Food Texture Design and Optimization

IFT Press

The *IFT Press* series reflects the mission of the Institute of Food Technologists – to advance the science of food contributing to healthier people everywhere. Developed in partnership with Wiley-Blackwell, *IFT Press* books serve as leading-edge handbooks for industrial application and reference and as essential texts for academic programs. Crafted through rigorous peer review and meticulous research, *IFT Press* publications represent the latest, most significant resources available to food scientists and related agriculture professionals worldwide. Founded in 1939, the Institute of Food Technologists is a nonprofit scientific society with 18,000 individual members working in food science, food technology, and related professions in industry, academia, and government. IFT serves as a conduit for multidisciplinary science thought leadership, championing the use of sound science across the food value chain through knowledge sharing, education, and advocacy.

WILEY Blackwell

Food Texture Design and Optimization

Edited by

Yadunandan Lal Dar
Ingredion Incorporated, Mogi Guaçu, São Paulo, Brasil

Joseph M. Light
Ingredion Incorporated, Bridgewater, New Jersey, USA

IFT Press

WILEY Blackwell

Registered Office
John Wiley & Sons, Ltd, The Atrium, Southern Gate, Chichester, West Sussex, PO19 8SQ, UK

Editorial Offices
9600 Garsington Road, Oxford, OX4 2DQ, UK
The Atrium, Southern Gate, Chichester, West Sussex, PO19 8SQ, UK
111 River Street, Hoboken, NJ 07030-5774, USA

For details of our global editorial offices, for customer services and for information about how to apply for permission to reuse the copyright material in this book please see our website at www.wiley.com/wiley-blackwell.

Library of Congress Cataloging-in-Publication Data

Food texture design and optimization / edited by Yadunandan Lal Dar and Joseph M. Light. – First edition.
pages cm. – (Institute of food technologists series)
ISBN 978-0-470-67242-6 (hardback)
1. Food texture. I. Dar, Yadunandan. II. Light, Joseph.
TX546.F665 2014
664′.072–dc23

2013022910

A catalogue record for this book is available from the British Library.

Wiley also publishes its books in a variety of electronic formats. Some content that appears in print may not be available in electronic books.

Cover image © iStockphoto/Artefy
Cover design by Meaden Creative

Set in 10.5/12.5pt Times by SPi Publisher Services, Pondicherry, India
Printed and bound in Malaysia by Vivar Printing Sdn Bhd

1 2014

Titles in the IFT Press series

- *Accelerating New Food Product Design and Development* (Jacqueline H. Beckley, Elizabeth J. Topp, M. Michele Foley, J.C. Huang, and Witoon Prinyawiwatkul)
- *Advances in Dairy Ingredients* (Geoffrey W. Smithers and Mary Ann Augustin)
- *Bioactive Compounds from Marine Foods: Plant and Animal Sources* (Blanca Hernández-Ledesma and Miguel Herrero)
- *Bioactive Proteins and Peptides as Functional Foods and Nutraceuticals* (Yoshinori Mine, Eunice Li - Chan, and Bo Jiang)
- *Biofilms in the Food Environment* (Hans P. Blaschek, Hua H. Wang, and Meredith E. Agle)
- *Calorimetry in Food Processing: Analysis and Design of Food Systems* (Gönül Kaletunç)
- *Coffee: Emerging Health Effects and Disease Prevention* (YiFang Chu)
- *Food Carbohydrate Chemistry* (Ronald E. Wrolstad)
- *Food Ingredients for the Global Market* (Yao - Wen Huang and Claire L. Kruger)
- *Food Irradiation Research and Technology, second edition* (Christoper H. Sommers and Xuetong Fan) *Foodborne Pathogens in the Food Processing Environment: Sources, Detection and Control* (Sadhana Ravishankar, Vijay K. Juneja, and Divya Jaroni)
- *Food Oligosaccharides: Production, Analysis and Bioactivity* (F. Javier Moreno and Maria Luz Sanz
- *Food Texture Design and Optimization* (Yadunandan Lal Dar and Joseph M. Light)
- *High Pressure Processing of Foods* (Christopher J. Doona and Florence E. Feeherry)
- *Hydrocolloids in Food Processing* (Thomas R. Laaman)
- *Improving Import Food Safety* (Wayne C. Ellefson, Lorna Zach, and Darryl Sullivan)
- *Innovative Food Processing Technologies: Advances in Multiphysics Simulation* (Kai Knoerzer, Pablo Juliano, Peter Roupas, and Cornelis Versteeg)
- *Mathematical and Statistical Methods in Food Science and Technology* (Daniel Granato and Gastón Ares)
- *Microbial Safety of Fresh Produce* (Xuetong Fan, Brendan A. Niemira, Christopher J. Doona, Florence E. Feeherry, and Robert B. Gravani)
- *Microbiology and Technology of Fermented Foods* (Robert W. Hutkins)
- *Multiphysics Simulation of Emerging Food Processing Technologies* (Kai Knoerzer, Pablo Juliano, Peter Roupas, and Cornelis Versteeg)
- *Multivariate and Probabilistic Analyses of Sensory Science Problems* (Jean - François Meullenet, Rui Xiong, and Christopher J. Findlay
- *Nanoscience and Nanotechnology in Food Systems* (Hongda Chen)
- *Natural Food Flavors and Colorants* (Mathew Attokaran)
- *Nondestructive Testing of Food Quality* (Joseph Irudayaraj and Christoph Reh)
- *Nondigestible Carbohydrates and Digestive Health* (Teresa M. Paeschke and William R. Aimutis)
- *Nonthermal Processing Technologies for Food* (Howard Q. Zhang, Gustavo V. Barbosa - Cánovas, V.M. Balasubramaniam, C. Patrick Dunne, Daniel F. Farkas, and James T.C. Yuan)
- *Nutraceuticals, Glycemic Health and Type 2 Diabetes* (Vijai K. Pasupuleti and James W. Anderson)
- *Organic Meat Production and Processing* (Steven C. Ricke, Ellen J. Van Loo, Michael G. Johnson, and Corliss A. O ' Bryan)
- *Packaging for Nonthermal Processing of Food* (Jung H. Han)
- *Practical Ethics for the Food Professional: Ethics in Research, Education and the Workplace* (J. Peter Clark and Christopher Ritson)
- *Preharvest and Postharvest Food Safety: Contemporary Issues and Future Directions* (Ross C. Beier, Suresh D. Pillai, and Timothy D. Phillips, Editors; Richard L. Ziprin, Associate Editor)

- *Processing and Nutrition of Fats and Oils* (Ernesto M. Hernandez and Afaf Kamal - Eldin)
- *Processing Organic Foods for the Global Market* (Gwendolyn V. Wyard, Anne Plotto, Jessica Walden, and Kathryn Schuett)
- *Regulation of Functional Foods and Nutraceuticals: A Global Perspective* (Clare M. Hasler)
- *Resistant Starch: Sources, Applications and Health Benefits* (Yong - Cheng Shi and Clodualdo Maningat)
- *Sensory and Consumer Research in Food Product Design and Development* (Howard R. Moskowitz, Jacqueline H. Beckley, and Anna V.A. Resurreccion)
- *Sustainability in the Food Industry* (Cheryl J. Baldwin)
- *Thermal Processing of Foods: Control and Automation* (K.P. Sandeep)
- *Trait - Modified Oils in Foods* (Frank T. Orthoefer and Gary R. List)
- *Water Activity in Foods: Fundamentals and Applications* (Gustavo V. Barbosa - Cánovas, Anthony J. Fontana Jr., Shelly J. Schmidt, and Theodore P. Labuza)
- *Whey Processing, Functionality and Health Benefits* (Charles I. Onwulata and Peter J. Huth)

WILEY Blackwell

Contents

Contributors

Marcela Alexander, Ph.D.
Department of Food Science
University of Guelph
Guelph
Ontario N1G 2W1
Canada

Jeffrey Banes
Technical Application Manager
Grande Custom Ingredients Group
302 Main Street
Lomira
Wisconsin 53048
USA

David A. Beattie, Ph.D.
Ian Wark Research Institute
University of South Australia
Mawson Lakes
SA 5095
Australia

Jacqueline H. Beckley
The Understanding & Insight Group, LLC
3 Rosewood Lane
Suite 103
Denville
New Jersey 07834
USA

Piet L. Buwalda, Ph.D.
AVEBE Innovation Centre
PO Box 15
9640 AA Veendam
The Netherlands

Gail Vance Civille
President
Sensory Spectrum, Inc.
554 Central Avenue
New Providence
New Jersey 07974
USA

Milena Corredig, Ph.D.
Department of Food Science
University of Guelph
Guelph
Ontario N1G 2W1
Canada

Yadunandan Lal Dar, Ph.D.
Director
Ingredient Applications, South America
Ingredion Incorporated
Rua Paula Bueno, 2935
Mogi Guaçu, Brasil, 13841-010

Els de Hoog, Ph.D.
Project Manager
Texture Design and Encapsulation
NIZO Food Research
Kernhemseweg 2, 6718 ZB Ede
PO Box 20, 6710 BA Ede
The Netherlands

Milda E. Embuscado, Ph.D.
Principal Scientist
Materials and Process Technology
McCormick & Company, Inc.
204 Wight Avenue
Hunt Valley
Maryland 21031
USA

Takahiro Funami
Deputy General Manager
Hydrocolloid & Texture Design Division
San-Ei Gen F.F.I., Inc.
1-1-11 Sanwa-cho
Toyonaka
Osaka 561-8588
Japan

Thomas Helm
Director of Food Technology
Grande Custom Ingredients Group
302 Main Street
Lomira,
Wisconsin 53048
USA

Sayaka Ishihara
Researcher
Texture Design Division
San-Ei Gen F.F.I., Inc.
1-1-11 Sanwa-cho
Toyonaka
Osaka 561-8588
Japan

Melissa Jeltema, Ph.D.
The Understanding & Insight Group, LLC
3 Rosewood Lane
Suite 103
Denville
New Jersey 07834
USA

Arun Kilara, Ph.D.
Principal
Nutri + Food Business Consultants
117 Westside Drive
Chapel Hill
North Carolina 27516-4431
USA

Kaoru Kohyama
Research Leader
National Food Research Institute
Project Leader of Food Flavor and Texture Evaluation
National Agriculture and Food Research Organization
2-1-12 Kannondai
Tsukuba
Ibaraki 305-8642
Japan

Marta Krasowska, Ph.D.
Ian Wark Research Institute
University of South Australia
Mawson Lakes SA 5095
Australia

Joseph M. Light
Vice President
Global Development
Ingredion Incorporated
10 Finderne Avenue
Bridgewater
New Jersey 08807
USA

Alessandra Marti, Ph.D.
Department of Food
Environmental and Nutritional Sciences (DeFENS)
Università degli Studi di Milano
via Celoria, 2 – 20133
Milan
Italy

Rajen S. Mehta, Ph.D.
Senior Director
Fiber Applications
SunOpta Ingredients Group
100 Apollo Drive
Chelmsford
Massachusetts 01824
USA

Howard R. Moskowitz, Ph.D.
President
Moskowitz Jacobs, Inc.
1025 Westchester Avenue, Suite 400
White Plains
New York 10604
USA

Maria Ambrogina Pagani, Ph.D.
Department of Food
Environmental and Nutritional Sciences (DeFENS)
Università degli Studi di Milano
via Celoria, 2 – 20133
Milan
Italy

Chandani Perera, Ph.D.
Senior Team Leader
Food and Industrial Applications
Roquette America Inc.
2211 Innovation Drive
Geneva
Illinois 60134
USA

Clive A. Prestidge, Ph.D.
Ian Wark Research Institute
University of South Australia
Mawson Lakes SA 5095
Australia

Koushik Seetharaman, Ph.D.
116, Department of Food Science
University of Guelph
Guelph
Ontario N1G 2W1
Canada

Joanne Seltsam
Program Director
Sensory Spectrum, Inc.
554 Central Avenue
New Providence
New Jersey 07974
USA

Tapashi Sengupta, Ph.D.
Senior Principal Engineer
Food Processing
PepsiCo Advanced Research
7701 Legacy Drive
Plano
Texas 75024
USA

David Taylor, Ph.D.
Senior Research and Development Scientist
Grande Custom Ingredients Group
302 Main Street
Lomira
Wisconsin 53048
USA

Jennifer Vahalik
The Understanding & Insight Group, LLC
3 Rosewood Lane
Suite 103
Denville
New Jersey 07834
USA

Fred van de Velde, Ph.D.
Group Leader
Protein Functionality
NIZO Food Research
Kernhemseweg 2, 6718 ZB Ede
P.O. Box 20, 6710 BA Ede
The Netherlands

Xin Yang, Ph.D.
Associate
Systems Design & Measurement
Customer Solutions & Product Innovation
Ingredion Incorporated
10 Finderne Avenue
Bridgewater
New Jersey 08807
USA

1 Introduction

Yadunandan Lal Dar[1] and Joseph M. Light[2]

[1]*Ingredion Incorporated, Mogi Guaçu, São Paulo, Brasil*
[2]*Ingredion Incorporated, Bridgewater, New Jersey, USA*

1.1 THE BASICS

Food texture defines the eating experience and drives consumer liking or preference of food products. This has always been a fundamental aspect of the human (and perhaps animal) eating experience and has been known to food preparers as well as discerning consumers since the very beginning of food preparation. However, the development of technical and business expertise in food texture optimization and design using structured multidisciplinary tools is a more recent development. As culinary arts, food science, materials science, sensory science, and the study of consumer behavior have evolved in scope and complexity; it is now possible to develop a view of how the ability to study and optimize texture can drive consumer choice. The science and technology of designing and optimizing food texture have evolved significantly in recent years and continue to do so at a rapid pace.

The current volume is being compiled as a showcase for the current state of the art in this field. There have been several volumes that have been written as introductory texts or overviews (Bourne, 2002; Moskowitz, 1987; Rosenthal, 1999) or comprehensive references (McKenna, 2003; Kilcast, 2004) for those who would like to initiate study in the field. The scope of this volume is for those new to the field as well as for those who have experience but perhaps would like to gain deeper and wider insight on food texture and gain familiarity with the state of the art at the beginning of the twenty-first century. From the list of contributors in this volume, it should be clear to any reader that the scope of this work is both global and diverse. Enjoyment of food and how it impacts social and cultural aspects of anthropological evolution is a shared influence for every culture on this planet and despite regional and cultural variations there are strong similarities in ingredients and formulation considerations that make a particular food from a region more appealing than another. Indeed, there is scope for collaboration on a more global scale in the pursuit of greater understanding of the evolution and impact of food texture.

'The Ketchup Conundrum' is a widely quoted article by Malcolm Gladwell (2004) and brings to light a basic dilemma with the perception of food. There is inherent duality

Food Texture Design and Optimization, First Edition. Edited by Yadunandan Lal Dar and Joseph M. Light.

in the way humans treat food – each cultural sub-population has some food products that are 'sacred', that is, there are perceived requirements for ingredients, flavor, texture, and even preparation techniques. However, for all these products, if one is able to gather perspective over a long enough time-frame and over different population bases, it is clear that there has been significant evolution even in these products and preferences change gradually but surely across regions and over time. Mr. Gladwell's article offers a sound perspective using experiments with condiments in the U.S. marketplace but once you expand scope to include Asian or other regional markets there is emergence of other factors that influence consumer choice.

The editors recently had the opportunity to be personally involved in a study of texture for ketchup and barbecue sauce and noted how this phenomenon is manifested in this category. Ketchup in different parts of the world, even for the same brand, can have marked differences in flavor and texture. Further, different populations use ketchup differently and hence it has led to an entire population of products that are tomato based sauces where any opportunity to differentiate flavor or texture leads to a new name for the product. In some regions (e.g. the United States), this is mandated by law due to standards of identity, but in others it is just a way of innovating in a marketplace that is constantly evolving. Ketchup products from different parts of the world have significant differences in thickness, cling, cohesiveness, and flavor aspects including sweetness vs. sourness vs. saltiness vs. spice content. The presence of such products over a long time in different marketplaces has driven consumer acceptance and familiarity with the local products and has evolved the definition for what is 'traditional'. Regional taste preferences as well as differences in quality and type of raw materials also shape these effects.

The other side of this duality is that there are certain food products that populations love experimenting with and will constantly explore new and different flavors and textures. Certain categories of confectionary and snack foods are perhaps the best examples of this in many parts of the world. Not every exploration produces positive results and certain flavors or textures are more enduring than others. However, the main observation here is that the same people will not want to change certain types of food but will actively look to change others and that the types of food and the extent of change that is acceptable varies widely. Thus, if one takes Mr. Gladwell's analysis of why ketchup does not have varieties, but extends that to barbecue sauce in the United States, the picture changes substantially. Barbecue sauce is made using similar basic ingredients and processes that are used for ketchup but is one of the most experimented with condiment category across the United States. There are hundreds of brands and strong preferences from state to state, sometimes even with states.

The other phenomenon that makes a global view of food texture somewhat complicated is that even for food products that are similar in evolution, and function, regional ingredients, supply quality, preparation methods, and storage methods/expectations vary so significantly that they can introduce substantial differences in product texture and quality.

A good example of ingredient variation is wheat flour – a basic ingredient that is perhaps used in almost every country in the world (Shewry and Khan, 2009) Despite the product being called wheat flour and labeling as such on the food product, there are significant variations in content and properties of protein, content/quality of starch, as well as other compositional differences that impact the processing and milling of these grains and ultimately the final texture of the wheat based products. Seasonal variations in wheat quality and properties add further complexity and all of these effects have led to a global industry that specializes in the standardization of grades of wheat flour to reduce the variation and uncertainty from region to region and year to year. This is especially relevant with the globalization of the food supply chain that leads to consumers who routinely eat food made with grains, fruits, vegetables, meat, and spices produced thousands of miles from where they live.

An interesting example of texture optimization from the last century is from Australia with the use of wheat gluten to optimize bread texture from the 1930s. (Technology in Australia, 2000) Fielder's Gluten Pty. Ltd was formed to commercialize the use of 'vital wheat gluten' to improve bread texture and allow bread of good quality to be made from grades of flour that were previously not used successfully. This is still a commonly used approach to optimize bread texture and counter seasonal or varietal variations in wheat flour.

The observation for wheat flour can be extended to other grains and tubers and is valid to different extents for meats, dairy products, fruits, vegetables, and other ingredients. Frequently, shortages in one region will require ingredient substitution from another. Government and trade regulations can limit the freedom of this exchange to some extent but the phenomenon has been growing in scope for many years. Produce and grains from agricultural economies are exported around the world and variations in trade and supply can lead to significant changes in ingredients.

The globalization of food brands from many corporations is another key trend that has led to additional complexity and challenge in addressing food texture related challenges. Important multi-national brands often have sales in billions of dollars and have signature texture and flavor profiles. The brand owners strive to maintain these profiles in different regions of the world. This leads to formulation challenges where locally sourced raw materials with differences in properties from other regions or legacy local processes with differences in temperature or shear profiles have to be optimized to provide consistent final product properties.

Supply and quality fluctuations have further impact on food and ingredient costs. This has a significant impact on food costs. The World Bank tracks food prices globally and the world is going through a series of crises as regards food inflation as shown in Figure 1.1. (World Bank Food Price Watch Report, November 2012). This has led to a constant search for approaches to maintain food quality while managing through changes in the prices of both raw and processed foods.

New developments in processing and packaging techniques have also had a significant impact in the study of food texture. New techniques that include using higher pressure and shear and consequently reliance on lower temperatures to achieve the desired food safety requirements lead to the ability to manufacture packaged foods with texture and flavor close to freshly prepared meals. The ability to use these new techniques while still achieving target food textures for packaged foods that were so far manufactured using traditional techniques often leads to reliance on formulation based approaches to optimize texture. High pressure processing (HPP) and pasteurization is one example of a recent technique that is leading to an evolution in how food is made safe for packaging and consumption (Doona and Feeherry, 2007). The use of very high pressures to achieve pasteurization allows foods to be processed at significantly lower temperatures, which can lead to large differences in texture when compared to traditional thermal pasteurization processes.

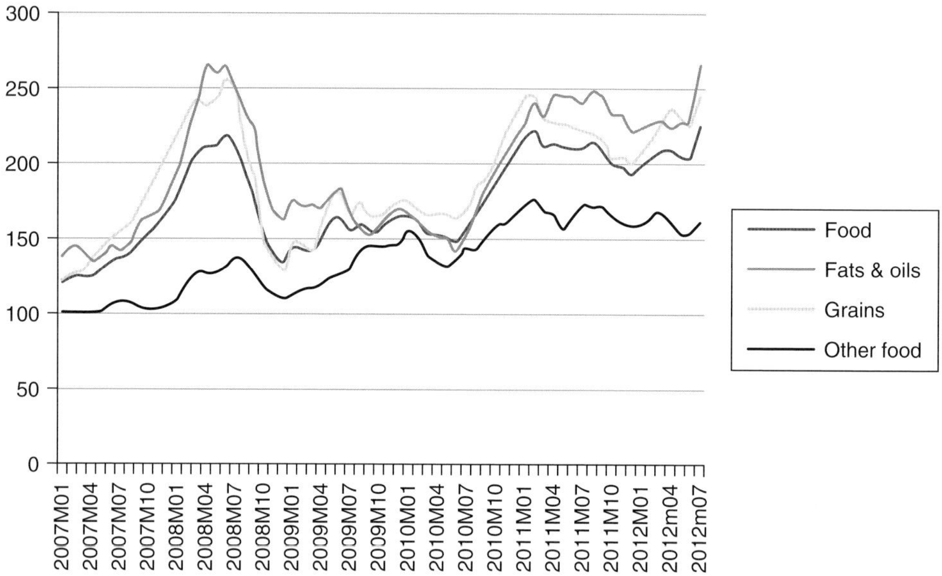

Figure 1.1 Food inflation (Food Price Index: 100=2005 Prices). (Source: World Bank Global Food Price Index. Reproduced with permission.)

Shelf life is a key concern for refrigerated, frozen, as well as shelf-stable packaged foods. Shelf life and stability criteria are different based on storage temperature for texture as well as microbiological stability. This has also led to the requirement for formulation based approaches that depend on functional ingredients to provide the desired texture and stability for the target product shelf life. Shelf life is a key consideration when considering the time taken from food production and distribution centers to local sales outlets and finally to consumers' homes or to restaurants and food service establishments.

What we know regarding texture today is that texture is a critical criterion in addressing food product design. The factors that make this important are summarized in Table 1.1. There is a strong penalty if food product developers are not able to hit the desired target texture. This is typically reflected in lower acceptance and consumption of the food product by the target consumers. This is true of a family consuming a home cooked meal, a customer ordering a favorite dish at a restaurant, as well as for an individual purchasing a packaged food product from the supermarket for a quick dinner.

Texture will continue to play a critical role for future food product development as the factors described in Table 1.1 as well as others will continue to evolve and impact how food texture preference and its impact on consumer choice will change. Texture needs to be addressed as part of the initial formulation before final flavor optimization and locking in the final recipe and formulation. This does not mean that flavor is less demanding but optimizing flavor before texture can lead to double work as changes in texture can lead to significant changes in flavor perception.

In order to understand and change texture, it is important to be able to measure it precisely. Measurement of texture is still evolving but there are two broad classes of techniques that are commonly used. Since texture is primarily a sensory concept and its perception is a combination of perceptions from several human senses, the ultimate measurement of

Table 1.1 Factors impacting texture choice or texture properties that impact choice.

Factor	Impact
Regional and cultural history	Individual preference toward familiar textures based on family and cultural history
Seasonal variation	Seasonal changes in grains, tubers, fruits, vegetables and other food products can change their composition and texture, and can in turn lead to changes in texture of the food products made with them
Regional variation	Different regions use the same names for products (e.g. wheat or rice) but varietal and other factors can lead to significant variation in texture
Supply constraints	Supply constraints for ingredients for formulated foods can lead to their substitution by others leading to changes in texture
Cost fluctuations	Cost increases in ingredients for formulated foods can lead to their substitution by lower cost ingredients leading to changes in texture
New processing techniques	Advances in processing techniques including the use of higher pressure or shear can lead to changes in texture
New packaging techniques	New packaging containers, equipment, and other factors can lead to changes in how the product is handled and how it changes over its shelf life leading to changes in texture
Regulatory developments	Regulatory changes can reduce or eliminate the use of certain ingredients requiring their substitution and changes in texture
Genetic makeup	Some genetic factors such as amount and potency of saliva or genetic predisposition to certain types of mouth behavior can lead to preference for particular textures

texture is through sensory measurements by humans. There are also several instrumental techniques that are used to measure physical properties that are either strongly correlated to or that strongly impact the sensory perception of texture. Having access to both types of techniques and being able to translate data into action to develop desired textures is a requirement for texture optimization.

1.2 DEFINING FOOD TEXTURE

Food texture has been defined by the International Standards Organization (ISO) in their standard vocabulary for sensory analysis as 'All the rheological and structure (geometrical and surface) attributes of a food product perceptible by means of mechanical, tactile, and where appropriate, visual and auditory receptors' (ISO 5492, 2008). The measurement of food texture using sensory and instrumental techniques has been the topic of considerable study over the last few decades (Rosenthal, 1999; Moskowitz, 1987). The entire process of food texture perception has been a topic of significant interest over the years. The overall process is illustrated in Figure 1.2. A more comprehensive description can be found in Sczezsniak (2002).

The overall perception of texture starts at first visual contact with the food product. The visual appearance includes color, shine, visual flow characteristics and other similar attributes. The next step in the perception of texture is the sense of touch. This is an opportunity to perceive surface properties like stickiness or roughness as well as bulk

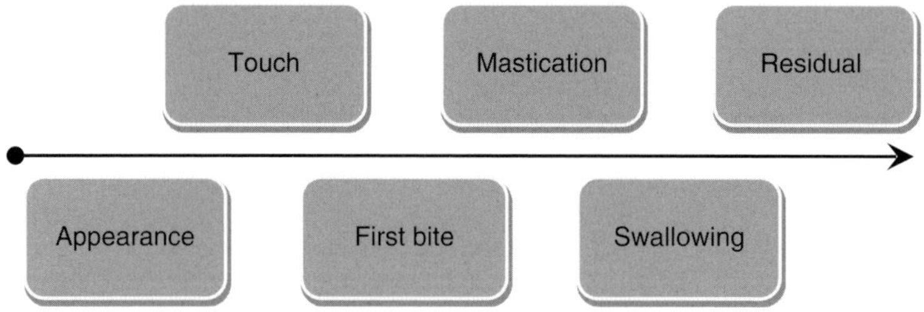

Figure 1.2 Eating experience: The perception of texture.

properties such as hardness or ease of flow. The next step in the consumption of food is the first sip, first bite or first compression of the food. This is a critical step where the senses of hearing and taste provide the first sensory input. The loudness as well as the frequency of sound has been shown to drive consumer interest and preference. The initial perception of hardness, flowability or other textural sensations in the mouth have a similar impact. This is also the first point where trigeminal sensations such as those related to temperature can be perceived.

The next stage in the perception of texture is the mastication or chewing of food. This stage provides perhaps the richest and most complex array of experiences in the textural perception of food products. The key texture attributes include resistance to chewing, the way the food product breaks down in the mouth, the extent to which it coats the palate, how the product sticks to the teeth or the tongue and eventually how long it takes to break down and dissolve in the mouth.

How the food feels as it is being swallowed is nearly the final stage in the perception of food texture. The characteristics can include the flow properties, roughness or smoothness, and how easy or difficult it is to swallow. Once the food has been swallowed, there is one final set of attributes related to texture. These are the residual textural effects of the food even after the consumer has swallowed it. These attributes include residual coatings of the mouth and the throat as well as any residual particulate elements left in the mouth.

As shown in Figure 1.2 and summarized above, the perception of food texture is a complex sensory process that involves the interplay of a variety of product characteristics and human senses. This is one of the reasons that it has taken so many years to measure and understand the texture of different food products and its impact on consumer perception and liking. The optimization of food texture requires the ability to measure it as well as the ability to make changes in the food formulation or preparation process to optimize the texture as well as the overall eating experience.

1.3 MEASURING FOOD TEXTURE

Food texture is primarily a collection of sensory properties and a variety of techniques have been developed to measure different sensory aspects of food texture. However, conducting sensory analysis can be time consuming and access to a suitable set of evaluators may not always be available. This has led to the utilization of a variety of instrumental techniques that can be used effectively to screen for certain critical properties that correlate strongly with sensory perceptions. In certain cases, the instrumental techniques can be used as a proxy for sensory analysis.

1.3.1 Sensory testing and analysis

The purpose of sensory testing and analysis is to employ humans to test food products and to use their perceptions to understand and optimize product texture. The techniques described here are used not only for texture but for flavor as well as overall evaluation of food products and the eating experience. Sensory tests can be used to determine differences in products as well as consumers' acceptance and/or preference for the food products being evaluated. The data generated through such analysis is typically used to address desired organoleptic characteristics to achieve business targets.

Amerine et al. (1965) still remains a good reference for sensory evaluation though many more recent references are available and used. A good recent reference is Moskowitz et al. (2006). Sensory evaluation is very commonly used in the food industry for new product development or reformulation, but is also used for a variety of other development activities. It can be used to compare products to a target and to qualify process changes to ensure that the final product meets expected organoleptic standards. It is used for quality assurance including the evaluation of storage stability and for shelf-life testing. A variety of different sensory tests are used based on the type of data that is required for decision making. These tests are summarized in Table 1.2.

The commonly used methods named in the last column of Table 1.2 are further described in Table 1.3. For consumer acceptance or preference tests, one of three common implementation methods is typically used. Central location testing involves recruiting consumers and bringing them to a specific location for testing. In-home placement involves recruiting target consumers and providing products to be used in their households for testing. Mall or store intercept involves stopping and interviewing consumers in shopping malls or stores. Each method has its own unique advantages as well as its own complexities.

Table 1.2 Commonly used methods for sensory analysis.

Test	Objective	Evaluators	Methods
Difference	To determine if differences exist between products	Experienced with test methodology	Triangle Duo-trio Directional
Acceptance	To determine the acceptability of the product	Target consumers	Monadic Paired
Preference	To determine if one product is preferred to another	Target consumers	Monadic Paired Ranking
Descriptive	To identify and quantify specific product characteristics of flavor and/or texture	Highly trained and calibrated to reference standards	Quantitative Descriptive Analysis Spectrum Descriptive Analysis

Table 1.3 Methods used for sample presentation and sensory testing.

Method	**Description**
Triangle	Two identical and one different sample are presented. Evaluators identify the different sample.
Duo-trio	Reference presented followed by one identical and one different sample. Evaluators identify the sample same as the reference.
Directional	Two different samples are presented. Evaluators are asked which one is stronger in specific sensory attributes (flavor or texture).
Monadic	One sample or series of single samples is presented for evaluation
Paired	Two samples are presented side-by-side or sequentially for evaluation
Ranking	Three or more samples are presented to be ranked for preference
Quantitative descriptive analysis	A panel of trained evaluators evaluates a set of samples for specified attributes to arrive at statistically derived product description of flavor and/or texture
Spectrum descriptive analysis	A panel of trained evaluators evaluates a set of samples using calibrated scales for specified sensory attributes to provide a complete and detailed descriptive characterization of the sensory attributes and their intensities

In order to obtain reliable and high quality sensory data, the facilities, setup, and training for sensory evaluation play a very important role. A good setup helps generate good quality data whereas inadequate attention to logistics and details can compromise the quality of the data and call into question the conclusions made from the data. The most common factor that impacts the quality of data is noise – changes in the factors being studied not caused by the parameters of interest. Noise can be random, as caused by random changes in the samples or environment or it can be systematic, as caused by factors other than those being studies influencing the outcome of the work. The recommendation below can significantly help reduce the impact of noise on the data.

The physical set-up for the analysis is important and can impact the outcome of the testing. Requirements for controlled testing include an area in which extraneous odors and distractions are minimized – individual booths are highly recommended. Laboratories or kitchens used for sample preparation should be equipped with a sink, food warmers, refrigerator/freezer, analytical balance, lab glassware, oven and stove. In addition any equipment needed to prepare and serve food products, such as blenders, mixers, and serving/sampling containers that are useful for the range of samples being evaluated.

Another area that needs to be taken into account is sample presentation. It is important to provide consistent sample sizes and portions for different samples and to different evaluators. The serving temperature needs to be regulated uniformly across samples and panelists. The samples should also be served in similar, preferably identical containers. Further, the samples should be coded so the evaluators cannot identify them by name and therefore will not be influenced by any prejudice towards the names or descriptions of samples. The serving order for the samples should also be changed using a suitable strategy for randomization to ensure that any effects due to order of presentation do not confound the data.

The selection of suitable evaluators is an area that requires significant expertise as well. For consumers being asked to evaluate a food product, it is important for most tests to use a selection or screening questionnaire to classify the consumers and eliminate any potential outliers. For example, for people evaluating yogurts to pick the most liked sample out of a set, it is important to identify people who consume yogurt regularly and actually have a liking for the product vs. those who do not like yogurt. If there are people in the set who are not frequent consumers and do not like yogurt at all, their views are not representative of mainstream yogurt consumers and will most likely need to be classified separately from people who like yogurt and consume it regularly. For trained panelists conducting difference of descriptive testing, it is important to recruit people to the panel who have the capability to differentiate between the attributes being evaluated. Certain people may not qualify based on their being overly sensitive to certain tastes or textures. In addition, food safety is a key factor to consider while conducting sensory analysis and testing. Proper safety guidelines for food preparation, storage, and serving need to be followed to ensure food safety and it is critical to ensure that people who are tasting food be informed about the ingredients used to ensure that there is no accidental exposure to any allergens for people who suffer from food allergies.

There is highly specialized training that is available and routinely conducted for expert evaluators for flavor and texture. There are three types of languages or terms commonly used to describe texture. These are basic/fundamental terms, integrated terms, and consumer terms. The ideal texture language used for sensory evaluation should be fundamental/basic and should translate integrated and consumer language into elements that can be measured and, if needed, altered through processing or formulation.

The basic/fundamental language terms constitute building blocks of a composite sensory perception and are developed from the basic elements of texture perception. They are discrete, concrete and clearly defined. Such elements can not only be measured reliably with high accuracy, but are also understood by a scientist or food formulator and are the fundamental elements of a texture experience. Examples of basic texture terms include thick, slippery, cohesive, firm, fracturable, loud and so on. Each term measures a fundamental property and though some of the terms may correlate to each other, each term measures a fundamental texture property of interest for the food.

Integrated language terms are so called because they are made of two or more fundamental terms. These are typically used by consumers or by experts when relating consumer terms to fundamental texture terms. These terms typically cannot be directly measured but have to be derived by either expert evaluation of the fundamental terms that comprise them or through consumer evaluation of the overall sensory impact. Examples include crispy, crunchy, creamy and chewy.

Consumer language terms are those typically used by non-expert consumers to describe the foods they eat. These are typically not well defined and can be difficult to translate into specific product characteristics. This is because these terms evolve from sensory, cultural, and even emotional responses to food. Even though the terms imply a consistent broad sense to most people, the same terms used by different people can often describe very different physical characteristics. A further complicating factor is that many consumers use similar terms to those used in fundamental or integrated languages to describe complex consumer experiences that convey meanings different from those conveyed by the same terms when used in the fundamental or integrated languages by expert evaluators. Examples include velvety, creamy, crunchy, chewy, and soupy.

A trained texture evaluation team needs to have the capability to be able to distinguish between the different terms and be able to conduct measurements and translate across the three different texture languages.

1.3.2 Instrumental measurements

Bourne (2002) provides a comprehensive overview of the history of texture measurement. One of the earliest documented examples of the systematic testing and optimization of food texture is from the mid-1800s in Germany where mechanical testing was used to measure the firmness of jellies or from the late 1800s in the United States where viscosity was measured to test the quality of milk or cream. Several key developments in Europe and the United States in the late 1800s and the early 1900s included systematic measurement and optimization of texture of dairy products, oil/butter, fruits, grains, and dough-based products.

There has been extensive research conducted in the use of instrumental measurements to measure physical properties of the food products that can be correlated to the sensory perception of texture. There is a large volume of information available on the use of these techniques in Rosenthal (1999) and Bourne (2002). These tests can be divided into three types of measurements.

The first type is the measurement of primary physical characteristics including size, shape, volume, porosity, and color. These properties are relatively straightforward to measure and provide data that is easy to compare between products and can sometimes be directly related to sensory perception. These characteristics can also be routinely measured during quality control and basic physical attribute testing.

The second type of measurements are based on mechanical deformation or flow behavior and are typically based on relating the rate or scale of deformation in the instrument to that in the mouth or in a processing unit. These measurements are related to rheological testing for liquids or semi-solids or to compressive or tensile testing for solids. A typical test involves preparation of samples of defined shape or geometry and deforming them using carefully measured forces. The test can be conducted by controlling the force and measuring the deformation or by controlling the deformation and measuring the required force. The force is measured as a function of deformation and/or temperature and/or rate of deformation. The magnitude of the force and deformation as well as the shape of the curve including shape and size of the peaks and valleys are related to different fundamental properties and in some cases can be directly related to sensory perception. The so-called texturometer was an instrument used for this type of measurement (Szezsniak, 2002) and there are a variety of instruments routinely used for liquid as well as semi-solid foods. Many examples are provided in McKenna (2003) and Kilcast (2004).

Rheology for semi-solid as well as solid foods is an important tool for texture evaluation. It requires significant expertise and training to generate as well as interpret reliable data and can be time as well as effort consuming. However, rheology enables precise evaluation of food texture in a way that few other techniques can. Rheological properties broadly correlate to sensory properties in the range where both types of evaluation provide reliable results. However, rheological testing can have advantages over sensory testing for samples or conditions where sensory testing is difficult to conduct. Examples could be samples that are too hot or too cold and the extreme temperatures can impact a normal trained panel's ability to conduct objective evaluation across samples. Another example is samples that are too spicy, for example, hot pepper sauces that some panelists would find difficult to taste and different panelists would have different degrees of sensitivities to the level of spiciness. Such trigeminal sensations can introduce significant noise and variation in the evaluation of other texture attributes. Another area where rheology excels is the ability to pick out small differences in the flow behavior of samples. These may or may not be important for the immediate eating experience but do play a role in influencing the stability of texture over time and ultimately play a role in influencing the product texture.

For these reasons, rheological testing finds frequent use in determining ingredient functionality in product development, intermediate or final product quality control, shelf life testing, as well as process engineering calculations for scale-up of processing operations.

The third type of measurements are custom fabricated to detect physical attributes that are specific to the human perception of texture. These attributes include devices that track jaw movement or measure the neurological activity of jaw muscles and relate them to properties of the food and the response of consumers. Some good examples are described in Cakir (2011).

1.3.3 The ultimate texture language

Research and product development efforts from different parts of the world have evolved the measurement and understanding of texture using a variety of approaches. Though these approaches are all useful in their own way, small yet significant differences make it difficult to translate work conducted across groups or locations and even the same samples and terms measured by different groups can lead to different results.

This is true for sensory and, to a lesser extent, for instrumental measurement and is due to the fact that texture measurement is inherently a complex undertaking. Our understanding of what comprises fundamental texture and how to best measure it is still evolving for several food products. The ultimate texture language would perhaps be more universal in its applicability and understanding and pull together the different groups working on solving similar problems in different parts of the world.

Figure 1.3 shows an example graphic illustrating the TEXICON™ or texture lexicon used by Ingredion Incorporated. This is an example graphic illustrating how fundamental instrumental measurements (outermost layer) could be related to fundamental sensory terms (next two layers) and ultimately to integrated consumer terms (innermost layer). The ability to understand this relationship and use it to change food texture by design can lead to the perfect texture (as illustrated at the core of the graphic). Another way to look at this interconnection is that understanding this relationship can help translate the eating experience (at the center of the graphic) into measureable and, ultimately, controllable attributes that can be optimized to design the perfect texture. This graphic and the concept of the TEXICON™ is used here as a tool to illustrate the importance of

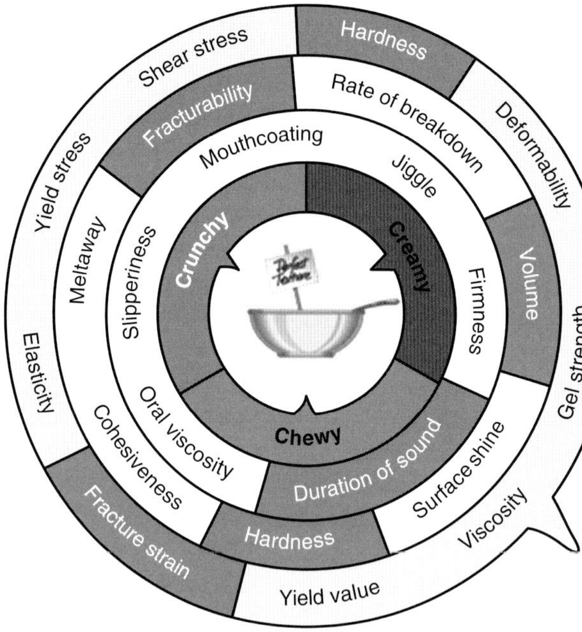

Figure 1.3 TEXICON: The Texture Lexicon. (Source: Copyright © 2013 Ingredion Incorporated. Reproduced with permission.)

understanding these relationships but the fundamental concept and approach here is broadly applicable to any team and any organization that endeavors to understand and optimize texture as well as the overall eating experience.

1.4 THE OPTIMIZATION OF FOOD TEXTURE

A variety of tools and techniques are used to develop optimum food textures. This is true for a person preparing a home cooked pasta dinner where a combination of the right cooking time and type of pasta is needed to get the desired *al dente* texture. It is also true for a food scientist designing the next popular meal choice for consumers in the food aisle in a grocery store. A skilled chef truly understands the synergy between food ingredients, flavor dynamics as well as manipulating cooking conditions to achieve highly desirable, even sensational food textures and taste. From a texture perspective, the tools available to optimize food texture can be divided into two sets: formulation and process optimization.

1.4.1 Formulation

Food formulation has traditionally been looked upon as an art as it had its roots in highly skilled chefs using experience and creativity to gain mastery of texture, and ultimately taste. As food science knowledge has increased and many research teams have developed and documented this knowledge, food formulation has evolved into a much more scientific tool to develop and optimize food texture. Different ingredients have traditionally been used to impart a variety of textural experiences. Traditional uses of ingredients include the use of fats to provide lubricity or mouthcoating, starch based ingredients to provide viscosity, or protein based ingredients to provide gel or set characteristics. Significant developments and advances have been made in the ingredients available that provide food formulators with a versatile toolbox to design and optimize texture. The advances made include a variety of ingredients based on carbohydrates, proteins and fats. Carbohydrate ingredients include starches, gums, fibers and a variety of derivative ingredients. Fat based ingredients from different sources with different levels of unsaturation, and refining can influence a variety of textures. Protein based ingredients from animal and vegetable sources including concentrates and isolates can also provide a range of texture and nutrition benefits. Functional ingredients provide a variety of texture properties including viscosity, gelling, and other textural attributes in many applications. For example, carbohydrate based ingredients can now be used to provide attributes related to indulgence and lubricity as well as gelation that were only previously possible using fat or protein based ingredients.

In addition to functional ingredients, a variety of computer-based applications and tools have evolved to enable formulation development and optimization. These tools include the ability to generate and optimize nutritional information, conduct design of experiments and optimize formulations based on driving preferred texture attributes as well as the ability to routinely review and statistically analyze multivariate texture data.

1.4.2 Process Optimization

The impact of processing including temperature, level of shear and pressure has been a topic of considerable study by food scientists and engineers for more than a hundred years. The time spent at a particular temperature, shear level, or pressure also plays a significant role in determining the texture of the food product. The changes induced by processing are due to changes in the physical characteristics of the ingredients due to the cooking process. From a texture perspective, this can include changes in viscosity as well as in the hardness of ingredients. The actual changes are strongly impacted by the physical breakdown of the native ingredients added to the recipe as they interact with each other and water. The actual changes can be quite complex and the order of ingredient addition plays an important role.

The primary impact of processing is based on temperature, shear, pressure and the length of time the product is subject to these parameters. Not only are the specific conditions important but the scale of the process as well as the type of apparatus used can have a significant impact. The intensity of cooking depends directly on time at an elevated temperature. Higher levels of shear or pressure typically accelerate the cooking process. Shear also independently acts to change texture through imparting mechanical force that can help with mixing as well as with breaking down certain ingredients.

The choice of equipment used is important in several ways. The application of heat plays a strong role in developing ideal texture. Whether the heat is applied through uniform exposure to moderately high temperature or through short exposure to very high temperature or through long exposure to a range of temperatures can lead to very different textures for the same substrate. For example, chicken that is boiled, fried or grilled has very different textural attributes that are primarily related to the cooking process and how that changes the texture of the food differently. Another factor that has strong impact is how the heat is transmitted to the food being cooked. Whether it is through being immersed in hot water, hot oil, or steam or being cooked in a shallow layer of water or oil can all lead to different textures.

For home cooking, there is specific equipment used for these different types of processes ranging from stove-tops, ovens, convection ovens, grills based on a variety of different fuels, and more recently microwaves. A variety of cooking utensils including pans, steamers, stock-pots, griddles, and pressure cookers further help in developing optimal textures. In a commercial production environment, there is a variety of equipment that is commonly used to get the same target textures at a larger scale, with greater reliability, and faster speed. The type of equipment can vary significantly by scale, source of heat, food safety regulations, and economic as well as cultural considerations. The choice of equipment has a strong influence on texture and changing equipment scale as well as type typically needs to be compensated with a change in processing parameters or formulation or both to maintain or optimize texture. Trained food formulators and applications scientists typically have the expertise to deliver target textures as well as overall taste while navigating the complexities of process development and optimization.

1.5 A HOLISTIC APPROACH TO INTEGRATED FOOD TEXTURE DESIGN

The optimization of food texture can be a complex undertaking as illustrated by the different aspects that can lead to changes in texture perception as well as the many different tools and techniques that are required. A systematic and data driven approach to food texture design can help lead to robust solutions faster. A good example based on the editors' experience is the DIAL-IN® Texture Technology, which is Ingredion's integrated approach to food texture design and optimization. This approach is unique to Ingredion in that it uniquely uses the knowledge and experience of the Ingredion team to provide robust and reliable texture solutions. However, most of the elements in this approach are based on a general framework that is applicable for anyone who seeks to develop strong capability in designing food texture, or for that matter, other aspects of food product design and optimization.

This approach is a rapid, robust, data-driven methodology that combines the art and science of food formulation and translates consumer needs and sensory preferences into food products that provide a superior eating experience. The approach has five steps as shown in Figure 1.4.

1. Define business and technical goals: This is perhaps one of the most critical steps in addressing any product development issue. It is critical to not only define the goals and objectives but also to define how to address these through the project being planned. This includes relevant cultural insights, label or nutritional requirements, desired package claims, the relevance of any benchmarks, as well as the desired timeline and measures of success.

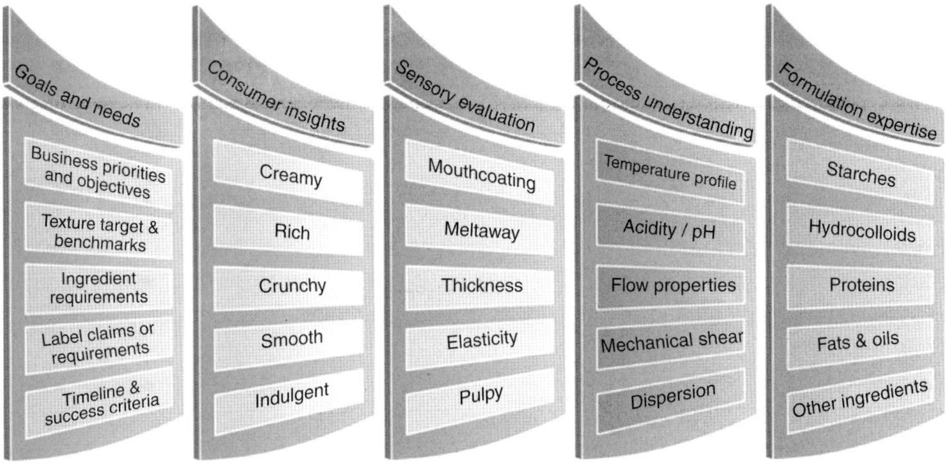

Figure 1.4 The systematic approach used for DIAL-IN® Texture Technology. (Source: Copyright © 2013 Ingredion Incorporated. Reproduced with permission.)

2. Gather consumer as well as market insights: Define consumer expectations for the product being designed and to define the requirements to gain consumer acceptance and liking. This step requires access to capabilities and expertise to be able to develop and utilize consumer insights.
3. Evaluate sensory attributes and set texture target: Ensure that there is adequate understanding of the sensory texture attributes that need to be measured to define the eating experience of the food product being developed or optimized. This may include characterization (e.g. descriptive sensory analysis) of a number of commercial products in the market that represent a range of textural diversity. This information is then used to set a texture target including a definition of the fundamental sensory texture terms that need to be optimized and the sensory and instrumental measurements that need to be conducted. It is important to have access to a capable sensory program with the capabilities summarized in Table 1.2.
4. Understand impact of process equipment and constraints on texture: The specific process equipment, process times and temperature, pH and the amount of shear need to be set and their impact on texture needs to be considered. Food processing expertise and compliance with current food regulations are critical to ensure success in this step.
5. Formulate to optimize texture: The final step is to use a scientific structured process to optimize formulations using the best functional and basic ingredients to develop the target texture. In order to be successful in this step, it is important to have expertise in formulation science. This includes several key capabilities including intimate knowledge of ingredients and their impact on texture, expertise in tools to rapidly design and test food formulations, as well as the ability to use measurements to detect gaps and address these gaps through formulation know-how. Use of capabilities including experimental design, statistical analysis and optimization as well as multivariate data visualization are all helpful in providing rigor and robustness to this step to optimize texture.

1.6 CURRENT MARKET TRENDS AND FUTURE CHALLENGES IN FOOD TEXTURE DESIGN AND OPTIMIZATION

Food texture has been one of the fundamental characteristics that consumers look for in their food products and drives consumer liking and purchase intent. This has always been important but as science and technology have developed, it is increasingly possible to optimize texture and to target new product texture by design. There are three main categories of texture related product development initiatives that are commonly encountered in the food industry.

1. Developing a new food product with a desired target for texture as part of the overall eating experience.
2. Building back the texture of an existing food product after changing some key ingredients, for example, reducing fat in a sauce/dressing or adding whole grains to a bread loaf. In either case, it is desirable to maintain texture and the eating experience while creating a more healthy and desirable nutritional profile.
3. Transforming the texture of an existing product to one that is more highly desirable for the target consumers.

One of the best ways to assess the attention on food texture is to look for texture claims on food products introduced to the marketplace. It is possible to do this using many different approaches. In recent years, front of package texture claims have risen so that almost 1 in 2 products launched has a front of package texture claim. It is possible to review these claims through using any suitable market insight tool. Figure 1.5 provides examples of texture claims on bakery and snack products manufactured in the United States from 2009–2011. Over 3000 products, more than 50% of all product launched, had texture claims on the front of their packages. The top texture terms are shown in Figure 1.5 broken down by category.

Based on current and emerging market trends, texture will continue to be a key factor in food product design and optimization for the foreseeable future. A key area of focus where texture will need to be addressed is cost optimization and affordability initiatives as fluctuations in price of food ingredients necessitate initiatives to optimize the cost of food formulations while maintaining a desirable eating experience. This could include replacement of costly sources of carbohydrates, fats or proteins with more cost-effective ones while minimizing undesirable changes in the eating experience.

Another area of focus continues to be the reduction of ingredients that could be harmful if consumed in excess. Current initiatives in different parts of the world include reduction of different types of fats, salt, and sugar. These ingredients have an impact on texture and flavor but can also have impact on other factors including shelf stability and food safety. The impact of reducing or eliminating the undesirable ingredients from the food formulations needs to be compensated for by using a combination of ingredients and processing.

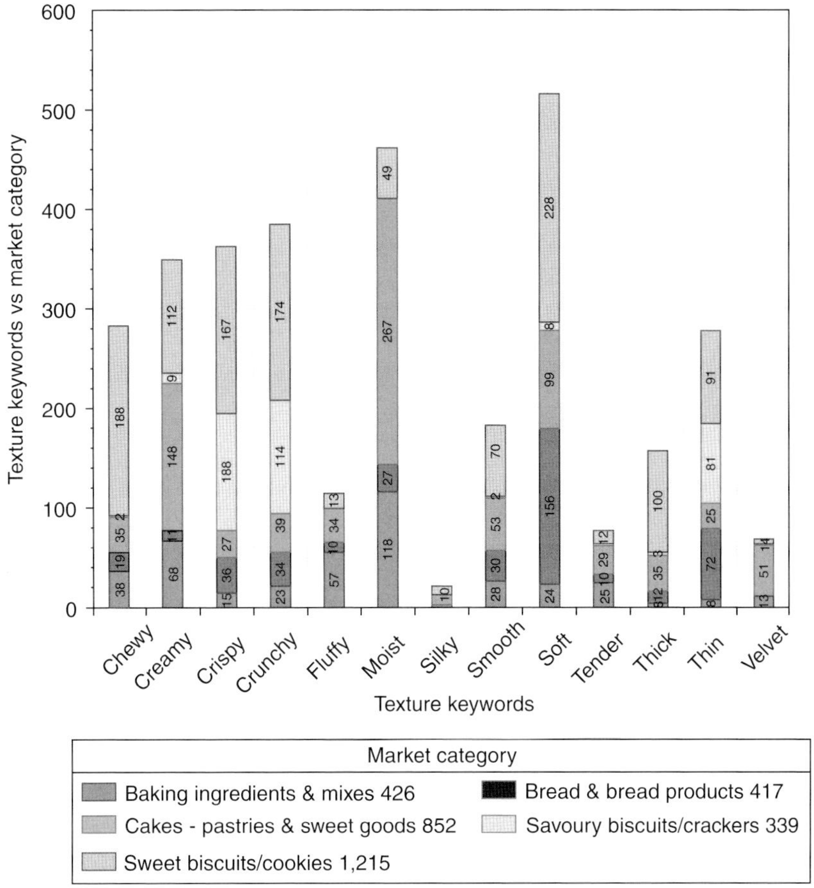

Figure 1.5 Top texture terms featured in texture claims on bakery and snack product launched between 2009–2011. (Source: Image generated using Innova Market Insights by S. Mutz-Darwell, Ingredion Incorporated. Reproduced with permission.)

The addition of healthy ingredients to food product to promote personal health and well-being will also continue to lead to texture challenges. The addition of wholegrain or multi-grain flours, soluble and insoluble fibers, and proteins lead to texture challenges due to the unique and sometimes undesirable texture attributes introduced by these ingredients in a variety of food products.

The need for allergen free foods for populations suffering from different food allergies will also continue to generate texture challenges that will need to be addressed. As an example, the development of gluten free baked goods have seen significant interest and advancement over the last 10 years but several challenges still remain to be addressed.

Processing and shelf-life optimization are also important challenges that will continue to provide new food texture challenges. New processing techniques such as high pressure processing (HPP) (Doona and Feeherry, 2007) provide new avenues to develop food products with desirable attributes. However, these techniques lead to texture related challenges as the difference in processing conditions lead to changes in texture from ingredients as compared to traditional cooking methods. High quality microwavable products still continue to be a challenge. Evolution in shelf life and food safety requirements can also lead to changes in formulation and processing and consequently changes in texture.

References

Amerine, M.A., Pangborn, R.M., and Roessler, E.B. (1965) *Principles of Sensory Evaluation of Food*, Academic Press, New York.

Bourne, M. C. (2002) *Food Texture and Viscosity: Concept and Measurement*, 2nd edn., Academic Press, San Diego CA.

Cakir, E. (2011) *Understanding Textural Properties of Protein-Based Soft-Solid Structures using Oral Processing*, PhD thesis, North Carolina State University, Raleigh NC.

Doona, C.J. and Feeherry, F.E. (eds.) (2007) *High Pressure Processing of Foods*, Wiley-Blackwell, Oxford.

Gladwell, M. (September 6, 2004), *The Ketchup Conundrum*, The New Yorker, New York.

ISO 5492 (2008) *Sensory Analysis – Vocabulary*.

Kilcast, D. (ed.) (2004) *Texture in Foods: Vol. 2: Solid Foods*, Woodhead, Cambridge.

McKenna, B.M. (ed.) (2003) *Texture in Foods: Vol. 1: Semi-Solid Foods*, Woodhead, Cambridge.

Moskowitz, H.R. (ed.) (1987) *Food Texture: Instrumental and Sensory Measurement*, Marcel Dekker, New York.

Moskowitz, H.R., Beckley, J.H., and Resurreccion, A.V.A. (2006) *Sensory and Consumer Research in Food Product Design and Development*, Wiley-Blackwell, Oxford.

Rosenthal, A.J. (1999) *Food Texture: Measurement and Perception*, Springer, New York.

Shewry, P. R. and Khan, K. (eds.) (2009) *Wheat: Chemistry and Technology*, AACC International, Minneapolis.

Szezsniak, A. S. (2002) Texture is a sensory property, *Food Quality and Preference*, 13: 215–225.

TECHNOLOGY IN AUSTRALIA 1788 – 1988 (2000), Available: http://www.gpatamworth.com.au/who-we-are/our-history

World Bank (2012) *Food Price Watch* Vol. 3, Issue 12 Washington DC.

I Product development challenges and texture solutions

This volume provides a current snapshot of the depth and breadth of efforts around the world in food texture design and optimization. It is divided into two parts. Part A provides examples of key product development challenges that product developers and researchers around the world are dealing with. Part B deals with advances in tools and techniques used for texture design and optimization. Chapter 2 provides an insightful personal perspective on the past, present and key future developments in this rapidly advancing field by an author who led several pioneering initiatives in advancing food texture design and optimization.

Chapters 3, 7 and 10 provide examples of developing food products for people who need to avoid certain ingredients due to allergies. Chapter 3 provides a good overview of the texture challenges and some solutions to develop egg-free food products. Chapter 7 provides an overview of the challenges and current options available to replace wheat, barley or rye flour to develop gluten free baked goods. Chapter 10 describes the unique challenges related to pasta texture and how these need to be addressed in the development of gluten free pasta while maintaining desired texture characteristics.

Chapters 4, 5, 6 and 11 describe approaches to replacing or adding ingredients in food products for health or labeling concerns. Chapter 4 addresses challenges with low-fat ice-cream and how to maintain the desired eating experience for this indulgent product while providing a healthier nutritional profile. Chapter 5 describes some of the issues related to gelatin in food products and describes options to replace gelatin without compromising in texture. Chapter 6 describes the use of specialty whey protein as a healthy ingredient with textural benefits in reduced and low fat foods. Chapter 11 describes approaches and options to address texture challenges in baked goods with fiber.

Chapter 8 addresses unique textural aspects of battered and breaded foods and how processing and functional ingredients can be used to optimize texture. Chapter 9 is a comprehensive exploration of the unique challenges related to multi-textured foods and how to design and optimize textures for products that have components with different textures.

Food Texture Design and Optimization, First Edition. Edited by Yadunandan Lal Dar and Joseph M. Light.

2 People, products, texture: a personal retrospective

Howard R. Moskowitz

Moskowitz Jacobs, Inc., White Plains, New York, USA

2.1 INTRODUCTION

In 1967, I sat down with my doctoral advisor, S.S. (Smitty) Stevens, for one of our informal meetings. Smitty, Harvard's legendary psychophysicist and a proponent of systematic measurement of sensory experience, would instruct by sitting with a student and patiently going over the student's data. It was frightening, the way graduate life should be, at least at Harvard.

Stanley Smith Stevens – 1906–1973. (Source: Eastern Psychological Association. www.easternpsychological.org/i4a/pages/Index.cfm?pageID=3379)

Looking back, what happened in that half hour was important for me, at least as the first sharp, clear memory of what would turn out to be a lifetime of professional study in texture. Smitty invited me to sit with him at his data analysis table, a large translucent

Food Texture Design and Optimization, First Edition. Edited by Yadunandan Lal Dar and Joseph M. Light.

rectangle of plastic, tilted about 30 degrees, with a light underneath. It was Smitty's lightbox, a home-made device that in the 1960s allowed him to trace data, compare different curves, and essentially become far more intimate with the data than we, gifted with computers, ever do today.

That morning Smitty had received a letter of invitation from the D. Reidel Publishing Company in Dordrecht, Holland, to join the editorial board of a new-to-the-world journal, the *Journal of Texture Studies*. Smitty was an expert in a then arcane field of experimental psychology, with the focus on relating physical stimuli to sensory perceptions. In his characteristically brusque, always memorable fashion, he grunted, muttered to himself about yet another journal, looked at me, smiled, and then said something to the effect that *I'm going to see whether Joe (Stevens, no relation) would be interested in this new area, texture. After all, Joe did a lot of the work in perceiving hardness. What do you think, Howie?* And that was it, a five minute digression, interrupting our sacred time together, giving me an opportunity to see how the grand man reacted to this new world of texture.

Well, Joe ended up joining the editorial board, I ended up getting a Ph.D. with Smitty, working on the sense of taste; texture was momentarily disposed of, but not for long. The romance just started with that off-hand introduction. Texture would remain of tangential interest to me for the year and a half, 1967–late 1968, when I did experiments on taste mixtures for my doctoral thesis. Texture would again rear its head, however, around November 1968. I had already submitted my written doctoral thesis and was waiting for oral exams, which would not take place for two months (early January 1969). What should I do? And now, enter texture once again, this time to stay with me for 42, now almost 43 years, longer than most marriages, happier than some, continually revealing new sides, new dimensions, like the best marriages do.

2.2 PART I – FORMATION

2.2.1 Phipps Arabie and Rose Marie Pangborn

During those tension-filled late fall and early winter months after I had handed in my thesis, I waited both for word from 'on high' about my fate, as well as the prospect an answer about my first real job, as a research scientist at the U.S. Army Soldier Systems Center. The laboratories were in Natick, 17 miles west of Cambridge, a straight shot along the Mass Pike, a world totally, utterly beguiling, which beckoned me away from Harvard into adulthood.

But what to do in those two months? I happened across papers by Rose Marie Pangborn in the Department of Food Technology at the University of California, Davis. She was then doyenne of the newly emerging world of sensory evaluation and co-author of the now iconic *Principles of Sensory Evaluation of Food*, and highly disciplined researcher. I took note of Pangborn's work, especially taste and texture. If sweetness, for example, could be suppressed by bitterness, what happens when we taste sugar say in a thickened product? It is the same rules. Could I apply the same procedures of my discipline, sensory psychophysics, to mixtures of sweetener agent? So, in some ways, it was Rose Marie Pangborn who pushed me into texture research.

But it wasn't just Rose Marie at Davis and the nervous moments between thesis and job that led to my interest in texture. It was Dr. Phipps Arabie of Rutgers University, then a mere Harvard undergraduate, who hung around the psychology department, took

Rosemarie Pangborn. (Source: American Chemoreception Research Society. www.achems.org/i4a/pages/Index.cfm?pageID=3296)

advanced courses, became my assistant, and as the years progressed, my lifelong friend who urged me. We spoke for many years, until Phipps's untimely passing in June, 2011, as this chapter was being written.

But just how did texture fit into this? Neither Phipps nor I knew anything about texture. But we were young, eager and the world was opening up. Phipps was always game to pursue new experiences in science. Texture, or really taste-texture relations, was about to hit us right in the face.

Phipps Arabie. (Source: www.multiresolutions.com/strule/tcd/57.jpg)

The reason was simple. As I explored the relation between sensory intensity and physical concentration, I kept getting power functions of the form (Sensory Intensity) = a(Concentration)^b. It was linear except for one repeated departure, high concentrations of sucrose in water, which flattened out. Much to my chagrin, the nicely linear sugar-sweetness function in log-log coordinates lost something at around 10% sucrose in water, and then started flattening out. Just what was going on? Why wasn't this sweet, gooey, sugar water solution obeying the power function, and continuing to go straight up? As much as I tried, it was impossible to straighten out. Here was sucrose sweetness, throwing me a 'curve' right at the start of my professional career.

And so, everything was nice for this almost finished graduate student. Everything except a nagging curvature. And then there was my hypothesis – maybe the very thick syrupy nature of the sugar solution interfered with the ability of the molecule to reach the tongue. Maybe, just maybe, it was texture, a 'disturbing secondary variable' that underlay the curvature!

It was Phipps who suggested the experiments relating perceived taste intensity jointly to the concentration of tastants, and to the apparent viscosity of the tastant-gum solution. This would end up being a psychophysical function comprising two variables. His suggestion was all that was needed to start the ball rolling. I invited Phipps to join me in the research, which, of course, he did. We performed our taste-texture experiments in the harsh winter of 1968, just before and just after the Christmas vacation. Inspired by Smitty Stevens, who looked on benevolently and encouragingly, we measured the exponent of the power function relating taste intensity to concentration.

At that time, a lot of our thinking was dictated by the analysis of relations between one independent variable (e.g., concentration) and one rating attribute (e.g., taste intensity). We looked for the exponent b, of the simple power function: Taste Intensity = a(Concentration)^b. It was a first for me, to relate taste intensity to the levels of two variables, concentration and apparent physical viscosity: Taste Intensity = a(Concentration)^b (Viscosity)^c.

In the end, Phipps and I published our findings. Our first publication reported the results of a series of such experiments, appearing in the aforementioned *Journal of Texture Studies* (Moskowitz and Arabie, 1970). Our second publication was in the experimental psychology journal, *Perception & Psychophysics*, a relatively new journal (Moskowitz,1972). Our romance with texture had started. The key lesson learned there was that psychophysics, my chosen field, the science relating perceived magnitude and physical magnitude, could encompass texture. Texture had started out as a minor, disturbing factor in taste; it would grow to be a great love.

And, what of the lesson in this history? Is there some wisdom to be gleaned? In early 1969, sensory researchers in psychology were just beginning to wake up to the relevance of texture. It had little to do with food. Their focus was on the sensation of touch and deep feel (kinaesthesis). The idea that such sensory research had anything to do with food and drink was, at best, a hint from the 'real world.' Looking back, I continue to be amazed at just how insulated we were, how little we knew, how rarely scientists from different fields knew what was going on in each other's domains.

The more important lesson, however, is 'why?' With hindsight, it seems that lack of knowledge came from at least two sources.

1. *Isolation:* We simply didn't read each other's literature. Google® had not yet been thought of, so to keep abreast of the 'outside world' we relied on *Current Contents*, a weekly photo-offset of tables of contents pages of a journal that we bought from Philadelphia's Institute of Science Information.

2. *Development:* The field of texture research had to come into its own, and was beginning to grab attention. Texture research was young and scientists had a more difficult time connecting. The timing simply wasn't right yet. But it would be right. The future was on its way.

And so, this personal story continues, now moving 17 miles, to the countrified old New England town of Natick, Massachusetts, home of the Natick Soldier RD&E Center at the U.S. Army Natick Soldier Systems Center, and the site of my next seven years of joyful scientific labor.

2.3 PART II – DISCOVERY

2.3.1 The Natick years and a whole new world

It's fair to say that the years after one receives the Ph.D. end up being the most formative ones in a career. During those years, the newly minted scientist doesn't carry around the deadening weight of experience or the galling taste of dashed hopes. The world is new; one has received the long sought-after goal, the Ph.D., allowing for attendance at conferences and the right to be addressed as Doctor.

But there's more. Those early years after the Ph.D. allow one to wander off a bit, to go into new areas, and to meet new people. It seems like one is never as alive as during that first magic year, when the world is fresh, the mind inquiring, the body ready to stay up all hours to discuss science, to challenge, to be challenged, and to simply live the life of the mind. They were heady times, these early years, and truly formative for an appreciation of texture.

Those words describe the first couple of years I spent at the U.S. Army Natick Laboratories, where I worked as a senior research scientist in the Behavioral Sciences Division. My 'boss,' really more of an older colleague, the late Dr. Harry Jacobs, had the wisdom to encourage his young scientists, Drs. Linda Bartoshuk and Herbert Meiselman, and me, to explore our own individual interests. And so began the second chapter of texture, this time in earnest, and not as the investigation of a so-called secondary disturbing phenomenon, as it had been when Phipps Arabie and I explored taste and texture, with the aid of sugar, water, and sodium carboxy methyl cellulose (CMC).

It's fair to say that my experience of the world of texture changed dramatically when I had the good fortune to meet Dr. John Kapsalis, a rheologist. He was a scientist who studied the effects of force and deformation on physical products, in his case, foods. In the hallways of the food science building, I met John's technician, Hillar Haan. He showed me their Instron Universal Testing Machine®, which was able to apply forces and measure all sorts of deformations. I was hooked. For a psychophysicist, there was an opportunity to apply basic forces to products, to measure their deformation, and to do the same type of measurement with people, this time asking them to chew and to rate hardness, or springiness, and so forth. For this then 25-year-old scientist, the prospect was exciting.

After I had had two or three meetings with Hillar, John Kapsalis introduced himself, feeling that his technician had 'vetted me,' and found something possibly of interest. Our first meeting was electric, as John and I sat together for what must have been at least

90 minutes, just chatting. That first 90 minutes would multiply over the years, becoming in time a deep friendship. John and I talked at first about rheology and science, then later about John's poetry (he was an accomplished poet at that time, writing in both Greek and English), and finally, about life in general. John had lived through the Nazi occupation of Greece, had come to the United States, married, raised a family, and was filled with life's wisdom. He was also a member of the informal GIFT (Greek IFT), a group of Greek colleagues in the world of food science who I would eventually get to know over the years from his introductions and attending professional meetings in the food industry.

My education in texture was beginning. With John, discussions of psychophysics took a new turn, as I became a mentor to food scientists. It turned out to be a heady time, truly filled with joy at learning, meeting, developing. The most important part of my education came through Friday lunches with John, where we talked texture and the potential of psychophysics to add to the knowledge of this newly emerging science. We did informal experiments on products and ended up publishing an article on the perception of hardness and crunchiness of space cubes (Moskowitz, H.R. and Kapsalis, J.G., 1975). The key thing about that research was its practicability; one could create equations that described how sensory perception varied with fundamental parameters of food texture.

2.3.2 Meeting and then entering into the circle of texture researchers

When recounting one's personal history of science, it's common to recall and recount significant scientific encounters. Perhaps the most significant for this author was the circa 1970 meeting in Tarrytown, New York, hosted by Dr. Alina Szczesniak, a senior scientist at General Foods, Inc., the doyenne of texture research, at least from the point of consumer researchers.

Alina Szczesniak '48 in 1984.

Alina Szczesniak. (Source: Bryn Mawr College. www.brynmawr.edu/alumnae/bulletin/szcz.htm)

As I was about to discover, in the late 1960s and early 1970s the field of texture research was quite small. To my delight, John Kapsalis had received an invitation from Alina to attend a meeting about texture, and the newly born field of psychorheology (the study of the perception of texture). Dr. Szczesniak and Amihud Kramer had been invited to edit what would become the first book on texture from practicing scientists. Ami, a prolific researcher in texture, immediately caught my fancy. He was born in the town of Drogobych, now the western Ukraine, then Galicia, a province of the Austro-Hungarian Empire. One of my great-great-grandfathers had been the Chassidic rabbi in Drogobych, a century and a half ago. John and I were put up at the Hilton Hotel near Tarrytown. On a personal note, I pass the hotel now and again on my way to the Tappan Zee Bridge. The hotel has changed a bit, gotten older as we all have, changed names four or five times. I can't help but smile when I think of that meeting held 41 years ago. Certain events just never seem to fade away, becoming ever more precious as one ages.

People from different fields, including Malcolm Bourne from Cornell, Philip Sherman from Unilever, Essex Finney from the USDA, Birger Drake from SIK in Gothenberg, and so on attended the meeting. Actually there were a few more; these individuals are just the ones that stand out. The most remarkable fact was just how little we all knew, but how excited we all were at the prospect of discovering the future in a new field.

The topics of conversation were varied. We each had to present a short précis of what we were doing, what we were thinking, and where we saw the future. The work product that came out of it was a very thin book, *Texture Measurements of Foods* (Kramer and Szczesniak, 1971). It was a beginning, as texture researchers were getting to know each other, creating bonds and linkages that would last years. At 26 years old, I became enthralled with texture and a member of the inner circle.

If I were to summarize the importance of that meeting, it is that small conferences, by invitation only, have been and continue to be, a very effective method for bringing 'new blood' into the circle. They revitalize a field of science by encouraging new ideas. You can't overestimate the value of experience, making connections with others, as a true foundation for good science. The soft science there was as powerful as the hard, and a great deal more inspiring.

There are other benefits as well, some of which I think every young scientist experiences when starting out in a career, meeting the leaders, the doyens of the field. This texture meeting was an exciting sharing of information. Looking back, however, it was something more, something bigger. It was the subtle molding of a mind. Talking to leaders in the field, listening to the chit-chat, hearing colleagues making references, sharing war stories, all created a wonderfully fertile bed that nourished my young professional mind. It was the soil of a young mind, being turned, fertilized, seeded with ideas. We don't even know it's happening until years later, when we might look at some experiments and realize the chain from a chance discussion to the experiment and data lying before us. These were the memorable times. They're important in the early days, experiences to be cherished. They don't happen as often later on, perhaps because after just a few years our heads are all filled with a lot of what we knew, protecting us from that which might threaten us, the new and unknown, that early friend of ours, later our mortal enemy.

2.3.3 Psychophysics, texture, Roland Harper and Smitty

During those heady years at the Natick Laboratories, I had a chance to get to know Smitty better. Smitty was more delightful when visited as one's professor, but I felt safer after I had my Ph.D. Discussions were then far more delightful than when I was advised by Smitty during those tough years in the mid-late 1960s. Smitty wasn't interested in texture, per se, but he maintained an avid interest in sensory perception. During the period after the Ph.D. was granted, and until his untimely death in January, 1973, I'd visit Smitty, often just to chat, sometimes to get some wisdom, sometimes just to touch base with Harvard psychophysics.

One of the wonderful times during these early years was when I was asked to chauffeur the late Professor Roland Harper. Smitty and Roland had collaborated, as Roland told me, in the late 1950s and early 1960s. Dr. Harper was a doyen/professor of applied sensory science at the University of Reading in the U.K., a psychophysicist, a gentle scholar, full of mirth, who delighted in talking with young scientists.

Roland Harper. (Source: Society of Chemical Industry. www.soci.org/About-Us/About-SCI/History/Notable-Chemists)

It was during Roland's visit to the U.S. that I drove him around. While we chatted, I got the sense of what texture science was like in the late 1930s and 1940s, when Roland was working with cheese graders, and studying the psychophysics of the touch and feel of foods.

I bring this memory up, because this meeting with Roland and Smitty made me realize that it was perfectly valid, acceptable, to move between the science of Smitty, the pure psychophysics of perception applied to texture, and the application of these principles to real food, the way Roland was doing it. Often, those of us schooled in the pure academic tradition feel that we are betraying our intellectual heritage and education by applying what we know, rather than attempting to maintain the purity of our thinking, the splendid isolation of our science.

It was occasions like these, 40 plus years ago, that are indelibly engraved on my mind, as treasures of a hybrid, academic/applied life in sensory psychophysics and texture, and why I write about that occasion. Writing this memory is important. In some ways, that one visit was the impetus for my own acceptance of applied work. After all, if Roland Harper could do it, and get Smitty's respect, it must be worthwhile doing.

2.3.4 Texture and the Natick Laboratories library

One of the very nice things about being young is that one has the energy to follow through on a lot of things. Today as I look back, I realize that I very rarely go to a library. All that I need is, at least in my opinion, at my fingertips. Google®, Google Scholar®, JSTOR digital archives, and other web based tools are available to search the literature. My fingers which originally walked in the Yellow Pages™, can now do so in journal pages. And so, I don't have to waste time or expend extra energy. I know what I'm going to write. More often than not I can go right away to the reference I need. References, articles previously published, are like bits of decoration and mortar, to be put in the wall, to decorate, or in some cases simply to establish that the paper is appropriate for the so-called scientific community.

Thinking back 42 years ago, I recall becoming interested in the world of texture while also thirsting to know more. Rose Marie Pangborn's aforementioned book *Principles of Sensory Evaluation of Food* (Amerine, Pangborn and Roessler, 1965) certainly was a good starting place. She was a meticulous collector of scientific information, a wonderfully disciplined practitioner.

But Amerine, Pangborn and Roessler just didn't do the trick; I wasn't satisfied. And so, lunch time at Natick would find me wandering the stacks of the laboratory library, a magical repository of the history of the world of food research, among other disciplines. I could get my hands on dairy journals, on meat journals, even *The Journal of Food Science*, from its founding days (late 1930s). The hunger for knowledge was there. I'd read about the Ridgliemeter, the Gelometer, put myself through some of the German food literature, and really just immersed myself. Turning the pages, looking at the articles, the ads, seeing what people were writing about, saying, doing, were all part of this early affair with texture. It was delightful, chatting with John Kapsalis on the one hand, going to texture meetings with other scientists and then, on the other hand, reading what happened 30, 40, 50 years before. I pieced together my own sense of how the world of texture evolved. It's one thing to read about methods and results. It's quite another to pore through the actual article and, in some ways, be transported back to the laboratory as the author of the article struggled to put the results of a scientific experiment on paper.

2.3.5 Space cubes

Over the next five years, John Kapsalis and I would become close friends, sharing ideas about science, poetry, life, people, and so forth. After about six months, John asked his assistant, Hiller Haan, to invite me to collaborate on the study of spice cubes, small hard, bite sized food cubes, about the size of a Herbox® cube, the type used for soups or seasonings. The topic was texture. I wasn't privy to the issues leading up to the exact study, and if truth be known, those issues weren't interesting. I was simply interested in getting my hands on data from John's Instron Universal Testing Machine®, and correlating the fundamental physical properties with perceived hardness. John, or more realistically Hillar, would make the measurements, and I'd have my panelists rate hardness. And so, we looked at a number of these cubes, getting basic physical measures from the Instron®; the modulus of elasticity and ultimate strength. From the consumer panelists, we'd get ratings of hardness and crunchiness, both of which co-varied with modulus of elasticity, but far more with ultimate strength. The question was 'which is the better independent variable'. I suggested the 'better' variable was the one that generated a power function with exponent nearer 1.0.

I often think about that experiment, not because the substantive results were so good, but rather because it revealed a fundamental issue with texture. Just what was the stimulus? We could measure the response, the hardness, the crunchiness, with a fair degree of accuracy. And, in those moments, I was transported back six years to Smitty Steven's same recurrent question – what is the stimulus? This is the first order question for texture – just what should you measure?

2.3.6 Ed Cussler, Jay Kadane, Joe Kokini ... and the new world of 'what is the real stimulus?'

Towards the end of my seven wonderful years at the US Army Natick Laboratories, I ran across Edward Cussler and Jay Kadane of Carnegie Mellon University. The phrase 'running across' is probably incorrect. Our paths met, because Ed called me one day to talk about texture, psychophysics, his work, and the possibility of working together. Would I care to come to visit him in Pittsburgh? Jay Kadane was a statistician, a close friend of Ed's, and as it turned out, the son of one of my late father's dearest colleagues. What could be more auspicious!

Ed's work lay exactly in the 'sweet spot' in texture, and was irresistible. Ed was a physicist/engineer, who had come across Stevens' work, and realized that a lot of what we knew in texture had to go further than the machines. Whether we were working with integrative machines that simulated the mouth (e.g., Kramer's Shear Press®, Szczesniak's GF Texturometer®, or any of the other machines trying to be analogue sensors), or whether we were working with fundamental physical measures such as stress and strains measured by the Instron Universal Testing Machine®, we were missing something important. That 'something' was a description of the forces possibly occurring at the site of the receptor. And it was that work, the effort to describe the local forces, which was so attractive to me, a sort of siren call. The particulars of the work, now approaching almost 40 years, are not as interesting as the path.

It's worth a short digression to elaborate on just how exciting I found the notion of measuring forces at the receptor. A decade before, in the mid 1960s, Smitty Stevens would often punctuate his monologues about data and experiments with a word of wisdom about science and the deeper issues. Many a time, he would opine something to the effect that the real task of psychophysics is to measure the response to the REAL stimulus, the stimulus at the receptor.

In texture the big word was 'subjective-instrumental' or 'subjective-objective' correlations. One need only look at *The Journal of Food Science*, for tables showing correlation coefficients between objective measures and sensory measures. To a world that had to be satisfied up to then with correlations, Cussler's work on the proximate physical stimulus was a gift from heaven. And this psychophysicist, this author, wasn't about to pass up that gift!

2.3.7 Final Natick moments

By the mid 1970s, I was beginning to get a bit of wanderlust. Natick Laboratories was my first job after the Ph.D. As jobs and scientific environments 'go,' I couldn't have started in a better place. Or, if I were to stay there for the next 40 years, as my dear friend and colleague Herbert Meiselman did, also in Behavioral Sciences, I could point to a life and career well and smartly spent.

But I needed more. My late father, Moses Moskowitz, challenged me on one visit with the notion that perhaps I ought to explore other opportunities, push myself, open up new vistas, and not just relax. Any other person but my father would have said those words to no avail. But when Dad said them, it made me stop and think, and then slowly, inevitably take action. The irony is that Dad was not at all scientific, not at all quantitative. An intellectual, writer, founder of a non-governmental organization (Consultative Council of Jewish Organizations), Dad was always pushing for more.

Perhaps another, more sensitive reason, now in retrospect, is that Dad worked closely in his organization with Mrs. Helane Kadane, Jay Kadane's mother. It was that sense that Dad was somehow part of my work. That eventuated my listening to him and, in fact, taking my science into the real world of application, a world that Dad could not hope to enter, but had always admired.

As a final effort, though, Dad encouraged me to move beyond simple experiments, to create a grand plan, an architecture, that would represent more than simple experiments. And, when he heard that I was working with Ed Cussler and, indirectly, with Jay Kadane, he pushed even harder. And so, my final effort was to attempt to synthesize an analog psychophysics of texture. I didn't get very far, but the notion was simple. If, for instance, $F=MA$ (force = mass × acceleration), then perhaps $PM=PF \times PA$, or (psychological force = psychological mass × psychological acceleration). A primitive notion, of course, but I began to think about the problem, wondering whether it could be applied to texture. Was there a physics of texture and a parallel psychophysics? It was with that problem, and the massive difficulties ensuing trying to solve it, that moved me out of the world of basic science and into the world of industry. I had met a topic area worthy enough to spend a lifetime, developing a basic science of the mind, a mental science of texture, and realized I could not solve the problem. And off I went into industry.

2.4 PART III – APPLICATION

2.4.1 By way of a second introduction

In 1974, I began to get restless. Natick was a wonderful place, but I had this nagging feeling inside. Was all that I had learned simply destined to end up in academic papers, some dry, some not so dry? Was psychophysics simply self-gratification, intellectual trips on highways and byways, leading nowhere?

One of the wonderful things about Natick was the encouragement given to me by my then supervisor, Harry Jacobs, Chief of the Behavioral Sciences Division. Harry believed in letting go, not restraining his cadre of young scientists, and giving them permission to roam the academic and business worlds, as long as the roaming could be justified.

During my seven years at Natick, I was fortunate to attend meetings of the Institute of Food Technologists (IFT), home to food scientists, food engineers and, of course, food ingredient suppliers, not to mention at that time research scientists and business people from the food industry.

It wasn't long before a number of the people I met, especially the late Victor V. Studer of Thomas J. Lipton, began to recognize that there might be something of value in psychophysics that could be applied to industry. Vic, and his colleague Arthur Hawkins, of Colworth House, Unilever's food research laboratory located in an idyllic English manor house just outside London, would meet me at the occasional scientific conference. Both delightful colleagues often invited me for a drink or for a dinner. These were heady times, and as I was to discover, they were the early siren calls of a new career, the next stage of my romance with texture.

2.4.2 A new view of texture

On October 31, 1975, I left Natick Laboratories to begin the next phase of my career, as a marketing researcher. During the preceding year I had been accepting small, part-time assignments to evaluate food and cosmetic products, working with MPI, a marketing research firm in New York. The work was interesting, but what was really exciting was the chance for a young scientist to validate himself, to test out his learning, to compete with others, to see whether his education and dreams really meant anything in the cold world of commerce.

Texture would play a key role here, because my maiden voyage into this new world of application was begun with a small, paid consulting engagement at Unilever's aforementioned Colworth House. Arthur Hawkins, deeply involved in Unilever's scientific outreach, had come across my work in texture. He invited me to consult for two days with his scientists. I was honored, excited, delighted, and just plain ready to tackle the world. And so, my life in business began with texture as my first 'real-world' gig.

When I returned to New York, moved down to my apartment on East 24th Street in Manhattan and started in earnest, I faced a new world. It was no longer the happy, simple, and idealistic world of the young research scientist. It was the world where questions were to be answered. I was to be the expert. Questions of taste and texture weren't going to be referred to a scientific group. I would be asked to solve problems, such as how do we make this sauce better?

The world-view changed the relevance for texture. In the applied world of market research, there isn't time to explore how the world works. Not when one wants to survive. Questions of texture were important, not only to discover the mechanisms, but rather texture as a sensory factor to be measured and fixed, if necessary.

The real world educated me about texture in a way that the research laboratory could never hope to. At Natick, the problems were scientific and the issues to be solved revolved around the search for 'how the world works.' Correctness, validity, or any of a dozen other such evaluations came from the solidity of the data, just as much as from the approbation of one's fellow scientists. Advancement in science and in the field of texture would come by being invited to contribute chapters to books, to go to conferences and meet other scientists and, of course, to chair sessions, review papers, and the numberless activities that fill the day of the scientist.

But not in business; it was different. Smitty had a phrase for it, 'validity is a matter of opinion.' In business, especially consumer evaluation of products, texture was one of the dimensions. There were some scientific issues, but mainly it was acceptance by the client that meant the most. What an education! I had come in with the dreams and fantasies of a young, 31-year-old scientist, just seven years after the Ph.D. I had hoped to bring my methods to the commercial world. Instead, the commercial world reshaped me, another smithy of my soul, just like Smitty had been that first smithy (no pun intended here).

2.4.3 Texture to the consumer researcher

At this point, I ought to confess that I came to the field of consumer research as a psychophysicist and not as a so-called sensory specialist. In the field of sensory and consumer research with products, a great divide had taken place in the early 1970s. The industrial world had to answer problems, to ensure that products would sell and ensure that their quality was maintained. Two main avenues of effort emerged out of these 'had to's.' One was psychophysics, studying the relation between physical stimuli and perceptions, with the goal of engineering the product. The other was descriptive analysis, the description of the sensory aspects, for example, the dimensions and nuances of texture. I was a firm follower of psychophysics, and not particularly a fan of descriptive analysis.

Just what did psychophysics do then, in these formative days? Well, I didn't spend my time profiling the different characteristics of texture as was then *au courant*. That was introduced to the field by scientific doyenne Alina Szczesniak and her loyal General Foods lieutenant and colleague, Elaine Z. Skinner. I moved in a different direction. I was interested in texture as a dimension of perception, a driver of perception that could be studied quantitatively.

Howard Moskowitz. (Source: Photo © Moskowitz Jacobs Inc.)

A short aside – It seems that in many fields, description rather than functional relations, constitute the first steps to a science. Texture, it seemed, was no different. Most of the science I had learned dealt with the search for functional relations between variables, with the goal being to understand how to change variables so as to create new perceptions. The sensory world that I would encounter during the years 1975–1992 rejected psychophysics. Instead, corporate time and money was spent training expert panels to describe the sensory 'notes' of products, whether flavor or texture. But over time, descriptive analysis receded. From being an absolute must in the arsenal of a sensory researcher, it reverted back into the tool that it is, without providing the magic that people sought. This pattern of wild adoption followed by disillusion followed by 'normalcy' occurred in the world of flavor description and, finally, in texture description. Scientists realized that they couldn't create new products by descriptive analysis, but only describe them. It's that realization which led product developers and marketing directors away from descriptive analysis and into what ended up being a version of psychophysical exploration and modeling (see Boring, 1929; Cairncross and Sjostrom, 1950; Szczesniak, Brandt and Friedman, 1963).

Most of the early commercial studies that I did tested a number of different products. Sometimes the products represented the range of available products in the commercial market (so-called category appraisal). At other times, the studies included systematically varied products, such as parboiled rice created by combining different treatments, or honey bran cereal formulated with different ingredients, and also under different process conditions.

These studies weren't focused on texture, not in the way I was accustomed to focusing when I was a scientist at Natick Laboratories. They were certainly not conducted in the manner that experts in descriptive analysis would work with products, trying to create their texture profile, to discover their texture signature.

How did I conduct my studies? My instructions to the respondent appear in Figure 2.1. My data typically looked like the data presented in Table 2.1. (See next page). The rows correspond to products, the columns correspond to products. The respondent was instructed to look, smell, taste, perhaps even swallow the product, rating the different characteristics on the scale provided. Texture, in this case, comes to the fore because the respondent is instructed to rate both liking of texture (how the product feels in the mouth), as well as sensory intensity of different attributes.

The data that we see in Table 2.1 are typical of the data that the business generated in its commercial assignments. When someone generates this type of data in academic settings, typically the goal is to publish the work, more or less in order to add to the store of human knowledge and, of course, get the coveted publication. The academic doesn't really have to do anything with these data, other than obtain the data in a reasonably rigorous, defensible way, show how the problem fits into some string of previous effort, and then present the data as a new, generally small, but worthy contribution to the field.

In business, however, the unspoken ground rules are a bit different. Certainly one has to acquire the data in a reasonable, scientifically acceptable way, so that the data are unbiased. But that's not the essence of the story. The data have to *do* something, they have to *lead* somewhere, produce some piece of knowledge worth paying for. Increasingly, the focus in business is 'WiiFM,' what's in it for me, what's in these data for me to move my business ahead, to get my business task done? One can't simply get away with saying that the data fill a hole in the literature. The data have to *do* something within the larger structure of the business.

As I progressed in business, working on projects as varied as the study of cereals, meats, yogurts, beverages, I realized that there had to be what one calls a 'value-add.'

Respondents are pre-recruited to a CLT (central location test) according to specific criteria set forth by the client. They are re-screened at the site and seated numerically in the interviewing room at individual computer stations.

The moderator conducts a brief orientation about the session's procedure. During this time respondents are told about the types of questions they will be answering and are instructed on how to use the rating scale.

Respondents are served their first product.

Once all respondents have their first product, the moderator will take the respondents through the first evaluation as a group. At this point, the moderator ensures that everyone comprehends the questions, rating scale and evaluation procedure.

Respondents answer a battery of questions on key appearance, aroma taste and texture attributes.

Figure 2.1 The instructions. (Source: © Moskowitz Jacobs Inc. Reproduced with permission.)

Table 2.1 The product (column) x attribute (row) matrix. (Source: © Moskowitz Jacobs Inc. Reproduced with permission.)

	Attribute	**1**	**2**	**3**	**4**
Like	**Total**	53	55	47	40
Like	**Appearance**	48	58	45	46
Sensory	**Dark**	64	67	31	44
Sensory	**Uniform**	60	65	67	61
Sensory	**Chewy**	38	69	74	33
Sensory	**Wide**	36	35	32	50

The data, texture information just as much as taste information, had to lead somewhere. But where? What would constitute a good strategy to convince clients that the data were worthwhile? It would be impossible to survive in business simply producing scientific data. Perhaps the descriptive analysis panels could do that, but those panels were inside the corporation, sheltered, given a more or less permanent job, no matter what they did. As an outsider, a supplier, a research 'vendor,' I had to sell my services, again and again, convincing the buyer that these data were worth spending money to acquire.

2.4.4 Necessity is truly the mother of invention – enter the 'Ernie Curve'

In the world of commerce when it comes time to earning a living one ought to have either rich parents, a secure job or, in my case, a 'bag of tricks,' methods by which I could add value to the client. As I was entering the business in the mid 1970s, I made two discoveries that would provide me with income for the rest of my life (or at least so far).

1. What clients really buy: Clients may buy consultants, but only once or twice. After that, people buy projects, that is, results, data, work-products. Consulting has to be part of the project, but the real purchase is for data, for solutions. That was the most important lesson. It told me that although I had a splendid experience with Arthur Hawkins at Colworth House where we talked science, not to mention having a great meal in the executive dining room, the real sustainability would be in projects, and not in those wonderful, ethereal, challenging, but alas all-too-rare experiences.
2. It's *actionable value* add that makes the difference. I watched how others were surviving in the field. When it came to texture, there were a few people consulting, but mainly on product applications. These people had to be generalists. The truth of the matter was that one could talk about a topic like texture, but only a few times, and only once to any of the few companies wanting to learning something. What could one bring to the world of texture that would make a product developer or a marketer want to pay money, more than once? I would have to develop a system that generated value, that produced various generations in its wake, that is, release 1.0, release 2.0 and so forth.

I mention this as a way of introducing the 'Ernie Curve,' which has nothing to do with science. The name Ernie Curve refers to the univariate relation between sensory intensity on the abscissa and degree of liking, or some image attribute on the ordinate. I had discovered, or rather rediscovered, the curve in the 1970s, when working with sugars, and instructing respondents to rate sweetness and liking. They were different curves. What seems so simple now, in 1971 was quite a revelation (Moskowitz, 1971).

The name Ernie Curve evolved from an afternoon's experience with Nestlé USA in 1982. I was invited by Ernst Schmid, a brave, caring and daring market researcher, to tell Nestlé about my insights about food, specifically coffee. In searching for something to share with him, and perhaps even to win a project from Nestlé, I developed the notion that we could plot liking versus sensory ratings and fit a quadratic function. Optimal liking lies somewhere in the middle of the sensory range, not at the extreme. It's simple now, but then it was a revelation to many in the scientific and industrial worlds. It was the start of a new world for me. Figure 2.2 shows the difference between the linear plot and the quadratic plot (see Moskowitz, 1981). (See next page).

By the way, the name Ernie Curve came from the security guard, who when I told him I was at Nestlé to visit Ernst Schmid said, 'Oh, Ernie Smith. He's in room XYZ.' And so was born the Ernie Curve, or at least that's the way I remember it.

So what's special about the Ernie Curve? When you look at Figure 2.2 you see the data, but the curve shows you the pattern or a possible pattern presenting a smoothed representation of the more granular reality that is nature. In some ways, that simple pattern was enough. In all of the applied work I had been doing for clients, generally on the

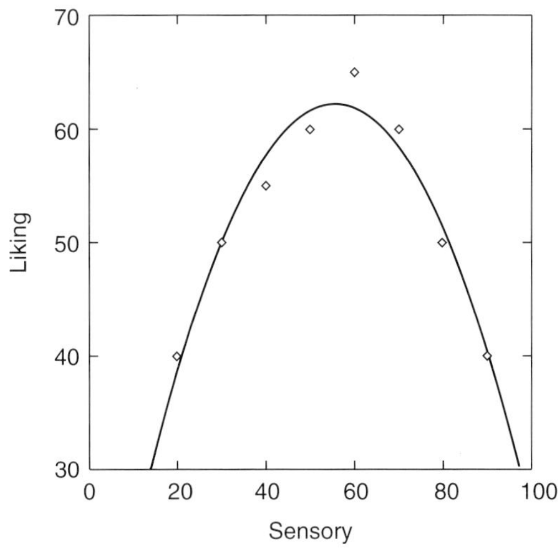

Figure 2.2 The Ernie Curve (relation between sensory attribute level and liking). (Source: © Moskowitz Jacobs Inc. Reproduced with permission.)

basis of a problem they posed and a 'proposal/bid' system, it was important to get the business, but to get it by bringing something new to the business engagement. The Ernie Curve was new and, indeed the notion of discovering a pattern, not just reporting data, was new. I realized that this Ernie Curve was important and perhaps foundational when the clients actually began to think about the implications of such curves and incorporate the curve in development efforts.

2.4.5 How does the Ernie Curve contribute to our knowledge of texture?

I had developed the Ernie Curve to study flavor, specifically of coffee. In the case of coffee, texture or mouthfeel doesn't come into play. Temperature of the coffee is important, but typically temperature was not a factor in any of the early product studies.

It would be a very short time, perhaps a matter of a month, until I would use the Ernie Curve in texture. The client was Vlasic, of pickle fame. The topic was to better understand what drives the acceptance of pickles. The specific issue was the then recent (1982) and disturbing finding that a competitor, General Foods' Claussen Pickles, scored better than Vlasic's for a number of regular pickle products. Claussen was making this claim which was undercutting Vlasic's market share. Nothing so worries a marketing professional in a company as the prospect that his product might be found to be inferior, lose market share because of this inferiority, and that the loss would be a blot on his record, a stumbling block in his future career. It was time for applied psychophysics; I was called in to help, Ernie Curve in hand.

Excitement about an opportunity to do applied research is one thing. Of course, the prospect of getting paid and even ensuring one's future makes that excitement even more delicious. The question remained – just how to make the Ernie Curve 'pay out?' The question was not a theoretical one. I had already learned that simply doing statistics, T tests, analyses of variance, even factor analyses, didn't add much value to projects, and that all the statistical legerdemain I could muster probably wouldn't add much. The Ernie Curve had to contribute, but 'how?'

Thinking about the Ernie Curve, that sensory-liking relation, revealed two prospective uses. These were directions for changing the product and segmentation on the basis of the Ernie Curve, respectively. Both would contribute to thinking about texture, not so much because they were focused on texture, but rather focused on using sensory data to guide product development; sensory, texture being just one of the sensory dimensions involved. And, to paraphrase the words of poet Robert Frost, that *made all the difference*.

2.4.6 Ernie Curves, sensitivity to sensory changes, and the location of the optimum product

The Ernie Curve gives us a tool by which to explore how liking changes with sensory attribute level. What's more, we can construct the Ernie curves by instructing respondents to evaluate a series of related products on overall liking, and then on sensory attributes, and afterwards use a curve fitting program to plot the curve of best fit. We already saw the data that would be used in Table 2.1, and the curve in Figure 2.2. Now, it would be empirical data that would round out the picture, telling us what happened.

Vlasic Pickles and its market fight with Claussen's Deli Dills®, taught us our first lesson in texture using the Ernie Curve. We found something very simple. As the pickle became crunchier, overall liking increased. It was a fairly straight line. People just didn't like limp pickles. A pickle had to hold its ground, maintain its posture, lie there with dignity, and simply not wilt, and not be so soft that insides spurted out as the pickle is bitten into. Pickles were interesting and easy. But there were more worlds to conquer, worlds that would teach a lot more about texture.

Around the same time, in the early – mid 1980s, I met Dennis Degeneffe, first when he was working with McDonald's and later when he joined Oscar Mayer, a division of General Foods (later of Kraft). Dennis had been encouraged by Robert Drane, head of the marketing research department at Oscar Mayer, to look at new ways to understand how consumers evaluate the quality of processed meat. Oscar Mayer, a major division of General Foods, was headquartered in Wisconsin, giving it freedom from General Foods' by-now-standardized processes. It was able to do consumer research without a lot of stifling controls that hobble a large corporation. It was with Oscar Mayer and the studies with all sorts of meats, that some of the applications of the Ernie Curve would reach their highest point, especially for texture. This isn't the place to go through all the wonderful learning about texture that took place during those years, but rather to acknowledge how Oscar Mayer provided the opportunity for texture research to advance significantly in the quest for better products.

The first task posed by researchers at Oscar Mayer was to guide the product developers to make better hot dogs and bolognas. During these years, Oscar Mayer employed food technologists, market researchers and marketers. There were no 'standard ways' of measuring responses, no expert panels and, indeed, very little in the way of what today we would call traditional sensory analysis. The opportunity was open to try new ideas from psychophysics, to uncover relations between variables in the way that the Ernie Curve does, rather than having to conform to sensory best practices, which at that time in the early 1980s was primarily descriptive analysis and tests of difference between products using the respondent as the judge.

The studies for Oscar Mayer are relevant to understanding of texture. All of the data were done with consumers. All of the Oscar Mayer studies were done using 0-100 rating scales, whether for liking, sensory attributes or more complex cognitive attributes. Occasionally, the liking and sensory attributes were augmented by JAR, just-about-right scales. The results from each study generated a treasure trove of data; each study typically worked with as few as five and as many as 20 different samples.

2.4.7 What the Oscar Mayer meat data taught us about texture

When we began our work with Oscar Mayer, our specific task was to address the business issue, the problem whose prospective solution was funded by marketing. Many of these issues began with the question of 'how is my product performing' and 'how do I improve my product.' Those questions were addressed, of course, the results reported, and actions taken. What is remarkable, however, is the residual value of those studies to teach us more, to show us how the world works. Studies with multiple products, multiple attributes, with each respondent evaluating a number of products give us that legacy.

By 1983, we and Oscar Mayer were deeply involved in texture. Oscar Mayer's meat products and, indeed most meat products, vary as much in texture as they do in flavor and appearance. Indeed, it would be fair to say that our work with Oscar Mayer products turned our focus to texture and then to appearance as a key driver. Flavor was pretty well fixed in most of the processed meats, except for the few cases where it was deliberately varied by the change in spice blend. What was not known was how consumers would respond to the different textures.

We now go into a few types of learning that we got from these studies. Our illustrative data comes from some of these meat studies. There are many more studies for texture from other projects, other foods, but we can learn a lot by staying in one area, meat.

1. How much overall liking is generated from a unit liking of texture versus a unit liking of taste, or a unit liking of appearance? The work with Oscar Mayer led us into new directions, mostly hedonics, with questions like this one. When we plotted overall liking as the dependent variable versus attribute liking as the independent variable, we continued to find a straight line. Correlations and factor analysis kept telling us that attribute liking and overall liking were the same. Yet, when we fit a straight line to the data, we ended up with lines having different slopes. Table 2.2 shows these results. Texture does drive liking. In meat, however, texture is far less impactful than taste and aroma, impressions from the chemical senses (see Moskowitz and Krieger, 1995).

Table 2.2 How unit increases in liking of an attribute co-vary with increases in overall liking. The equation is: Overall liking = Constant + M(sensory attribute liking). (Source: © Moskowitz Jacobs Inc. Reproduced with permission.)

	Constant	**M = Slope**
Like Texture	7.06	0.22
Like Appearance	44.9	0.26
Like Aroma	36.5	0.42
Like Taste	−9.98	1.15

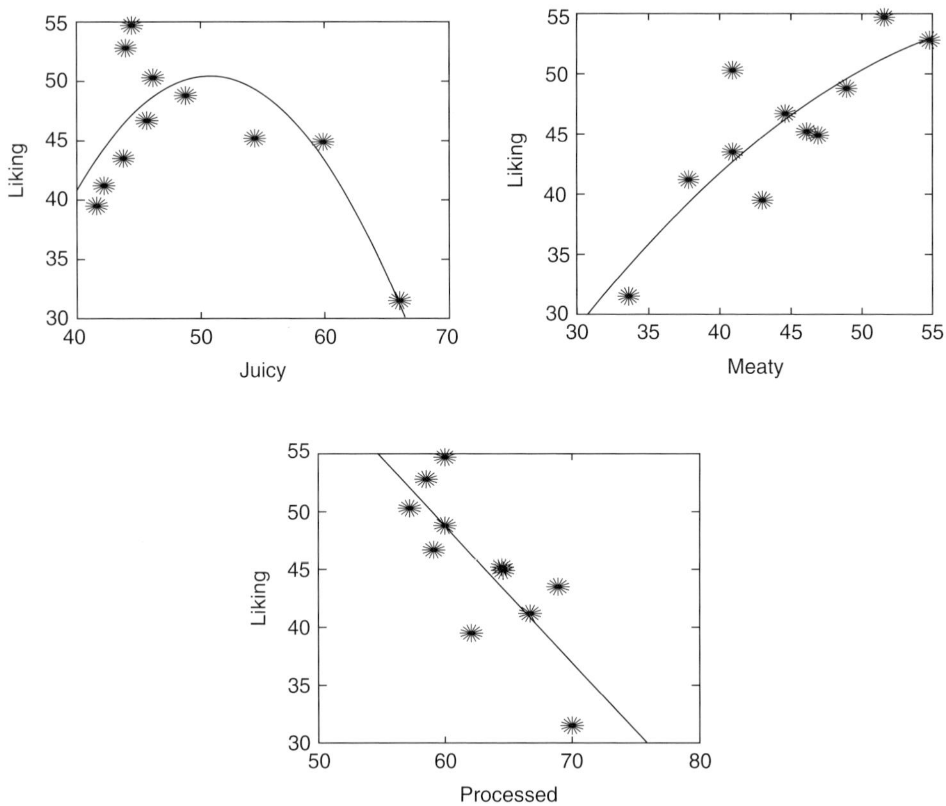

Figure 2.3 Ernie Curves different attributes of hot dogs. (Source: © Moskowitz Jacobs Inc. Reproduced with permission.)

2. What is the shape of the Ernie Curve (steep versus flat)? Often the steepest curves come from the sensory attributes of flavor, including taste and complex flavor characteristics sensed by mouth. We don't respond hedonically to changes in appearance attributes in the way that we respond to changes in taste/flavor. Figure 2.3 shows three Ernie Curves; for 'juicy,' a traditional texture attribute, for 'meaty,' a flavor/image attribute, and for 'processed,' a pure image attribute.
3. The self designed ideal – do respondents assign to their optimal product a level lying within an achievable range? At the end of the evaluations, we often instructed the respondents to assign a rating to their 'optimal product' using the same 0–100 rating scale that they had just used to evaluate the actual products. We wanted to learn whether the respondents could identify the optimal level, and whether this optimal level lay within the range of the products that they tested. For many sensory attributes, including texture attributes, we discovered that the self-defined ideal level lay within the range. There are, however, a number of attributes including a texture attribute (greasiness), whose self-designed ideal lies outside the range of products tested. Table 2.3 shows results from the hot dog study. The respondents evaluated the different hot dogs, rating them on liking and attributes, and then profiled the ideal as if it were another product.

Table 2.3 Does the self designed ideal lie within the range of products actually tested. The data from the hot dog study suggests that respondents may design a product with an unachievable level of some emotionally-tinged attributes, such as meatiness (unachievable, ideal high) and greasiness (unachievable, ideal too low). (Source: © Moskowitz Jacobs Inc. Reproduced with permission.)

	Maximum achieved by a product in the study	Minimum achieved by a product in the study	Self Designed Ideal
Dark	67	19	45
Long	74	26	67
Processed Appearance	**69**	**57**	**30**
Aroma	54	41	46
Smoke Flavor	51	31	36
Meaty	**55**	**34**	**76**
Salty	**51**	**30**	**25**
Aftertaste	**61**	**40**	**32**
Firm	60	29	46
Juicy	60	42	57
Greasy	**34**	**21**	**13**

Table 2.4 The JAR (just about right) ratings for the eight hot dogs (P1 – P8) on five sensory attributes. Numbers near 0 mean that the products are 'on target.' We look for a pattern where all products under-deliver (e.g., meatiness) or over-deliver. These two patterns suggest that respondent's don't really understand the sensory attribute, and that product developers can be misled. Respondents do understand 'greasiness.' (Source: © Moskowitz Jacobs Inc. Reproduced with permission.)

Attribute	P1	P2	P3	P4	P5	P6	P7	P8
Dark	6	7	−17	−10	−21	1	−31	−4
Smoky	−1	−4	−8	−11	−16	−8	−16	−7
Meaty	−8	−10	−16	−16	−14	−11	−13	−10
Salty	2	1	2	2	5	4	−5	−2
Greasy	−1	−3	−1	1	−2	5	2	−4

4. The JAR (just about right) scale, and texture. In these Oscar Mayer studies, the respondents rated each sensory attribute, rated the liking of the attribute, and then rated the attribute on JAR (just about right). The center point, just about right was 50, far too little was 0 and far too much was 100. Before analysis, we subtracted 50 from each rating, so negatives mean too little, and positives mean too much. For texture, Table 2.4 tells us that the products were rated on 'greasiness' close to 50. The bigger learning for texture is that respondents 'know' when an actual product delivers too much or too little of a textural attribute, at least greasy. Respondents are probably fairly well-anchored in the perception of texture attributes. In contrast, when it comes to the over-delivery or under-delivery of actual products on an attribute such as 'meatiness,' the respondents feel that no product actually delivers enough 'meatiness.'
5. How well does the self designed ideal or the JAR actually correspond to the product liked most, and how well do texture ratings 'hit the mark?' This was the most important question of all. In 1988 I had moved the Ernie Curve beyond one-dimensional

analysis of the type discussed above to looking at the more general case. Without going into the agonizing detail, the organizing idea was that the sensory attributes interacted, and that one had to create a 'model' relating the sensory attributes, including texture, to liking. The problem was that up to the late 1980s, most of the effort was concentrated on multiple linear regression to relate sensory attributes to liking, without, however, accounting for curvature. The approach I developed first created a 'basis space,' a parsimonious set of independent variables by factor-analyzing the sensory attributes, and then used the factor scores, their squares, and their interactions in order to predict liking. The specifics aren't important, but rather what it implied for texture. I ended up creating a constrained optimization and reverse engineering algorithm, that identified optimal products in three ways:

a. the sensory profile of the optimal product (sensory optimization),
b. the sensory profile of the optimal product by setting the 'goal' to the sensory profile of the self-designed ideal,
c. the sensory profile of the optimal product by setting the goal of all JAR values to '0' (reverse engineering by goal fitting).

The bottom line to the actual experiments was that self-design ideal and the product with the 'best' JAR profile (0's) were 'on target' for most texture and appearance attributes, and some taste/flavor attributes. But there were exceptions, those texture words that are emotion-laden, such as 'greasy,' and 'fatty,' where it was easy to find the sensory optimum (approach 'a' above), but difficult to when one had to rely upon the 'direction' provided by the respondent, who was asked directly to guide the developer (see Moskowitz, 1994, 1999; Moskowitz and Jacobs, 1989).

2.4.8 Sensory preference segmentation – does it occur with texture?

The other part of the story is also important. Differences in sensory preferences seem to be the rule of the day, even for water solutions of sugar or of salt (see Ekman & Akesson, 1964; Pangborn, 1970). Sensory segmentation was dramatic for flavor. It couldn't be escaped. When we did the segmentation with Vlasic® pickles, with Prego® pasta sauce, or Oscar Mayer® hot dogs, again and again the segments were defined by clear differences in the ***taste/flavor*** that people liked. And the flavors were polarizing; the high impact, very strongly flavored, spicy pickle or pasta sauce, smoky hot dogs, was down-rated, virtually rejected by those who liked bland products (see Moskowitz, Jacobs and Lazar, 1985).

For texture, the picture was entirely different. Yes, there were segments for texture, but it was a matter of degree. There wasn't the dramatic polarization, the crossing of two curves (Figure 2.4, left panel), but rather a change in the location of the optimum sensory impression. People didn't reject one texture and love another (Figure 2.4, right panel). They liked both textures, some liking the firmer texture more and the softer texture less, others showing the opposite pattern.

Sensory preference segmentation ended up being important for product development. The primary segmentation tended to be on flavor. In fact, best-selling author

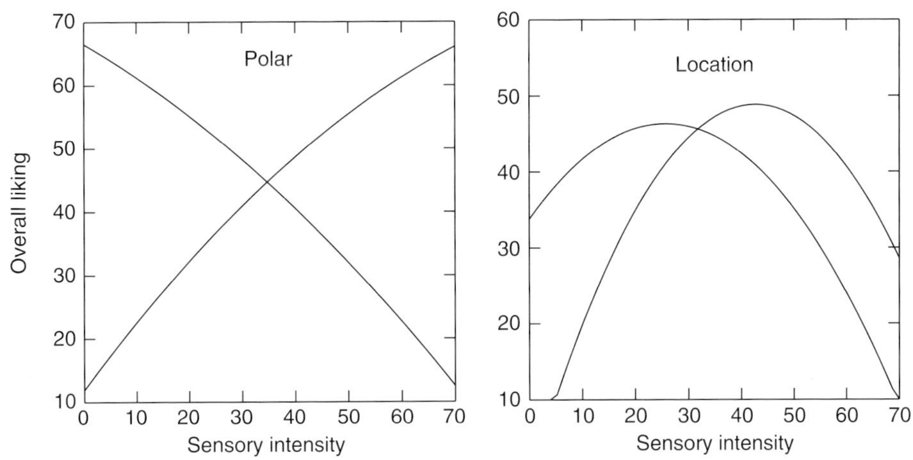

Figure 2.4 Schematic of sensory polarization versus location. The left panel shows polarized liking segments for the same sensory attribute; what one segment likes the other dislikes. This type of segmentation occurs for flavor. The right panel shows segmentation that typically occurs for texture. Both segments are fairly similar in their general sensory-liking pattern. It's only the location of the optimum that differs. (Source: © Moskowitz Jacobs Inc. Reproduced with permission.)

Malcolm Gladwell talks about the development of different pasta sauce products for Prego, using sensory segmentation (Gladwell, 2009). The story is primarily taste/flavor, and not texture, but texture does play a role. For pasta sauce there is a segment that loves chunky, thick sauce.

Over the past 30 years working with sensory preference modeling and segmentation, both in science and in application, it strikes me that segmentation is something that cannot be avoided when we deal with taste/flavor, but may be imposed, intellectualized, and thus be of only secondary importance in texture, as well as in appearance. Some of us love spicy products, others hate spicy products. We don't have that same fierce 'gut reaction' when we talk about hard versus soft, thin versus thick, chewy versus melts in your mouth. We do have emotional reactions to texture, to be sure, but these are secondary to our emotional reactions to taste/flavor, where we talk about hate/love, without having to explain WHY. When it comes to segmented preferences for texture, we have differences among people, often accompanied by a need to explain the rationale, the reason behind the segmented preferences that a person enjoys.

2.4.9 Summing up

Writing a personal history of texture, as this chapter attempts, means going back into one's memory, one's experiences, reliving occasions, and in the typically clichéd phrase, 'trying to make sense of it all.' The history that I have written is not the grand intellectual history of a field. Indeed, texture is not yet old enough as a domain of inquiry to merit this grand history. Rather, the history is one of people, scientists trying to make sense of how we interact with foods, how our sensory systems react to the complex stimuli that we eat and swallow, stimuli to which we react almost mindlessly every day as we go about our business.

There is a wonderful world yet to be discovered in texture. We really don't know the true lexicon of texture perceptions. Nor do we know the true proximal stimulus, the 'what's going on at the receptor.' But most of all, we don't know the way the mind works about texture. We know that texture is important, but we haven't truly probed how the mind integrates the stimulus in front of it, the food or drink, to emerge with a texture percept. And beyond that stretch even more unexplored continents, such as the role of texture in product acceptance, and even how to systematically create new textures that will delight the generations of consumers yet to come.

This psychophysicist's 42-year journey is one attempt to 'make sense of it all.'

References

Amerine, M., Pangborn, R.M., and Roessler, E. (1965) *Principles of Sensory Evaluation of Food,* New York, Academic Press.

Arabie, P. and Moskowitz, H.R. (1972) The effects of viscosity upon perceived sweetness. *Perception & Psychophysics*, 9, 410–412.

Boring, E.G. (1929) *Sensation and Perception in the History of Experimental Psychology*, New York: Appleton Century Crofts.

Cairncross, S.E., Sjostrom, L.B. (1950) Flavor profiles – a new approach to flavor problems. *Food Technology*, 4, 308–311.

Ekman, G., and Akesson, C.A. (1964) Saltiness, sweetness and preference: A study of quantitative relations in individual subjects. Report 177, *Psychological Laboratories*, University of Stockholm, Sweden.

Gladwell, M. (2009) The ketchup conundrum. In: *What the Dog Saw*. Little Brown & Company, New York.

Kramer, A., and Szczesniak, A. (1971) *Texture Measurements of Foods*, Reidel, Dordrecht.

Moskowitz, H.R. (1999) Inter-relating data sets for product development: the reverse engineering approach. *Food Quality and Preference*, 11, 105–119.

Moskowitz, H.R. (1994) *Food Concepts and Products: Just in Time Development*. Food and Nutrition Press, Trumbull.

Moskowitz, H.R. (1971) The sweetness and pleasantness of sugars. *American Journal of Psychology*, 884, 387–405.

Moskowitz, H.R. and Arabie, P. (1970) Taste intensity as a function of stimulus concentration and solvent viscosity. *Journal of Texture Studies*, 1, 502–510.

Moskowitz, H.R., and Jacobs, B.E. (1989) Using in-market products to generate target sensory profiles in early stage development. In: *Product Testing With Consumers For Research Guidance* (ed. L. Wu), American Society for Testing and Materials, Philadelphia, STP 1035, 64–74.

Moskowitz, H.R. and Kapsalis, J.G. (1975) The texture profile: its foundations and outlook. *Journal of Texture Studies* 6, 157–166.

Moskowitz, H.R., and Krieger, B. (1995) The contribution of sensory liking to overall liking: An analysis of six food categories. *Food Quality and Preference*, 6, 83–90.

Moskowitz, H.R. (1981) Sensory intensity vs hedonic functions: Classical psychophysical approaches, *Journal of Food Quality*, 5, 109–138.

Moskowitz, H.R., Jacobs, B.E., and Lazar, N. (1985) Product response segmentation and the analysis of individual differences in liking, *Journal of Food Quality*, 8, 168–191.

Pangborn, R.M. (1970) Individual variations in affective responses to taste stimuli. *Psychonomic Science*, 21, 125–128.

Szczesniak, A.S., Brandt, M.A., and Friedman, H.J. (1963) Development of standard rating scales for mechanical properties of texture and correlations between the objective and sensory methods of texture evaluation. *Journal of Food Science*, 28, 397–403.

3 Optimizing textural properties of soft solid foods: replacing eggs

Xin Yang

Ingredion Incorporated, Bridgewater, New Jersey, USA

3.1 INTRODUCTION

Egg is an important ingredient in the food industry, providing unique texture and preferable flavor in various products. Egg is also well known for its nutritional values, including perfectly balanced proteins, easily digestible lipids, and certain vitamins and minerals. However, there are several reasons why ingredients designed to replace eggs in food formulations may be desired. First, the high content of cholesterol (250–300 mg/egg) may be concerning to consumers. Second, allergenic concerns of egg source exist. In addition to the ingested allergy, several egg proteins, both from egg white and yolk, have been described as a cause of inhalant allergy. Third, egg substitutes provide more food options for people with specific dietary needs or restrictions, such as vegans and vegetarians. The major challenge to replacing egg products in foods is to maintain the desirable texture. Understanding egg functionality is a key step to successful replacement of eggs in food products. In this chapter, we will firstly summarize the texture functionalities of egg ingredients, then consider the application of egg and egg substitutes in various food systems, and finally discuss the current development of commercial egg substitutes.

3.2 EGG COMPOSITION AND FUNCTIONALITY

An egg includes about 58% egg white (albumen) and 31% yolk by weight. The chemical compositions of egg white and yolk are summarized in Table 3.1. Due to different compositions, the functionalities of egg white and yolk are dissimilar. Isolated egg white and yolk as well as the whole egg are used as food ingredients to provide the desired functional properties. To successfully replace egg, ingredients should be based on understanding of their functional contributions to food products and identifying the key components to substitute these functionalities.

Food Texture Design and Optimization, First Edition. Edited by Yadunandan Lal Dar and Joseph M. Light.

Table 3.1 Chemical composition of the hen's egg by percentage. (Source: Data from USDA.)

Component	%	Water	Protein	Fat	Ash
Whole egg	100	65.5	11.8	11.0	11.7
Egg white	58	88.0	11.0	0.2	0.8
Egg yolk	31	48.0	17.5	32.5	2.0
Shell	11				

3.2.1 Egg white

The egg white consists mainly of water and protein, and a few minerals. From a simplified point of view, egg white is similar to an 11% protein solution. However, the protein composition is quite complicated and the functionalities of egg white are provided by a combination of several proteins (Table 3.2). Extensive studies have been conducted to investigate the specific functionalities associated with each protein component and reviews are found in the literature (Li-Chan and Nakai, 1989; Mine, 1995). The two most important functional properties of egg white in food products are foaming ability and gel formation (Mine, 1995).

The foaming ability of egg white has been known for a long time and used in many aerated food products. Egg white proteins are good for foam formation and show excellent stability to resist liquid drainage in a foam structure. The egg white proteins function as surfactants by adsorbing at the air/water interface during the foaming process. This adsorption lowers the interfacial tension and promotes bubble formation. The adsorbed proteins tend to form a continuous film at interface via intermolecular interactions, which shows viscoelastic characteristics and promotes foam stability by slowing down bubble destabilization changes, such as coalescence and disproportionation (Dickinson, 1992; Wilde, 2000).

The foaming properties of individual proteins in egg white have been investigated. The formation of a stable foam is mainly attributed to ovoglobulins, which alone can generate foam as good as egg white (Nakamura, 1963; Johnson and Zabik, 1981a, b). Ovomucin has high viscosity, which limits its foaming capacity but contributes to foam stability by preventing liquid draining (Kato et al., 1985). Enzymatic hydrolysis increases solubility and surface hydrophobicity of ovomucin, leading to higher foaming capacity (Hammershøj et al., 2008). The foaming capacities of ovalbumin, ovotransferrin, and lysozyme are similar but less than those of ovoglobulin and the egg white control (Johnson and Zabik, 1981b). Lysozyme is a basic protein and can interact with other acid proteins, such as ovomucin, through electrosatic interactions. The formation of lysozyme-ovomucin complex in egg white liquid reduces its foaming capacity (Johnson and Zabik, 1981a, b). The dissociation of lysozyme-ovomucin complex during storage leads to better foaming capacity of older eggs compared to the fresh eggs (Ternes, 2001; Hammershøj and Qvist, 2001).

The foaming properties of egg white proteins are influenced by various factors, such as pH, ionic strength, co-solutes, whipping time, temperature and egg source, and therefore can be optimized at a certain condition (Dickinson, 1992; Hammershøj et al., 1999; Pernell et al., 2002a; Hammershøj et al., 2006). Room temperature egg white has a

Table 3.2 Composition and functional properties of egg white. (Source: Data from Linden & Lorient 1999 and Hammershøj et al. 1999.)

Proteins	% of total protein	Molecular weight (Da)	PI[1]	T_d[2] (°C)	Carbohydrate moiety	Surface tension (mN/m)	Texture functional properties
Ovalbumin	54	46,000	4.6	84	Yes	51.8	Gelling
Conalbumin (ovotransferrin)	12	76,000	6.5	61	Yes	42.4	–
Ovomucoid	11	28,000	4.0	70	Yes	39.0	–
Ovoglobulins	4+4[3]	36,000–45,000	5.6	92	Yes	45.4	Foaming
Lysozyme	3.5	14,300	10.5	75	No	42.0	Electrostatic interaction with acid proteins
Ovomucin	3	$5.5–8.3\times10^6$	4.5–5.0	–	Yes	–	Viscosity
Ovoinhibitor	1.5	49,000	5.1	–	Yes	–	–
Ovoglycoprotein	1.0	24,400	3.9	–	Yes	–	Viscosity
Flavoprotein	0.8	34,000	4.0	–	Yes	–	–
Ovomacroglobulin	0.5	7.7×10^5	4.5	–	Yes	–	–
Ficin inhibitor (cystatin)	0.05	12,700	5.1	–	No	–	–
Avidin	0.05	68,300	10	–	Yes	–	–

1. PI: Isoelectric point
2. T_d: Temperature of denaturation
3. There are two types of ovoglobulins (G2 and G3) and each compose of 4%.
4. –: Data are not available.
5. Data from Linden and Lorient (1999) and Hammershøj et al. (1999)

lower viscosity than egg white at refrigerated temperature and generates higher foam volume because air bubbles are more easily incorporated into a less viscous phase. Sugar is added after foam generation due to its enhancing effect on viscosity, which increases foam stability but decreases foam volume. Positive relationships were observed between foam drainage stability and sugar content for egg white proteins (Lau and Dickinson, 2005; Yang and Foegeding, 2010). Cream of tartar and trace amounts of salt are added into egg white foam to achieve a stiff texture (soft peak) by adjusting pH and ionic strength. Egg white foam prepared at pH 4.8, where both positive and negative net charge proteins are present and interact through electrostatic attractions, shows the largest foaming capacity, smallest bubble size and most foam stability compared to other pH conditions (Hammershøj et al., 1999). Increasing salt content enhances protein adsorption at interface and increases foaming capacity and foam stability, possibly due to screening of protein charges at high ionic strength (Raikos et al., 2007a). Beating egg white in a copper vessel has been noted to give better foam stability. The hypothetical principle for this effect is the formation of a copper-conalbumin-complex, which denatures at the air-water interface more easily than the iron-conalbumin-complex (Ternes, 2001).

Gelling property is another important functionality of egg white and contributes to the textures of a range of food products, such as custards and scrambled eggs. The egg white solution transits to a thermo-irreversible gel after heating due to the denaturation and aggregation of proteins. The mechanism of heat-induced protein aggregation has been proposed as the following transition: native monomer → denatured monomer → soluble aggregate → gel (Mine, 1995). Hydrophobic interactions among denatured proteins are involved in the formation of a gel. The sulfhydryl-disulfide interchanges lead to disulfide bonds between proteins and reinforce the gel network. Ovalbumin is the major contributor to gel formation (Mine, 1995). Synergistic effects on heat-induced gel strength was found between ovalbumin and lysozyme at neutral pH, due to electrostatic interactions (Arntfield and Bernatsky, 1993). The formation of large protein aggregates causes an opaque and white appearance of egg white gels. The turbidity and hardness of egg white gels can be adjusted by pH, ionic strength and sugar (Hatta et al., 1986; Raikos et al., 2007b). Aggregation of egg white proteins occurs more readily when close to pI and at a higher ionic strength due to less net charge of proteins and easier interactions among them. The pH of egg white from a newly laid egg is between 7 and 8.5, which generally rises during storage due to the loss of carbon dioxide. Increasing pH from 6.5 to 9 leads to higher egg white gel firmness, corresponding to a transition from more porous to denser gel structure (Chang and Chen, 2000). Handa et al. (1998) observed a fine ordered network of egg white gel at pH 11, which contributes to the highest gel hardness and maximum water-holding capacity in the range of pH 3–11. The higher gel firmness at alkaline pH could be due to the accelerated sulfhydryl-disulfide exchanges. Sugar and salt tend to protect the native structure of proteins and to increase denaturation temperature, leading to higher gelling temperature for egg white (Raikos et al., 2007b). However, salts also have a screening effect on protein electrostatic repulsions and promote protein interactions. Decreased gelling temperature was observed for bovine serum plasma in the presence of salt (Baier and McClements, 2003). Addition of sugar or salt reduces the gel firmness of egg white (Raikos et al., 2007b). The decreasing effect of salt could be due to its influences on protein aggregation and gel network formation. The sugar effects on gel properties depend on several factors. Sugar increases viscosity and reduces the diffusing rate of protein molecules, causing a slower gelling rate. On the other hand, the presence of sugar reinforces protein-protein hydrophobic interactions due to a preferential interaction mechanism, resulting in higher gel strength if enough time is allowed for gel formation. Baier and McClements (2003) reported that sucrose decreases the strength of bovine serum albumin gels prepared at 90 °C/15min but increases the gel strength if prepared at 90 °C/120min.

3.2.2 Egg yolk

Egg yolk contains approximately 50% solids, including 17.5% proteins and 32.5% lipids (Table 3.1). It is a mixture of proteins and lipoproteins and exists as a dispersion of granules in a continuous aqueous phase or plasma (Table 3.3). The individual constituents of yolk are hard to separate; while the granules and plasma can be easily fractionated by a mild centrifugation in industrial scale (McBee and Cotterill, 1979). The sediment granules consist of 70% high-density lipoproteins (HDL), 16% phosvitin, and 12% low-density lipoproteins (LDL); and the supernatant plasma contains 85% LDL and 15% livetins (McBee and Cotterill, 1979). The major composition of granules, HDL, contains about 22 to 26% lipid (61% phopholipids, 35% triglycerides, and 4% cholesterol and cholesterol esters) (Martin et al., 1963). The HDL is held together with a phosphoprotein, phosvitin, through phosphocalcic bridges, forming an insoluble granule complex. At high NaCl concentrations, the divalent Ca^{+} in the phosphocalcic bridges are substituted by monovalent Na^{+}, leading to dissociation of the complex and soluble HDL and phosvitin (Causeret et al., 1991). The main component in plasma, LDL, contains about 86–89% lipids (27% phospholipids, 69% triglycerides, and 4% cholesterol and cholesterol ester) and is in a form of a lipid core surrounded by an interfacial layer of phospholipids and proteins (Martin et al., 1963). The functional properties of egg yolk are mostly attributed to the proteins, including the apoproteins from lipoproteins. On the other hand, the yolk lipids should always be considered together with proteins for yolk functionality as they are all associated with proteins. Twelve dominant protein

Table 3.3 Composition of egg yolk. (Source: Data from Linden and Lorient 1999.)

Fractions	% of total yolk	% of yolk proteins	Lipid content %	MW	Textural functional properties	Location
Phosvitin	4	10	0	36,000	–	Granules
HDL	16	36	20	400,000	Emulsifying	
LDL	68	24	88	L_1:10×10^6 L_2:3×10^6	Gelling, emulsion stability, viscosity	Granules & Plasma
Livetins	10	30	0	α:80,000 β:45,000 γ:150,000	Emulsifying	Plasma
YRBP	1.5	0.4	0	36,000	–	

1. HDL: High density lipoprotein (Lipovitellin)
2. LDL: Low density lipoprotein (Lipovitellenin)
3. YRBP: Yolk riboflavin binding proteins
4. Data from Linden and Lorient (1999)

compositions with MW of 20 to 240 kDa are determined in egg yolk by electrophoresis and mass spectrometry (Nilsson et al., 2007). The yolk lipids include 65–70% triglycerides and 25–30% phospholipids (Linden and Lorient, 1999). The phospholipids of HDL and LDL are similar, including 74% lecihins, 18% cephalins, and 8% minor phospholipids (Martin et al., 1963). The fatty acid compositions depend on hen's feed.

Two major functional properties of egg yolk are emulsifying ability and gel formation. Emulsifying properties of egg yolk are important for the texture and stability of many food products, such as mayonnaise, salad dressings, egg-based sauces and creams. The egg yolk constituents have the capability to adsorb at oil-water interface and reduce interfacial tension, enabling oil droplets to form during emulsification. The emulsifying properties of yolk and its two major fractions, plasma and granules, were compared in several studies. Yolk granules show lower emulsifying activities than yolk plasma in distilled water, where the granules are in complex forms of HDL-phosvitin and cannot dissolve in water; however, granules exhibit no less emulsifying activities than yolk or yolk plasma in 0.5 M NaCl solution, in which the complex forms dissociate to HDL and phosvitin and dissolve in solution (Dyer-Hurdon and Nnanna, 1993; Anton and Gandemer, 1997). Le Denmat et al. (2000) evaluated the emulsifying properties of yolk, yolk plasma and granules in a range of pH and salt concentrations. They found yolk and yolk plasma have similar emulsifying activities, which are independent of pH and salt concentration; however, the emulsifying activity of yolk granules are comparable to the other two at pH 7 and much lower at pH 3. This is also explained as a result of the poor solubility of yolk globules at pH 3, when they are in an aggregated state. At pH 7, even a low salt concentration leads to disruption of the HDL-phosvitin complex and partially dissolves the yolk granules into the solution, providing more efficient surface activity. Therefore, yolk plasma plays an important role in emulsion formation when yolk granules are in an insoluble state; however, globules can exhibit comparable emulsifying activity when the native aggregates disrupt and dissolve in solution.

When compared for emulsion stability, yolk and yolk plasma show similar creaming rates; however, yolk granules exhibit more stability against creaming process at neutral pH and have less stability at acidic pH (Anton and Gandemer, 1997; Le Denmat et al., 2000). The creaming rate is associated with emulsion destabilization changes, including flocculation, coalescence and creaming. The adsorption of egg yolk constituents at the oil/water interface leads to a film formation surrounding oil droplets that offers protection against coalescence and other destabilization changes during long term storage. The emulsion stabilities are largely dependent on the composition and structure of the adsorbed films. The pH can influence the competition between yolk constituents for interfacial adsorption due to its effects on the solubility, conformation and/or surface charges of proteins. At neutral pH, lipoproteins (HDL and LDL) were found to have higher adsorption affinity at the oil/water interface than the globular proteins (phosvitin and livetin) because of their more flexible structure and greater surface hydrophobicity (Anton and Gandemer, 1999; Le Denmat et al., 2000). It is postulated that the lipoproteins break down at the interface; and apoproteins and phospholipids stay at the interface and form a film while the neutral lipids coalesce with oil droplets (Kiosseoglou and Sherman, 1983). The HDL in yolk globules has a higher protein percentage than LDL in yolk plasma and possibly forms a thicker interfacial film that offers better prevention for oil droplet destabilizations. At acid pH, the HDL maintains an aggregated structure and shows less emulsifying activity and lower emulsion stability (Anton and Gandemer, 1999; Le Denmat et al., 2000). Although the phosvitin has more absorption at the interface at acid pH due to reduced surface charges and creates small oil droplets by itself, the HDL are more interfacial active and displace phosvitin at the interface in the mixtures, leading to larger droplet size and lower emulsion stability (Aluko and Mine, 1997).

The competitive adsorption of egg yolk constituents at oil/water interface has been analyzed by identifying protein compositions at interfaces using electrophoresis. There are conflicting results concerning the protein composition at interfaces. Mine (1998a) measured protein components in creams and serums after centrifuging the emulsions prepared with 1–3% egg yolk and reported that the HDL fraction from yolk globules adsorb preferentially both at pH 3 and 7, suggesting the LDL fraction is not a major contributor to emulsification. Aluko and Mine (1997) evaluated the interface compositions for emulsions prepared with 0.5% protein mixtures of globule lipoprotein and phosvitin and indicated lipoproteins are more interfacially active and displace phosvitin more efficiently at neutral pH than pH 4. In another study, Anton and Gandemer (1999) analyzed the emulsions prepared with 0.5% egg yolk and found the protein composition of the interface varies with pH. Phosvitin is the major protein at pH 3; while almost all main proteins in yolk are adsorbed at pH 6. Selected apoproteins from both LDL and HDL and no phosvitin are adsorbed at pH 9. The lowest interfacial tension was observed at pH 6, which is close to the isoelectric region of yolk proteins, suggesting the protein charges influence yolk adsorption (Anton and Gandemer, 1999; Mel'nikov, 2003). Le Denmat et al. (2000) observed a preferential adsorption of plasma proteins at pH 3 due to the poor solubility of yolk globules and almost equal adsorptions of proteins from plasma and globules at the interface at pH 7 (using emulsions made with 2.5% egg yolk). This suggests that plasma proteins show similar adsorption to yolk globules. Mine (1998b) evaluated the protein composition of interfaces prepared with LDL from yolk plasma and found most LDL-polypeptides adsorbing on the oil droplet except for three polypeptides (64, 43, and 19 kDa) remaining in serum. Minor changes were observed in the protein compositions of the LDL interface at pH 3.5, 6.5 and 8.5. Nilsson et al. (2007) evaluated the yolk protein fraction at interfaces using electrophoresis followed by mass spectrometry and found that the proteins showing strong interfacial adsorption capability always have long average hydrophobic and shorter hydrophilic domain lengths. They suggested that the hydrophobic domains of the protein molecules rather than the lipids from the protein-lipid complex are the major contributors to interfacial adsorption of lipoproteins. Although dissimilar adsorption behavior was observed for egg yolk constituents, possibly due to the complex compositions and different emulsifying method, such as pH, salt concentration, buffer substances, homogenization conditions, and so on, it can be concluded that most components in egg yolk show good interfacial adsorption and the lipoproteins play an important role in the development of interfacial films around the oil droplets.

Removal of triglycerides and cholesterol in egg yolk by extracting with organic solvents or supercritical CO_2 may influence the emulsifying properties. Compared to globular proteins, the lipoproteins are the major adsorbing components at the oil/water interface in egg yolk due to their flexible structures. Extraction of lipids leads to concentrated yolk proteins and results in thicker films around the oil droplet due to protein aggregation (Kiosseoglou, 2003). The adsorption behavior of low cholesterol yolk depends on the solvent used for lipid extraction, which may alter the structure of yolk lipoproteins and their interfacial properties (Paraskevopoulou et al., 1997a). The low cholesterol yolk prepared using petroleum ether shows similar adsorption behavior and comparable emulsifying properties as egg yolk; while the cholesterol extraction using a mixture of petroleum ether and ethanol leads to less interfacial activity and reduces emulsion stability (Paraskevopoulou and Kiosseoglou, 1994; Paraskevopoulou et al., 1997a). The extraction using supercritical CO_2 results in comparable emulsification and emulsion stability although it leads to less interfacial activity (Bringe et al., 1996; Paraskevopoulou et al., 1997a, 1997b). These complicated behaviors are attributed to the variations in compositions and structures of yolk lipoproteins after removing lipids using different methods.

Another important functionality of yolk is gel formation, which contributes to the texture of many food products, such as cakes, omelet, confectionery, and so on. The formation of the gel network is attributed to the denaturation and aggregation of yolk proteins upon heating. The lipids associated with proteins are also included in this process and influence gel texture. For example, the aggregation of proteins dominating on the surface of LDL or HDL particles leads to an inter-particle yolk gel network, in which the lipid cores are buried in the particle interior and modify the gel texture as fillers (Kiosseoglou, 2003; Kiosseoglou and Paraskevopoulou, 2005). This structure is similar to protein-stabilized concentrated emulsion gels. The LDL in yolk plasma appears to dominate yolk gelation as similar gelling patterns were observed between yolk and yolk plasma; while the yolk granules show slower and less effective gel formation (Anton et al., 2001). This difference is attributed to a relatively higher denaturation temperature of yolk granule proteins (above 75°C) compared to the plasma LDL apoproteins (above 70°C), which is postulated due to the complex structure of HDL-phosvitin (Le Denmat et al., 1999; Dixon and Cotterill, 1981). Anton et al. (2001) dissociated the native structure of yolk granules using a high NaCl concentration and investigated the heat resistance of disrupted yolk granules. They observed no difference in protein denaturation temperature between native and disrupted yolk globules; however, the native structure tends to restrict aggregation and avoids gelation of globule solutions during heating. Although the yolk granules require a higher temperature to develop a gel network, Kiosseoglou and Paraskevopoulou (2005) found, upon heating at 90°C/30 min, the yolk granules form a much stronger gel than yolk or yolk plasma at equal solids basis. They hypothesized that the plasma constituents in yolk might form an initial gel network at a lower temperature, which keeps the granules intact, and is not reinforced by the granule constituents even after they denature at higher temperatures. The intact granules are entrapped in the yolk gels and act as gel structure weakening points or inactive fillers, leading to slightly lower gel strength of liquid yolk compared to yolk plasma. However, it was also suggested that the disintegrated yolk globules at high ionic strength can interact with plasma constituents and reinforce the gel network structure (Kiosseoglou and Paraskevopoulou, 2005; Raikos et al., 2007b).

The gelling properties of egg yolk are influenced by several factors, including processing conditions, removal or addition of components, and pH. Heat treatments during spray drying may destabilize the supermolecular structure of yolk granules and lead to weak gel formation of yolk dispersions at room temperature (Kiosseoglou, 2003).

Removal of lipids, such as cholesterol, increases the protein percentage in yolk and causes more efficient gel formation and stronger gel texture at the same solid level (Paraskevopoulou and Kiosseoglou, 1997). However, the full fat yolk requires lower critical protein concentration to form a gel network compared to the yolk protein concentrates with removal of lipids, suggesting the presence of lipids is important for gel structure formation in dilute dispersion (Paraskevopoulou et al., 2000). Incorporation of oil droplets leads to slower gelling rate and weaker gel strength of emulsion gels prepared from yolk and yolk plasma, suggesting the oil droplets coated with plasma constituents act as inactive fillers (Anton et al., 2001; Koidis et al., 2002). A possible explanation of this effect is that parts of LDL in plasma tend to adsorb at the oil-water interface, due to their flexibility, and cannot be involved in hydrophobic protein interactions for the yolk gel network formation as their hydrophobic patches are oriented towards the oil phase (Kiosseoglou, 2003). The weakening effect of oil droplet to yolk protein gel strength is less effective when Tween 20 is used as an emulsifier, possibly due to the surfactant-lipoprotein interactions (Koidis et al., 2002). On the other hand, the emulsion gels prepared with yolk granules show faster gelling rate and higher gel strength compared to the solution gels, suggesting the oil droplets coated with granules behave as active fillers (Anton et al., 2001). Addition of salt slightly increases gel strength of egg yolk, probably due to the dissociated HDL-phosvitin complex contributing to gel structure formation (Raikos et al., 2007b). The strength of heat-set yolk gels increases with increasing pH in a range of pH 2–8, which is explained due to increased yolk protein solubility (Raikos et al., 2007b).

3.2.3 Whole egg

The functional properties of whole egg are contributed by the constituents from egg white and yolk. The emulsifying properties are mainly provided by yolk fractions due to their efficient interfacial adsorbing capability. Aeration of whole egg is achieved by egg white proteins and the interfacial active constituents from yolk. Due to the high affinity to hydrophobic phases, the lipoproteins from egg yolk tend to replace egg white proteins at air/water interfaces, leading to reduced foaming capacity of whole egg. Whole egg form gels at lower temperatures and show higher gel firmness compared to egg white and yolk (Raikos et al., 2007b). Incorporation of yolk or emulsified oils coated with yolk shows enhanced effects on the egg white gels, suggesting interactions between egg white proteins from gel matrix and the yolk lipoprotein from oil droplet surfaces (Kalkani et al., 2007). These results suggest that gel formation of whole egg involves protein aggregations from egg white and yolk as well as interactions between the two components.

3.3 EGG SUBSTITUTION IN FOOD PRODUCTS

The textural functionalities of egg are mainly contributed by proteins and lipoproteins. Potential substitutes are suggested according to the required functionality and include proteins from other sources, polysaccharides, and emulsifiers (Table 3.4). Egg ingredients usually provide multiple functionalities in a food system. For example, egg white

Table 3.4 Texture functionality of egg ingredients in different types of food products.

Food products		**Egg white**					**Egg yolk**		
		Foaming capacity	**Foam stability**	**Foam stiffness**	**Heat Coagulation**	**Thickening**	**Emulsifying**	**Heat gelation**	**Thickening**
Aerated products based on egg white foam (lipid free)	Angel food cake	**	**	**	**	*	–	–	–
	Meringues	**	**	**	**	*	–	–	–
	Marshmallows	**	**	**	–	*	–	–	–
	Egg white frosting	**	**	**	–	–	–	–	–
Aerated products using whole eggs (lipid contained)	Mousses	**	**	**	–	**	**	–	**
	Nougats	*	*	–	–	*	*	–	–
	Cakes containing whole eggs	**	*	*	**	*	**	*	–
	Soufflé	**	**	**	**	*	*	**	*
	Cookies	–	–	–	*	*	**	*	*
	Biscuits	–	–	–	*	*	*	*	*
	Egg bread	–	–	–	*	*	*	*	*
Products based on egg gelation	Pan fried egg	–	–	–	**	–	–	**	–
	Boiled egg	–	–	–	**	–	–	**	–
	Scrambled egg	–	–	–	**	–	**	**	–
	Custard and Puddings	–	–	–	**	–	**	**	–
	Egg salad	–	–	–	**	–	**	**	–
	French toast	–	–	–	**	*	**	**	*
	Omelettes	–	–	–	**	*	**	**	*
	Casserole	–	–	–	**	*	**	**	*
	Egg noodle	–	–	–	**	–	–	**	–
	Meat products	–	–	–	**	*	**	**	*
Emulsion products	Mayonnaise and dressings	–	–	–	–	*	**	–	**
	Egg nog	–	–	–	–	*	**	–	**
	Ice cream	–	–	–	–	–	**	–	–
Potential Substitutes		Whey proteins, bovine plasma, white lupine protein, soapwort extracts, xanthan gum			Polysaccharids, proteins, modified starch		Soy lecithins, proteins	Polysaccharids, proteins, modified starch	

**Very important   *Important   –Not important

provides a foam base while the egg yolk assists blending of different components in a cake batter. It is important to be aware of the egg functionalities and their interactions with other ingredients in a food system in order to select proper ingredient or their combinations for egg substitution. In the following part, the applications and substitution of egg ingredients will be discussed based on their functionalities in several food systems (Table 3.4).

3.3.1 Aerated products based on egg white foam (lipid free)

Egg white foam is a traditional ingredient to create a light texture of many aerated food products, such as angel food cakes and meringues. With no leavening agent, the foamy structure of angel food cake is completely created by egg white foam. A typical composition of angel food cake includes egg white, sugar, cake flour, trace of cream of tartar and salt. During the whipping process, egg white proteins adsorb at the air/water interface and produce a foam structure with increasing six to eight times in volume. Cream of tartar and salt are added into the foam to create a stiff texture (soft peak stage of foam) by adjusting pH and ionic strength. Sugars are gradually added with low speed whipping and stabilize the foam by increasing continuous phase viscosity and enhancing interfacial viscoelastic properties of egg white proteins (Yang et al., 2010). After gently folding in the cake flour, the cake batter is transferred into a cake pan and baked. During heating, egg white proteins aggregate and starches from cake flour gelatinize, leading to a transition from wet foam (cake batter) to a stable dry foam structure (angel food cake). Large quantities of air bubbles in the egg white foam are expanding during heating and this increases cake volume. Egg white is unique for its capabilities to generate foams of high volume and excellent stability, especially during mixing and heating, and to aggregate during heating.

Several studies were conducted to investigate the functional properties of individual egg white proteins and their interactions in angel food cake. The ovoglobulins alone has good foaming capacity and produces angel food cake with equal or greater volume than the egg white control and excellent texture (Johnson and Zabik, 1981a). The mixture of ovoglobulins and ovomucin shows good foamability and makes a nice angel food cake batter; however the batter expands normally but then collapses during baking, leading to a lower cake volume (MacDonnell et al., 1954; Johnson and Zabik, 1981b). Addition of ovalbumin into this mixture can increase cake volume to normal level, indicating ovalbumin is critical for cake structure formation during heating (MacDonnell et al., 1954). Ovalbumin alone can generate a cake with similar volume to egg white if given a long enough whipping time; however, the cake structure is coarser (MacDonnell et al., 1954; Johnson and Zabik, 1981b). This suggests ovalbumin cannot provide a fine bubble formation or it lacks stability to maintain small bubbles. Angel food cakes made with lysozyme and conalbumin show smaller volume than the egg white control, which is attributed to less air incorporation and lower foam stability (Johnson and Zabik, 1981a). Raeker and Johnson (1995) indicated that lysozyme and conalbumin have lower denaturation temperatures than other egg white proteins and tend to denature early at the bubble surface during heating, which probably reduces the film elasticity and prevents further cake expansion. Ovomucoid and ovomucin show low foaming capacity and produce angel food cakes with least volumes (Johnson and Zabik, 1981a). Lysozyme was found to increase angel food cake volume when combined with ovomucin, probably due to electrostatic interactions (Johnson and Zabik, 1981a, 1981b). In general, ovoglobulins and ovalbumin are the major contributors to foam formation, foam stability and fine cake structure formation during baking. Ovomucin adds viscosity of pre-foam solutions, which reduces the foaming capacity but prevents foam drainage during cake batter preparation. The key properties required from egg white proteins to produce good angel food cakes are formation of protein films with rheological properties, that allow for foam formation and stabilization, and at the same time heat aggregation at appropriate temperatures to allow for maximum cake batter expansion during baking.

With a comparable foaming capacity, whey protein isolates (WPI) have been used to substitute egg white protein (EWP) in angel food cakes (Arunepanlop et al., 1996; Pernell et al., 2002b; Berry et al., 2009; Yang and Foegeding, 2010). Although WPI can generate the same foam volume as EWP, substitution of EWP with WPI in angel food cake results in lower cake volume and coarser cake structure, suggesting lower foam stability. The lack of foam stability, especially during heating, is the major obstacle for WPI foams in applications involving heating or baking. Significantly lower foam drainage stability was observed for WPI compared to EWP, even when the solutions have similar viscosities (Yang et al., 2009). Foam stability is associated with the interfacial properties of air bubbles. The proteins adsorbed at the air-water interface tend to denature and undergo a further rearrangement and/or interaction to form an elastic film, which can prevent bubble destabilization changes, including disproportionation and coalescence (Bos and van Vliet, 2001). The WPI forms a less elastic film at the air-water interface than EWP, leading to lower bubble stability and therefore poor foam stability (Davis and Foegeding, 2007; Yang and Foegeding, 2010). Pernall et al. (2002b) measured cake expansion during baking and found cakes containing WPI expand to an equal or higher volume than EWP in the initial phase but collapse in the later stage. They found the cake batters made with WPI stop expansion at 75°C but that of EWP continue growing until 85°C. The two temperatures correspond to the denaturation temperatures of WPI and EWP solutions, which are associated with their major components – β-lactoglobulin in WPI and ovalbumin in EWP. The viscoelastic properties of cake batters were also measured during baking. The WPI cake batter becomes more viscous (increased phase angle) at β-lactoglobulin denaturation temperature range, leading to the collapse of cake structure. However, the cake batter containing EWP shows continuously increased elastic characteristic (decreased phase angle), corresponding to continuously cake volume expansion until starch gelatinization 'sets' the cake structure (Pernell et al., 2002b).

Partially replacing 25% EWP with WPI can create angel food cake with almost the same volume (Figure 3.1); however, a coarse structure with big air bubbles is formed, indicating bubble coalescence during baking (Figure 3.2). Berry et al. (2009) observed the microstructures of cake batters during heating and found big bubbles quickly appear in the cake batters using the mixture of WPI and EWP, in a way similar to WPI (Figure 3.3). The EWP cake batter maintains a homogeneous fine structure through heating to 85°C, where starch gelatinization starts and assists cake matrix formation (Figure 3.3). They indicated that WPI tends to dominate at the air bubble interface and determines the bubble stability and the cake structure formation, although EWP in the

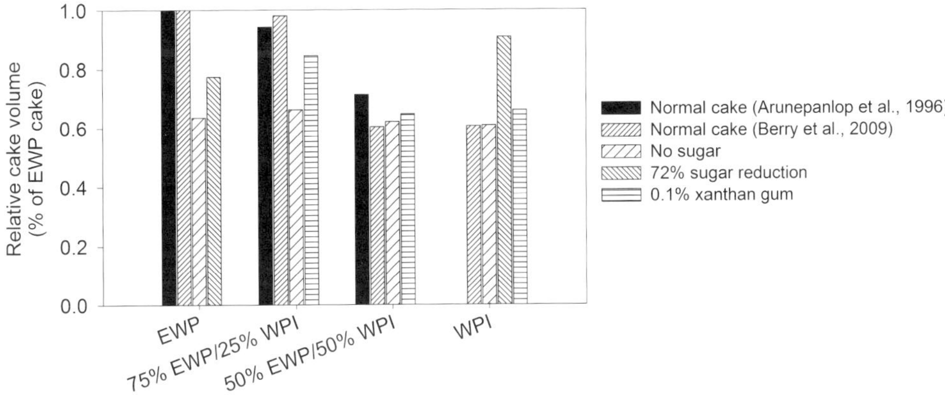

Figure 3.1 Relative volume of angle food cakes prepared from egg white proteins (EWP) and whey protein isolates (WPI). The cake volumes are relative values compared to the angel food cakes containing 100% EWP controls. (Source: The data for no sugar cakes are from Berry et al. 2009. The data for 72% sugar reduction cakes are from Yang and Foegeding 2010. The data for 75%EWP/25%WPI and 50%EWP/50% WPI with 0.1% xanthan gum are from Arunepanlop et al. 1996. The data for the WPI cake with 0.1% xanthan gum are from Pernall et al. 2002b.)

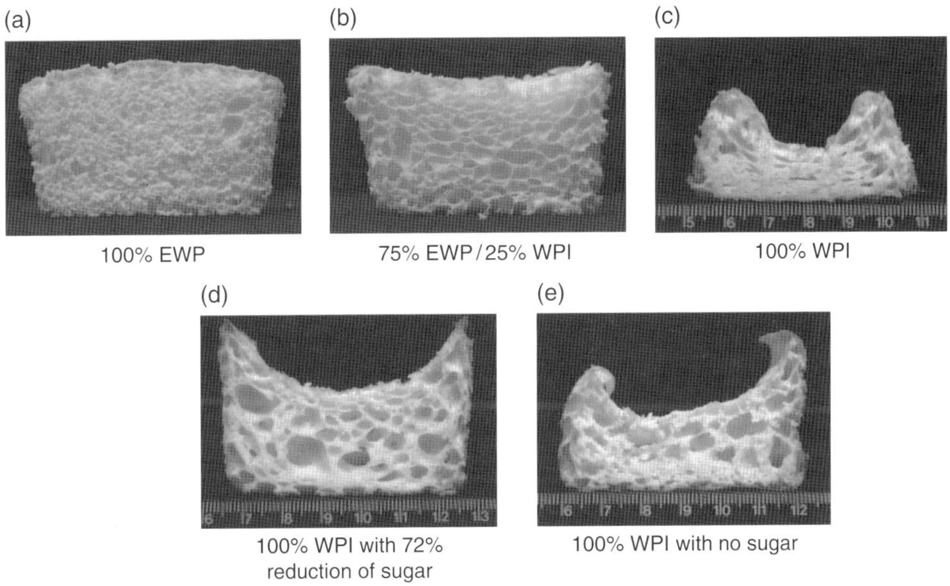

Figure 3.2 Structure of angel food cakes prepared from 100% EWP foam (a), 75% EWP/25% WPI (b), 100% WPI (c), 100% WPI with 72% reduction of sugar (d) and 100% WPI with no sugar (e).
(Source: Data for (a) and (b) from Berry et al. 2009. Data (c), (d) and (e) from Yang and Foegeding 2010.)

continuous phase of cake batters assists cake matrix formation during baking. This is confirmed by a dominating effect of WPI to the interfacial properties of the WPI and EWP mixture solutions, even at a low WPI ratio (Yang et al., 2009). Addition of xanthan gum improves the fine structure of 75%EWP/25%WPI cake but shows no increase in

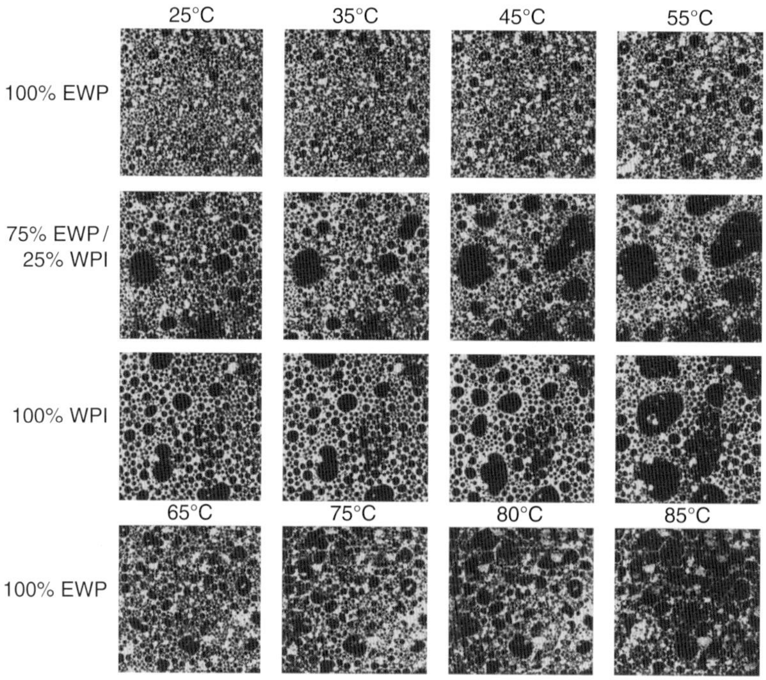

Figure 3.3 Air bubble expansion during heating process in angel food cake batters prepared from 100% EWP foam, 75% EWP/25% WPI foam, and 100% WPI foam. (Source: Data from Berry et al. 2009.)

cake volume (Arunepanlop et al., 1996; Figure 3.1). Xanthan gum slightly increases the volume of WPI cake although it increases the solution viscosity by more than 9 times, suggesting the viscosity of foam solution may lower cake collapse but it is not the critical factor (Pernell et al., 2002b). On the other hand, reducing sugar content decreases foam solution viscosity but leads to an increased cake volume and less collapse in angel food cakes containing WPI (Yang and Foegeding, 2010; Figure 3.1 and 3.2). Reduced protein film elasticity at air-water interface was observed for WPI with sugar addition, possibly contributing to less bubble stability and more collapse of cake structure during heating (Yang and Foegeding, 2010). In a further evaluation of the individual protein components in WPI, the detrimental effect of sugar on interfacial film elasticity was found to be mainly associated with α-lactoalbumin, which tends to dominate at the air-water interface in the mixtures with β-lactoglobulin even at a lower percentage (Yang, 2008). In summary, substitution of EWP with WPI leads to faster growth of the air bubble size and the collapse of foam structure before cake matrix formation, which are associated with lower elasticity of the protein film at air-water interface and lower protein denaturation temperature.

Beating egg foam to a soft peak stage is recommended for cake preparations. At this stage, the foam shows a solid-like property or fluffy texture and exhibits a yield stress value. This characteristic is unique for EWP foam and contributes to the textures of many aerated food products, such as mousses and soufflé. The foam yield stress corresponds to a high air phase fraction in foam, usually above 0.74, and formation of a 'polyhedral' structure, in which bubbles can compress each other against the Laplace pressure and behave like a soft solid (Edwards and Wasan, 1996; Pernell et al., 2000). This rheological property has been theoretically predicted using geometrical models of bubbles and can be attributed to several factors, including air phase fraction, solution viscosity, interfacial tension, interfacial rheology, bubble size distribution, and bubble size shape. The high yield stress of EWP foam is not reproduced by WPI, which cannot be explained using air phase fraction, interfacial tension, or bubble size (Pernell et al., 2002a). Proportional relationships were observed between foam yield stress and interfacial film elasticity for the whey protein based materials; however, this relationship cannot be extended to explain the high yield stress of EWP (Foegeding et al., 2006; Yang et al., 2009; Yang and Foegeding, 2011). Many modifications were applied to WPI and its major component, β-lactoglobulin, to modulate the interfacial rheology and foam yield stress, including heat polymerization, enzyme hydrolysis, addition of salt, glycine and lactose, and combinations with other proteins (Luck et al., 2002; Davis and Foegeding, 2004; Davis et al., 2004, 2005; Yang et al., 2009; Yang and Foegeding, 2010). Although the foam yield stress and/or interfacial rheology can be improved, using the modified whey proteins still cannot produce comparable angel food cake as egg white, with less cake volume and poor texture (Zhu and Damodaran, 1994; Pernell et al., 2002b; Abu-Ghoush et al., 2008). Heat polymerized WPI leads to better foam stability and improved cake volume than WPI, which is attributed to increased viscosity; however, the cake volume and quality are still less than EWP cake (Pernell et al., 2002b). The hydrolyzed WPI shows excellent foaming capacity and stability; however, it generates poor cake batters and fails to prepare angel food cake (Abu-Ghoush et al., 2008). Abu-Ghoush et al. (2008) applied proteins from various sources to replace egg white in angel food cakes, including WPI, hydrolyzed WPI, gelatin, fish proteins, pea protein, rice protein, soy protein, corn protein and casein, and concluded that WPI performs best among these egg alternatives.

Bovine blood plasma protein isolates show good foaming capability and are used to replace egg white in angel food cakes (Khan et al., 1979). The compatibility of plasma protein isolates with other cake ingredients is not as good as egg white, resulting in significantly reduced cake height when replacing more than 30% egg white in angel food cakes. The cake with 25% substitution of EWP with plasma protein isolates has slightly lower volume and shows a coarse structure, suggesting bubble destabilization during baking. Raeker and Johnson (1995) indicated that the bovine plasma protein isolates have comparable foaming capability but less foaming stability than EWP.

Miller and Setser (1983) used mixtures of wheat starch, water and xanthan gum to prepare egg white reduced angel food cake. Removal of 35% egg white from the cake formula leads to a more condensed batter and increased cake density, due to reduced moisture content. Addition of water increases air incorporation into cake batter but decreases foam stability. Xanthan gum enhances the cake batter viscosity and increases foam stability; however, it does not improve the angel food cake volume, suggesting no contribution to cake matrix formation. Wheat starch adds viscosity to the cake batter and contributes to cake structure formation during heating. A combination of wheat starch and water produces angel food cakes with 35% reduction of egg white and comparable volume and no significant difference in sensory characteristics from the egg white control.

Addition of basic proteins, such as lysozyme, can improve foamability and foam stability for acidic food proteins, such as β-lactoglobulin and bovine serum albumin (Phillips et al., 1989; Dickinson, 1992). This enhancing effect is because of the electrostatic interactions between lysozyme and acidic proteins (Dickinson, 1992). A basic soy protein was found to improve foam stability of egg white, possibly by acting in a similar way as lysozyme to interact with positively charged proteins in foams (Wang and Wang, 2009). This effect suggests combinations of basic proteins and acid proteins may provide potential foaming agents in angel food cakes to substitute egg white.

3.3.2 Aerated products using whole eggs (lipid contained)

In cakes using whole eggs, such as sponge cake and butter cake, egg yolk and/or lipid ingredients (butters, shortenings and vegetable oils) are added and influence the cake texture. The presence of lipophilic ingredients, such as lipoproteins from yolk, can significantly reduce the egg white foaming capacity and lead to lower foam stability because lipids tend to replace egg white proteins at the air/water interface, accelerating air bubble coalescence (Walstra, 2003). On the other hand, the presence of emulsifiers, such as phospholipids from yolk, can lower the air/water interfacial tension and assist air being whipped into the batters. The monoglycerides in plastic shortening also ensures numerous small air cells are created during the beating or creaming process of shortening. Unlike angel food cake, the chemical leavening agent – baking powder – is frequently used to generate gas (CO_2) during heating process in order to create the foamy structure in cakes. The dissolved CO_2 can merge into air cells to increase bubble size and cake volume but does not spontaneously form bubbles. Therefore, formation of a large amount of air bubbles in the original cake batters is critical for the final cake volume. Typical ingredients in a sponge cake include whole egg, cake flour or wheat flour, sugar, and baking powder. Butters or shortenings are added in butter cake to 'shorten' or tenderize the texture. During cake preparation, egg white can be separated from whole egg, generates a foam individually, and then is folded with other ingredients to make cake batter. This technique is used in preparing Chiffon cake. In this case, the egg white foam functions in a similar way as in angel food cake but the effects from egg yolk and lipids to foam stability should be considered. In the 'all in' cake process, all ingredients are mixed together at the same time and the air stabilization is mainly achieved by emulsifiers. Although proteins may still play a functional role in cake batters, the emulsifier or lipophilic ingredients are the dominating surface active components around air bubbles. Therefore, the aeration by egg white proteins is less important than in angel food cake. The functionality of egg ingredients include: 1. Foam formation and stabilization by egg white proteins in the case that egg white foam is generated in an individual step; 2. Emulsification by emulsifiers from egg yolk – phospholipids and lipoproteins, which assist in blending of ingredients in cake batters and create more air bubbles by reducing air/water interfacial tension; 3. Heat aggregation of egg white and yolk to modulate cake batter rheological properties, which influence air bubble expansion and cake matrix formation.

Dairy and vegetable proteins have been used for partial or total substitution of eggs in cake formulations and can influence the textural and physical properties of cakes.

Mohamed et al. (1995) substituted the whole egg with proteins from various sources in an 'all in' sponge cake based on equal protein content and found the final cake volumes in an order of egg yolk > whole egg > egg white = skim milk > soy protein isolates > whey powder (12% w/w protein). The cake volume has a weak correlation with cake batter density. They suggested that the emulsifiers, such as lecithin from egg yolk, plays a more important role to create and stabilize air bubbles in cake batters than egg white proteins. The low density lipoprotein (LDL) of yolk is a major contributor for air incorporation to cake batters; while the yolk high density lipoprotein (HDL) granules inhibits aeration but assists in stabilizing air bubbles during whipping and baking (Kamat et al., 1973). Removal of cholesterol from egg yolk does not change its functionalities in cakes, resulting in comparable volume and structure as the control (Awad et al., 1997); however, the reduced lipid content leads to a firmer texture (Paraskevopoulou and Kiosseoglou, 1997). Although aqueous solutions of egg white proteins can generate excellent foam, their ability to form and stabilize air bubbles in a lipid-containing system is limited. The β-casein from milk proteins have surfactant-like properties due to the asymmetric distribution of hydrophobic and hydrophilic residues, leading to a higher cake volume compared to soy and whey proteins (Mohamed et al., 1995).

Incorporation of whey protein concentrates to substitute 20% wheat flour in eggless cake results in a higher volume and better quality, which are comparable to the cake made with full egg (Jyotsna et al., 2007). The authors indicated that addition of whey proteins provides enough surface active components and enhances air incorporation in eggless cake batters. Whey protein isolate, wheat starch, gums and their blends were used as egg replacers in a yellow cake system (Kohrs et al., 2010). Whey protein isolate or wheat starch alone leads to reduced cake volume when substituting 50% whole egg; however, their blends with guar gum or xanthan gum result in equal or greater cake volumes and comparable texture characteristics as the whole egg control. Increased cake volumes with addition of hydrocolloids were observed in other studies and attributed to increases in batter viscosity and interactions between gums and starches that alter the setting temperature of cake matrix (Miller and Hoseney, 1993; Gomez et al., 2007). More surface stickiness was observed for cakes containing guar gum and xanthan gum, due to liquid migration to the crust surface. Addition of hydrocolloids was found to increase the moisture retention in cakes during baking (Gomez et al., 2007; Ashwini et al., 2009).

Combinations of hydrocolloids and emulsifiers were applied to prepare eggless cakes (Ashwini et al., 2009). Addition of hydrocolloids leads to a higher cake batter viscosity, which reduces air incorporation but slows down air bubble destabilization changes during the early stage of baking. The increased viscosity of continuous phase also changes the starch gelatinization profile and alters cake matrix formation during the later stage of baking. When added into cakes containing no emulsifier, hydroxypropylmethycellulose (HPMC) increases cake volume, whereas arabic gum and xanthan gum cause no significant change and guar gum and carrageenan decrease it. However, when used in cakes containing emulsifier – sodium steoryl-2-lactylate, all hydrocolloids result in increased cake volumes even with less aeration in cake batters compared to the control, and HPMC leads to the highest improvement. In the cakes containing another emulsifier – glycerol monostearate, HPMC, xanthan gum, and carrageenan

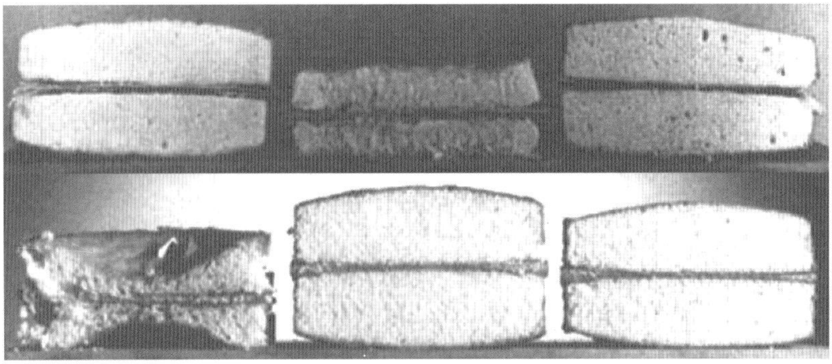

Figure 3.4 Structure of white layer cakes (top) prepared from egg white control (left), No egg white negative control (middle), Bovine blood plasma (right) and yellow cakes (bottom) prepared from No egg negative control (left), Fresh whole egg control (middle), Bovine blood plasma (right). The protein contents of controls and bovine blood plasma cake are equivalent. (Source: Data from Johnson et al. 1979.)

increase cake volume, while arabic gum and guar gum decrease it. Microscopic images suggest the interaction between hydrocolloids and starch granules in the cake matrix plays a role to determine the cake volume and quality, which is dependent on the types of hydrocolloids and emulsifiers (Ashwini et al., 2009).

The white lupine protein (WLP) was used to replace whole egg in yellow cake when combined with baking powder, mono- and diglycerides, and soy lecithin (Arozarena et al., 2001). The cake made with WLP expands during baking due to the leavening action of baking powder but collapses when taken out from the oven, indicating an unstable foam structure. This is explained as the heat aggregation of egg proteins is not accomplished by the denaturation of WLP, leading to the structure collapse when the cake is allowed to cool. Addition of xanthan gum significantly increases the WPL cake height and prevents the vertical collapse of the cake structure; however, increasing xanthan gum content thickens the cake batter and limits the cake expansion. After combining WLP, xanthan gum and other ingredients in an optimum range, a maximum cake volume can be achieved, which is still 9% smaller than the whole egg control and corresponds to a harder cake texture (Arozarena et al., 2001). In summary, combinations of proteins, emulsifiers, chemically leavening agent and/or hydrocolloids provide options to substitute egg solids in a cake system; however, their influence on cake texture characteristics should be considered.

Soapwort (*Gypsophila arrostii*) is a genus of *Gypsophila* and contains saponins, which have surfactant properties. Soapwort extract shows excellent foam formation property because of its saponin content. Celik et al. (2007) used soapwort extract to replace up to 75% egg white in sponge cakes and found no significant difference in cake volumes and equal likeness of the cake sensory characteristics, including crumb cell structure, color, smell, flavor, chewiness, sweetness and overall likeness, when compared to the control. However, the author indicated that 100% substitution of egg white with soapwort extract results in cakes with inferior physical properties.

Bovine blood plasma was used to substitute egg white or whole egg in layer cakes and generated comparable volume and structure as the controls (Johnson et al., 1979; Figure 3.4). During the preparation of layer cakes, the egg white foam was generated individually and then combined with other components. The foam formation of proteins is important for aeration; although the presence of emulsifiers helps air incorporation during beating of shortenings and yolks. The 100% substitution of egg white with bovine blood plasma protein isolate results in half reduction of volume in angel food cake; however, only 8% smaller volume is obtained in white layer cakes when replacing egg white with bovine blood plasma solids based on an equal protein level (Khan et al., 1979; Johnson et al., 1979). Bovine blood plasma foam fails in angel food cakes due to less foam stability but functions almost as well as egg white in white layer cake, because emulsifiers and baking powders also assist in maintaining air bubbles in cake matrix during baking. Slightly less efficient foaming capacity was observed for bovine blood plasma compared to egg white (Johnson et al., 1979). Raeker and Johnson (1995) evaluated the individual protein fractions of bovine blood plasma in white layer cakes and found a positive correlation between protein denaturation temperature and cake volume. The γ-globulin has the highest denaturation temperature and produces cakes with larger volumes and better crumb structures than other protein fractions and the whole bovine plasma. On the other hand, fibringogen, the most heat-sensitive protein in bovine blood plasma, generates cakes with smallest volumes and flat shapes. This is explained as the early denaturation of protein films around air bubbles prevents them from further expansion during baking (Raeker and Johnson, 1995). When substituting whole egg, bovine blood plasma leads to about 10% reduction in cake volume; while addition of 1% lecithin enhances the cake volume (Johnson et al., 1979). The added lecithin compensates the natural phospholipids in egg yolk. A blend of hydrolyzed bovine plasma, beef stock, lecithin and xanthan gum can replace 50% of egg in Devil's food cake with only a slight decrease of cake volume and no significant difference in cake symmetry and shrinkage (Lee et al., 1993).

Cookies have more condensed structure than cakes and usually contain less egg solids. The foaming ability of egg white is not important in this system, as the aeration is mainly achieved by emulsifiers during blending of shortenings and yolks and enhanced by a chemical leavening agent during heating. The emulsifiers from egg yolk assist aeration and mixing of ingredients in cookie dough. A good emulsification can prevent spreading of cookie dough during baking. The heat aggregation of egg proteins influences the cookie structure formation during heating. The proteins and lipids from egg contribute to cookie texture, flavor and appearance. Few studies were conducted to replace egg components in cookies, in which more attention has been placed on reducing the high amount of butter and sugar.

Fermentation is an important step for yeast leavened products, such as breads and biscuits. Substitution of egg proteins with other ingredients changes the nutrition sources for yeast growth and may lead to a difference in carbon dioxide generation during fermentation and therefore influence the texture. Mohamed et al. (1995) evaluated egg white, whey and soy proteins in both chemically leavened and yeast leavened rice cakes and found egg white proteins led to higher cake volume than whey and soy proteins in yeast leavened systems while no big difference was observed in chemically leavened systems. They suggested that egg white probably provides an excellent amino acid source for the yeast growth, which is lacking in proteins from other sources.

3.3.3 Emulsion products

Egg and egg yolk are important ingredients for emulsion products, including mayonnaise, salad dressings, egg-based sauces, and egg nog. The egg ingredients provide functionalities of emulsifying, texturizing and thickening. Although egg white proteins are included in many food emulsions and influence their texture, the major emulsifying functionality is provided by egg yolk and attributed to the lipoproteins and phospholipids. Most egg based emulsion food products contain a high amount of oil fraction and form concentrated emulsions. For example, a typical mayonnaise contains about 75% oil. Concentrated emulsions show semi-solid like properties, due to the oil droplets compressing each other in a way similar to air bubbles in a polyhedral foam structure (Princen and Kiss, 1989). The rheological properties depend on oil phase fraction, size and geometry of oil droplets, interfacial tension and rheological characteristics of the films around oil droplets and continuous phase properties. Substitution of egg or egg yolk with other components significantly changes the composition and structure of the interfacial film and alters oil droplet formation during emulsification, leading to different food textures.

Soy lecithin shows similar emulsifying capability and creates comparable oil droplet sizes in emulsions as egg yolk; however, the emulsion stability is much lower compared to egg yolk, corresponding to a faster increase of oil droplet size over storage time (Bringe et al., 1996). The apoproteins from egg yolk lipoproteins contribute to interfacial film formation, which protect oil droplets from destabilization changes and increase emulsion stability. Soy lecithin adsorbs at interface nicely but lacks protein components to form an elastic interfacial film, corresponding to excellent emulsifying capacity but low emulsion stability.

Proteins from animal and vegetable sources and their derivatives exhibit good emulsifying properties and have potential to replace egg ingredients, including whey proteins, caseinate, soy proteins, tomato seed proteins, white lupine proteins, Arrowtooth flounder proteins, and so on. (Velev et al., 1993; Aluko et al., 1998; Mine and Keeratiurai, 2000; Raymundo et al., 2002; Diftis et al., 2005; Sathivel et al., 2005). However, these proteins always show less efficient interfacial adsorption than egg yolk, leading to larger droplet size and lower stability of emulsions. Polysaccharides were added to enhance the emulsion texture and to improve oil droplet stability by increasing continuous phase viscosity and interacting with protein films (Xie and Hettiarachchy, 1997; Benichou et al., 2002). The excellent emulsifying properties of egg yolk lipoproteins are due to their unique protein-lipid structures, which provide higher hydrophobicity compared to other proteins. It is also suggested the high hydrophobicity of the apoproteins themselves contribute to the strong interfacial adsorption capacity (Nilsson et al., 2007). Nevertheless, the yolk proteins, including apoproteins from lipoproteins, always have stronger hydrophobic affinity than proteins from other sources and contribute most efficiently to emulsion formation and stabilization.

Modified starches with emulsifying capacity were used to prepare emulsions (Tesch et al., 2002). The octenyl-succinate starch (OSA starch) has hydrophobic side groups attached to the hydrophilic polysaccharide chain and can adsorb to the oil/water interface to stabilize an emulsion. The emulsifying capacity of OSA starch was found to be independent of pH and ionic strength (Tesch et al., 2002). When adsorbed at the oil droplet surface, the OSA starch provides steric stabilization for emulsions due to its large size and branched structure. In addition to emulsifying activity, the OSA starches increase continuous phase viscosity and may act as a combination of emulsifier and stabilizer. The physical properties of OSA starch, such as viscosity and solubility, can be adjusted by use of various starch bases, including native starch, acid converted starch, or dextrin. When applied at acid pH, the OSA starch is negatively charged and may interact with the positively charge proteins in emulsion systems. Magnusson and Nilsson (2011) observed the strong interaction between OSA starch and yolk livetin protein at pH 4, which leads to complex formation in solution and reduced available OSA starch for interfacial adsorption. They also found that addition of OSA starch causes no significant difference in droplet size or stability of yolk livetin stabilized emulsions, although some starch-protein interactions are suggested at the interface.

It is challenging to find ingredients that can stabilize small oil droplets and create a thick texture in full fat emulsion foods like egg yolk. Fat mimetics or replacers are added in low fat or fat free emulsions to enhance the texture. In these systems, less or no egg yolk is required to stabilize the lipid phase. Thickening agents include modified starch, xanthan gum, locust bean gum, alginate and other polysaccharides (Mancini et al., 2002; Dolz et al., 2006). These ingredients can mimic oil droplets by acting as fillers or increase continuous phase viscosity as soluble polymers.

Low cholesterol yolk (yolk protein concentrates) and yolk granules were used to prepare mayonnaises in order to reduce the cholesterol content (Paraskevopoulou et al., 1997c; Laca et al., 2010). The high protein content of low cholesterol yolk leads to more elastic structures in salad dressings when compared to high cholesterol egg yolk at equal yolk solid basis (Paraskevopoulou et al., 1997c). Lower texture stabilities during storage were also observed for salad dressings made with low cholesterol yolk, corresponding to fast decreases in network consistency and rapid increases of oil droplet sizes. The yolk granules have lower cholesterol content than raw yolk and maintain good emulsifying functionality. The mayonnaises prepared with yolk granules contains only 1/6 the amount of cholesterol as a commercial sample and shows comparable rheological characteristics and sensory scores when using similar yolk protein levels (Laca et al., 2010).

Egg yolk has been used as an emulsifier in traditional ice cream to assist in blending of ingredients and to improve texture and appearance of ice cream during melting. Several egg alternatives were applied in French vanilla ice cream to replace egg, including modified corn starch, whey protein concentrate and soy protein isolate (Herald et al., 2008). Lower whiteness was observed in ice creams using egg substitutes compared to the control. Using modified corn starch results in a much higher viscosity but no significant differences in texture mouthfeel. Modified corn starch and whey protein isolate improve the melting property and provide closest sensory traits of ice cream compared to the egg yolk control.

3.3.4 Products based on egg gelation

Egg ingredients are critical for the texture of gel type foods, such as custards, puddings and scrambled eggs. Many of these products contain both egg white and yolk and have emulsion gel structures. The yolk lipids and added lipid ingredients, such as cream, not

Table 3.5 Proximate composition of egg products and egg substitutes according to supplier specifications.

Product	Water	Protein	Carbohydrates	Lipids	Others	Ingredients
Liquid products						
Liquid whole egg	65.5	11.8	–	11.0	11.7	–
Liquid egg white	88.0	11.0	–	0.2	0.8	–
Liquid egg yolk	48.0	17.5	–	32.5	2.0	–
Egg Beaters® Regular	88.2	9.8	1.6	0.0	0.3	99% Egg white
Egg Beaters® Whites	86.8	10.9	2.2	0.0	0.2	100% Egg white
Second Nature® Eggs	88.5	9.6	1.6	0.0	0.3	99% Egg white
Better'n Eggs®	87.1	10.7	1.8	0.0	0.4	98% Egg white
Powdered products						
Whole egg powder	5	43	5	40	3.65	–
Egg white powder	11.8	82.2	4.7	0	1.2	–
Egg yolk powder	–	31.1	0.5	65.1	4.5	–
Ener-G® Egg Replacer	0.0	0.0	99.9	0.0	0.1	Potato starch, tapioca starch flour, leavening, sodium CMC, Methylcellulose
NATIONAL® FRIGEX	11.0	0.0	89.0	0.0	0.0	Modified corn starch
Pasta Power	7.0	85.0	5.0	1.0~2.0	6.5	Wheat protein isolate
Soy Flour 220T	8.0	46.0	26.0	15~19	–	Defatted soy flour, soybean oil, soybean lecithin
Soy protein isolate	6.1	82.1	7.1	3.6	1.0	Soy proteins
Eggstend® 300	3.5	57.0	26.0	5.5	5.0	Whey protein concentrate
Biozate1®	5.5	90.0	1.0	1.0	6.0	Whey protein isolate
Bipro®	<5.0	95.0	1.0	1.0	3.0	Whey protein isolate
NFDM	4.0	35.1	52.2	0.7	8.0	Non-fat dry milk powder
Other options						
Ground flaxseeds	9.8	18.5	29.2	42.3	0.3	Carbohydrates, lipids
Silken® soft tofu	89.9	4.8	2.4	3.0	0.0	Water, Soybeans, Gluconolactone, Calcium Chloride
Banana	75.0	0.8	23.8	0.0	0.4	Water, carbohydrates
Applesauce	80.0	0.0	19.9	0.0	0.0	Water, carbohydrates, pectin

only bring the desired flavor in food products but also contribute to the soft gel texture by acting as fillers in the protein gel network. Egg substitutions in these products require functionalities of both gelling and emulsifying for emulsion gel formation as well as enough lipids for a desired texture. Egg substitutes containing egg white, vegetable oil and soy or milk proteins were used to prepare scrambled eggs in substitution of whole eggs (Leutzinger et al., 1977). The texture of scrambled eggs was found to depend on the solid content in egg substitutes. The egg substitute with similar solid content as whole egg results in comparable texture of scrambled eggs but inferior aroma and flavor.

Eggs have water binding, lipid retention and adhesion properties and function as a binding agent in meat products, such as croquettes, and in some culinary cooked egg dishes, such as omelettes and French toast. These properties are mainly associated with the emulsifying, thickening and gelling functionalities. Many non-egg proteins, such as gelatin, whey proteins and soy proteins, were also used as binding agents in meat products and provide similar functionalities (Colmenero, 1996).

Egg noodles or pastas are made from high gluten flour, usually semolina, durum flour, farina, wheat flour or their combinations, and egg ingredients. The gluten network development in dough contributes to a firm noodle texture and provides resistance to overcooking. Egg proteins can improve the cooking quality of noodles by maintaining starch in the dough and strengthening the texture during cooking. Egg yolk also provides the yellow color of pasta. Several egg substitutes were used to prepare egg noodles, including proteins from vegetable, dairy and animal sources (Khouryieh et al., 2006; Chang et al., 2011). The lipids in whole egg were replaced by either proteins or carbohydrates in egg substitutes (Khouryieh et al., 2006). Egg substitutes with more carbohydrates show more cooking loss and lower noodle firmness than whole eggs. The noodle firmness increases with protein content when using different types of whey protein to replace 50% whole egg. Cuttlefish paste was used to substitute liquid egg in noodles and generates no significant difference in protein contents and slightly lower lipid contents (Chang et al., 2011). Increasing Cuttlefish paste substation leads to less cooking loss and higher noodle firmness.

3.4 COMMERCIAL EGG PRODUCTS AND SUBSTITUTES

Various forms of egg products have been developed for the applications in food industry and home consumption. The liquid egg products can be directly used in food preparation and provide a convenient option for consumers. Frozen egg products have similar compositions as liquid egg products and can be stored for a longer time. Concentrated and powdered products have reduced volume and save the cost for storage and transportation. A variety of egg substitutes have been commercialized with adaption to consumer dietary needs or restrictions and allergenic concerns of egg ingredients (Table 3.5).

The majority of liquid egg products or 'egg substitutes' are based on egg white for a low fat and low cholesterol alternative for whole eggs. These products contain more than 98% egg white and are mainly used to replace egg functionality in home culinary dishes, such as gelling and foaming. In some products, minor other ingredients, such as flavorants, colorant, vitamins and minerals, are added to mimic the flavor and yellow color of yolk and to compensate for the nutritional values lost with removal of yolk. Polysaccharides or gums may also be added in these products, usually at very low levels, in order to enhance the viscosity and to provide more similar characteristics as whole eggs. However, these products cannot function well when emulsification is required as the lipoproteins and phospholipids from egg yolks are missed. In these cases, it is recommended to partially replace whole egg ingredients with egg white based products for reduced fat and cholesterol content and still allowing some yolk constituents to function as emulsifiers.

The egg yolk is a mixture of proteins and lipids (Table 3.3) and can be substituted using combinations of vegetable oil and proteins from other sources. Phospholipids, such as soy lecithin, may also be added to enhance the emulsifying properties. In two old versions of egg substitute products, Egg Beater® (82% egg white, 10% corn oil and 7% nonfat dry milk) and Second Nature® (84% egg white, 7% water, 3% partially hardened soybean oil, 2.5% isolated soy protein and 1.8% egg white solids), egg yolk is replaced by combinations of vegetable oil and milk or soy proteins and used to prepare scrambled eggs (Leutzinger et al., 1977).

The powdered egg substitutes are mainly based on dairy and vegetable proteins and polysaccharides. Proteins denature and form aggregates and/or gels during heating and can be used to mimic egg functionality in gel type foods and bakery products. Proteins also provide emulsifying properties and can be applied to prepare emulsion foods. The Ener-G® Egg Replacer is recommended for use in baking. It includes starches and gums, which provide thickening effect and increase the viscosity of the bakery product batters. The leavening agent can assist foam structure formation during the baking process.

For culinary cooking, several options are recommended for egg replacement, including carbohydrate-based ingredients, such as ground flaxseeds, banana and apple sauces, and protein-based ingredients, such as tofu. A combination of flour, cooking oil and salt is suggested to replace egg yolk and used as a binding agent in meat loaf. In bakery products, starch-based ingredients may provide egg functionalities. Recommendations include potato, pumpkin or squash, pureed prunes, banana, and apple sauces and so on. Tofu is a soy protein gel and suggested as egg substitutions to create egg gel type texture in food systems, such as custards and scrambled eggs.

In summary, selection of egg substitutes depends on the food product characteristics and the egg functionalities. Partial or even complete replacement of egg ingredients can be realized by using proper alternative ingredients and may lead to minor changes in certain food products. It should be considered that most food systems are dispersions of two or more phases, including foam, emulsions, and emulsion gels. Egg ingredients play important roles in blending different components, creating desired texture and maintaining stability. Although many proteins and small molecular emulsifiers have comparable foaming and emulsifying capacity as egg white and yolk, the excellent stability of dispersion systems prepared using egg ingredients is still a challenge to match using other ingredients or their combinations.

References

Abu-Ghoush, M., Herald, T.J., and Aramouni, F.M. (2008) Comparative study of egg white protein and egg alternatives used in an angel food cake system. *Journal of Food Processing and Preservation.* 34(s2), 411–425.

Aluko, R.E., and Mine, Y. (1997) Competetive adsorption of hen's egg yolk granule lipoproteins and phosvitin in oil-water emulsions. *Journal of Agricultural and Food Chemistry.* 45(12), 4564–4570.

Aluko, R.E., Keeratiurai, M., and Mine, Y. (1998) Competitive adsorption between egg yolk lipoproteins and whey proteins on oil-in-water interfaces. *Colloids and Surfaces B: Biointerfaces.* 10(6), 385–393.

Anton, M., and Gandemer, G. (1997) Composition, solubility and emulsifying properties of granules and plasma of egg yolk. *Journal of Food Science.* 62(3), 484–487.

Anton, M., and Gandemer, G. (1999) Effect of pH on interface composition and on quality of oil-in-water emulsions made with hen egg yolk. *Colloids and Surfaces B: Biointerfaces.* 12(3–6), 351–358.

Anton, M., Le Denmat, M., Beaumal, V., and Pilet, P. (2001) Filler effects of oil droplets on the rheology of heat-set emulsion gels prepared with egg yolk and egg yolk fractions. *Colloids and surfaces B: Biointerfaces.* 21(1–3), 137–247.

Arntfield, S.D., and Bernatsky, A. (1993) Characteristics of heat-induced networks for mixtures of ovalbumin and lysozyme. *Journal of Agricultural and Food Chemistry.* 41(12), 2291–2295.

Arozarena, I., Bertholo, H., Empis, J., Bunger, A., and Sousa, I. (2001) Study of the total replacement of egg by white lupine protein, emulsifiers and xanthan gum in yellow cakes. *European Food Research and Technology.* 213(4–5), 312–316.

Arunepanlop, B., Morr, C.V., Karleskind, D., and Laye, I. (1996) Partial replacement of egg white proteins with whey proteins in angel food cakes. *Journal of Food Science*, 61(5), 1085–1093.

Ashwini, A., Jyotsna, R., and Indrani, D. (2009) Effect of hydrocolloids and emulsifiers on the rheological, microstructural and quality characteristics of eggless cake. Food *Hydrocolloids.* 23(3), 700–707.

Awad, A.C., Bennink, M.R., and Smith, D.M. (1997) Composition and functional properties of cholesterol reduced egg yolk. *Poultry Science.* 76(4), 649–653.

Baier, S.K., and McClements, D.J. (2003) Combined influence of NaCl and sucrose on heat-induced gelation of bovine serum albumin. *Journal of Agricultural and Food Chemistry.* 51(27), 8107–8112.

Benichou, A., Aserin, A., and Garti, N. (2002) Protein-polysaccharide interactions for stabilization of food emulsions. *Journal of Dispersion Science and Technology.* 23(1–3), 93–123.

Berry, T.K., Yang, X., and Foegeding, E.A. (2009) Foams prepared from whey protein isolate and egg white protein: 2. Changes associated with angel food cake functionality. *Journal of Food Science.* 74(5), E269–E277.

Bringe, N.A., Howard, D.B., and Clark, D.R. (1996) Emulsifying properties of low-fat, low-cholesterol egg yolk prepared by supercritical CO_2 extraction. *Journal of Food Science.* 61(1), 19–23.

Bos, M.A., and van Vliet, T. (2001) Interfacial rheological properties of adsorbed protein layers and surfactants: a review. *Advances in Colloid and Interface Science.* 91(3), 437–471.

Causeret, D., Matringe, E., and Lorient, D. (1991) Ionic strength and pH effects on composition and microstructure of yolk granules. *Journal of Food Science.* 56(6), 1532–1536.

Celik, L., Yilmaz, Y., Isik, F., and Ustun, O. (2007) Effect of soapwort extract on physical and sensory properties of sponge cakes and rheological properties of sponge cake batters. *Food Chemistry.* 101(3), 907–911.

Chang, Y.I., and Chen, T.C. (2000) Functional and gel characteristics of liquid whole egg as affected by pH alternation. *Journal of Food Engineering.* 45(4), 237–241.

Chang, H.C., Chen, H.H., and Hu, H.H. (2011) Textural changes in fresh egg noodles formulated with seaweed powder and full or partial replacement of cuttlefish paste. *Journal of Texture Studies.* 42(1), 61–71.

Colmenero, F.J. (1996) Technologies for developing low-fat meat products. *Trends in Food Science and Technology.* 7(2), 41–48.

Davis, J.P., and Foegeding, E.A. (2004) Foaming and interfacial properties of polymerized whey protein isolate. *Journal of Food Science.* 69(5), C404–C410.

Davis, J.P., Foegeding, E.A., and Hansen, F.K. (2004) Electrostatic effects on the yield stress of whey protein isolate foams. *Colloids and Surfaces B: Biointerfaces.* 34(1), 13–23.

Davis, J.P., Doucet, D., and Foegeding, E.A. (2005) Foaming and interfacial properties of hydrolyzed β-lactoglobulin. *Journal of Colloid and Interface Science.* 288(2), 412–422.

Davis, J.P., and Foegeding, E.A. (2007) Comparisons of the foaming and interfacial properties of whey protein isolate and egg white proteins. *Colloids and Surfaces B: Biointerfaces*. 54(2), 200–210.

Dickinson, E. (1992) *An Introduction to Food Colloids*. Oxford: Oxford University Press. Pp.135 and 216.

Diftis, N.G., Biliaderis, C.G., and Kiosseoglou, V.D. (2005) Rheological properties and stability of model salad dressing emulsions prepared with a dry-heated soybean protein isolate-dextran mixture. *Food Hydrocolloids*. 19(6), 1025–1031.

Dixon, D.K., and Cotterill, O.J. (1981) Electrophoretic and chromatographic changes in egg yolk proteins due to heat. *Journal of Food Science*. 46(4), 981–983, 990.

Dolz, M., Hernandez, M.J., and Delegido, J. (2006) Oscillatory measurements for salad dressings stabilized with modified starch, xanthan gum, and locust bean gum. *Journal of Applied Polymer Science*. 102(1), 897–903.

Dyer-Hurdon, J.N. and Nnanna, I.A. (1993) Cholesterol content and functionality of plasma and granules fractionated from egg yolk. *Journal of Food Science*. 58(6), 1277–1281.

Edwards, D.A., and Wasan, D.T. (1996) Foam rheology: the theogy and role of interfacial rheologycial properties. Chapter 3 in "Foams: theory, measurements and applications" Edited by Prud'homme, R.K. and Khan, S.A. New York: Marcel Dekker, Inc., p.189.

Foegeding, E.A., Luck, P.J., and Davis, J.P. (2006) Factors determining the physical properties of protein foams. *Food Hydrocolloids*. 20(2–3), 284–292.

Gomez, M., Ronda, F., Caballero, P.A., Blanco, C.A., and Rosell, C.M. (2007) Functionality of different hydrocolloids on the quality and shelf-life of yellow layer cakes. *Food Hydrocolloids*. 21(2), 167–173.

Hammershøj, M., Prins, A., and Qvist, K.B. (1999) Influence of pH on surface properties of aqueous egg albumen solutions in relation to foaming behaviour. *Journal of the Science of Food and Agriculture*. 79(6), 859–868.

Hammershøj, M., and Qvist, K.B. (2001) Importance of hen age and egg storage time for egg albumen foaming. *LWT – Food Science and Technology*. 34(2), 118–120.

Hammershøj, M., Nording, J.A., Rasmussen, H.C., Carstens, J.H., and Pedersen, H. (2006) Dry pasteurisation of egg albumen powder in a fluidised bed. I. Effect on microbiology, physical and chemical parameters. *International Journal of Food Science & Technology*. 41(3), 249–261.

Hammershøj, M., Nebel, C., and Carstens, J.H. (2008) Enzymatic hydrolysis of ovomucin and effect on foaming properties. *Food Research International*. 41(5), 522–531.

Handa, A., Takahashi, K., Kuroda, N., and Froning, G.W. (1998) Heat-induced egg white gels as affected by pH. *Journal of Food Science*. 63(3), 403–407.

Hatta, H., Kitabatake, N., and Doi, E. (1986) Turbidity and hardness of a heat-induced gel of hen egg ovalbumin. *Agricultural and Biological Chemistry*. 50(8), 2083–2089.

Herald, T.J., Aramouni, F.M., and Abu-Ghoush, M.H. (2008) Comparison study of egg yolks and egg alternative in French vanilla ice cream. *Journal of Texture Studies*. 39(3), 284–295.

Johnson, L.A., Havel, E.F., and Hoseney, R.C. (1979) Bovine plasma as a replacement for egg in cakes. *Cereal Chemistry*. 56(4), 339–342.

Johnson, T.M., and Zabik, M.E. (1981a) Response surface methodology for analysis of protein interactions in angel food cakes. *Journal of Food Science*. 46(4), 1226–1230.

Johnson, T.M. and Zabik, M.E. (1981b) Egg albumen proteins interactions in an angel food cake system. *Journal of Food Science*. 46(4), 1231–1236.

Jyotsna, R., Manohar, R.S., Indrani, D., and Rao, G.V. (2007) Effect of whey protein concentrate on the rheological and baking properties of eggless cake. *International Journal of Food Properties*. 10(3), 599–606.

Kalkani, A., Paraskevopoulou, A., and Kiosseoglou, V. (2007) Protein interactions and filler effects in heat-set gels based on egg. *Food Hydrocolloids*. 21(2), 191–197.

Kamat, V.B., Lawrence, G.A., Hart, C.J., and Yoell, R. (1973) Contribution of egg yolk lipoproteins to cake structure. *Journal of the Science of Food and Agriculture*. 24(1), 77–88.

Kato, A., Oda, S., Yamanaka, Y., Matsudomi, N., and Kobayashi, K. (1985) Functional and structural properties of ovomucin. *Agricultural and Biological Chemistry*. 49(12), 3501–3504.

Khan, M.N., Rooney, L.W., Dill, C.W. (1979) Baking properties of plasma protein isolate. *Journal of Food Science*. 44(1), 274–276.

Khouryieh, H., Herald, T., and Aramouni, F. (2006) Quality and sensory properties of fresh egg noodles formulated with either total or partial replacement of egg substitutes. *Journal of Food Science*. 71(6), S433–S437.

Kiosseoglou, V. (2003) Egg yolk protein gels and emulsions. *Current Opinion in Colloid & Interface Science*. 8(4–5), 365–370.

Kiosseoglou, V., and Paraskevopoulou, A. (2005) Molecular interactions in gels prepared with egg yolk and its fractions. *Food Hydrocolloids.* 19(3), 527–532.

Kiosseoglou, V., and Sherman, P. (1983) The influence of egg yolk lipoproteins in the rheology and stability of o/w emulsions and mayonnaise: 2. Interfacial tension-time behavior of egg yolk lipoproteins at the groundnut oil–water interface. *Colloid & Polymer Science.* 261(6), 502–507.

Kohrs, D., Herald, T.J., Aramouni, F.M., and Abughoush, M. (2010) Evaluation of egg replacers in a yellow cake system. *Emirate Journal of Food and Agriculture.* 22(5), 340–352.

Koidis, A., Paraskevopoulou, A., and Kiosseoglou, V. (2002) Fracture and textural properties of low fat egg yolk gels containing emulsion droplets. *Food Hydrocolloids.* 16(6), 673–678.

Laca, A., Saenz, M.C., Paredes, B., and Diaz, M. (2010) Rheological properties, stability and sensory evaluation of low-cholesterol mayonnaises prepared using egg yolk granules as emulsifying agent. *Journal of Food Engineering.* 97(2), 243–252.

Lau, C.K., and Dickinson, E. (2005) Instability and structural change in an aerated system containing egg albumen and invert sugar. *Food Hydrocolloids.* 19(1), 111–121.

Le Denmat, M., Anton, M., and Gandemer, G. (1999) Protein denaturation and emulsifying properties of plasma and granules of egg yolk as related to heat treatment. *Journal of Food Science.* 64(2), 194–197.

Le Denmat, M., Anton, M., and Beaumal, V. (2000) Characterization of emulsion properties and of interface composition in O/W emulsions prepared with hen egg yolk, plasma and granules. *Food Hydrocolloids.* 14(6), 539–549.

Lee, C.C., Love, J.A., and Johnson, L.A. (1993) Sensory and physical properties of cakes with bovine plasma products substituted for egg. *Cereal Chemistry.* 70(1), 18–21.

Leutzinger, R.L., Baldwin, R.E., and Cotterill, O.J. (1977) Sensory attributes of commercial egg substitutes. *Journal of Food Science.* 42(4), 1124–1125.

Li-Chan, E., and Nakai, S. (1989) Biochemical Basis for the Properties of Egg White. *CRC Critical Review in Poultry Biology.* 2(1), 21–58.

Linden, G., and Lorient, D. (1999) Egg products in "New ingredients in food processing: biochemistry and agriculture". CRC Press LLC. Pp.121–138.

Luck, P.J., Bray, N., and Foegeding, E.A. (2002) Factors determining yield stress and overrun of whey protein foams. *Journal of Food Science.* 67(5), 1677–1681.

MacDonnell, L.R., Feeney, R.E., Hanson, H.L., Campbell, A., and Sugihara, T.F. (1954) The functional properties of the egg white proteins. *Food Technology.* 8, 49–53.

Magnusson, E., and Nilsson, L. (2011) Interactions between hydrophobically modified starch and egg yolk proteins in solution and emulsions. *Food Hydrocolloids.* 25(4), 764–772.

Mancini, F., Montanari, L., Peressini, D., and Fantozzi, P. (2002) Influence of alginate concentration and molecular weight on functional properties of mayonnaise. *LWT – Food Science and Technology.* 35(6), 517–525.

Martin, W.G., Tattrie, N.H., and Cook, W.H. (1963) Lipid extraction and distribution studies of egg yolk lipoproteins. *Canadian Journal of Biochemisry and Physiology.* 41(3), 657–666.

McBee, L.E., and Cotterill, O.J. (1979) Ion exchange chromatography and electrophoresis of egg yolk. *Journal of Food Science.* 44(3), 656–667.

Mel'nikov, S.M. (2003) Effect of pH on the adsorption kinetics of egg yolk at the triacylglycerol-water interface and viscoelastic properties of interfacial egg yolk films: a dynamic drop tensiometry study. *Colloids and Surfaces B: Biointerfaces.* 27(2–3), 265–275.

Miller, R.A., and Hoseney, R.C. (1993) The role of xanthan gum in white layer cakes. *Cereal Chemistry.* 70(5), 585–588.

Miller, L.L., and Setser, C. (1983) Xanthan gum in reduced-egg-white angel food cake. *Cereal Chemistry.* 60(1), 62–65.

Mine, Y. (1995) Recent advances in the understanding of egg white protein functionality. *Trends in Food Science & Technology.* 6(7), 225–232.

Mine, Y. (1998a) Emulsifying characterization of hens egg yolk proteins in oil-in-water emulsions. *Food Hydrocolloids.* 12(4), 409–415.

Mine, Y. (1998b) Adsorption behavior of egg yolk low-density lipoproteins in oil-in-water emulsions. *Journal of Agricultural and Food Chemistry.* 46(1), 36–41.

Mine, Y., and Keeratiurai, M. (2000) Selective displacement of caseinate proteins by hens egg yolk lipoproteins at oil-in-water interfaces. *Colloids and surfaces B: Biointerfaces.* 18(1), 1–11.

Mohamed, S., Lajis, S.M.M., and Hamid, N.A. (1995) Effects of protein from different sources on the characteristics of sponge cakes, rice cakes (APAM), doughnuts and frying batters. *Journal of the Science of Food and Agriculture.* 68(3), 271–277.

Nakamura, R. (1963) Studies on the foaming property of the chicken egg white. VI. Spread monolayer of the protein fraction of the chicken egg white. *Agricultural and Biological Chemistry.* 27(6), 427–432.

Nilsson, L., Osmark, P., Fernandez, C., and Bergenstahl, B. (2007) Competitive adsorption of proteins from total hen egg yolk during emulsification. *Journal of Agricultural and Food Chemistry.* 55(16), 6746–6753.

Paraskevopoulou, A., and Kiosseoglou, V. (1994) Cholesterol and other lipid extraction from egg yolk using organic solvents: Effects on functional properties. *Journal of Food Science.* 59(4), 766–768.

Paraskevopoulou, A., and Kiosseoglou, V. (1997) Texture profile analysis of heat-formed gels and cakes prepared with low cholesterol egg yolk concentrates. *Journal of Food Science.* 62(1), 208–211.

Paraskevopoulou, A., Kiosseoglou, V., Alevisopoulos, S., and Kasapis, S. (2000) Small deformation measurements of single and mixed gels of low cholesterol yolk and egg white. *Journal of Texture Studies.* 31(2), 225–244.

Paraskevopoulou, A., Kiosseoglou, V., and Pegiadou, S. (1997a) Interfacial behaviour of egg yolk with reduced cholesterol content. *Journal of Agricultural and Food Chemistry.* 45(10), 3717–3722.

Paraskevopoulou, A., Pegiadou, S., and Kiosseoglou, V. (1997b) Functional properties of egg yolk concentrate with a reduced cholesterol content prepared by treatment with supercritical CO_2. *Food Hydrocolloids.* 11(4), 385–391.

Paraskevopoulou, A., Kiosseoglou, V., Alevisopoulos, S., and Kasapis, S. (1997c) Small deformation properties of model salad dressings prepared with reduced cholesterol egg yolk. *Journal of Texture Studies.* 28(2), 221–237.

Pernell, C.W., Foegeding, E.A., and Daubert, C.R. (2000) Measurement of the yield stress of protein foams by vane rheometry. *Journal of Food Science.* 65(1), 110–114.

Pernell, C.W., Foegeding, E.A., Luck, P.J., and Davis, J.P. (2002a) Properties of whey and egg white protein foams. *Colloids and Surfaces A: Physicochemical and Engineering Aspects.* 204(1–3), 9–21.

Pernell, C.W., Luck, P.J., Foegeding, E.A., and Daubert, C.R. (2002b) Heat-induced changes in angel food cakes containing egg-white protein or whey protein isolate. *Journal of Food Science.* 67(8), 2945–2951.

Phillips, L.G., Yang, S.T., Schulman, W., and Kinsella, J.E. (1989) Effects of lysozyme, clupeine, and sucrose on the foaming properties of whey protein isolate and β-lactoglobulin. *Journal of Food Science.* 54(3), 743–747.

Princen, H.M., and Kiss, A.D. (1989) Rheology of foams and highly concentrated emulsions IV: An experimental study of the shear viscosity and yield stress of concentrated emulsions. *Journal of Colloid and Interface Science.* 128(1), 176–187.

Raeker, M.O., and Johnson, L.A. (1995) Thermal and functional properties of bovine blood plasma and egg white proteins. *Journal of Food Science.* 60(4), 685–690.

Raikos, V., Campbell, L., and Euston, S.R. (2007a) Effects of sucrose and sodium chloride on foaming properties of egg white proteins. *Food Research International.* 40(3), 347–355.

Raikos, V., Campbell, L., and Euston, S.R. (2007b) Rheology and texture of hen's egg protein heat-set gels as affected by pH and the addition of sugar and/or salt. Food *Hydrocolloids.* 21(2), 237–244.

Raymundo, A., Franco, J.M., Empis, J., and Sousa, I. (2002) Optimization of the composition of low-fat oil-in-water emulsions stabilized by white lupin protein. *Journal of the American Oil Chemist Society.* 79(8), 783–790.

Sathivel, S., Bechtel, P.J., Babbitt, J.K., Prinyawiwatkul, W., and Patterson, M. (2005) Functional, nutritional, and rheological properties of protein powders from Arrowtooth flounder and their application in mayonnaise. *Journal of Food Science.* 70(2), E57–E63.

Ternes, W. (2001) Egg proteins. In *Chemical and Functional Properties of Food Proteins.* Edited by Sikorski, Z.E., Lancaster, PA: Technomic Publishing Co. Inc., pp.335–371.

Tesch, S., Gerhards, C., and Schubert, H. (2002) Stabilization of emulsions by OSA starches. *Journal of Food Engineering.* 54(2), 167–174.

Velev, O.D., Nikolov, A.D., Denkov, N.D., Doxastakis, G., Kiosseoglu, V., and Stalidis, G. (1993) Investigation of the mechanisms of stabilization of food emulsions by vegetable proteins. *Food Hydrocolloids.* 7(1), 55–71.

Walstra, P. (2003) *Physical Chemistry of Foods.* New York: Marcel Dekker. p355.

Wang, G., and Wang, T. (2009) Improving foaming properties of yolk-contaminated egg albumen by basic soy protein. *Journal of Food Science.* 74(8), C581–C587.

Wilde, P.J. (2000) Interfaces: their role in foam and emulsion behavior. *Current Opinion in Colloid & Interface Science.* 5(3–4), 176–181.

Xie, Y.R., and Nettiarachchy, N.S. (1997) Xanthan gum effects on solubility and emulsification properties of soy protein isolate. *Journal of Food Science.* 62(6), 1101–1104.

Yang, X. (2008) PhD Dissertation. North Carolina State University. Effect of sucrose on the foaming and interfacial properties of egg white proteins and whey protein isolates.

Yang, X., Berry, T.K., and Foegeding, E.A. (2009) Foams prepared from whey protein isolate and egg white protein: 1. Physical, microstructural, and interfacial properties. *Journal of Food Science*. 74(5), E259–E268.

Yang, X., and Foegeding, E.A. (2010) Effects of sucrose on egg white protein and whey protein isolate foams: Factors determining properties of wet and dry foams (cakes). *Food Hydrocolloids*. 24(2–3), 227–238.

Yang, X., and Foegeding, E.A. (2011) The stability and physical properties of egg white and whey protein foams explained based on microstructure and interfacial properties. *Food Hydrocolloids*. 25(7), 1687–1701.

Zhu, H., and Damodaran, S. (1994) Protease peptones and physical factors affect foaming properties of whey protein isolate. *Journal of Food Science*. 59(3), 554–560.

4 Low fat ice cream

Arun Kilara

Nutri+Food Business Consultants, Chapel Hill, North Carolina, USA

4.1 INTRODUCTION

Despite the myths surrounding the origins of ice cream, we do not know for sure how the evolution of ice cream resulted. Snow and ice have been popular during warmer seasons because of the cooling properties of these materials. Perhaps first snow and ice were mixed with fruit juices and later with milk or yogurt and this resulted in a gradual evolution of these products as we know them. Until the nineteenth century harvesting ice and storing it for use during summer was a labor intensive process and therefore, ice cream was a food for the rich only. With the invention of the hand cranked freezer and a ready availability of ice, ice cream moved down the social ladder and towards the end of the nineteenth century it was sold on the streets of metropolitan areas.

Ice cream and frozen desserts are popular throughout the world. Although ice cream is a popular frozen dessert in all parts of the world, the major consumer countries of ice cream are New Zealand, the United States, Canada, Australia, Belgium, Finland and Sweden. In some areas of the world, unavailability of appropriate ingredients, lack of refrigerated distribution chain, economics or other cultural factors may deter the manufacture (and therefore consumption) of ice cream. There are several different names for ice cream in the world. In Norwegian it is *iskrem*, in Portuguese *sorvettes*, in Spanish *healdos*, in French *glace*, in Italian *gelato*, in Hebrew *glidah*, in German *eis*, in Finnish *jatelo*, in Greek *pagoto* and in Chinese *bing qi lin* or *sou go*.

The popularity of ice cream results from several characteristics such as partial freezing, cooling and refreshing sensation produced when the product is consumed, its sweet taste and the lack of a pre-conditioning aroma. In all these products a sugar solution is the common denominator. The characteristics of the syrup are manipulated by the addition of other materials to obtain desired taste, texture, consistency and appearance.

Ice cream is manufactured as regular, custard/French, reduced fat, light, low and no fat versions. Other frozen desserts include frozen yogurt, sherbet, water ice, mellorine, frozen dairy dessert, frozen confection, frozen dairy confection, milk shake, smoothies, shake and slush. The nomenclature varies from country to country depending on the prevailing legislation. Two manufacturing practices that affect the characteristics of

Food Texture Design and Optimization, First Edition. Edited by Yadunandan Lal Dar and Joseph M. Light.

Table 4.1 Typical composition of several frozen dairy desserts.

Component	No-fat	Low-fat	Reduced fat	Regular	Premium	Super Premium
				%		
Milk fat	0.50	3.0	6.0	10.0	12.0	16.0
Milk solids not fat	13.5	13.0	12.5	8.5	9.5	9.0
Whey solids	–	–	2.0	2.0	–	–
Sugar	10.0	9.0	10.0	11.0	11.0	16.0
Corn Syrup Solids	10.0	9.0	8.0	6.5	6.0	–
Stabilizer/Emulsifier	1.0	0.8	0.6	0.4	0.3	–
TOTAL SOLIDS	35.0	34.8	39.1	38.4	38.8	43.0
WATER	65.0	65.2	60.9	61.6	61.2	57.0

frozen desserts are the freezing technique and degree of freezing. The freezing technique may involve stirring (agitation) during freezing, or without stirring (quiescent) or a combination of the two. The extent of freezing results in products that are hard frozen, or designed for dipping or scooping, or used as soft serve or a milk shake.

According to United States Food and Drug Administration (US FDA) ice cream must contain a minimum of 10% milk fat, 20% total milk solids, must be frozen under agitation and should weigh not less than 4.5 lbs per US gallon (Code of Federal Regulations, 2011).

Due to changing dietary needs the US Congress passed the Nutritional Labeling and Education Act (NLEA) in 1992 and USFDA promulgated regulations to enforce this law. Among the numerous changes brought about by this legislation were the introductions of new categories of foods labeled as Light, Low-fat and No-fat. The term low fat as applied to all foods stipulates that a serving of that food should contain no more than 3 g of fat. Therefore, low fat ice cream contains no more than 3 g milk fat in a 4 fluid ounce serving of the product. Typical composition of several ice cream products is provided (Table 4.1).

Reduction of any component in a formulation requires the substitution of other ingredients. For example, when fat is reduced or eliminated other solids have to be added to achieve a total solids minimum of 35%. The reason for such substitution is evident when the function of each component is examined in detail.

4.2 ROLE OF COMPONENTS IN AN ICE CREAM MIX

The main components in an ice cream mix are fat, milk solids but not fat, sweeteners and stabilizers and emulsifiers. Milk solids not fat includes components such as whey protein concentrate and whey powder and sweetener solids encompassing sugar, corn syrup solids, maltodextrins and so on.

4.2.1 Milk fat

Milk fat ice cream plays an important role in the enjoyment of the product by consumers. Milk fat imparts lubrication, warmth and creaminess to a product. It also ameliorates the perception of ice crystals in the mouth. It is a major contributor of solids to an ice cream mix. Since most flavor volatiles are fat soluble it is an excellent carrier of characterizing added flavors. Milk fat agglomeration plays a key role in the texture of ice cream.

Milk fat is a complex mixture of many types of organic compounds. Fat in milk is secreted as tiny droplets called globules. A drop of milk contains millions of such globules. Each globule is surrounded by a membrane called milk fat globule membrane. This fat is made up of triglycerides, traces of di- and mono-glycerides, cholesterol and phospholipids among many other substances. In fact milk fat is composed of some 3600 different compounds. The triglycerides which are the main component are synthesized by the cow by linking three molecules of fatty acids to one molecule of glycerol, hence the name tri-glycerides. These fatty acids consist of carbon, hydrogen and oxygen atoms. Fatty acids can have as few as four carbon atoms or as many as 26 carbon atoms. In milk fat, we have a number of different fatty acids including those containing four, six, and eight carbon atoms. This is significant because the characteristic flavor of milk fat is in large part due to the presence of these lower chain fatty acids butyric, caproic, caprylic and capric acids, carbon numbers 4, 6, 8, and 10, respectively. Further there are fatty acids that are unsaturated and thus we have fatty acids with one double bond (monoenoic) fatty acids, two double bonds (dienoic) or with three fatty acids (trienoic). The unsaturated fatty acids with multiple double bonds are functional, healthy and essential for human beings.

Milk fat melts and crystallizes, is unctuous, depresses the cold sensation, contributes desirable flavor, is a solvent for added flavors, adds structure to ice cream and is of great importance in extrusion properties of ice cream. Extrusion helps shape ice cream in novelties.

4.2.2 Milk solids not fat

The term milk solids not fat refers to all components in milk solids other than fat and includes proteins, lactose and minerals (ash). Proteins play an important role in emulsification of the fat. Milk proteins also help in developing overrun (aeration). The combination of emulsification and foaming create desirable texture. Proteins also contribute to the viscosity of the mix. Proteins are surface active agents (surfactants), and are responsible for desirable interfacial behavior. Lactose, a disaccharide consisting of glucose and galactose, is the predominant carbohydrate in milk. Lactose is only approximately 20% as sweet as sugar (sucrose). Lactose contributes to the depression of freezing point of an ice cream mix. In mixes with a very high content of milk solids not fat, lactose crystalization results in a textural defect known as sandiness. Milk salts exist in both free and bound forms. Calcium and phosphorus are integral to the casein micelle. Approximately 66% of these two elements occur in a bound form in the casein micelles. Other milk salts in a soluble form such as citrate, sodium, potassium and chloride are free. Soluble or free salts contribute to lowering the freezing point of the mix.

Standards of identity for ice cream and frozen desserts allow the substitution of 25% of the milk solids not fat content of a mix with whey derived ingredients. Whey derived ingredients include whey powder, partially demineralized whey powder, partially delactosed whey powder, partially demineralized and delactosed whey powder, whey protein concentrates and whey protein isolates. Whey powders are low in protein and high in lactose and minerals. Removal of both lactose and minerals results in elevating the protein content resulting in whey protein concentrates. Whey protein concentrate with 34% protein is used as an economic substitute for skim milk powder. Its gross composition is similar to skim milk powder with the exception that the protein content is devoid of caseins. Therefore textural control is more challenging in ice cream mixes in which milk solids are substituted by whey solids.

Milk proteins are good foaming agents and play a key role in the development of overrun. Overrun is a term used to measure the volume expansion from mix to ice cream. The density of ice cream is always lower than the density of the mix from which this product was made.

4.2.3 Nutritive sweeteners

In addition to lactose contributed by milk solids, sugar and other sweeteners are also added in the manufacture of ice cream mix. For economic reasons, when sugar prices are high, high fructose corn syrup may be used as a partial substitute for sucrose. Honey has also been considered as an ingredient in mix manufacture. To improve the body and texture of certain types of ice cream maltodextrins and corn syrup solids are added to mixes. Nutritive sweeteners contribute 4 kcal/g of the substance and hence the classification as nutritive sweeteners.

Since sugar is crystalized it indicates that it is a pure chemical, and the source of sucrose, whether beet or cane, does not influence sweetness or other physicochemical properties. It provides sweetness, depresses freezing point, affects freezing performance, affects body and texture, enhances flavor, and contributes bulk or total solids and impacts on economics. Generally the equivalent of 15% sucrose is considered optimal sweetness in ice cream. In making no-sugar added ice cream and frozen desserts, the bulk contributed by sugar is absent and therefore bulking agents, such as polydextrose, are used. Sugars depress the freezing point of the ice cream mix. On a mole basis, monosaccharides depress the freezing point approximately two-fold when compared with disaccharides. If ice cream was manufactured using only monosaccharides, the product would be soft and 'soupy' because of the depression of freezing point of the mix.

Often corn syrup solids or maltodextrins are added to ice cream mix formulations. Corn sweeteners are derived from the modification of corn starch. Low conversion corn syrups are 23–38 dextrose equivalents (DE). DE is a measure of the extent of hydrolysis or modification of starch. Regular conversion syrups are 23–38 DE, intermediate conversion syrups are 48–58 DE and high conversions are 58–68 DE. These products are not as sweet as sugar but they contribute total solids to the mix. By increasing total solids to the mix, heat shock protection is provided.

In certain low fat and no fat mixes as well as no sugar added mixes, maltodextrins of 5 or 10 DE are used to provide solids in the mix without adversely affecting the freezing point of the mix. These products are also derived from corn starch and the modification of the starch is to a much lesser extent than with corn syrup solids. Typically the DE ranges from 5–15 for maltodextrins. Maltodextrins contribute to the texture and chewiness of ice cream.

4.2.4 Stabilizers

The term stabilizer is used for a group of substances that help stabilize the structure of ice cream. Other names include colloids, hydrocolloids, and gums, which indicate that these substances are large molecules (macromolecules) that are capable of interacting with water. Interaction with water also lets some of these compounds interact with proteins and lipids in the mix. A variety of materials are used as stabilizers. These include gelatin, guar gum, sodium carboxymethylcellulose, microcrystaline cellulose, locust bean gum (carob) and carrageenan. During mix processing, presence of gums affects mix viscosity and homogeneity. During freezing gums exert secondary effects in dryness and stiffness of ice cream, the finished frozen desserts, it controls the properties of the water that is unfrozen. This last point means that ice cream is smoother and ice crystals take longer to grow in the presence of stabilizers especially during storage and distribution of these products. Usually, stabilizers are used at 0.1–0.5% levels in the mix but the actual amount depends on the type of stabilizer, strength of the stabilizer, total solids and fat level of the mix, duration and temperature of storage of ice cream and the method of pasteurization. High fat and high total solids mixes require a lesser level of stabilizers. More stabilizer is needed for ice cream that is stored for a long period of time or if the temperature fluctuation during storage is frequent. If the mix is pasteurized by the high temperature short time method, more stabilizers may be needed than if the same mix were pasteurized by the batch method or by the ultra-high temperature method.

A good stabilizer should be nontoxic, readily dispersible in the mix, not cause excessive viscosity, separation or foam in the mix, not clog strainers and filters, provide ice cream with good meltdown, be economical and not impart off flavor to the mix. Some of the common stabilizers used in ice cream are guar gum, sodium carboxymethylcellulose, locust bean gum and carrageenan.

4.2.5 Emulsifiers

As opposed to stabilizers, emulsifiers exert their action on the fat phase of ice cream. Emulsifiers are surface active agents (surfactant). Fat and water do not mix. Emulsifiers facilitate the mixing of fat and water because these molecules have two domains, one that likes water (hydrophilic) and another that likes fat (hydrophobic). When the hydrophobic part of a surfactant interacts with the fat, the water-loving part of the molecule can interact with water, thus facilitating the suspension of fat in water. Generally mono- and di-glycerides and ethoxylated esters of sorbitol (polysorbates) are the commonly used emulsifiers. Mono and di-glycerides are derived from fatty acids and glycerol. Therefore, emulsifiers are fatty substances. They also show fat like properties of melting point, crystallinity, and they can be composed of saturated or unsaturated fatty acids. Presence of emulsifiers in ice cream leads to smoother texture and better shape retention, while improving the ability of the mix to incorporate air.

Mono- and di-glyceride mixtures are obtained by the chemical treatment of fats such as lard, palm kernel or soybean oil. Most of the mono- and di-glycerides are solid at room temperature and are added to the mix prior to pasteurization at a level of 0.1–0.2%. Emulsifiers with high monoglyceride content are also effective drying agents.

Polysorbates are polyoxyethylene compounds. These synthetic chemicals are the most effective drying agents. Polysorbate 80 is an oleic acid derivative. It is a very powerful drying agent and is used at 0.04–0.07% levels. Polysorbate 65 is helpful as a whipping agent, that is, it helps in air incorporation. To obtain comparable stiffness, more polysorbate 65 has to be used than polysorbate 80. At high levels, polysorbate 80 imparts off-flavors whereas polysorbate 65 does not. Polysorbates are generally liquids and cause churning of the mix.

Dried or frozen egg yolks are used to produce dry, stiff ice cream. Dried egg yolks are harder to incorporate into a mix than frozen and sugared egg yolks. The general use level of egg yolks is 0.3–0.5%. If a French-style or custard is required, a minimum of 1.4% egg yolk solids are necessary. Lecithin, a phospholipid present in egg yolks is thought to act as the emulsifier. Lecithin can also be derived from soybean oil. Buttermilk powder provides phospholipids which can act as emulsifiers.

4.3 PROCESSING

The components mentioned above are mixed in proportions defined by the desired composition of the mix. The proportions are arrived at by calculations that use matrix algebra. Such calculations are necessary because a dairy ingredient like cream contributes not only fat to an ice cream mix but also provides milk solids not fat. Sweetened condensed milk, if used in a formulation, contributes, fat, milk solids not fat and sugar. Calculations are necessary to achieve the required composition of the mix.

Processing of the mix is important for mitigating public health concerns, achieving accuracy in obtaining the desired composition and for facilitating the building of some structural elements in ice cream. A schematic of the steps involved in the processing of an ice cream mix are shown in Figure 4.1.

Liquid ingredients and dry ingredients suspended in a liquid are mixed together in pre-determined proportions to form the raw ice cream mix. The raw mix is pasteurized. Pasteurization can be a batch process termed low temperature-long time (LTLT) or continuous process. Continuous processes are either conventional pasteurization or ultra-high temperature (UHT) processes. The primary purpose of pasteurization is to render the mix safe from a public health safety viewpoint. Every particle of the mix is heated to a specified temperature and held for a specified time prior to homogenization and cooling. These time and temperatures are specified by the US Public Health Service and enforced by the Food and Drug Administration.

The pasteurized mix is homogenized. If a HTST process is used for pasteurization, it is customary to homogenize the mix as it exits the raw regeneration section prior to entering the heating section (Figure 4.2). In some factories the homogenization step is carried out after the pasteurized mix exits the flow diversion device and prior to entering the pasteurized regeneration side of the plate heat exchanger.

The approved pasteurization temperatures for ice cream mix are provided (Table 4.2). Because of the high fat, high viscosity and high sweetener concentrations the temperatures shown are 3 °C (5 °F) higher than that used for milk (US Dept. Health & Human Services, 2009).

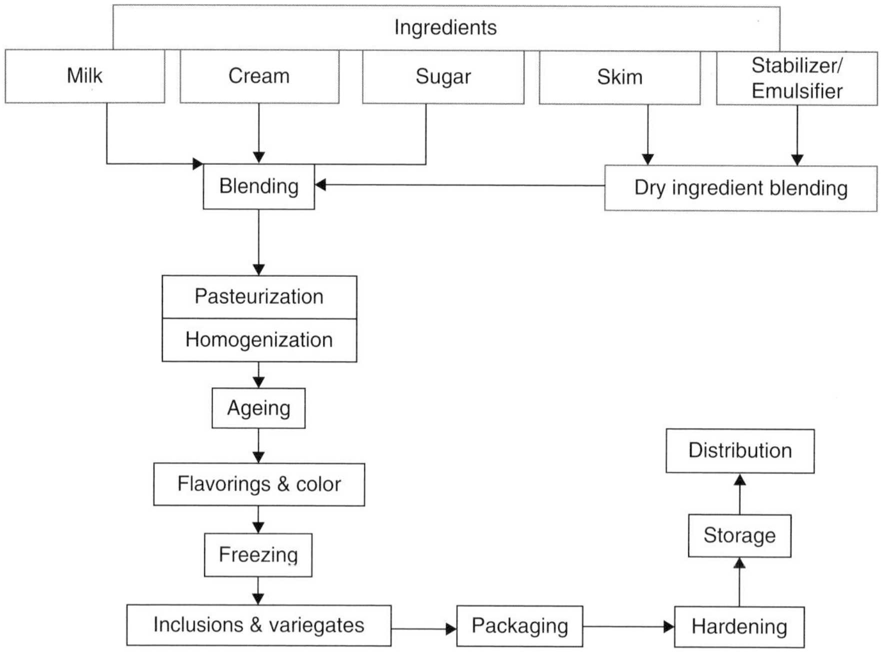

Figure 4.1 Process flow diagram for the manufacture of ice cream.

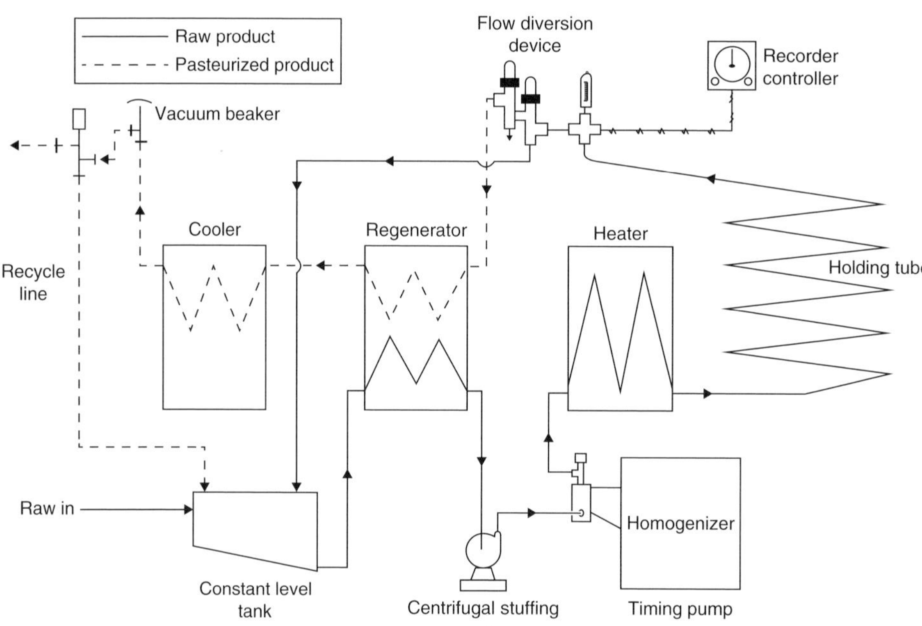

Figure 4.2 Schematic of a pasteurizer in which the homogenizer acts as a timing pump. (Source: Food and Drug Administration. Pasteurized Milk Ordinance 2009.)

Table 4.2 Time and temperature combinations approved by the US Food and Drug Administration for the pasteurization of ice cream mix. (Source: Food and Drug Administration. Pasteurized Milk Ordinance 2009.)

Temperature		**Time**
°C	**°F**	
66	150	30 min.
75	166	15 sec.
92	196	1.0 sec.
93	199	0.5 sec
97	206	0.1 sec.
99	209	0.05 sec.
103	213	0.01 sec.

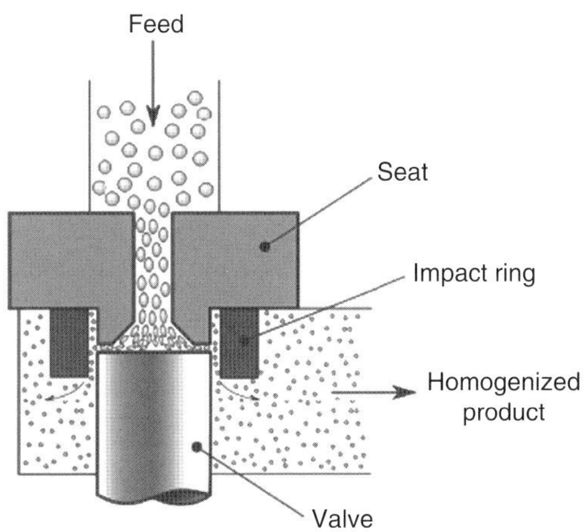

Figure 4.3 Schematic of a homogenization valve.

Pasteurization is a heat treatment accorded to the ice cream mix in which every particle of mix is heated to a specified temperature and then held for a specified period. The purpose of this step is to render the mix safe from a public health viewpoint by destroying disease causing vegetative microorganisms. The heating step also facilitates the dissolution of sugars and other carbohydrates, melts the milk fat and activates the added stabilizers.

Homogenization is a step performed in conjunction with pasteurization. This is a unit operation that results in fat particle size reduction. It improves the rheological properties of the mix while also preventing separation of fat from the serum and provides stability against partial fat coalescence. A high pressure pump forces the mix through a very narrow opening (homogenizer valve) causing fat particle size reduction. The time for the mix to pass through this narrow opening is less than a millisecond (Figure 4.3). Breakage of large fat droplets to smaller sizes results in a tremendous increase in surface area (Table 4.3). Proteins adsorb on to the newly created surface in approximately one microsecond. Adsorption of the proteins partially prevents the coalescence of the fat droplets.

Table 4.3 Changes in droplet size and surface area as a result of homogenization. (Source: Barfod et al. 1991.)

Property	Pre-homogenization	Post-homogenization
Average fat globule size (μm)	3.6	0.6
Number of globules /L	4×10^{12}	8.6×10^{14}
Surface area (m^2/L)	162.8	976.8

The factors that affect fat globule size are: type of homogenizer, pressure of homogenization, stages of homogenization (single or two stage), fat content, fat to protein ratio, temperature of homogenization and proper maintenance and operation of the equipment. Generally, as the pressure increases fat droplet size decreases. Efficient homogenization is possible with a two stage homogenizer and a mix temperature of around 55–60 °C (130–140 °F). The average fat droplet size should be < 1 μm post homogenization.

The pasteurized, homogenized mix is stored for a period of 4–14 hours prior to being frozen. This step is known as ageing and the temperature of the mix during ageing is 2–4 °C (35–40 °F). During ageing the fat and hydrocolloid components are undergoing changes. A majority of the fat crystallizes during ageing and the stabilizers complete their hydration. Modern emulsifiers have obviated the need for ageing but in practice 2–4 hours elapse in an ice cream factory prior to the flavoring and freezing of the mix.

The aged mix is flavored, colored and frozen in a continuous ice cream freezer. During this process approximately 50% of the water in the mix is frozen under agitation. Air is also incorporated into the semi-frozen mélange of mix and ice. The exit temperature of the ice cream from the freezer ranges from −5.5 to −7.8 °C (18–22° F). At this stage the ice cream is flowable and can be easily packaged into containers of different sizes. The packaged ice cream is then placed in cardboard trays and overwrapped in plastic prior to being hardened.

During hardening, the core temperature of the ice cream is dropped to −30 °C (−22 °F) as rapidly as possible so that the remaining water that can freeze is converted to ice. The water freezes on to the large number of small ice crystals created in a properly frozen ice cream. If ice cream is handled properly and hardening is efficient, no new ice crystal nuclei should form. If hardening is inefficient ice crystalization can result in the creation of two sizes; a large number of small crystals in the ice cream freezer and a smaller number of larger ice crystals. This is highly undesirable.

The packaged ice cream is then stored and transported to warehouses. From there it is distributed to the points of sale.

4.4 STRUCTURAL ELEMENTS OF ICE CREAM

View ice cream as a sugar syrup into which all other ingredients are either suspended or dissolved and then water is frozen in this mélange while incorporating air. It is easy to visualize this mélange as consisting of ice crystals and a concentrated sugar syrup which contains proteins, fat, milk salts, air cells, stabilizers, emulsifiers, colors, flavors

Table 4.4 Sizes of structural elements in ice cream as it exits the freezer. (Source: Walstra 1997.) Ice cream has 100% overrun and lactose crystals are generally not present in ice cream.

Structural element	Average size (μm)	Volume fraction
Ice crystals	50	0.3
Air bubbles	100	0.5
Lamellae thickness	15	–
Fat globules	0.6	0.06 (Fat globules and Fat clumps)
Fat clumps	3	
Lactose crystals	(15)	(0.005)

and other particulate ingredients that may have been added post freezing. As freezing progresses in the ice cream freezer, more water is converted to ice, and the concentration of all other constituents in the mix is increased. This process is iterative until the ice cream exits the freezer. At this point, approximately 50% of the water is frozen into very small ice crystals. These small crystals serve as nuclei for the remainder of the water to freeze on during hardening. No new ice crystals are created once the ice cream exits the freezer. Depending upon the mix composition the exit temperature of ice cream from the freezing cylinder is between –5.5 to – 7.7 °C (18–22 °F). The relative size of various particles in ice cream has been described by Walstra and Jonkman (1998) (Table 4.4).

Milk fat exists as small droplets coated with milk proteins and low molecular weight emulsifiers and due to the agitation process in the freezing cylinder the frequency of the collisions of fat droplets against one another increases. This results in a partial coalescence of fat droplets. If all or a majority of the fat droplets coalesce butter is formed and this is undesirable. The low molecular weight emulsifier provides an energy barrier to prevent extensive fat droplet coalescence. Milk fat is a mixture of triglycerides of varying composition, and at the exit temperature of ice cream, some of the fat is in solid state and some of it is in liquid state. Even upon hardening, some triglycerides in milk fat never solidify. Fat has a profound influence on the properties of ice cream (Table 4.4) (Kilara, 1998).

In low fat ice cream the fat content of the mix is between 3–4%, that is, the volume fraction of fat is lower than regular ice cream. Therefore, there are fewer numbers of fat droplets post homogenization in low fat ice cream and the newly created surface area of fat is also much less. The protein load in such mixes is not altered when compared with regular ice cream mix and an excess of protein is available to coat the nascent fat droplets post-homogenization. This may also suggest that there is a need for lowering the emulsifier concentration in a low fat mix. However, it has been observed in practice that not altering the emulsifier concentration (from that present in regular ice cream mix) results in a better eating experience of the low fat ice cream. The overall role of fat in perception of ice cream properties has been discussed (Kilara, 1998) (Table 4.5).

As shown in Table 4.1, low fat ice cream mix has a higher milk solids-not fat, sweetener solids and stabilizer/emulsifier concentrations. The total solids of a low fat mix are still less than those of regular ice cream mix even after the increase in other solids in the mix. Lower total solids are indicative of more water in the mix and therefore

Table 4.5 Influence of fat on the properties of ice cream. (Source: Kilara 1998.)

Physical State	Frozen	Meltdown	Post-melt down
Events leading to melt	Scooping and first lick/bite	Chewing, compression & manipulations as product melts	Manipulation of melt and swallowing
Consumer quality attributes	Scoopability, firmness	Denseness, smoothness, chewiness, iciness, creaminess	Gumminess, thickness, stickiness
Rheological measurements	Firmness, adhesion, dynamic measures	Thermal analyses	Viscosity vs. shear rate, dynamic measures
Structural aspects	Ice crystal, fat dispersion, air cell		Melted ice crystal, fat dispersion, air cell

control of this extra moisture is critical for a low fat ice cream with good texture and consumer appeal.

Air is another important contributor to the texture of ice cream. Air content (measured as overrun) is one aspect while another is the size of a dispersed air cell. Overrun measures the content of air; greater air incorporation is measured by increased overrun. Air is an insulator as it has a very low dielectric constant. In high fat ice cream not only does the fat serve as an insulator but so does the overrun. In low fat ice cream the amount of fat is not sufficient to act as an insulator. This, in the case of low fat ice cream, often translates to a colder eating ice cream. The feeling of coldness can somewhat be offset by increasing the overrun and to manipulate the size of the air cells.

Incorporation of air in ice cream is facilitated by the dasher assembly in the freezing barrel. Generally, two types of dashers are available, classified as low displacement and high displacement. It refers to the volume of the freezing cylinder occupied by the dasher assembly and in a low displacement dasher the assembly occupies a relatively lower volume of the barrel. Low displacement dashers are favored in the manufacture of bulk ice cream items, that is, 250 mL, 500 mL and larger volume packages, and are thought to produce ice cream with relatively large air cell dimensions. High displacement dashers occupy a large volume in relation to the volume of the freezing cylinder so less mix is frozen per unit time in comparison to a low displacement dasher. The air cells in ice cream frozen by a high displacement dasher are smaller and therefore provide greater texture to the product. High displacement dashers are favored in the manufacture of extruded ice cream novelties and for low and no-fat ice creams. The air cell structure of ice cream manufactured from a high displacement dasher is generally dryer (lack of surface wetness) and is amenable to be shaped without melting.

Air is hydrophobic and therefore the surface of the air cell attracts fat globules and proteins. Also present around the air cell is some of the lipid that is in the molten state. Air cell sizes have been estimated to be 30–60 μm and a minimum thickness of 9 μm (Berger, 1978). One gram of ice cream is estimated to have 8×10^6 each of air cells and ice crystals. In a given volume of ice cream approximately 80% of it consists of air cells and ice crystals.

Air is introduced into the rear of the freezing cylinder at the point where ice cream mix also enters. Prior to its introduction, air is compressed and filtered so as to render it clean and sterile and free of oil. At the rear of the cylinder the mix is liquid and compressed air blows through it. As freezing progresses, the viscosity of the mix is such that air can no longer bubble through the product and is incorporated into the product as bubbles. In the freezing cylinder, ice is also being formed and scrapped and fat is also destabilizing and the scrapper blades are breaking all of these physical structures down continuously. This complicated collision and particle size reduction results in smaller air cells with destabilized fat and protein coating the surface of both the air cells and the fat droplets. Since compressed air is introduced into the freezing cylinder the contents of the cylinder are under pressure. A drop in pressure results when the ice cream exits the freezing cylinder. This drop in pressure results in a slight volume expansion or a drop in density of the product.

A third structural element of ice cream is the ice itself. The freezing process results in the formation of ice along the walls of the freezing cylinder. The formed ice is scrapped by the blades of the dasher assembly and mixed into the product. When the product temperature is above the freezing point of the mix the scrapped ice crystals melt. As freezing progresses, the scrapped ice crystals are retained but the crystaline structure examined under a microscope show them to have rounded, irregular shape. Shape irregularities in ice crystals results from free energy consideration as it takes greater free energy to maintain sharp right angled structures. Additionally, the collisions of numerous phases in the freezing cylinder may also facilitate the rounding off of ice crystals.

The exit temperature of the ice cream is between −5.5 and −7.5 °C (18–22 °F). Approximately 50% of the water in the mix is frozen into small crystals. The largest numbers of small ice crystals are formed in the freezing cylinder. During hardening of the packaged ice cream, the remainder of the water that can freeze does so onto the existing ice crystals. During storage and distribution of ice cream, temperature abuse occurs resulting in increased rate of Ostwald ripening; a process in which larger ice crystals grow at the expense of the small crystals. Eventually this process of Ostwald ripening results in a coarse and icy texture in ice cream which is highly undesirable for the enjoyment of eating of ice cream.

The last component that is important for texture is the unfrozen suspension referred to as matrix. This matrix is a supersaturated suspension of sugars, proteins and hydrocolloids (stabilizers) and milk salts. The volume fraction of the matrix is temperature dependent because the state of the ice crystals at a given temperature determines the amount of moisture in the matrix. The relative volume fractions of structural elements in ice cream show that the matrix comprises 15% for regular ice cream as opposed to that of low fat ice cream which has a value of 20% (Table 4.6). This is to be expected as the total solid of low fat mix is lower than that of regular ice cream.

The physical state of the matrix has been described as a gel (Regand and Goff, 2003). Using model ice cream systems, these authors concluded that interactions between protein and stabilizers result in retardation of ice crystal formation. However, some non-gelling hydrocolloids such as xanthan gum were more efficacious in preventing ice recrystalization than gelling hydrocolloids. They hypothesize that steric blocking of the ice crystal-matrix interface prevented solute transport to and from the matrix. Presence of locust bean gum or guar gum in ice cream mix as a stabilizer was

Table 4.6 The relative volume fractions of structural elements in ice cream and low fat ice cream. (Source: Clarke 2005. Reproduced with permission of Royal Society of Chemistry.)

Structural element	Regular ice cream	Low fat ice cream
		Volume fraction
Ice	0.30	0.31
Air	0.50	0.48
Fat	0.05	0.01
Matrix	0.15	0.20

Parameters of structural elements in one liter of regular ice cream

	Size (μm)	Number	Surface Area (m^2/L)
Ice	50	5×10^9	40
Air	60	4×10^9	50
Fat	1	1×10^{14}	300

Table 4.7 Relationships between microstructure, physical and sensory aspects of ice cream. (Source: Clarke 2005. Reproduced with permission of Royal Society of Chemistry.)

Microstructure	Physical	Sensory
Ice content & contiguity, overrun, matrix viscosity	Young's modulus, hardness, yield stress	Firmness on spooning and in mouth, crumbliness, scoopability
Ice content & contiguity	Latent heat & thermal conductivity	Coldness, amount of cooling
Ice crystal size & air bubble size		Smoothness, coarseness, iciness
Overrun	Young's modulus	Lightness
Fat content, fat coalescence, matrix viscosity	Meltdown rate	Rate of melt
Matrix volume fraction, stabilizer & fat content	Viscosity when melting	Thickness, mouth coating, gumminess
Lactose crystals		Sandiness

observed to slow the rate of ice crystal growth in ice cream stored under isothermal conditions or was subject to temperature cycling (Sutton and Wilcox, 1998). After storage of ice cream at −10 °C (14 °F) for two months the rate of crystalization increased slightly for unstabilized ice cream with increase in ice phase volume. In general, presence of stabilizers in the matrix facilitate a mechanism whereby ice crystals melt and then regrow instead of a mechanism in which the ice crystals melt, diffuse and grow (Regand and Goff, 2002).

The relationships of structure, physical and sensory properties are complex. However, Clarke (2004) points out the utility of some measurements in relating microstructure with the other properties (Table 4.7). The material science approach of relating composition, structure and properties of ice cream (Figure 4.4) to the overall performance characteristics of ice cream is another aspect of trying to design ice cream with specific characteristics.

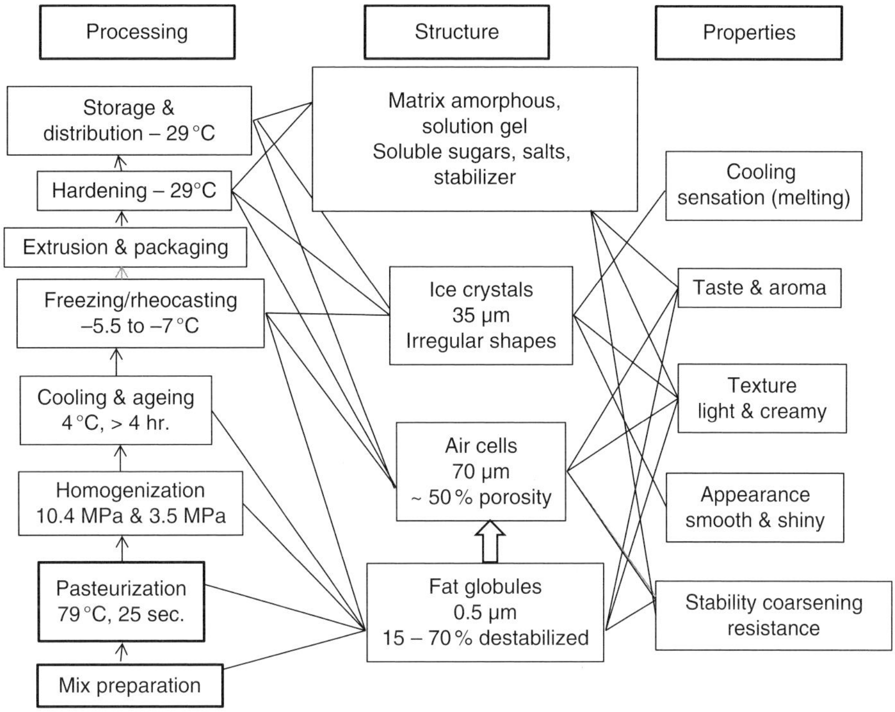

Figure 4.4 Material science approach to ice cream tries to relate composition, structure and properties to the performance of ice cream. (Source: Kilara and Sharkasi 1994.)

4.5 CONTROLLING ICE CREAM STRUCTURE

Mix formulation (including total solids and individual components) plays an important role in the texture of the frozen product. In addition, processing and proper maintenance and operation of the freezer is also critical.

4.5.1 Mix formulation

4.5.1.1 Total solids

Higher total solids of the mix translate to less water to freeze and manage which in turn is responsible for smoother texture. Experience demonstrates that small differences in solids have a major impact on texture control. For example, it is more difficult to control the texture of a 34% total solids ice cream than it is for a 36% total solids product. Differences can be noticed between 38% and 40% total solids products.

4.5.1.2 Milkfat

Milkfat provides lubrication, imparts warmth and creaminess to ice cream. It also mitigates ice crystal perception. Moreover, milkfat provides a substantial amount of solids to the mix and thus, reduces the amount of water to be controlled.

4.5.1.3 Milk solids not fat

These are critical because of their contribution of proteins. Proteins not only bind water but also act as surface active agents in facilitating emulsion and foam formation. Emulsions and foams contribute greatly to the texture of ice cream as discussed above. It is not surprising that ice cream with a high level of milk solids not fat is capable of withstanding greater temperature cycling (heat shock) without adverse effects to the texture. Protein content in ice cream is reduced when skim solids are substituted with whey solids. The legal maximum for such substitution in the U.S. is up to 25% of the milk solids not fat. Whey solids contain 13% protein in comparison to skim solids which contain approximately 35% protein. In commercial practice such substitution may not be readily apparent because the manufacturers may increase the type and levels of other texture control agents (e.g., stabilizers).

Milk solids not fat also contribute to lowering the freezing point of ice cream by virtue of the concentration of lactose. However, lactose is not as soluble as other di-saccharides and has the potential to crystallize upon storage. This results in a textural defect called sandiness. Sandiness can be exacerbated by thermal cycling or heat shock and by improper introduction of particulate inclusions into ice cream post freezing. Sandiness is not a common textural defect because modern day stabilizer systems prevent lactose crystalization. Lactose crystals of > 10 μm can be sensed in the mouth as if sand was being felt in the mouth and hence the name.

4.5.1.4 Sweeteners

Commonly used sweetener solids are sucrose, corn syrup solids (glucose syrup solids) and high fructose corn syrup. The latter is used mainly in economy brands of ice cream. Besides imparting sweet taste these solids also control the freezing point of the mix and are responsible for the freeze/thaw characteristics of ice cream. Freezing point is a colligative property and depending upon the number of molecules in solution the initial freezing point is altered. The higher the concentration of sweetener solids the greater the depression in the initial freezing point of water. Ice/water equilibrium contributes to the pliability of ice cream and therefore, at some concentration of sweetener solids, it may become very difficult to freeze an ice cream mix.

Sucrose is the most widely used sweetener, but due to economic reasons, a part of the sucrose may be substituted by high fructose corn syrup. Substitution of sucrose with high fructose corn syrup results in products that deteriorate more rapidly when subjected to heat shock. When high fructose corn syrup is used it results in approximately 30% less use of solids to achieve equivalent sweetness because fructose is sweeter than sucrose. This is especially true if 55% fructose high fructose corn syrup is used. In addition, since high fructose corn syrup contains monosaccharides, the depression of freezing point is greater than just using sucrose. If adequate care is not exercised in formulation of products containing high fructose corn syrup, texture deterioration during storage can become a major issue.

While sucrose may be the dominant sweetener solid in ice cream, a partial substitution of sucrose with corn syrup solids is practiced. Low dextrose equivalent corn syrup solids (36–42 D.E.) are marginally sweet and can substitute 25–50% of sucrose on a weight basis. Due to this substitution the total solids of the mix are increased and the texture of the ice cream will be changed from a rapidly melting product to one with a chewy texture. Corn syrup solids therefore provide heat shock protection to ice cream (Keeney, 1979).

4.5.1.5 Stabilizers and emulsifiers

Stabilizers bind water, act as viscosifiers and control the ice crystal size by manipulating the properties of the matrix. This is especially true during the storage and distribution of ice cream. Sodium carboxymethyl-cellulose, guar and locust bean gums, xanthan gum, alginates and carrageenans are commonly used as stabilizers and as their name implies, they stabilize the texture of ice cream. As concentrations of stabilizers increase the protection of textural damage due to temperature fluctuations is increased. If, however, too much stabilizer is used in ice cream a texture defect referred to as gumminess can result. Gumminess is a defect in which after the ice cream melts in the mouth a viscous (slimy) feeling is perceived by the consumer.

Emulsifiers also affect texture by controlling the degree of fat destabilization. In an ice cream mix, homogenization results in the formation of a stable emulsion in which the lipid droplet is coated by a layer of milk protein. Some of this milk protein is replaced by low molecular weight emulsifiers during ageing of the ice cream mix. In the freezing process, where agitation is essential fat can form butter because in butter manufacture frequent collisions of fat droplets results in agglomeration and eventually to phase inversion. In ice cream, the presence of butter droplets is undesirable and the product tastes greasy and unpleasant. Low molecular weight surfactants act as an energy barrier to prevent butter formation (Berger and White, 1976).

4.5.1.6 Other texture control agents

Microcrystaline cellulose is often used as a texture control agent that behaves differently than conventional stabilizers. This additive is finely ground cellulose that is co-dried with sodium carboxymethyl cellulose or xanthan gum to facilitate drying and dispersion of the ingredient. Microcrystaline cellulose is not a stabilizer because of its poor water binding. Its mode of action is thought to simulate small ice crystals. During the heat shock process, refreezing of the water takes place on the cellulose nuclei hence delaying the appearance of coarse ice crystals. Usage levels of microcrystaline cellulose vary between 0.2–0.5% in ice cream mix formulations.

4.5.2 Processing factors

The role of blending, pasteurizing, homogenizing, cooling and ageing the mix in the manufacture of good quality ice cream has been discussed elsewhere in this chapter and cannot be over-emphasized. In addition, freezing, hardening and storage have also been referred to in preceding sections. The operation of a continuous ice cream freezer demands care and management on the part of the operator. Some points to consider here are:

i. Keep the ammonia jacket clean, free from oil, water and non-volatile ammonia fractions.
ii. Keep scrapper blades sharp and straight.

iii. Keep mix pumps in proper working condition paying particular attention to the lubrication and tightness of belts.
iv. Make sure there is a plentiful supply of ammonia at the freezer. This is crucial to obtaining the desired draw temperatures.
v. Provide a steady suction pressure at all times which is normally 0.45 kg below that at which the freezer must operate.
vi. Ensure proper cleaning procedures are used and avoid bending parts of the dasher/mutator assembly during cleaning.
vii. During operation make sure the mix lines are tight to prevent mix from leaking out or air from leaking in.

Ice itself is an insulator and attempts have been made to dip ice cream novelties into water to obtain a thin film of ice on the surface, especially when these items are non-aerated or lightly aerated. This film of ice may provide heat shock protection to novelties. Its effect on consumer perception and acceptability remains unclear.

4.6 STORAGE AND DISTRIBUTION

Fast freezing, small ice crystals and low storage temperatures are quality vectors common to most frozen foods. Ice cream in particular is more susceptible to changes during storage. Ostwald ripening, growth of large ice crystals at the expense of small crystals, ultimately results in textural changes in ice cream. Too many large ice crystals lead to an icy, coarse texture that detracts from the smooth creaminess expected by the consumer. Temperature abuses of the packaged ice cream result in these undesirable textural alterations. Temperature abuse is also termed heat shock. Heat shock is inevitable during storage of ice cream as temperature cycling is an integral part of refrigerated/frozen storage. Every storage container (domestic freezer, warehouse freezers etc.) have a thermostat which controls the mean desirable temperature. When the refrigerant removes heat to achieve the set temperature the compressor turns off. In the absence of the refrigerant temperature rises until the cut in temperature is reached. At the cut in temperature the compressor is turned on and heat is removed until the set temperature is reached once again. Under the best of circumstances ice cream experiences heat shock or temperature cycling. As the product warms up some of the ice melts and when cooling commences water is reconverted to ice which has to freeze on available crystal nuclei. Over a period of time Ostwald ripening leads to a coarsening of texture. There are two important aspects to this issue, the first is the amplitude of the temperature fluctuation and second the frequency of cycling. Larger amplitudes and/or increased frequency of temperature cycling can lead to faster rates of texture deterioration.

In practice, it is observed that when the product is under the control of the factory or manufacturing facility every precaution is exercised to ensure least heat shock to ice cream. When ice cream is transported from the factory warehouse to intermediate distribution warehouses and from these warehouses to the grocery chain distribution center and from there to individual stores, heat shock becomes progressively worse. In the individual grocery, the rapidity with which the ice cream is moved from the receiving dock to the freezer and then to the display cabinet is critical. Once the product is properly stacked in the display cases the situation becomes uncontrollable. Consumers open merchandising cabinet doors and take their time in deciding which product to purchase all the while keeping the door open. When the door is open warm air from the store rushes in. Subsequently when the consumer closes the door the refrigeration cycle takes time to cool the cabinet down to the set point. If the warm air that enters the cabinet is humid as well frost build up may lead to frequent defrost cycles in the merchandising cabinet. Defrost cycles subject the ice cream to heat shock as well. An additional consideration is the type of merchandising cabinet used in a store. Open type cabinets, colloquially referred to as 'coffin style,' cause more heat shock than closed merchandising cabinets. Under similar environmental conditions open style merchandising cabinets undergo more frequent defrosting than closed cabinets.

4.7 SUMMARY

In this chapter the ingredients, processes and storage of ice cream and their impact on texture has been discussed. Low fat ice cream differs from regular ice cream in the total fat content. Low fat ice cream, by U.S. regulations, must contain 3 grams or less of fat per serving. A serving of ice cream in the U.S. is 4 fluid ounces or half a cup (approximately 120 mL) and the measure is volumetric. Therefore, overrun of the product has an influence on how much mix is actually present in that volume of product. For example, if the low fat product has 100% overrun, then in a serving approximately 50% by volume is air and the remainder is ice cream mix. If the overrun is less than 100% more mix is present in a serving.

Low fat ice cream mix has to be formulated differently to compensate for the lower fat content. If this is not done, the total solids of the mix will be low and management of more water through the freezing, storage and distribution can pose problems of texture deterioration. The principles involved in managing the ice/water equilibrium are the same regardless of the fat content of the frozen dessert. Typically adjustments in milk solids not fat content, types of sweetening agents and amount and type of stabilizers used are the means by which total solids are altered in a low fat mix.

The structural elements of ice cream consist of milk fat emulsion, air cells, ice crystals and the matrix in which these other elements reside. Destabilized fat is critical for shape retention of ice cream and the fat network stabilizes air cells. The role of the matrix is also critical as this element contains sugars, salts, proteins and hydrocolloids.

The role of temperature during storage, transportation and distribution is another critical factor in determining texture of ice cream and frozen desserts. Minimizing temperature fluctuations helps prolong the textural attributes of ice cream and cannot be overemphasized.

References

Barfod, N.M., Krog, N., Larson, G. and Buchheim, W. (1991) Effects of emulsifiers on protein-fat interaction in ice cream mix aging.1. Quantitative analysis. *Fat Science Technology*, 93, 24–29.

Berger, K.J. (1978) The ice cream fat globule membrane. *Confectionery Manufacturing* Marketing February issue, pp 23–27.

Berger, K.J. and White, G.H. (1976) The fat globule membrane in ice cream. Part II. Emulsifiers and high melting triglycerides. The mosaic model of the membrane. *Dairy Industry International*, 41, 237–241, 243.

Keeney, P.G. (1979) Confusion over heat shock. *Food Engineering* 51(6), 116–118.

Kilara, A. (1998) Fat mimetics in ice cream and frozen dessert manufacture. In: *Ice Cream*. Ed. Buchheim, W. International Dairy Federation, Brussels, Belgium. Pp. 65–74.

Kilara, A. and Sharkasi, T.Y. (1994) Microstructure and processing of ice cream. *MRS Bulletin* XIX (7), 51–55.

Regand, A. and Goff, H.D. (2002) Effect of biopolymers on the structure and ice recrystallization in dynamically frozen ice cream model systems. *Journal of Dairy Science*, 85, 2722–2732.

Regand, A. and Goff, H.D. (2003) Structure and ice crystallization in frozen stabilized ice cream model systems. *Food Hydrocolloids*, 17, 95–102.

Sutton, R.L. and Wilcox, J. (1998) Recrystallization in ice cream as affected by stabilizers. *Journal of Food Science*, 63, 104–107.

United States Department of Health and Human Services, Public Health Service, Food and Drug Administration. 2009. Grade 'A' Pasteurized Milk Ordinance. [Online] Available:http://www.fda.gov/downloads/Food/FoodSafety/Product-SpecificInformation/MilkSafety/NationalConferenceonInterstateMilkShipments NCIMSModelDocuments/UCM209789.pdf

United States Federal Food and Drug Administration, 2011. [Online] Available:http://ecfr.gpoaccess.gov/cgi/t/text/text-idx?c=ecfr&sid=c15e3d6e94fbd87eebe70f968ea82b77&rgn=div5&view=text&node=21:2.0.1.1.23&idno=21

Walstra, P. and Jonkman, M. (1998) Role of milk fat and protein in ice cream. In: *Ice Cream*. Ed. Buchheim, W. International Dairy Federation, Brussels, Belgium. Pp. 17–24.

5 Formulating gelatin free products

P.L. Buwalda

AVEBE Innovation Centre, Veendam, The Netherlands

5.1 GELLED FOODS

Gels are applied in many systems in particular in foods. The appreciation of a food item is governed by many factors. One of the most important factors is the texture which is that part of the food structure that is perceived by feeling with teeth, fingers, tongue, and/or palate (Guinard and Mazzucchelli, 1996). Other factors including eyesight, anticipation and hearing play a role, too. The ISO norm for sensory analyses for textures (5492/3) stipulates:

> All rheological and structure attributes of a food product perceptible by means of mechanical, tactile, and, where appropriate, visual and auditory receptors.

This structure is governed by many influencing factors and components: fats, proteins, viscosity and so on. One of the most dominating factors is the gel character of the food item as is suggested by mentioning of the world rheology in the ISO-norm. Although the word gel is frequently used, many definitions exist, which are sometimes inconsistent with each other. For this overview we will stick to the definition suggested by Almdal and co-workers (Almdal et al., 1993) which is basically a rheological description.

> A gel is a diluted system in water with substantial solid properties.

The solid properties of the watery system are imparted by a three dimensional network of a diluted agent dissolved in the water. These agents are usually polymeric in nature and are referred to as gelling agents. Hydrocolloids derived from plants, algae, fungi, animals and bacteria are well known gelling agents in day to day life.

The reason why the gel is formed from the solvent is because the solute forms cross-bondings, however weak and whatever in nature, and these crossbonds are kept together by longer non-interacting chains. In the end, these alternating crossbonds and longer segments 'embrace' the solvent (see Figure 5.1) thus imparting a solid character. The nature of the gel depends on a large number of variables: strength/rigidity of the crossbonds, the length of the segments, the solvent-solute interaction, and so on. There is some mix up of terminology in the food industry. For the sake of clarity in this chapter we will use

Food Texture Design and Optimization, First Edition. Edited by Yadunandan Lal Dar and Joseph M. Light.

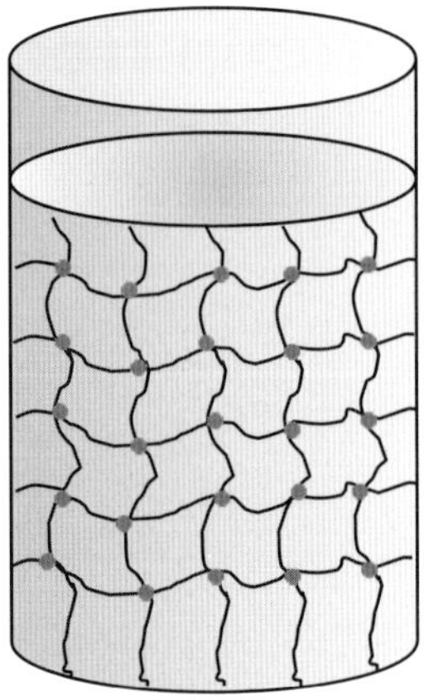

Figure 5.1 Artist impression of crossbonds and segments embracing a solvent.

the term gelling agent for a product that enhances solid properties and thickener for a product that merely enhances viscosity (a product may have very high viscosity with no solid properties at all).

Many examples of gels exist in food and non-food systems. Bakery creams, mayonnaise style products, and scent blocks, are all systems that predominantly consist of water but maintain shape during the period of use. In these water based systems (Saha and Bhattacharya, 2010), hydrocolloids including starches play a pivotal role to create and maintain this shape.

5.1.1 Gelatin

Gelatin is the most common of all gelling agents. The total world market accounts to approximately 1.5 billion US dollars with roughly 300 000 metric tonnes of sales (N.N. 2013). Gelatin is used in many foods such as desserts, aspics, confections, sauces, cheese style products and low fat margarines. For a relatively recent review see (Boran and Regenstein, 2010). Its characteristic properties are undistinguishable and unique. Gelatin is of animal origin and derived from pork, horse, cattle, sheep, and fish. Gelatin is obtained by hydrolytical treatment of the byproducts of the meat industry. Bones, hides, and so on are treated with acid or alkali and after processing gelatin is isolated.

Gelatin forms clear, elastic gels that melt on heating to temperatures above 30 °C. The viscosity of the resulting molten solutions is very low. These features make gelatin well suited for the food industry. At elevated temperatures viscosities are low allowing processing and, upon cooling, the desired shaping can be applied. Gelatin is sold mentioning the bloom value. This value indicates the gel strength obtained per gram of gelatin given a specific measurement in water. High bloom values indicate that a low amount of gelatin induces a strong gel in water. In some tropical countries, ambient temperatures are above the melting point of gelatin gels leading to melting of food products thus destroying texture and appearance of the food item making them unpalatable.

In the food industry there is a long quest for replacing gelatin. Apart from the physical properties, which are inherent, a large and growing consumer group disfavors products with gelatin. Obviously, vegans and vegetarians don't want to eat products of animal origin.

Furthermore, special qualities are necessary to meet Kosher and/or Halal standards imparting high prices on the gelatins. A further perceived drawback is the scare of food scandals, although the food industry has implemented many procedures to ensure safe supply.

It has to be mentioned that the ever present competitive pressure on pricing is really the biggest driver of replacing gelatin with a cheaper, readily available raw material and without a doubt starch is the material of choice.

Many attempts have been made to replace gelatin in a generic way, but the versatility of gelatin has prevented this approach. This chapter will therefore focus on a more application by application approach where a dominant property of gelatin is matched with a specific starch derivative.

5.1.2 Starch and starch gelling

Starch is an abundant carbohydrate polymer that is used in many food products, paper, adhesives and so on. The production is well over 75 million metric tonnes.

Starch is used to thicken food products, but also to impart texture and gel character in food products.

Starch gelling is a complex phenomenon that is driven by many factors such as granules, molecules, impurities, cooking process, botanical origin of the starch, granular remnants, and so on. The generally accepted model has been described elegantly by Morris (1990). Crystalized amylose and amylopectin, as well as granular remnants called ghosts play a pivotal role in the formation and rigidity of the crossbonds; the non-crystalized amylose and amylopectin form the longer chains.

Excellent reviews on starches (Hoover, 2001; Jane et al., 1999), their components (Pérez and Bertoft, 2010) and their botanical origin (Fredriksson et al., 1998) have been written. In Table 5.1 some selected data from these papers are given which support the paragraphs to follow.

As can clearly be seen from the table, although to the untrained eye one starch may seem like another starch, on a molecular as well as on a granular level starches are very different. The amount of amylose in the starch, the molecular weight and the branch chain length of the amylopectin differ. Potato starch and amylopectin potato starch are unique in that they carry bound phosphate groups in significant quantities, giving these

Table 5.1 Characteristic data of selected starches. (Source: Data from Jane et al. 1999 and Frederikson et al. 1998.)

Starch	Amylose %	Amylopectin %	CL (amylop)	Bound Phosphate mg/g	Granule size µm	Crystal
Potato	20	80	31.4	0.6	44	B
Tapiooa	18	83	27.6	0.07	17	A
Corn	23	77	24.4	0.03	15	A
High amylose corn	40	60	30.7	NA	15	B
Waxy corn	2	98	23.5	0.02	15	A
Wheat	28	72	25.7	0.04	12	A
Amylopectin potato	0	100	34.8	0.8	42	B

Data from Jane et al. 1999 and Frederikson et al. 1998, Granular size to be published (volume average)

starches a mild polyelectrolyte character in solutions in water. Furthermore, high amylose corn, potato and high amylopectin potato starch crystallize in a so-called B-isomorph, whereas other starches crystallize in an A isomorph. Hence all starches will have different properties and will generate different types of gelling agents.

Starch is modified in order to induce or enhance certain properties (Kaur et al., 2012; Singh et al., 2007). Gelling too can be enhanced and modulated using starch modification. For an excellent review see (Wurzburg, 1986). In particular, acid degradation, enzyme degradation, dextrinization and oxidation of amylose containing starches are intended to enhance the gelation characteristics.

Gelling starches are those starches that are specifically intended to be used as the *sole* gelling ingredient. Recently, some elegant reviews have been published (BeMiller, 2011; Banerjee and Bhattacharya, 2012) where starch is used to support the gelling properties of other hydrocolloids. For the sake of simplicity this approach will not be discussed here.

Amylose gelling is thought to be the motor of most starch gelling, linked to the strong tendency of amylose to gel rapidly from solutions (Gidley, 1989; Clark et al., 1989). Gidley and co-workers have shown that a minimal concentration of approximately 1% is needed, although interaction of individual amylose molecules already occurs at lower concentration. The gelling speed depends on the molecular weight, with an optimal speed at approximately DP = 1000. The melting points of the resulting gels depend on the degree of polymerization of the amylose. This was shown in a series of elegant studies by a.o. S.G. Ring. (Moates et al., 1997). In Figure 5.2 the melting temperature Tm is depicted as a function of degree of polymerization of the amylose. It is clear that there is a limit value for the melting of a gel at approximately 140 °C if the degree of polymerization goes to infinity. Or, to put it in the right perspective, in order to dissolve amylose completely temperatures of above 140 °C are necessary.

Many publications describe the crystalization of amylose which is considered to be the driver behind the crossbonding process during gelation. One of the most elegant approaches has been described by Schierbaum and co-workers (Gernat et al., 1987). In this paper it is shown that the amylose like maltodextrins crystallize in a small disk

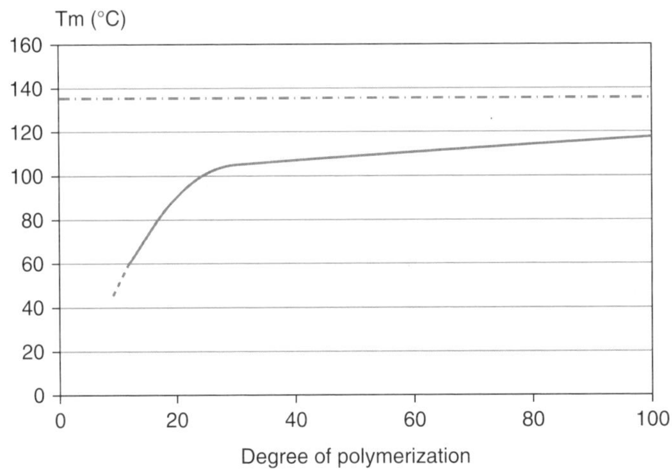

Figure 5.2 Melting point vs degree of polymerization of amylose. (Source: Data from Moates et al. 1997.)

280 nm wide and 28 nm high, which are then connected through longer parts of the molecules that may contain the branch points. The crystalization occurs in the so-called B-isomorph similar to the isomorph of potato and high amylose corn starches. These disks scatter the visible light leading to the typical white or opaque color of starch gels.

Although it is well known that part of the starch gelling is caused by amylopectin gelling (Miles et al., 1985; Goodfellow and Wilson, 1990), not a lot of literature describes the gelling of amylopectin as part of an isolated gelling system; certainly no applications have really been suggested in the scientific literature. Huizenga and co-workers (Huizenga et al., 1997) have shown that amylopectin potato starch (virtually 100% amylopectin) tends to gel when long chain aliphatic molecules are added. The authors suggest that an interaction between the amylopectin and the aliphatic chain is the driver for gelation. In this study it was shown that waxy corn starch does not gel under the same conditions, suggesting that the outer chain interact with the aliphatic chain and that the differences occur because of different average chain length. Other waxy starches were not investigated. Buwalda and co-workers (Buwalda and Meima, 2003) have shown that degraded waxy potato starch gives gels without adding the long chain aliphatic molecules also in comparison with waxy corn, again suggesting that the outer branch chain length is a dominating factor in amylopectin gelation. A special case arises when amylopectin gets elongated by the amylomaltase enzyme (van der Maarel et al., 2005; Kaper et al., 2004; Hansen et al., 2009). In the above publications it is mentioned that gels all melt at temperatures at or around 60 °C, suggesting small outer branches form the crystalized crossbonds for gelling. In Figure 5.3 the gel strength as a function of the temperature (cool-heat-cool) is depicted for amylomaltase treated starch.

The above case is mentioned as special, because, although on molecular level it is a branched molecule and therefore an amylopectin, its behavior can be better explained as an amylose as described by Clark (Clark et al., 1992).

The amylomaltase treated starches are in this sense quite a new class of starch derivatives (Kaper et al., 2004; van der Maarel et al., 2005; van der Maarel et al., 2002; Michael Riis Hansen et al., 2009; M. R. Hansen et al., 2009; Hansen et al., 2008). Amylomaltase

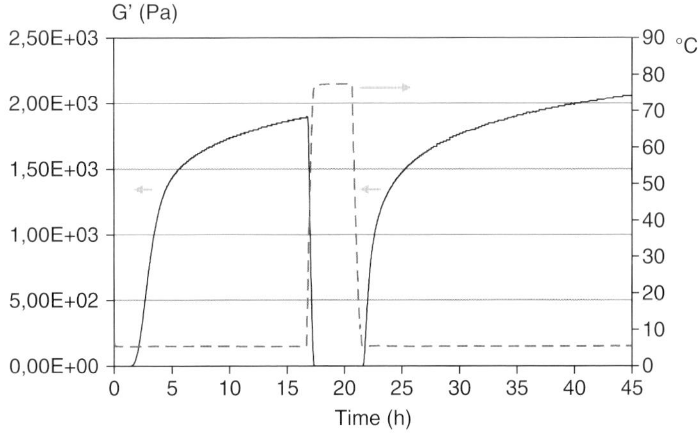

Figure 5.3 Gel strength of amylomaltase treated starch as function of temperature (5% w/w in demineralized water).

is an enzyme that transfers glucan chains from one glucan molecule onto another. Hence, the name disproportionating enzyme or D-enzyme. The more scientific name is 4-α-glucanotransferase. The transferring of glucans from one chain to another in the end leads to a full consumption of amylose which is 'welded back' onto the amylopectine back bone. This leads to chain elongated amylopectin, products that in water form strong thermo-reversible gels, with melting temperatures at approximately 60 °C and above. AVEBE has entered the market with a product based on this technology under the brand name ETENIA™ 457. This class is a special case of maltodextrins. In the past, maltodextrins were prepared using degrading enzymes (amylases) leading to starches with strongly reduced molecular weights. The amylomaltase enzyme is intended to maintain the maltodextrin in the high molecular weight levels, therefore increasing its gelling capacity (van der Maarel et al. 2005).

Apart from amylose and amylopectin, a third governing element for gelling is the formation of so-called granular ghost or remnants (Fannon and Bemiller, 1992). These ghosts are what remain as an organized item of the starch granules after cooking. In a gel they form their own micro-domains contributing to the overall network. The real mechanism in which these remnants contribute to the overall gel formation in rheological understanding is not described fully. In a special treatment combining acid degradation and crosslinking, a gelling agent was prepared making use of enhanced amylose gelling and granular ghosts to impart increased gel strength in one product (Chiu, 1984).

Apart from amylose, amylopectin and granules, minor components such as protein and fats contribute to one or more of the gelling components mentioned above.

Given these elements, starch gels are hard to understand and impossible to predict in their behavior given a certain food application and process. Much remains in the hands of the craftsmen and is trial and error.

5.2 REPLACING GELATIN BY STARCH

In Table 5.2 a comparison is made between the gel strength of a number of modified starches compared to gelatin. The comparison is made by making a gel with a gel strength of approximately 2000 Pa with the appropriate concentration for that particular gelling agent.

As can be seen, gelatin gels at a low to moderate concentration. Starches that are modified using the more classical acid degradation or enzyme treatments require much higher concentrations. Amylomaltase treated starch gels at similar concentration levels

Table 5.2 Gel strength in water of a number of starches as compared to gelatin.

Gelling agent	Concentration % (w/w)	G′ Pa
α-amylase degraded starch DE 2	20	2200
Acid degraded starch	9	1850
Pullulanase treated starch	8.5	2000
Gelatin240 bloom	3	2100
Amylomaltase treated starch	5	2050

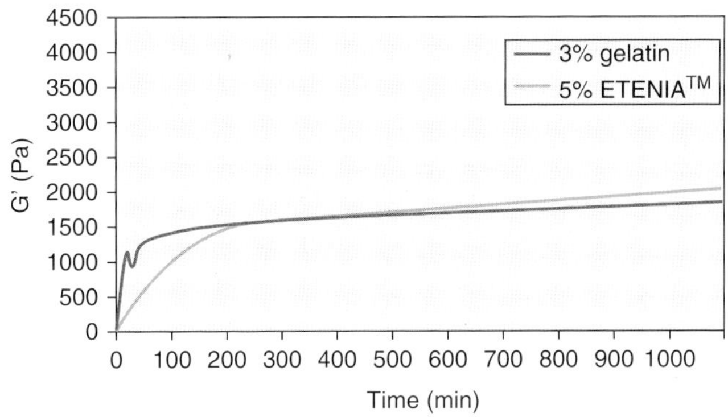

Figure 5.4 Gel strength of amylomaltase treated starch vs gelatin 240 bloom (ETENIA™ is amylomaltase treated starch).

as gelatin. However, it should be stipulated that not for all applications are low concentrations needed. For instance, in soft confections a high dry matter is needed for taste and regulating crystalization and water activity.

In Figure 5.4 the gel developed strength over time is depicted for gelatin and amylomaltase treated starch. From this figure and Table 5.2, it can be seen that approximately 3% gelatin 240 bloom gives similar gel strengths and gelling speed as amylomaltase treated starch using 5%. From this data it is tempting to suggest a 3 to 5 replacement ratio in any given application. This strategy has been tested and has failed many times. The reason is indicated in the definition of texture. Whilst the rheology parameter of gel strength may be similar, other parameters such as palatability, visual, clarity and so on, may be very different. This then leads to disappointing results in replacing gelatin in a given application. The strategy of choice is replacing gelatin on an application by application strategy.

Obviously, this can be attributed to the underlying difference in gelling mechanism between starches and gelatin. The starch gelling mechanism has been explained before. For gelatin the mechanism is quite different (Pezron et al., 1990; Pezron et al., 1991; Boran and Regenstein, 2010). The often postulated formation of collagen type fibrils, is an over simplification of the real underlying mechanism. From solutions the random coils of the hydrolyzed proteins change to a helix followed by transition to triple helix formation and subsequent alignment in a supramolecular structure comparable (but not the same) as collagen fibrils (Pezron et al., 1991). However, the physical nature depends on concentration and on cooling rate, salt and sugars present, and so on (Bohidar and Jena, 1993). The molecular architecture of the gelatins depends on the animal source and on the process conditions employed in the process of collagen hydrolysis and subsequent recovery. From this it becomes clear that the mechanism of crosslinking connected to the gelling of gelatin is not driven by crystalization, as is starch, but by molecular alignment in fibril type of crosslinks. This in turn implies that given an application, with its specific process conditions (including cooling for gel formation), gelatin replacement by starch to form a gel is only valid in one single point.

Figure 5.5 may serve as an example of the concentration dependency. In this figure amylomaltase treated starch is included for comparison. At high concentration (7–8%),

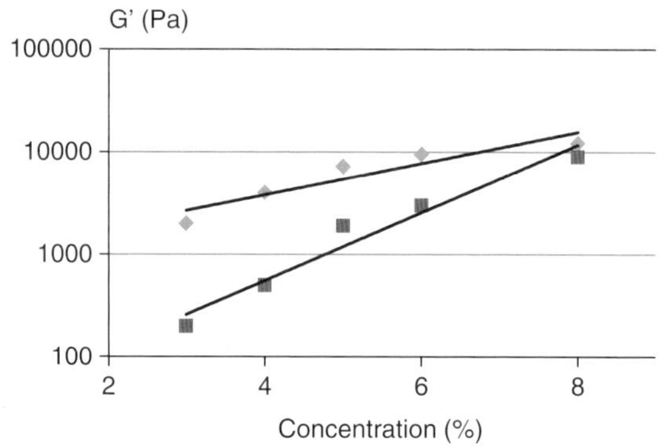

Figure 5.5 Concentration vs G′ of gelatin (♦) and amylomaltase treated starch (■).

gelatin and starch give very similar gel strengths. In the low concentration area (3%) gelatin forms much stronger gels.

5.3 YOGURTS

By far the most popular fermented dairy product is yogurt. Yogurt is made by heating milk, adding culture and subsequently fermenting the product to the desired taste and texture. The culture may consist of specific strains of Lactobacilli, Streptococi and/or Lactococci. During the fermentation the pH drops and the flocculated proteins form a delicate network typical of the yogurt texture. Next to that, the culture produces, during fermentation, specific exopolysaccharides (Robijn et al., 1996) that contribute to the network as well. The fat is captured in the protein network as specific fat globules. Upon eating, the network yields in the mouth, the fat melts and lubricates the tongue and platelet imparting a specific creamy sensation typical of full fat or cream style yogurt. The rheology and perception depends therefore on the milk (protein and fat content), the process and the culture. When lowering protein and/or fat it is obvious that texture needs to be repaired and gelatin is the hydrocolloid of choice. Gelatin in these applications is applied between 0.2 and 1 %. In low fat yogurt, often next to gelatin, a stable thickener is used to impart viscosity. In one of the first examples of replacing gelatin with starch described in US6093439 (Whaley et al., 2000), the use of high amylose starch is described imparting gelling in cultured dairy formulation to replace fat and/or gelatin. Actually, a combination of high amylose starch and a thickener was described. In recent years amylomaltase treated starch has been mentioned for the same purpose (Buwalda and Sein, 2006; Kanning et al., 2012; Alting et al., 2009). The organoleptic as well as physical characteristics have been described (Alting et al., 2009; Kanning et al., 2012) in two types of yogurt. Using confocal scanning light microscopy it could be proven that the amylomaltase treated is embedded in the network in domains. In Figure 5.6 an artist's impression of this is given.

In Table 5.3 a model recipe is given for 0% fat and 1% for stirred yogurts using an industry standard process. The viscosities of the resulting yogurts have been measured after one day, one week and two weeks. The results are reflected in Table 5.4.

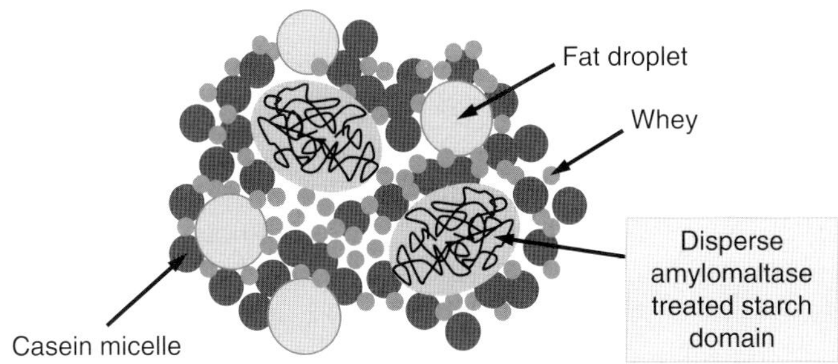

Figure 5.6 Artist impression of amylomaltase domain in the yogurt network. (Source: Drawing by Dr A. Sein (DSM).)

Table 5.3 Composition of 0% and 1% fat stirred yogurt.

yogurt	0% fat	1% fat
Ingredients	Percetage	Percentage
Milk (0% fat)	92.4	25.8
Milk (1.5 %fat)	–	65.3
Sugar	5	5
Whey Powder	0.9	1.1
Crosslinked HP starch	1.4	1.4
Gelling agent	0.35	0.35

Gelling agent either gelatin 240 bloom or amylomaltase treated starch

Table 5.4 Viscosity development over time of stirred yogurt.

Product	Day 1 mPas	Day 8 mPas	Day 15 mPas
0% fat gelatin	7500	13000	13000
0% fat amylomaltase starch	7000	14500	15100
1% fat gelatin	11500	25000	21100
1% fat amylomaltase starch	8000	21500	18000

As can clearly be seen from the table the viscosities are similar, although on day one, the yogurts containing amylomaltase treated starch are somewhat lower in viscosity. It is also clear that fat contributes to the overall viscosity. Further sensory analyses have shown that the yogurts containing gelatin have a shinier surface; the yogurts containing amylomaltase treated starch are judged as much creamier. In an exaggeration one could state that amylomaltase treated starch is replacing gelatin which in turn is replacing fat. However, amylomaltase treated starch imparts a nice creamy sensation due to the alfa amylase assisted break down of bulk viscosity. On the other hand, one might argue that amylomaltase treated starch is not exactly the same as gelatin in this application albeit that it is closer to the original texturizer: fat. The replacement ratio is approximately 1:1–1.5. This deviates from Figure 5.5 where at this low concentration more amylomaltase treated starch was to be expected.

The gelling of amylomaltase treated starch in this application is not obvious. Clark states (Clark et al., 1992) that amylose gels in concentrations above 1% (w/w) in water.

The physical characterization of the yogurts, however, shows that the amylomaltase treated starch does contribute to the overall gel strength at concentrations lower than 1%. From the CSLM pictures (Alting et al., 2009) it is clear that the starch is not the continuous gel network, but serves to support the protein network. This may implicate that the mechanism of gelling in this matrix is different from concentrated solution in water, although the authors do not comment on the issue. Therefore the replacement ratio does not follow the graph of Figure 5.5.

The conclusion of the above paragraph is that not so much gelatin replacement is the issue but the replacement of fat by either gelatin or starches. Important properties are the melting characteristics, the yield stress of the microgels (in the mouth) and the overall texture of the yogurt. An exact match for gelatin is not a requirement, but is part of the function.

5.4 LOW FAT MARGARINE AND BUTTERS

Margarine is originally intended to replace butter for baking and spreading on bread. The driving force behind this is the replacement of expensive animal fats by cheaper readily available plant based fats and oil. Given the global trend of increasing obesity (Ogden et al., 2012) from the early 1970s of the past century, low fat margarines and butters (low fat spreads) have become increasingly important items for spreading on bread, in particular, in Europe.

The replacement of gelatin in low fat margarine spreads and butters is arguably different from the replacement of gelatin in yogurt. In yogurt, gelatin is the established benchmark and recently some alternatives based on starch have been introduced. In low fat spreads, there has been competition between gelatin and starches from the beginning of the introduction of this food proposition.

Low fat spreads are water-in-oil emulsions meaning a continuous phase of oil and fats in which small droplets of water are interspersed. In order to prevent coalescence of the water droplets, they need to be stabilized by a gelling agent. The gelling agent has to meet certain criteria. The yield stress of the gel phase has to be similar to the fat. Although the droplets are very small and would not interfere with the spreading, the tongue is sensitive enough to register different rheology of the water phase as an off texture. The watery gel has to melt in the mouth at a similar temperature as the fat. For gelatin this criterion is met with ease. For starch this is less obvious. However, as with ATS in yogurt, the salivary alfa amylase induces a quick drop in viscosity mimicking fat melting. Other hydrocolloids such as alginates, pectins, and galactomanans lack this particular feature.

In patent US5904949 (Reddy et al., 1999) an overview is given that will help the reader with some good literature entries. Therefore, we will focus on the functionality of starches vs gelatin and take a perspective on future developments. In patent US5279844 (Wesdorp et al., 1994) for instance the advantageous formulation using gelatin is described. Special consideration is paid to the gel strength of the water phase. In US5904949 the same strategy has been applied to a starch based solution. Several modified starches were tested, including acid conversion and crosslinking. The starch base was waxy corn, high amylose corn and tapioca. The rheology was modified in such a way that the gel particles had similar rheological plasiticity as the fat component.

Table 5.5 Composition of low fat spreads.

Recipe	1	2	3	4
Oil phase				
Blend of Soy/Palm oil	39.3	39.3	39.3	39.3
Monoglycerides	0.7	0.7	0.7	0.7
ß-Carotene	0.02	0.02	0.02	0.02
Flavor	0.02	0.02	0.02	0.02
Water phase				
Amyomaltase treated starch	2.4	–	–	1.2
Gelatin	–	1.5	–	–
α-amylase treated starch DE2	–	–	3.72	–
Alginate	–	–	0.6	0.6
Salt	0.5	0.5	0.5	0.5
Potassium Sorbate	0.1	0.1	0.1	0.1
Skimmed milk powder	0.5	0.5	0.5	0.5
Flavor	0.02	0.02	0.02	0.02
Water	55.6	55.6	55.6	55.6

In recent years the emphasis has swung from gelatin replacement persé to a replacement of gelatin with starches that do not require labeling with an E-number or can be positioned as 'natural' or 'Eco'. In patent US2003099755 (Huizinga et al., 2003), the combination of unmodified starch and other hydrocolloids is described. In these systems the other hydrocolloids impart the gel character. In patent US2011104354 (Sein et al., 2011) the use of amylomaltase treated starch is described. In Europe, where 'E-number free' is an issue, enzyme treated starch can be labeled 'starch'. In Table 5.5 a comparison is made of low fat spreads using different gelling systems.

The resulting spreads were comparable to commercial samples and no bottle necks during processing were observed. This proves that for low fat spreads (40% fat) gelatin or combinations of starch and gelatin can be replaced by total starch solutions that can be labeled 'starch' in Europe.

A special case arises in low fat baking margarines (Sein et al., 2011). Here one cannot speak of gelatin replacement because gelatin is not widely employed to this end.

In contrast to yogurt, for low fat spreads the rheology of the starch gel is more important, preferably the melting temperature is similar to fat, although salivary alfa amylase is sufficient to generate melt in mouth characteristics.

5.5 CONFECTIONS

The use of gelatin in soft molded confections is well known. It generates clear candy, imparts low viscosity during molding, quick drying (in fact set) and the drying does not require extra drying equipment. The disadvantages are similar as in the previous applications: Kosher/Halal issues and non-vegetarian. The high use levels generate cost issues too. For the use of gelatin see (Boran and Regenstein, 2010). The use of starch for this type of confection is long known (Lacourse and Zallie, 1988). Gelatin imparts a relatively long elastic texture with non-stickiness to the confection product. This can be measured using a texture analyzer (see Figure 5.7). The stickiness correlates to relative (to maximum positive force) negative force needed to release the product from the probe (model for theeth). The texture is the relative negative (relative to the maximum positive force) area. This is an indication for the elongation of the product and connected to the well-known long ← → short texture attributes.

Using acid degraded potato starch reduces the quality with approximately 50% (unpublished results). The replacement ratio ranges from 1:1.5 to 1:2. Therefore, in the marketplace apart from full replacement, partial replacement is often observed.

The elasticity can be improved by adding crosslinked starch (Woltjes et al., 2004), although the molding viscosity may go up. No explanation of this phenomenon was given (see Table 5.6).

From this table it becomes clear that if only elasticity is important, it is viable to formulate all starch products that have similar elasticity. The typical springy bite of

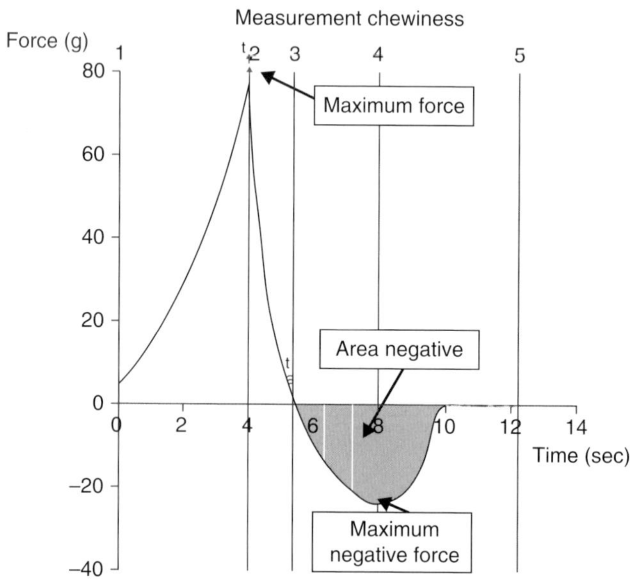

Figure 5.7 Stickiness and elasticity of gum confections.

Table 5.6 Composition and elasticity of soft confection with gelatin and starch.

Ingredients	Trial 1	Trial 2	Trial 3
Oxidized starch	12	–	8
HP crosslinked starch	–	3	–
Oxidized acetylated starch	–	9	–
Gelatin	–	–	4
Sugar	34	34	30
Glucose syrup DE42	34	34	38
Water	20	20	20
Elasticity	1	5	4

gelatin and the non-sticky bite (not in the table) could not be matched. So it is fair to say that good product can be made but differences are there.

In order to make products that can be molded at lower viscosities or cooked at higher dry matter amylopectin (Buwalda and Meima, 2003) starches or amylomaltase treated starches can be used (Buwalda and Tomasoa, 2011).

The conclusion is that this application is the most difficult one for replacing gelatin. The gelatin imparts, gel strength, elastic texture, springiness, low molding viscosity, and ease of process. Although some progress has been made and starch is widely used, the quality product still requires gelatin as the gelling agent, because not one starch can bring all of these properties in one product. There is one advantage that starch does have and that is the melting temperature of the gel. In hot areas or enclosures (i.e. a car in the sun) the gelatin may melt leaving a bag with one single fruity piece upon cooling.

5.6 MISCELLANEOUS PRODUCTS

Gelatin is used in soft and hard capsules to contain pharmaceutical, soap, paint, or any other liquid or powder composition. Many attempts have been made to replace gelatin in these applications without considerable commercial success. Medical capsules have been described as far back as 1988 (Wittwer et al., 1988). Soft capsules for soaps and paints have been mentioned (Muller and Innererber, 2011). Although there are market drivers to implement change, the bulk of the material still contains gelatin (or other non-starch colloid).

Further applications of gelatin are photographic paper and specialty adhesives. No structured approach is described to replace gelatin in these applications. Starch has its own functionality in these applications without the necessity to replace gelatin in its niche applications.

5.7 CONCLUDING REMARKS

Looking from the market backwards to the ingredient, it is obvious that starch and gelatin are competing over similar applications ánd functionalities in those applications. Where in the 1950s and 1960s many applications came about that employed gelatin as the gelling agent of choice, it was readily available and easy to use; now the balance is shifting towards gelling starches. The reasons are obvious: animal origin of gelatin, availability, religious groups, and so on. With new functionalities of starch employing 'friendly' modification techniques, starch is catching up in the traditional gelatin arena of gelling and texturizing.

References

Almdal, K., Dyre, J., Hvidt, S. and Kramer, O. (1993) Towards a phenomenological definition of the term 'gel', *Polymer Gels and Networks*, 1, 5–17.

Alting, A.C., van de Velde, F., Kanning, M.W., Burgering, M., Mulleners, L., Sein, A. and Buwalda, P. (2009) Improved creaminess of low-fat yogurt: The impact of amylomaltase-treated starch domains, *Food Hydrocolloids*, 23(3), 980–987.

Banerjee, S. and Bhattacharya, S. (2012) Food gels: gelling process and new applications, *Critical Review of Food Science Nutrition*, 52(4), 334–346.
BeMiller, J. N. (2011) Pasting, paste, and gel properties of starch-hydrocolloid combinations, *Carbohydrate Polymers*, 86(2), 386–423.
Bohidar, H.B. and Jena, S.S. (1993) Kinetics of sol–gel transition in thermoreversible gelation of gelatin, *The Journal of Chemical Physics*, 98(11), 8970.
Boran, G. and Regenstein, J.M. (2010) Fish gelatin, in *Advanced Food Nutrition Resources* 60, 119–143.
Buwalda, P.L. and Meima, H.R. (2003) *Reversible gel formation*, US2003157232A1.
Buwalda, P.L. and Sein, A. (2006) *Cream substitute*, US2010143542
Buwalda, P.L. and Tomasoa, D.T.B. (2011) *Gum confection*, US2011033600A.
Chiu, C.-W. (1984) *Instant Gelling Starches*, EP0108833.
Clark, A.H., Gidley, M.J., Richardson, R.K. and Ross-Murphy, S.B. (1989) Rheological studies of aqueous amylose gels: The effect of chain length and concentration on gel modulus, *Macromolecules*, 22(1), 346–351.
Fannon, J.E. and Bemiller, J.N. (1992) Structure of corn starch paste and granule remnants revealed by low-temperature scanning electron-microscopy after cryopreparation, *Cereal Chemistry*, 69(4), 456–460.
Fredriksson, H., Silverio, J., Andersson, R., Eliasson, A.C. and Aman, P. (1998) The influence of amylose and amylopectin characteristics on gelatinization and retrogradation properties of different starches, *Carbohydrate Polymers*, 35(3–4), 119–134.
Gernat, C., Reuther, F., Damaschun, G. and Schierbaum, F. (1987) Real structure of a partially crystalline maltodextrin gel as revealed by wide-angle X-ray-scattering, *Acta Polymerica*, 38(11), 603–607.
Gidley, M.J. (1989) Mechanisms of amylose aggregation and gelation, *Macromolecules*, 22(1), 351–358.
Goodfellow, B.J. and Wilson, R.H. (1990) A Fourier-Transform Ir study of the gelation of amylose and amylopectin, *Biopolymers*, 30(13–14), 1183–1189.
Guinard, J.X. and Mazzucchelli, R. (1996) The sensory perception of texture and mouthfeel, *Trends in Food Science & Technology*, 7(7), 213–219.
Hansen, M.R., Blennow, A., Farhat, I., Nørgaard, L., Pedersen, S. and Engelsen, S.B. (2009) Comparative NMR relaxometry of gels of amylomaltase-modified starch and gelatin, *Food Hydrocolloids*, 23(8), 2038–2048.
Hansen, M.R., Blennow, A., Pedersen, S. and Engelsen, S.B. (2009) Enzyme modification of starch with amylomaltase results in increasing gel melting point, *Carbohydrate Polymers*, 78(1), 72–79.
Hansen, M.R., Blennow, A., Pedersen, S., Norgaard, L. and Engelsen, S.B. (2008) Gel texture and chain structure of amylomaltase-modified starches compared to gelatin, *Food Hydrocolloids*, 22(8), 1551–1566.
Hoover, R. (2001) Composition, molecular structure, and physicochemical properties of tuber and root starches: a review, *Carbohydrate Polymers*, 45(3), 253–267.
Huizenga, R.H., Alrich, H.G. and de Groot, I.L. (1997) *Aqueous compositions comprising amylopectin-potatostarch and process for their manufacture* EP0799837.
Huizinga, H., Immerseel, A.R. and Pelan, E.W. (2003) *Edible emulsion spread*, US2003099755.
Jane, J.L., Chen, Y.Y., Lee, L.F., McPherson, A.E., Wong, K.S., Radosavljevic, M. and Kasemsuwan, T. (1999) Effect of amylopectin branch chain length and amylose content on the gelatinization and pasting properties of starch, *Cereal Chemistry*, 75(5), 629–637.
Kanning, M.W., van de Velde, F., Alting, A.C., Mulleners, L., Sein, A., van den Akker-Bleumink, B. and Buwalda, P. (2012) Improved creaminess in stirred yogurt through amylomaltase-treated starch domains, *International Dairy Journal*, 27(1–2), 86–91.
Kaper, T., van der Maarel, M.J.E.C., Euverink, G.J.W. and Dijkhuizen, L. (2004) Exploring and exploiting starch-modifying amylomaltases from thermophiles, *Biochemical Society Transactions*, 32, 279–282.
Kaur, B., Ariffin, F., Bhat, R. and Karim, A.A. (2012) Progress in starch modification in the last decade, *Food Hydrocolloids*, 26(2), 398–404.
Lacourse, N.L. and Zallie, J.P. (1988) *Starch based jelly confections*, US4726957.
Miles, M.J., Morris, V.J., Orford, P.D. and Ring, S.G. (1985) 'The Roles of Amylose and Amylopectin in the Gelation and Retrogradation of Starch', *Carbohydrate Research*, 135(2), 271–281.
Moates, G.K., Noel, T.R., Parker, R. and Ring, S.G. (1997) The effect of chain length and solvent interactions on the dissolution of the B-type crystalline polymorph of amylose in water, *Carbohydrate Research*, 298, 327–333.
Morris, V.J. (1990) Starch gelation and retrogradation, *Trends in Food Science & Technology*, (July), 2–6.
Muller, R. and Innererber, F. (2011) *Soft capsules based on starch and a method and device for the production thereof*, US2011319503.

N.N. (2013) [online], available: http://en.wikipedia.org/wiki/Gelatin [accessed

Ogden, C.L., Carroll, M.D., Kit, B.K. and Flegal, K.M. (2012) Prevalence of obesity in the United States, 2009–2010, *NCHS Data Brief*, No 82, 1–8.

Pérez, S. and Bertoft, E. (2010) The molecular structures of starch components and their contribution to the architecture of starch granules: A comprehensive review, *Starch – Stärke*, 62(8), 389–420.

Pezron, I., Djabourov, M., Bosio, L. and Leblond, J. (1990) X-ray-diffraction of gelatin fibers in the dry and swollen states, *Journal of Polymer Science Part B-Polymer Physics*, 28(10), 1823–1839.

Pezron, I., Djabourov, M. and Leblond, J. (1991) Conformation of gelatin chains in aqueous-solutions .1. A light and small-angle neutron-scattering study, *Polymer*, 32(17), 3201–3210.

Reddy, P.R., Trzasko, P., Kasica, J.J. and Jackson, J. (1999) *Water-in-oil emulsion spread*, US5904949.

Robijn, G.W., Gutierrez Gallego, R., van den Berg, D.J., Haas, H., Kamerling, J.P. and Vliegenthart, J.F. (1996) Structural characterization of the exopolysaccharide produced by Lactobacillus acidophilus LMG9433, *Carbohydr Res*, 288, 203–218.

Sein, A., Mastenbroek, J., Buwalda, P.L., Metselaar, R., Van Rooijen, C. and Visser, K.A. (2011) *Low fat shortening* US201110435.

Singh, J., Kaur, L. and McCarthy, O.J. (2007) Factors influencing the physico-chemical, morphological, thermal and rheological properties of some chemically modified starches for food applications – A review, *Food Hydrocolloids*, 21(1), 1–22.

van der Maarel, M.J.E.C., Capron, I., Euverink, G.J.W., Bos, H.T., Kaper, T., Binnema, D.J. and Steeneken, P.A.M. (2005) A novel thermoreversible gelling product made by enzymatic modification of starch, *Starch-Starke*, 57(10), 465–472.

van der Maarel, M.J.E.C., van der Veen, B., Uitdehaag, J.C.M., Leemhuis, H. and Dijkhuizen, L. (2002) Properties and applications of starch-converting enzymes of the alpha-amylase family, *Journal of Biotechnology*, 94(2), 137–155.

Wesdorp, L.H., Madsen, R.A., Kasica, J.J. and Kowblansky, M. (1994) *Edible plastic dispersion having a rapid gel-setting starch* US5279844.

Whaley, J.K., Kasica, J.J., Senkeleski, J.L., Foss, J.W. and Heigis, J.R. (2000) *Hydrocolloid composition for use as a gelling agent viscosifier and stabilizer* US6093439.

Wittwer, F., Tomka, I., Bodenman, H.U., Raible, T. and Gillow, L.S. (1988) *Method for forming pharmaceutical capsules from starch compositions*, US4738724.

Woltjes, J.R., Meima, H.R. and Buwalda, P.L. (2004) *Composition based on cross-linked starch and depolymerized starch suitable as gelatine replacement*, US6749880.

Wurzburg, O.B. (1986) *Modified Starches, Properties and Uses*, CRC Press Inc., Boca Raton, USA.

6 Modified whey proteins as texturizers in reduced and low-fat foods

Jeffrey Banes, Thomas Helm and David Taylor

Grande Custom Ingredients Group, Lomira, Wisconsin, USA

6.1 WHEY AND WHEY PROTEINS

Whey is the milk-derived liquid left over from the production of cheese and casein products. Whey is either classified as 'sweet whey' or 'acid whey' depending upon the method used to form the cheese or casein curd during manufacturing. Sweet whey is the liquid produced from cheeses that use rennet and other protease enzymes to form the curd. Acid whey is produced when acid, either from direct acid addition or from lactic acid bacteria, is added to milk until the pH of the milk lowers to the isoelectric point of the casein proteins. Both sweet and acid whey contain water, lactose and other sugars, organic acids, minerals, lipids, and proteins. The majority of this whey is then further processed and converted into spray-dried, non-hygroscopic powders that can be used as food ingredients.

While proteins comprise a minor component of the total solids found in whey, they are extensively researched. Numerous references to their biological, chemical, and physical properties are published in literature. Whey proteins' extensive research is due to their unique food functional and nutritional properties. As a result of that research, whey proteins are found in many foods across the world.

Like most food protein products, whey proteins are a mixture of several different individual protein components. These components include β-lactoglobulin (β-Lg), α-lactalbumin (α-Lb), bovine serum albumin (BSA), immunoglobulins, enzymes, and protein fragments known as peptides. These proteins comprise only about 0.8% (wt/wt) of liquid whey and about 12% (wt/wt) in dry whey powder. Lactose and ash make up most of the remaining solids in both liquid and dry whey.

The relative amount of protein in whey products is routinely altered through the processes of membrane filtration, ion exchange chromatography, and centrifugation. These processes remove lactose, ash, and other minor components in liquid whey and increase the percentage of protein. Whey protein ingredients that are up to 93% protein (wt/wt) can be manufactured using these processes. Table 6.1 shows typical composition of various whey protein ingredients.

Food Texture Design and Optimization, First Edition. Edited by Yadunandan Lal Dar and Joseph M. Light.

Table 6.1 Typical whey protein ingredient composition. (Source: Courtesy of Grande Custom Ingredients Group, Lomira WI.)

Whey Product	Protein (%)	Carbohydrate (%)	Fat (%)	Ash (%)	Moisture (%)
Liquid Whey	0.8	5.1	>0.1	0.5	93.5
Sweet Whey Powder	12.0	75.8	0.8	7.4	4.0
Reduced Lactose Whey	23.0	58.0	1.0	15.0	3.0
Whey Protein Concentrate (35%)	35.5	53.0	2.5	5.0	4.0
Whey Protein Concentrate (80%)	76.7	11.1	4.7	2.8	4.7
Whey Protein Isolate (90%)	89.4	3.2	0.4	2.5	4.5

Food formulators can use whey protein ingredients for a number of purposes. Low protein ingredients, such as dry whey (sweet or acid) and reduced lactose whey, are typically used for their bulking, sweetening, or Maillard browning properties. Higher protein ingredients, such as whey protein concentrates and isolates, are commonly used for their nutritional, functional, and textural attributes. In general, the higher the protein content of the whey based ingredient, the more functional the ingredient is considered. Along with the increased function comes increased cost because the operational resources required to produce higher protein ingredients are much higher than lower protein ingredients.

Whey proteins have many functional properties that can be utilized by formulators to improve the quality of finished food products. These properties include gelation, solubility, surface active properties for emulsification and foam formation, viscosity, water holding, and fat holding.

6.2 FOOD FUNCTIONAL PROPERTIES OF WHEY PROTEINS

6.2.1 Gelation

One of the most important functional properties of whey proteins is gelation. Gelation is a process in which proteins interact with one another and form a solid or semi-solid structure. Whey proteins will form gels when heated or exposed to ionic stresses (Morr and Ha, 1993). The type of gel formed by whey proteins is influenced by factors such as pH, ionic strength, the presence of specific ions (such as calcium), and protein concentration (Morr and Ha, 1993). Other components present in the system, such as fat and sugars, also influence the type and strength of whey protein gels.

Heat induced gelation is the most common method used to form whey protein gels in food systems. Typically, whey proteins must be heated to 70–90 °C and held for longer than 15 seconds for heat-induced gelation to occur. During heating, individual whey proteins denature (unfold) and aggregate (polymerize) to form a three-dimensional network (Langton and Hermansson, 1992). The formation of this network is driven by hydrophobic protein interactions and further stabilized during cooling through hydrogen bonding and disulfide exchanges between proteins (Chou and Morr, 1979). This network physically entraps the surrounding solution, increasing the viscosity of the

Table 6.2 Effect of pH and protein concentration on gelation of WPC 80 solutions. (Source: Courtesy of Grande Custom Ingredients Group, Lomira WI.)

	pH				
% Protein	**8**	**7**	**6**	**5**	**4**
4	Translucent Fluid	Translucent Fluid	White Fluid	White Loose Gel	White Loose Gel
6	Translucent Fluid	Translucent Fluid	White Firm Gel	White Firm Gel	White Loose Gel
8	Translucent Fluid	Translucent Fluid	White Firm Gel	White Firm Gel	White Firm Gel
10	Translucent Firm Gel	Off White Firm Gel	White Firm Gel	White Firm Gel	Off White Firm Gel
12	Translucent Firm Gel	Off White Firm Gel	White Firm Gel	White Firm Gel	Off White Firm Gel

solution. When the concentration of protein is high enough, a rigid gel will form. Gels formed in this manner are not reversible. At pH 7.0, whey protein concentrations above 7.5% (wt/wt) are generally required for gelation (Morr and Ha, 1993).

The pH of the whey protein solution also influences the characteristics of heat induced whey protein gels (see Table 6.2). When β-Lg, the major protein found in whey, is heated at low pH (<4.0) it typically forms fine stranded gels (Langton and Hermansson, 1992). These gels are characterized by their translucent appearance and the presence of fine protein strands that can be seen under a microscope. Gels formed at this pH have low stress and strain values and are fragile compared to gels formed at higher pH levels. Fine stranded gels are also formed when β -Lg is heated between pH 6.5 and 9.0, although the strands are larger and more flexible (Langton and Hermansson, 1992). These gels also have higher stress and strain values compared to their low pH counterparts.

Between pH 4 and 6.5, whey proteins form gels that are opaque and particulate when heated. These gels have spherical aggregates that are linked together through hydrophobic interactions and disulfide bonds (Langton and Hermansson, 1996). Mixtures of whey proteins, like those found in most commercial whey protein products, form soft and creamy gels when heated between pH 4.5 and 5.4 (Langton and Hermansson, 1996) but become more elastic and rubbery above pH 5.4. The elasticity is attributed to the stabilizing effects of disulfide interchanges between proteins (Langton and Hermansson, 1992).

Ions, such as sodium, calcium and phosphates, also have a significant effect on the textural and physiochemical characteristics of thermally induced whey protein gels (Bottcher and Foegeding, 1994; Verheul and Roefs, 1998). The addition of calcium and sodium increases gel coarseness, permeability, and decreases elastic modulus (G'). The addition of specific ions to solutions of whey proteins can also cause the formation of protein gels without heating (Ju and Kilara, 1998, Marangoni et al., 2000). These cold-set gels are characterized by an opaque appearance and large protein aggregates. Cold-set gels have limited use in food systems because of the specific conditions needed for their formation.

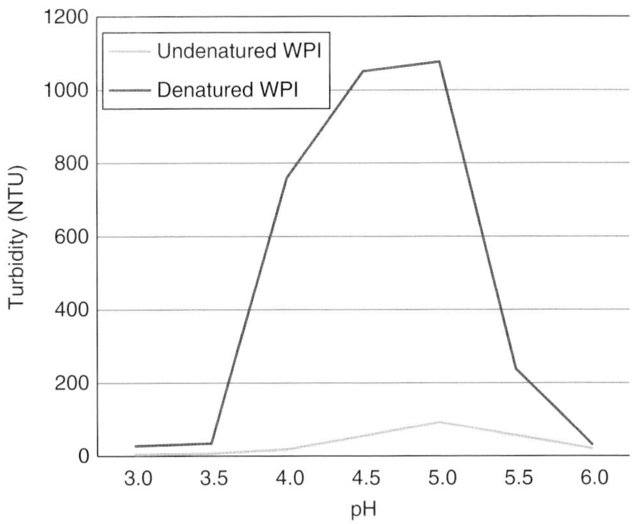

Figure 6.1 Turbidity of 2.3% whey protein isolate solutions at various pH. (Source: Courtesy of Grande Custom Ingredients Group, Lomira WI.)

6.2.2 Solubility

Undenatured whey proteins are highly soluble over a wide range of protein concentration, pH, ionic conditions, and temperature (Morr and Ha, 1993). Because commercial whey protein ingredients contain a mixture of proteins, the isoelectric point of these ingredients is very broad and typically ranges from pH 4.0 to 5.3 (Morr et al., 1973). Unlike other food proteins that aggregate at or near their isoelectric point, whey proteins remain soluble throughout this range provided they are not denatured. Denatured whey proteins will precipitate and aggregate between pH 4.0 and 5.3. In solutions, this aggregation is observed as an increase in turbidity (see Figure 6.1). A similar reaction occurs when solutions of undenatured whey proteins are heated in this pH range. The proteins will denature and form insoluble aggregates that contribute to chalky and grainy textures in food.

Below pH 4.0, and especially at pH levels below 3.5, whey proteins are highly soluble, even when heated. The low pH causes the whey proteins to take on a net positive electrical charge. The positive charge on individual proteins causes electrostatic repulsion between other proteins, preventing aggregation. The chalky, grainy texture associated with whey proteins heated at higher pH levels is greatly reduced.

6.2.3 Surface properties-emulsification and foaming

Whey proteins possess the physiochemical properties to form and stabilize both emulsions and foams. The amphiphilic nature of whey proteins allows them to unfold and interact with both polar and non-polar components of food systems. This is especially true when mechanical energy, in the form of shear, is applied to the system. As mechanical energy is applied to a liquid system containing whey proteins, the energy causes the proteins to unfold and orient themselves at the interface between the polar and non-polar substances.

Native, undenatured whey proteins typically are more effective at stabilizing emulsions and foams than denatured whey proteins, however, some heat denatured whey proteins do show improved surface active qualities (Dickinson and Parkinson, 2004; Bernard et al., 2011). Blends of whey protein ingredients containing differing amounts of denatured protein were shown to be very effective emulsifiers in cold emulsion systems such as mayonnaise and salad dressing (Holst et al., 1996).

It is hypothesized that undenatured whey proteins are responsible for the formation of a foam or emulsion, and the denatured whey proteins contribute to increased foam or emulsion stability. Undenatured whey proteins are able to more effectively unfold at the interface and denatured whey proteins prevent coalescence or draining between air or oil droplets, thus enhancing stability.

In addition to the protein components of whey ingredients, many concentrated whey ingredients contain high levels of milk phospholipids (Morr and Ha, 1993). These phospholipids have similar structures to commercially available lecithin, a common food emulsifier. It is reported that the phospholipid components in whey ingredients contribute to increased foaming and emulsification abilities (Morr and Ha, 1993).

6.2.4 Modifying whey proteins for improved functionality

The functional attributes of whey proteins can be enhanced through use of various processing techniques. Limited hydrolysis with proteolytic enzymes has been shown to improve the surface active properties of whey proteins (Konrad et al., 2005). Degrees of hydrolysis of 1–1.5% and 5.8% were found to be optimal for emulsification and foaming properties respectively. Hydrolysis results in increased surface hydrophobicity. A balance between hydrophobic and hydrophilic residues is required for good solubility and interfacial properties (Huang et al., 1996). Extensive hydrolysis results in large increases in surface hydrophobicity and reduced interfacial properties (Konrad et al., 2005). Hydrophobic peptides generated during hydrolysis cause bitterness which can be mediated by proper choice of enzymes and conditions (Spellman et al., 2005). Whey protein hydrolysates are readily available in the marketplace.

Whey proteins form aggregates in solution when heated under certain conditions. The water holding and viscosifying properties of these aggregates can be manipulated through control of concentration, heat, pH and shear in industrial processes (Spiegel, 1999, Hinrichs et al., 2004). Spray dried powders of these aggregates are useful as food ingredients to enhance viscosity, texture and mouth feel in a variety of applications including soups, sauces, dressings, dips, spreads and meat. Ingredients of this type are referred to herein as *modified whey proteins* (MWP).

Microparticulated whey proteins are a form of MWP produced under conditions of high shear. Microparticulation results in small aggregates that hold less water than larger aggregates but are said to mimic the lubricity of fat in the mouth. These MWP have a particle size in the range of 0.1 to 2.0 microns, which is similar to that of homogenized dairy fat. Because of the specific particle size, the mouth and tongue sense these MWP in similar ways that fat is sensed (Singer et al., 1988).

Several examples of commercial MWP ingredients exist in the marketplace today.

6.3 USING WHEY PROTEINS AS TEXTURE MODIFIERS IN FOODS

Whey proteins are used extensively in the food industry as texture modifiers. They are used in dairy and other food products to improve texture, viscosity, and mouthfeel after fat is removed from the food system.

6.3.1 Whey proteins in hard and semi-hard cheeses

Hard and semi-hard cheeses are among the most consumed of all dairy products. They are used as table cheeses, toppings, and ingredients in many other foods. Many of these cheeses are relatively high in fat (see Table 6.3), and as a result are a target for fat reduction. Many approaches are used to reduce the fat level in these cheeses, including the incorporation of whey proteins into the cheese during manufacturing and processing.

The incorporation of denatured whey proteins into hard and semi-hard cheese is extensively researched and published. The majority of these studies involved reduced-fat and low fat cheese products. In cheeses with typical amounts of fat, fat globules are dispersed throughout the casein matrix. These dispersed fat globules help to break up the casein matrix, creating an open microstructure in the cheese (Bryant et al., 1995). The result of this open microstructure is an ideal cheese texture that has the proper firmness, cohesive structure, and melting characteristics. Without the fat to break up the casein matrix, the overall texture of reduced-fat and low fat cheeses is characterized as being too firm and rubbery (Fife et al., 1996; McMahon et al., 1996; Mistry, 2001). The fat globules also improve the level of creaminess and lubricity detected in the mouth during mastication (Romeih et al., 2002). Poor meltability and an overall lack of flavor are also common defects in these cheeses.

One approach for improving the texture and mouth feel of low fat and reduced-fat cheeses is to increase the moisture content in the cheese. This is usually done by modifying the cheese making procedures to increase the moisture in the cheese curd (Merril et al., 1994). Increased moisture helps texture, but often leads to other defects such as poor flavor development and bitterness. The increased moisture often changes the

Table 6.3 Composition of various hard and semi-hard cheeses. (Source: From USDA National Nutrient Database for Standard Reference [Online], Available http://ndb.nal.usda.gov/ [10 October 2011].)

Type of Cheese	Fat (%)	Moisture (%)	Protein (%)	Carbohydrate (%)	Ash (%)
Cheddar	33.14	36.75	24.90	1.28	3.93
Cheddar, Low-Fat	7.00	63.10	24.35	1.91	3.64
Colby	32.11	38.20	23.76	2.57	3.36
Mozzarella, Low Moisture	24.64	48.38	21.60	2.47	2.91
Mozzarella, Low Moisture Part Skim	20.03	46.46	25.96	3.83	3.72
Mozzarella, Fat Free	0.00	60.20	31.70	3.50	4.60
Parmesan	25.83	29.16	35.75	3.22	6.04
Swiss	27.80	37.12	26.93	5.38	2.77
Swiss, Low-Fat	5.10	59.60	28.40	3.40	3.50

cheese microenvironment enough that the microorganisms responsible for flavor development during cheese ripening can be affected. Adding too much moisture to the cheese can also cause the cheese to become excessively soft and pasty.

Another approach to improve the texture of low and reduced-fat cheeses is to incorporate modified whey protein (MWP) ingredients into the cheese at various stages of the processing. When used, MWP increases moisture retention in the cheese and often functions like fat globules.

The use of MWP in cheese manufacturing is extensively studied and reported in the literature. In reduced-fat mozzarella cheese, MWP was shown to increase meltability and moisture level of the cheese although the extent of the improvement was largely dependent upon the degree of microparticulation (Lobato-Calleros et al., 2001). In this study it was hypothesized that the MWP particles become embedded within the casein matrix which allowed for greater moisture retention and less syneresis from the curd during the cheese making process. This also allowed for greater meltability, as meltability and moisture content are strongly correlated in cheeses.

The use of MWP is reported to improve the texture of low fat, fresh Kashar cheese. Kashar cheese is a popular cheese produced in Turkey and is used as an ingredient in sandwiches and pizza (Koca and Metin, 2003; Sahan et al., 2008). Typical low fat Kashar cheese is characterized as having a rubbery texture with poor meltability. Bitterness, off-flavors, and undesirable color are also problems in low fat versions of this cheese (Sahan et al., 2008). To overcome these textural and flavor challenges, various fat replacers are used. As with reduced-fat mozzarella cheese, incorporating MWP into Kashar cheese increase moisture retention, softens texture, and improves meltability (Koca and Metin, 2003; Sahan et al., 2008). Using texture profile analysis (TPA), Kashar cheeses containing MWP were shown to be improved over the low fat controls in the areas of hardness, cohesiveness, and chewiness. The texture still, however, was significantly different than the full fat controls. Excessive use of MWP can result in a cheese that retains too much moisture and becomes excessively soft (Punidadas et al., 2000).

The size of the particles within MWP ingredients is hypothesized to influence how effectively they work in low fat and reduced-fat cheese applications. MWP ingredients with smaller particle sizes often do not hold water as effectively as larger particles (Merril et al., 1994; Koca and Metin, 2003). The resulting texture of cheeses containing MWP ingredients also is dependent upon the types of heat treatments that the cheese or cheese milk experiences during processing (Koca and Metin, 2003). Higher heat treatments are often required for the water holding abilities of MWP to be maximized.

An alternate method of incorporating whey proteins into reduced-fat and low fat cheeses is to specially treat the cheese milk prior to cheese making. These techniques do not require the addition of any special dried whey ingredients. The simplest method to do this is to ultrafilter milk and then heat the milk at 85 °C for 17 s (Lo and Bastian, 1998). The heating denatures whey proteins, causing them to become entrapped within the casein during curd formation. This incorporation results in higher moisture retention and a less rubbery texture. Mechanical openings within the cheese, however, increased using this method.

Liquid whey from a previous batch of cheese can be similarly treated and incorporated back into a subsequent batch of low-fat cheese. Cheese whey heated to 95 °C,

Table 6.4 Summary of textural properties of reduced and low-fat cheese containing added whey. (Source: Courtesy of Grande Custom Ingredients Group, Lomira WI.)

Textural Attribute	Effect of Reducing Fat Content	Effect of Adding Whey Proteins	Reference
Hardness	Increase	Decrease	1, 2, 3, 4
Adhesion	Decrease	Increase	2
Cohesion	Increase	Decrease	1, 3, 5
Springiness	Increase	Decrease	1, 3, 5
Elasticity	Increase	NA	1
Meltability	Decrease	Increase	1, 3, 6, 7
Stretch	Decrease	Increase	6, 7

1 - Bryant *et al.* (1995), 2 - Koca and Metin (2003), 3 - Sahan *et al.* (2008), 4 - Lo and Bastian (1998), 5 - Lobato-Calleros *et al.* (2001), 6 - Punidadas *et al.* (1999), 7 - Punidadas *et al.* (2000)

Table 6.5 Typical composition of various cream cheese-type products. (Source: From USDA National Nutrient Database for Standard Reference [Online], Available http://ndb.nal.usda.gov/ [10 October 2011].)

Type of Cream Cheese	Fat (%)	Moisture (%)	Protein (%)	Carbo-hydrate (%)	Ash (%)
Cream Cheese, Full Fat	34.24	54.44	5.93	4.07	1.32
Cream Cheese, Low-Fat	15.28	66.72	7.85	8.12	2.03
Cream Cheese, Fat Free	1.00	71.87	15.69	7.67	3.77
Cream Cheese Spread	28.60	58.50	7.10	3.50	2.30
Neufchatel Cheese	22.78	63.11	9.15	3.59	1.37

homogenized, and then added into a new cheese vat resulted in higher cheese yields and improved texture (Punidadas et al., 1999, Punidadas et al., 2000). Cheese stretch and meltability were both improved by using heated and homogenized whey proteins. A summary of the textural effects of adding whey proteins to reduced and low-fat cheese is found in Table 6.4.

6.3.2 Whey proteins in cream cheeses and spreads

Cream cheese is a dairy product made from the acidification of milk and cream. According to the U.S. standards (21CFR133.133), cream cheese must have a fat level of at least 34% and a maximum moisture level of 55%. Neufchatel cheese, a lower fat version of cream cheese, must have a fat level between 20 and 33% and a moisture level less than 65% (21CFR133.162). These cheeses have soft texture and are primarily used for slicing, spreading, and baking. To meet consumer demand for lower fat products, manufacturers continue to develop reduced and low fat versions of cream cheese (see Table 6.5). As with hard and semi-hard cheeses, the reduction of fat in cream cheese creates many textural challenges for food manufacturers. As fat levels decrease, textural attributes, such as spreadability, mouth coating, and mouth meltaway often are affected in negative ways (Muir et al., 1997).

Spreadability is one of the most important textural properties in these types of products and thus it is critical to maintain good spreadability as the fat level is decreased. In full fat cream cheese and dairy spreads, total fat content is typically correlated to measured spreadability. But spreadability is also dependent upon the ratios of solid to liquid fat within the food system. Moisture, salt, and protein levels also affect spreadability (Wendin et al., 2000).

In order to reduce the fat in cream cheese and cheese spreads and maintain spreadability, a combination of moisture, starches, emulsifiers, and gums is typically added to the product during manufacturing (Finocchiaro, 1997; Tang, 1996). Cellulose and microcrystaline cellulose are also frequently used ingredients in the manufacture of reduced and low fat cream cheese products (Smith et al., 1992).

Dairy proteins, particularly modified whey proteins (MWP), are often added along with these carbohydrate based ingredients to improve other texture attributes. MWP that are microparticulated are used to improve mouth coating and fat-like texture of reduced-fat cream cheese products (Singer et al., 1992). The overall oral perception in these reduced-fat cream cheeses is similar to that of a full fat cheese. The usage level of MWP in cream cheeses and other spreadable foods can be as high as 50% (Singer et al., 1992a).

6.3.3 Whey proteins in other dairy based foods

Whey proteins are used in many other dairy foods for their ability to impact texture in positive ways. Whey protein isolate (WPI) is being used in reduced-fat and low-fat ice creams to modify many textural attributes. The addition of WPI to ice cream mixes prior to freezing significantly increases viscosity, hardness, and resistance to melting (Akalm et al., 2008). WPI is used in combination with other stabilizers to mimic the modified churning processes and produce reduced-fat ice creams with the texture of full-fat versions (Gerdes, 2009). The use of WPI increased overrun and did not negatively affect mouth or texture of the product.

Whey proteins are used extensively in the production of yogurt and yogurt-type products. They are commonly added to increase the solids of the yogurt, often helping to increase the firmness of the product. More recently they are used to provide fat-like texture to lower fat yogurt products. Whey protein concentrates (WPC) are often used because of their gelling properties (Sandoval-Castilla et al., 2004). Modified whey proteins (MWP) with microparticulated properties can be used to replace fat in yogurt, but often lead to a soft and loose yogurt texture (Singer et al., 1992b). MWP disrupt the casein structure of yogurt and contribute to a loose and open microstructure and thus are not as effective as WPC in improving yogurt texture.

6.3.4 Modified whey proteins in other food products

The texture of salad dressings can be modified by MWP. Singer (Singer et al., 1992b) demonstrated that combinations of MWP, water, and stabilizers can replace the majority of oil in ranch, French, and blue cheese dressings. MWP contributed to the overall texture and mouth feel typical of higher fat versions of the dressings. MWP levels of 5–28% were used in order to produce the desired textures and mouthfeel.

Use of MWP in other food products such as soups and sauces is commonly practiced commercially but not reported in the scientific literature.

6.4 FOOD APPLICATION CASE STUDIES

6.4.1 Cheese reduction and replacement in cheese sauces

The use of modified whey proteins (MWP) in reduced fat cheeses is extensively researched and was discussed earlier in this chapter. Natural cheese, however, is an inherently variable raw material, both within and between style and variety. Manufacturers of finished products, such as a cheese sauce, either solely, or for inclusion in a frozen entrée or dry mix often do not have control over the processing of the cheese ingredients they are using. Oftentimes, due to availability and/or purchasing restraints, developers must rely on limited styles or varieties of cheese ingredients in their formulation work. It is therefore essential to have other ingredients that enable the manipulation of texture and mouthfeel when, for example, only a high fat cheese is available in inventory for a low fat sauce formulation, or cost reductions are necessary through the reduction in the amount of cheese in the formula.

We can begin by discussing the role of cheese in a cheese sauce. Aside from the obvious cheese flavor delivery, cheese provides texture, body and a fat-like or creamy mouthfeel. It adds structure to the sauce through its protein and fat contributions and helps to provide cling on whatever substrate it is being served on. When the amount of cheese in a formula is reduced, whether for nutritional or cost reasons, the resulting loss of fat and protein has a negative impact on the texture of the sauce, reducing viscosity, which makes it feel thinner in the mouth. There are numerous ingredients on the market which claim to have fat mimetic properties, designed to counter formulation issues such as this.

MWP are useful for replacing cheese in sauces. As mentioned earlier, MWP are formed by heating or heating and shearing whey proteins in solution. They are used to increase moisture retention and function like fat globules. In addition to acting as a fat mimetic, they also contribute to overall body and texture. Through the use of MWP, the amount of cheese in a sauce can be reduced while maintaining, or even increasing the overall body and texture. MWP function differently than many carbohydrate- or hydrocolloid-based fat mimetic systems. While a carbohydrate or hydrocolloid system can increase viscosity and mouth coating, MWP, either on its own or in conjunction with carbohydrate based viscosifiers, contributes a creamy oral perception that more closely resembles the dairy fat that has been removed. Table 6.6 contains the formulas for several cheese sauces with different levels of cheese. It also shows viscosity impacts of cheese reduction and MWP addition.

The flavor and mouthfeel impact of the formulation changes in Table 6.6 were evaluated by an informal panel of experts. In addition to the viscosity differences, the creamy, fat-like perception in the mouth was lower in CR1 than CH1. The addition of 1% MWP in CRW1 changed the finished sauce dramatically, increasing viscosity and improving mouthfeel beyond just a thickening effect. It provided the perception of creaminess that became lost when the cheese level was reduced.

MWP should not be confused with true viscosifying ingredients like starches and hydrocolloids and should not be considered to replace these ingredients completely. The increase in body and texture from inclusion of MWP can be read on a viscometer as an increase in viscosity, but because of the physical nature of how each ingredient holds water, viscosity of a finished sauce should be controlled via starch and/or hydrocolloids.

Table 6.6 Formulas of cheese sauces containing cheese reductions. (Source: Courtesy of Grande Custom Ingredients Group, Lomira WI.)

Ingredient	% (wt/wt) in CH1	% (wt/wt) in CR1	% (wt/wt) in CRW1
Milk, reduced-fat	50.00	50.00	50.00
Water	22.45	22.45	22.45
Cheese, pasteurized process American	23.50	20.00	20.00
Starch*	3.25	3.25	3.25
Modifed whey protein**	—	—	1.00
Maltodextrin, M180	—	3.50	2.50
Salt	0.80	0.80	0.80
Totals	100.00	100.00	100.00
Viscosity (PaS) measured at 50 s^{-1} and 74 °C	0.412	0.213	0.589

*National 465, National Starch Food Innovation, Bridgewater NJ
**Grande Bravo 550, Grande Custom Ingredients Group, Lomira WI

Table 6.7 Basic cheese sauce formulas. (Source: Courtesy of Grande Custom Ingredients Group, Lomira WI.)

Ingredient	% (wt/wt) in C1	% (wt/wt) in C2	% (wt/wt) in W1	% (wt/wt) in W2	% (wt/wt) in W3
Milk, reduced-fat	50.00	50.00	50.00	50.00	50.00
Water	24.20	24.20	24.20	24.20	24.20
Cheese, pasteurized process American	20.00	20.00	20.00	20.00	20.00
Starch	3.25	3.75	—	3.25	2.50
Modified whey protein	—	—	3.25	1.00	2.50
Maltodextrin, M180	1.75	1.25	1.75	0.75	—
Salt	0.80	0.80	0.80	0.80	0.80
Total	100.00	100.00	100.00	100.00	100.00
Viscosity (PaS) measured at 50 s^{-1} and 74 °C	0.308	0.352	0.010	0.317	0.356

MWP is added to provide incremental viscosity, leading to an increase in body, texture, creaminess, and oral viscosity.

Table 6.7 shows some example formulations for a basic, reduced cheese, cheese sauce to evaluate viscosity and mouthfeel using modified corn starch and MWP.

The viscosity results verify that MWP, while contributing to viscosity, cannot be relied upon solely for increasing viscosity in a cheese sauce. When starch is completely replaced one to one with MWP, the viscosity decreases dramatically (C1 vs. W1), but when a small addition of MWP is added in addition to the starch, the viscosity is incrementally increased (C1 vs. W2).

All samples were evaluated organoleptically by an informal panel of experts to determine differences in mouth coating, fat-like perception and oral viscosity. Results showed that an increase in measureable viscosity alone does not change fat-like or creamy perception in the mouth. The differences between C1 and C2, having different levels of starch and different measureable viscosity, was minor when consumed. Both samples were described as thick in the mouth, but with quick release, meaning it did not coat the mouth and linger. Sample W2, with a 1% addition of MWP had a considerably longer hold time/coating in the mouth, presenting creaminess and more fat-like perception even at a similar viscosity. Sample W3 was compared to sample C2, with nearly identical viscosity readings. Altering the ratio of starch to MWP can yield different organoleptic results, while having the same measurable viscosity. Sample W3, with a higher level of MWP had a greater mouth coating component than sample C2, but was perceived as powdery. This can be understood if you refer back to the solubility section of this chapter. Sample W1 was not rated due to its obvious lack of viscosity. In addition to the perceived textural differences in the mouth, it should also be noted that the sauce samples with MWP also presented an increase in cheese flavor versus the control samples using only starch.

The results in this case study show that the use of MWP in cheese sauce formulations can lead to increased quality in lower fat or reduced cheese sauces by incrementally increasing viscosity and improving creaminess and mouthfeel. MWP are an excellent tool for product developers to use in sauces of all types to modify texture by increasing the creamy character or adding a fat-like mouthfeel component. As an added bonus, in most cases, the use of MWP to replace cheese will also lower the cost of the formula.

6.4.2 Low fat creamy salad dressing

Creamy salad dressings, such as ranch style dressings, are creamy from the oil in water emulsion that forms their base. A regular, full fat, salad dressing will contain from 40–50% oil to achieve this mouthfeel. Oil content is also directly related to viscosity, with higher oil contents producing more viscous emulsions. At the end of the spectrum lies mayonnaise, with a minimum oil content of 65%, typically produced at 77–82%. At these oil contents, when stabilized correctly, mayonnaise acts like a solid. Mayonnaise is actually considered a 'Bingham Plastic', meaning it acts like a solid until its yield value, the amount of shear stress needed to cause a solution to flow, is exceeded.

When the oil amount in a dressing is reduced, the number of oil droplets in the emulsion is reduced as well as the viscosity. The result is a change in the oral sensory characteristics. Low fat dressings can be thickened by starches and/or gums, and while the overall viscosity may be the same, oral perception may not be similar to the full fat dressing. In addition to potential differences in oral perception, thickening from starches and/or gums alone can lead to other attributes which detract from the quality of the low fat dressing. When introduced into the mouth, the presence of salivary amylase induces breakdown of starch containing products, which leads to a rapid melt-away sensation in the mouth. The resulting affect on the low fat dressing is that it feels thin in the mouth and will not linger, leading to the flavor and mouthfeel disappearing quickly. Gums, although excellent thickeners, at higher levels can produce a mouthfeel that is sometimes described as slippery or slimy. MWP are an excellent ingredient to use in conjunction with starches and gums to counter both of these negative attributes and lead to a full, creamy, smooth, and lingering mouthfeel. The following example formulation shows a low fat ranch dressing and the accompanying viscosity measurements via a Bostwick reading at 30 seconds (Table 6.8).

The use of a MWP in this example shows multiple benefits. The MWP allows this cold processed dressing's viscosity to be held consistent, while reducing the amount of starch and hydrocolloid used. In reducing the amount of starch and hydrocolloid, the

Table 6.8 Low-fat ranch dressing formulas. (Source: Courtesy of Grande Custom Ingredients Group, Lomira WI.)

Ingredient	% (wt/wt) in Control	%(wt/wt) in MWP
Water	61.12	61.12
Soybean oil	10.00	10.00
Maltodextrin	13.40	7.00
Modified whey protein*	0.00	8.00
Starch**	3.00	1.50
Sugar	6.00	6.00
Vinegar, distilled white, 5% acidity	2.00	2.00
Salt	2.50	2.50
Phosphoric acid, 40% (wt/wt)	0.75	0.75
Onion powder	0.60	0.60
Garlic powder	0.25	0.25
Xanthan gum***	0.20	0.10
Black pepper, ground, 22 mesh	0.08	0.08
Potassium sorbate	0.05	0.05
Sodium benzoate	0.05	0.05
Total	100.00	100.0
Bostwick Viscometer (cm), 30 s, 22 °C	11.0	10.3

*Grande Bravo 300, 500, and 550, Grande Custom Ingredients Group, Lomira WI
**Ultrasperse M, National Starch Food Innovation, Bridgewater NJ
***TIC Rapid 3 Powder, TIC Gums, White Marsh MD

perception of quick melt-away and/or slippery, slimy mouthfeel is reduced. At the same time, the viscosity and mouthfeel provided by the MWP are more consistent with a full fat dressing which is a stabilized emulsion of 40–50% oil. In an informal panel evaluation, the sample with MWP was preferred due to its heavier creamy feel and mouth coating aspects. In addition, the MWP sample was whiter and more opaque, without the use of coloring ingredients, such as titanium dioxide. A whiter, more opaque low fat dressing is more similar to a high fat emulsion, which is a side benefit to using a MWP, beyond the textural aspects that it provides.

In addition to MWP's use in low fat dressings, MWP can be used in other ways in the dressing industry. In regular dressing formulations, egg yolk is the most common emulsifier in use. Egg yolk can be purchased in various forms: liquid refrigerated or frozen, dried, salted, or sugared. Each form has its drawbacks from an emulsification capability standpoint, in addition to the refrigerated or frozen storage issues and costs. MWP act as an emulsifier by forming a film around dispersed oil droplets. The resulting emulsion has a very similar viscosity and mouthfeel to an egg yolk based emulsion. MWP, labeled as 'whey protein concentrate', is also more label-friendly than some alternative emulsifiers. The industry is making a push towards cleaner labeling, and an all natural ingredient such as whey protein concentrate is viewed as more label-friendly than ingredients such as mono- and diglycerides, carrageenan, carboxy methylcellulose, and propylene glycol alginate. In addition to labeling, some of these emulsifiers require a certain amount of heat to be activated, while MWP can be used in a cold process system. Table 6.9 shows a regular dressing formulation with viscosity measurements by Bostwick.

In the above example, egg yolk was replaced by an equivalent protein level of MWP. The balance of the egg yolk was replaced with water. The total solids of the two

Table 6.9 Full fat ranch dressing formulas. (Source: Courtesy of Grande Custom Ingredients Group, Lomira WI.)

Ingredient	% (wt/wt) in Control	%(wt/wt) in MWP
Soybean oil	46.00	46.00
Water	35.00	36.60
Maltodextrin	5.50	5.50
Buttermilk	2.00	2.00
Egg yolk	3.00	–
Modified whey protein*	–	1.40
Sugar	3.00	3.00
Salt	2.10	2.10
Vinegar, distilled white, 5% acidity	1.00	1.00
Starch**	0.75	0.75
Onion powder	0.60	0.60
Phosphoric acid, 40% (wt/wt)	0.50	0.50
Garlic powder	0.25	0.25
Xanthan gum***	0.12	0.12
Black pepper, ground, 22 mesh	0.08	0.08
Potassium sorbate	0.05	0.05
Sodium benzoate	0.05	0.05
Total	100.00	100.00
Bostwick Viscometer (cm), 30 s, 22 °C	7.9	7.6

*Grande Bravo 300, 500, and 550, Grande Custom Ingredients Group, Lomira WI
**Ultrasperse M, National Starch Food Innovation, Bridgewater NJ
***TIC Rapid 3 Powder, TIC Gums, White Marsh MD

formulations are equal. A very similar viscosity was maintained as well as a stable emulsion. The MWP sample also had the added benefit of a fuller, creamier mouthfeel than the egg yolk formula, as evaluated by an informal panel of experts.

In conclusion, MWP are found to be a useful tool in salad dressing formulation, for textural modification and as an emulsifying ingredient. Depending on the type of dressing being manufactured, MWP can be used to enhance, stabilize or modify in a positive way.

6.4.3 Reduced fat cream cheese spread

Spreadability and texture in a reduced fat cream cheese spread are of great importance to developers and marketers alike. For manufacturers of end products utilizing cream cheese products, control of spreadability and texture can be a challenging component. From a cost perspective, purchasing of a standard/stock cream cheese is financially prudent versus a customized blend, provided the manufacturer can modify the standard product to the desired fat level while maintaining spreadability and texture. The use of MWP has been shown to aid in the manipulation of cream cheese products to a desired textural end point.

Neufchatel cheese, a reduced fat version of cream cheese, can have a fat level between 20 and 33%, versus a minimum of 34% for standard cream cheese. However, the industry have settled on a fat level for Neufchatel of ~23%, which is 1/3 less fat than regular cream cheese, as is evident on retail store shelves. With this lower fat level, moisture content is greatly increased. Regular cream cheese has a maximum moisture content of 55%, whereas Neufchatel can contain up to 65%. In order to control moisture content,

Table 6.10 Cream cheese firmness profile as measured by TA. XTplus texture analysis (g of force needed to penetrate 24mm into a sample using a 60° conical Perspex probe). (Source: Courtesy of Grande Custom Ingredients Group, Lomira WI.)

Type of Cheese	Firmness (g)
Cream cheese, Full fat	13.95
Cream cheese, Neufchatel	11.95
Cream cheese, reduced-fat w/MWP	14.31

Figure 6.2 MWP (Grande Bravo 300) and water at a 2:3 Ratio (wt/wt). (Source: Courtesy of Grande Custom Ingredients Group, Lomira WI.)

spreadability and texture, a number of stabilizers are often used, including xanthan gum, carob bean gum, locust bean gum, guar gum, carrageenan, cellulose, and microcrystaline cellulose, in combinations or singly. Texture can vary greatly depending on the stabilizer and Neufchatel is not a direct substitute for regular cream cheese because of its softer texture (see Table 6.10). As can be seen in this table, there is a 15% decrease in firmness from regular cream cheese to Neufchatel. A combination of certain MWPs and water can create a textural equivalent to cream cheese (Figure 6.2).

The following formulation represents a reduced fat cream cheese spread, using standard cream cheese as the base and augmenting with MWP and water (Table 6.11). One important aspect to look at in the above example is moisture content. As mentioned previously, the maximum moisture content of standard cream cheese is no more than 55%, typically just under, while Neufchatel moisture content is typically around 63%, with a maximum of 65%. The moisture content of the above formulation is just under 55%, similar to standard cream cheese. The combination of the moisture content, along with the particle size of the MWP, leads to an oral perception more closely resembling homogenized dairy fat, making the reduced fat spread feel more like standard cream cheese in

Table 6.11 Cream cheese spread formula. (Source: Courtesy of Grande Custom Ingredients Group, Lomira WI.)

Ingredient	% (wt/wt) in Control	% (wt/wt) in Reduced-fat
Cream Cheese	100.00	67.00
Water	—	17.00
Modified whey protein	—	16.00
Total	100.00	100.00

the mouth. Firmness and texture of the reduced fat spread is also more similar to standard cream cheese than Neufchatel, while both are 33% reduced fat products (Table 6.10).

Organoleptic analysis was conducted on the samples. Firmness in the mouth mirrored what the texture analyzer showed, the reduced fat spread with MWP feeling firmer in the mouth than the Neufchatel and more similar to the regular, full fat cream cheese. In addition to the firmness difference, the oral viscosity of the samples was different. The Neufchatel was less cohesive in the mouth and had a mouthfeel consistent with the overuse of hydrocolloids, which can tend to feel a bit rubbery or slimy on the tongue. The reduced fat spread with MWP was more cohesive and had a mouthfeel similar to the standard full fat cream cheese.

As the above results show, control of fat content, firmness, spreadability, and texture can be managed within the finished product manufacturer's facility. This can be a very important developmental tool when making spreads and other products which utilize cream cheese, such as cheesecake, frostings, appetizer or snack fillings, and so on. The ability to 'dial in' the desired firmness in the cream cheese component can pay dividends in the final texture of the product being developed. Texture of a cheesecake can be improved, as well as cutability and clean slicing. Low fat fillings can benefit from a firmer texture, which can help prevent the filling from 'blowing out' during microwaving. In addition, the ratio of MWP and water used to replace cream cheese, in most instances, will reduce the overall cost of the formula. The use of MWP to extend cream cheese for reduced fat items represents an excellent use of whey proteins for modifying texture.

6.4.4 Light cream soup

Health and wellness initiatives have pushed manufacturers to continue to improve upon existing low fat, low calorie, and low sugar offerings, or in the case of new products, develop light offerings that raise the bar on what has previously been introduced to the marketplace. Light versions of creamy soups focus primarily on the low fat aspect of health and wellness, removing traditional cream or half & half, leaving a void in terms of mouthfeel, texture, opacity, flavor delivery, and satiety. The industry offers many carbohydrate, hydrocolloid, or protein based solutions as fat mimetics, ingredients with different chemical structures, but functional properties that mimic the characteristics of fat. In the case of dairy fat, it is well documented that MWP can mimic the oral perception of homogenized dairy fat due to the physical form of the protein and its particle size, which causes the tongue to sense the MWP in much the same way it senses dairy fat. According to the U.S. Dairy Export Council (Johnson 2000), 'reduced fat soups and sauces benefit from the efficient dispersion of fat and the excellent emulsifying properties of whey proteins. (They) impart creaminess and a superior texture to low-fat soups and sauces. Another key function in low-fat applications is increased opaqueness, which otherwise is significantly reduced with decreasing fat content.'

While traditional retail cream based soups can contain anywhere from 9 g to 15 g of fat per 1 cup serving, most light versions only contain 2 to 3 g of fat per 1 cup serving. This 70–80% reduction in fat can lead to drastic changes in the eating quality of the soup, as well as affecting the overall appearance. MWP can increase the smooth, creamy mouthfeel in these low fat offerings as well as contributing opacity. One common industry practice in retail soup production is to manufacture a homogenized blend of cream, vegetable oil, and water. This is a cost savings tool that extends the use of cream by incorporating a less expensive fat source, in the form of vegetable oil. If these ingredients were added to a soup formulation without additional processing, however, the texture would be oily and phase separation may occur. In order to hold these ingredients together in one homogeneous blend, an emulsifying agent is necessary and a homogenization step is required. MWP is a natural ingredient and commonly used as an emulsifier in this application. In addition to the emulsifying properties, the MWP also contributes lubricity and fat-like mouthfeel to the emulsion, as mentioned prior. Table 6.12 shows a low fat soup formula utilizing MWP as an emulsifying agent and mouthfeel enhancer. The soup contains 3 g of fat per 1 cup serving. Compared to an equivalent fat content from a cream only source, the use of the cream, oil, water emulsion, because of the presence of the MWP, creates a creamier, more opaque soup, leading to consumer preference.

Some soups are higher in fat to begin with than the retail soups just discussed. These are typically found in restaurants, schools, or other foodservice outlets, sup-

Table 6.12 Cream of mushroom soup formula. (Source: Courtesy of Grande Custom Ingredients Group, Lomira WI.)

Ingredient	% (wt/wt)
Emulsion Phase	
Water	5.20
Modified whey protein*	0.20
Cream	0.27
Vegetable oil	1.12
Thickeners/Flavors	
Water	61.73
Mushrooms, diced, IQF	10.00
Maltodextrin	5.00
Marsala cooking wine	4.13
Nonfat dry milk	4.00
Starch, modified	2.90
Chicken broth concentrate	2.00
Mushroom powder	1.00
Salt	1.00
Flour, all-purpose	0.60
Garlic, minced	0.50
Onion powder	0.30
Thyme	0.05
Total	100.00

*Grande Bravo 510, Grande Custom Ingredients Group, Lomira WI

Table 6.13 Reduced fat cream of mushroom soup formula. (Source: Courtesy of Grande Custom Ingredients Group, Lomira WI.)

Ingredient	Control % (wt/wt)	Reduced Fat % (wt/wt)
Cream	15.00	10.00
Modified whey protein*	–	0.45
Water	62.72	67.27
Mushrooms, diced, IQF	10.00	10.00
Marsala cooking wine	4.13	4.13
Mushroom powder	1.00	1.00
Chicken broth concentrate	2.00	2.00
Starch modified	2.70	2.70
Salt	1.00	1.00
Flour, all-purpose	0.60	0.60
Onion powder	0.30	0.30
Thyme	0.05	0.05
Garlic, minced	0.50	0.50
Total	100.00	100.00

*Grande Bravo 510, Grande Custom Ingredients Group, Lomira WI

plied by manufacturers in a frozen, refrigerated or dehydrated form. Restaurants that purchase ready-made or quick-scratch soups for their operations sometimes do not have the same nutritional consideration that retail products do. These soup formulas can, and do, contain higher amounts of fat and calories from many sources, but in the case of cream soups, the main source is heavy cream. Restaurants are feeling the pressure to offer healthier items, just as the retail sector is, so alternatives are needed to reduce the amount of fat in these soups, without sacrificing taste and texture. MWP can fulfill this need as well. Table 6.13 shows a different version of a cream of mushroom soup, one with a much higher amount of cream. If a restaurant or a soup manufacturer wanted to reduce the amount of fat slightly, say by 1/3, the cream amount must be reduced by 1/3. This reduction results in a loss of fat-like mouthfeel, opacity, and overall creaminess. A small addition of MWP can counteract this 1/3 loss quite easily, leading to a soup very similar to the higher fat 'control'. Table 6.13 also shows this reduced fat version.

MWP have been shown to have many applications in lower fat, creamy soups. From opacity and fat-like mouthfeel to emulsification capabilities, they provide benefits in their use. The use of MWP also offers a significant cost savings when directly replacing higher cost ingredients such as heavy cream. The clean, bland dairy flavor of MWP fits seamlessly in creamy soups, complimenting or enhancing the desired dairy flavor profile. The functional properties of MWP make them a valuable tool for product developers who need to improve texture in low fat cream soups.

6.5 SUMMARY

Whey proteins are important food ingredients that are used for their exceptional organoleptic, functional and nutritional properties. They are soluble over a wide pH range, form stable gels and have good foaming and emulsification properties. Modified Whey Proteins (MWP) have excellent water holding properties and are used to enhance viscosity and texture in a variety of food applications. MWP are commonly used in reduced and low fat cheese products, spreads, dips, soups, sauces and dressings to reduce cost while improving texture and mouthfeel.

References

Akalm, A.S., Karagözlü, C., and Ünal, G. (2008) Rheological properties of reduced-fat & low-fat ice cream containing whey protein isolate and inulin. *European Food Research and Technology*, 227, 889–895.

Bernard, C., Regnault, S., Gendreau, S., Charbonneau, S. and Relkin, P. (2011) Enhancement of emulsification properties of whey proteins by controlling spray-drying parameters. *Food Hydrocolloids*, 25, 758–763.

Bottcher, S.R. and Foegeding, E.A. (1994) Whey protein gels, fracture stress and strain and related microstructural properties. *Food Hydrocolloids* 8(2), 113–123.

Bryant, A., Ustunol, Z., and Steffe, J. (1995) Texture of cheddar cheese as influenced by fat reduction. *Journal of Food Science* 60(6), 1216–1219.

Chou, D.H.E. and Morr, C.V. (1979) Protein-water interactions & functional properties. *Journal of the American Oil Chemists Society* 56, 53A.

Dickinson, E. and Parkinson, E. (2004) Heat-induced aggregation of milk protein-stabilized emulsions, sensitivity to processing & composition. *International Dairy Journal* 14, 635–645.

Fife, R.L., McMahon, D.J., and Oberg, C.J. (1996) Functionality of low fat mozzarella cheese. *Journal of Dairy Science* 79, 1903–1910.

Finocchiaro, E.T. Opta Foods Ingredients, Inc. (1997) Fat free, reduced fat, & low fat margarine-like spreads & cream cheese. U.S. Pat. 5,679,395.

Gerdes, S. (2009) Whey proteins & slow churning. *Dairy Foods* 3, 82.

Hinrichs, R., Gotz, J., Noll, M., Wolfschoon, A., Eibel, H. and Weisser, H. (2004) Characterization of different treated whey protein concentrates by means of low-resolution nuclear magnetic resonance. *International Dairy Journal* 14, 817–827.

Holst, H.H., Christensen, A., Albertsen, K., Jensen, L.D., Pedersen, M.C., Thomsen, B., Rasche, H. & Hartmann, U. Danmark Protein A/S. (1996) Partially denatured whey protein product. U.S. Pat. 5,494,696.

Huang, X., Catignani, G., and Swaisgood, H. (1996) Improved emulsifying properties of β-Barrel peptides obtained by membrane-fractionation of a limited tryptic hydrolysate of β-lactoglobulin. *Journal of Agricultural & Food Chemistry* 44, 3437–3443.

Johnson, B.R. (2000) Whey Protein Concentrates in Low-fat Applications, [Online], Available: http://www.innovatewithdairy.com/SiteCollectionDocuments/G4ApplicMonogLowfat.pdf [30 Jan 2012].

Ju, Z.Y. and Kilara, A. (1998) Textural properties of cold-set gels induced from heat-denatured whey protein isolates. *Journal of Food Science* 63(2), 288–292.

Koca, N., and Metin, M. (2003) Textural, melting and sensory properties of low-fat fresh kashar cheeses produced by using fat replacers. *International Dairy Journal* 14, 365–373.

Konrad, G., Kleinschmidt, H., Rohenkohl, H. & Reimerdes, E. (2005) Peptic partial hydrolysis of whey protein concentrate for modifying the surface properties of whey protein. II. Effects on the emulsifying & foaming properties. *Milchwissenschaft* 60 (2), 195–198.

Langton, M. and Hermansson, A. (1996) Image analysis of particulate whey protein gels. *Food Hydrocolloids* 10(2), 179–191.

Langton, M. and Hermansson, A. (1992) Fine-stred and particulate gels of β-lactoglobulin & whey proteins at varying pH. *Food Hydrocolloids* 5(6), 523–539.

Lo, C.G. and Bastian, E.D. (1998) Incorporation of native and denatured whey proteins into cheese curd for manufacture of reduced fat, Havarti-type cheese. *Journal of Dairy Science* 81, 16–24.

Lobato-Calleros, C., Robles-Martinez, J.C., Caballero-Perez, J.F., Aguirre-Mandulano, E., and Vernon-Carter, E.J. (2001) Fat replacers in low-fat Mexican manchego cheese. *Journal of Texture Studies* 32, 1–14.

Marangoni, A.G., Barbut, S., McGauley, S.E., Marcone, M. and Narine, S.S. (2000) On the structure of particulate gels—the case of salt-induced cold gelation of heat-denatured whey protein isolate. *Food Hydrocolloids* 14, 61–74.

McMahon, D.J., Alleyne, M.C., Fife, R.L., and Oberg, C.J. (1996) Use of fat replacers in low fat mozzarella cheese. *Journal of Dairy Science*, 79, 1911–1921.

Merril, R.K., Oberg, C.J., and McMahon, D.J. (1994) Physical properties of mozzarella cheese. A method for manufacturing reduced fat mozzarella cheese. *Journal of Dairy Science*, 77, 1783.

Mistry ,V.V. (2001) Low fat cheese technology. *International Dairy Journal* 11, 413–422.

Morr, C.V. and Ha, Y.W. (1993) Whey protein concentrates and isolates, processing and functional properties. *Critical Reviews in Food Science & Nutrition* 33(6), 431–476.

Morr, C.V., Swenson, P.E. and Richter, R.L. (1973) Functional characteristics of whey protein concentrates. *Journal of Food Science* 38, 324–330.

Muir, D.D., Williams, S.A.R., Tamine, A.Y., and Shenana, M.E. (1997) Comparison of the sensory profiles of regular and reduced-fat commercial processed cheese spreads. *International Journal of Food Science and Technology* 32, 279–287.

Punidadas, P., Feirtag, J., and Tung, M.A. (1999) Incorporating whey proteins into mozzarella cheese. *International Journal of Dairy Technology* 52(2), 51–55.

Punidadas, P., Tung, M.A., and Feirtag, J. (2000) Potential use of homogenized whey protein dispersions & process modification for the manufacture of low fat and reduced fat cheddar type cheeses. *International Journal of Dairy Technology* 53(2), 45–50.

Romeih, E.A., Michaelidou, A., Biliaderis, C.G., and Zerfiridis, G.K. (2002) Low-fat white-brined cheese made from bovine milk and two commercial fat mimetics, chemical, physical and sensory attributes. *International Dairy Journal* 12(6), 525–540.

Sahan, N., Yasar, K., Hayaloglu, A.A., Karaca, O.B., and Kaya, A. (2008) Influence of fat replacers on chemical composition, proteolysis, textural profiles, meltability and sensory properties of low-fat Kashar cheese. *Journal of Dairy Research* 75, 1–7.

Sandoval-Castilla, O., Lobato-Calleros, C., Aguirre-Mandujano, E., and Vernon-Carter, E.J. (2004) Microstructure and texture of yogurt as influenced by fat replacers. *International Dairy Journal* 14, 151–159.

Singer, N., Yamamoto, S. and Latella, J. John Labatt Limited. (1988) Protein Product Base. U.S. Pat. 4,734,287.

Singer, N.S., Latella, J., and Yamamoto, S. John Labatt Limited. (1992) Reduced fat salad dressing. U.S. Pat. 5,102,681.

Singer, N.S., Latella, J., Yamamoto, S., John Labatt Limited. (1992) Reduced fat food product. U.S. Pat. 5,098,728.

Smith, D.A., Clark, K.M., and Firary, M.F. Kraft General Foods. (1992) Method of making a non-fat cream cheese product. U.S. Pat. 5,108,773.

Spellman, D., O'Cuinn, G. and Fitzgerald, J. (2005) Physicochemical & sensory characteristics of whey protein hydrolysates generated at different solids levels. *Journal of Dairy Research* 72, 138–143.

Spiegel, T. (1999) Whey protein aggregation under shear conditions –effects of lactose & heating temperature on aggregate size & structure. *International Journal of Food Science Technology* 34, 523–531.

Tang, P.S. Griffith Laboratories Worldwide, Inc. (1996) Method of making and using a fat mimetic in reduced fat/cholesterol foodstuffs. U.S. Pat. 5,589,215.

Verheul, M. and Roefs, S.P.F.M. (1998) Structure of whey protein gels, studied by permeability, scanning electron microscopy & rheology. *Food Hydrocolloids* 12, 17–24.

Wendin, K., Langton, M., Caous, L., and Hall, G. (2000) Dynamic analysis of sensory and microstructural properties of cream cheese. *Food Chemistry* 71, 363–378.

7 Texture design for breaded and battered foods

Chandani Perera[1] and Milda E. Embuscado[2]

[1]Roquette America Inc., Geneva, Illinois, USA
[2]McCormick & Company, Inc., Hunt Valley, Maryland, USA

7.1 FUNDAMENTALS OF COATING SYSTEMS

Batters and breadings provide value added visual, textural and sensory appeal to a variety of food products. Upon frying, coatings deliver a crisp golden brown exterior that acts as a shield to preserve moisture and flavor. Coated food adds an extra dimension of texture and flavor. Traditional breaded products include chicken, fish, and onion rings and these have been the popular items in the breaded category for many years. Currently, the variety of breaded products have grown to include cheese sticks, okra, corn, jalapeno poppers, mushrooms, zucchini, broccoli, cauliflower, chicken skin and many others. Batters and breading provide flavor, texture and act as a thermal buffer between the hot oil and the food.

In deep fat frying, food is totally submerged into hot oil. Pan frying is typically characterized by minimal use of oil. In general the oil level would be half of that of the food or merely to lubricate the pan. Before foods are deep fat fried or pan fried, they are normally coated with a dry or wet coating. Dry coatings are often referred to as 'breadings' and the wet coatings are called 'batters'. Breadings are either used in deep fat frying or pan frying. Batters are solely used in deep fat frying. In deep fat frying, foods are completely immersed in oil that is kept around 350° to 375°F. As the food hits the hot oil, moisture evaporates and a crunchy outer crust forms which helps to retain flavor and moisture of the meat, fish or vegetables. If batters are used in pan fried foods, the coating would not set as it would in the deep fat frying.

Cooking methods impact the flavor potential of coated foods. Huge variation in flavor profiles can be obtained when the coated food is fried, baked or microwaved. Typically deep fat fried foods deliver better flavor and texture profile than microwaved products.

7.2 TRADITIONAL COATING SYSTEMS

The first step in the traditional method of breading is to season the food (meat, fish or vegetable) with salt, pepper or any other seasoning. The food is then coated with flour followed by dipping in the beaten eggs and finally coating the food with breadcrumbs,

Food Texture Design and Optimization, First Edition. Edited by Yadunandan Lal Dar and Joseph M. Light.

to cover it completely. The flour acts as a base for the eggs to adhere to and the liquid eggs in turn form a base for the dry breadcrumbs to attach to. The entire system contributes to the overall flavor and texture of the final crust.

Today, commercial coatings are customized depending on the substrate that is being coated. Commercial coating system for a given food product may consist of predust, batter and breading applied in single or multiple passes. Formulation variables may include type of food, cooking methods, as well as moisture level, temperature and surface characteristics of the substrate. Processing variables are also important when designing a commercial coating system. Variables that play a role in adhesion, quality of the coating and finished product include line speed, water quality and breading pick-up.

7.2.1 Predust

A predust is used to enhance the adhesion between the substrate and the subsequent coating. Enhancing adhesion is especially important in the case of frozen fish in which coating is very difficult to adhere to. Typically, predust consists of plain flour, crackermeal or fine breadcrumbs. It may also contain starches and powdered eggs that help optimize adhesion. In some cases, salt, flavors and seasonings are added in the predust to build up the flavor profile. Since the predust is protected by the layers of the batter and the breader, the flavors added to predust are well protected during frying. As a rule of thumb, the predust has to be a fine powder for maximum adhesion.

7.2.2 Batter

Batters are liquid dough systems designed to be applied to food immediately before breading or frying. They are a mixture of fluid (e.g. milk, buttermilk, water, or beer) and flours and/or starches. Batters also contain other ingredients such as salts, seasonings, eggs, gums, and leavenings which provide expansion or puffing upon frying. Wheat flour is commonly used in batters. Typically wheat flours with 10–11% proteins are used in batters. Wheat flours with higher protein levels tend to increase batter viscosity and provide crisper texture. The viscosity of the batter plays an important role in determining a uniform thickness of the layer. Depending on the viscosity, the batter may or may not spread uniformly over the product. If the layer is not uniform, it may lead to poor adhesion. The level of adhesion varies depending on the fried product.

Batter systems can be classified into two groups; adhesion batter and puff or tempura batters (Loewe, 1990).

7.2.2.1 Adhesion batters

The main function of adhesion batter is to provide a binding medium between substrate and breading. Table 7.1 shows a typical adhesion batter recipe. Proper adhesion prevents surface voids through loss of breading and reduces crumb fall off during frying. This is achieved by adhesion of the batter to hydrophilic as well as hydrophobic areas of the substrate. The amount of protein, fat and moisture on the surface of the product affects adhesion. Depending on the product to be coated and the type of crumb, the adhesion batters can be classified as regular or high adhesion batters. High adhesion batter

Table 7.1 Adhesion batter recipe.

Ingredient	Percent
Wheat flour	50–65
Modified starch (oxidized)	10–15
Corn flour	8–14
Salt	2–4
Seasonings	3–6

Table 7.2 Tempura batter recipe.

Ingredient	Percent
Wheat flour	50–60
Corn flour	22–28
Modified starch (dextrin)	5–10
Salt	2–4
Dextrose	1–3
Baking soda	0.8–1.4

systems are used for the difficult to coat substrates, for example, fish or other seafood. Adhesion batters are generally high in adhesion starch which can be wheat-based or corn-based. The system may consist of modified starch, flour, and salt with or without leavening system. High adhesion batters for seafood usually do not contain a leavening system. Viscosity of these batters can be modified to deliver the desired pickup. Viscosity influences the uniformity and the thickness of the batter coating. Batters with high viscosity in general results in picking up higher percentage of breading compared to a less viscous batter. Modified starches are commonly used in adhesion batters to improve their viscosity and adhesive properties. These include mildly oxidized, thin boiled, cross-linked and acetylated starches. Oxidized starches are known to provide good adhesion. Acetylated starches are commonly used to provide the freeze thaw stability and to retain the crunchiness of the coating upon frying. Therefore, acetylated starches are used in batters applied in foods that are pre-fried, frozen and subsequently heated in a microwave or in conventional oven before consumption.

7.2.2.2 Tempura batters

Puff or tempura batters provide a protective outer coating with a variety of textures which can range from smooth, light airy, crunchy, crispy to cake-like texture. Tempura batter recipe is shown in Table 7.2. These batters also impart flavor and color to the product. Tempura batters are stand-alone systems that are not breaded. These batters are formulated to minimize the oil pick up and reduce greasiness. Typical tempura batter composed of water, flour, eggs and a leavening system that allows the coating to puff when exposed to oil at high temperature. Often the substrates for tempura batters are dusted with flour to provide better tempura batter adhesion. Figure 7.1 shows Tempura battered seafood and vegetables.

Figure 7.1 Tempura battered seafood and vegetables.

Tempura batters are more viscous than adhesion batters; therefore, they are able to form a thicker outer layer of the finished product without the use of crumbs. Since crumb is not used, the product must be fried to set the batter once the batter is applied. The use of tempura batters is quite successful in coating precooked products.

7.2.2.3 Specialty batters

7.2.2.3.1 Corn dog batter

This batter resembles a cake or donut formula which contains a combination of soft and hard wheat flours, corn meal and/or corn flour. Corn meal and corn flour in the formula contribute to the golden color and crispiness. Corn meal does not get completely cooked upon frying thus provides the cake-like texture. According to USDA regulations, corn dogs can carry up to 65% coating (Figure 7.2).

7.2.2.3.2 Clear coatings / French fry batters

Clear coatings provide the added value, convenience and functionality to products such as French fries, coated vegetables and seafood. These coatings are able to create products that are less soggy or greasy after frying. For example, clear coated French fries retain their crispiness for about 30 min after frying. The coating is usually applied on the fries before freezing the product. Potato starch, rice flour and dextrin are commonly used in clear coating formulations (Figure 7.3).

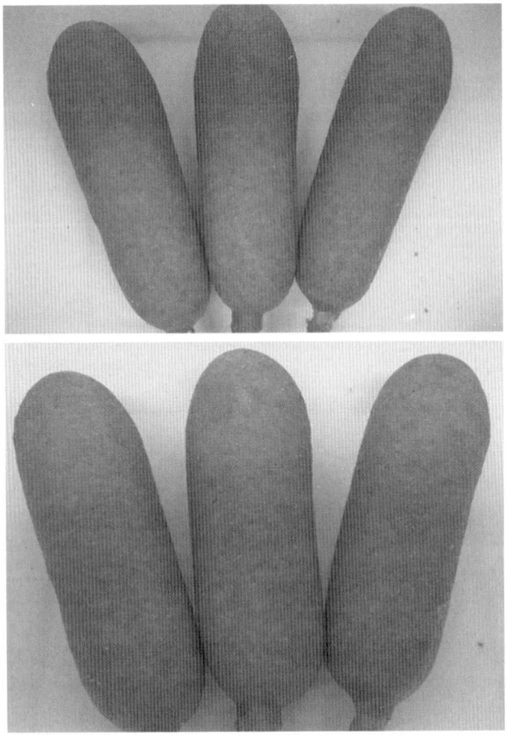

Figure 7.2 Corn dog batter.

Figure 7.3 Clear coated French fries.

7.2.3 Breadings

Breadings or crumbs are grain-based coatings which traditionally consist of cereal flours and many other ingredients. Common cereal-based ingredients in breadings include flour, breadcrumb, and cornmeal. However, breadings made up of breakfast cereals, pretzels, corn chips or whole grains are also becoming trendy in coated products. The type of breading influences the texture, color, flavor and oil pick up by the fried food.

Types of breading can be divided into a few groups: flour-based breading, sheeted breading (crackermeal, white bread crumbs), American bread crumbs (ABC) and Japanese bread crumbs (JBC).

7.2.3.1 Flour-based breading

The simplest breading is the flour breading. The most typical flour for breading is wheat flour. Various types of flours such as rice, corn and soy flours can also be used. Depending on the flour used, the final product may end up with a variety of texture, color and greasiness. Examples of flour-based breaded food products are shown in Figures 7.4 to 7.5.

7.2.3.2 Sheeted breadcrumbs

Typical ingredients in sheeted breadcrumbs include flour, shortening, salt, sugar, coloring agent and optionally a leavening agent. These ingredients are mixed with water to form denser dough, which is then forced through a series of rollers to form sheeted dough. The dough is conveyed to the oven for baking. The baking time varies depending on the dough thickness, temperature and moisture content. The baked sheet of dough is

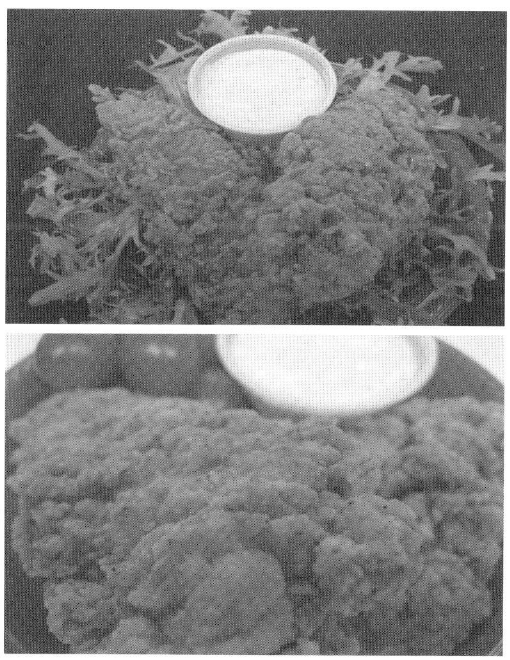

Figure 7.4 Chicken tenders using flour breading.

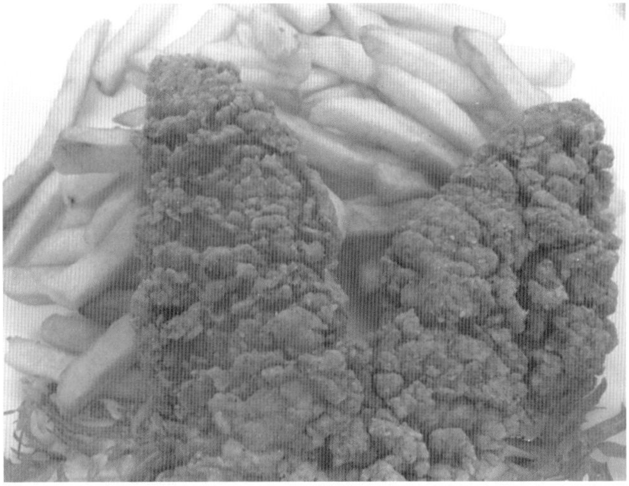

Figure 7.5 Fried chicken with flour breading and French fries with clear coating.

run through a granulating mill to achieve the desired particle size. The types of sheeted breadcrumbs include cracker meal and white breadcrumbs.

7.2.3.2.1 Cracker meals

Cracker meal breader is a simple system from a cracker-type formulation and common in the food industry. These are considered as the first generation of bread crumbs due to the simple formulation. They contain only two ingredients which are flour and water. Salts and sugars are added in some formulations. The dough may be fermented or non-fermented, and baked to approximately 30–35% moisture. When ground, these produce flat dense grains with a firm texture and the final moisture content of around 8–10%. Cracker meal breadings can be used alone or blended with other types of crumb to produce the desired texture. When used on fish products, it creates a hard, crunchy texture. As a functional ingredient, cracker meal is used as a binder, a thickener, and an extender. Cracker meal helps absorption of free moisture when applied to meat and vegetable products. Cracker meal in various applications is shown in Figures 7.6 to 7.7.

7.2.3.2.2 White breading

White breading gives a softer texture and is considered as the second generation bread crumbs due to its advanced formulation compared to the cracker meal breading. In addition to flour and water, they contain leavening agents, flavors, browning agents and salt. White beadings are commonly used as predust. In some applications they can be used in outer coatings with the addition of leavening agents, coloring agents and flavors to give crispy texture and more refined appearance.

7.2.3.3 American breadcrumbs (ABC)

American breadcrumbs come from fermented dough that is similar to white pan bread. The entire loaf is processed into crumbs and thus offers a gourmet-type appearance. The

Figure 7.6 Ravioli stuffing with cracker meal breading.

crumb is resistant to breakdown during processing; generally softer, porous and absorbs more oil and moisture than cracker meal. American breadcrumbs create crunchy texture and perform well on raw breaded, partially fried, oven-reconstituted and fryer-reconstituted products. It is available in fine to coarse particles with the average density of 20 to 28 pounds per cubic foot. Fine crumbs can be used in a variety of stuffing crumbs. American breadcrumbs are commonly used in seafood as a binder to retain moisture and to improve product quality. Examples of food products using ABC are shown in Figures 7.8 to 7.11.

7.2.3.4 Japanese-style breadcrumbs (JBC) / Panko

Japanese style breadcrumbs, also known as Panko breadcrumbs, originated in Japan in or about 1970. Panko crumb offers light appearance with crispy texture. Today, the gourmet breadcrumb of choice in many of the world's finest restaurants is the Japanese Bread Crumbs.

JBCs are produced by a process known as dielectric baking in which an electric current is passed through the dough to generate heat within the dough. The dielectric baking cooks the dough rapidly and leaves no crust, producing crumbs that are free from golden brown color. Proper formulation with a special mixture of flours and specific levels of salt is necessary to carry the current through the dough when cooked by dielectric technique. Insufficient salt levels do not cook the dough properly whereas higher salt levels will overcook (burn) the loaf. After baking, the loaf is shredded, dried and sorted according to particle size. The crumb resulting in this process is tender,

Figure 7.7 Meatballs with cracker meal breading in the binder.

Figure 7.8 Cheese sticks with American Breadcrumbs (ABC).

Figure 7.9 Chicken nuggets with American Breadcrumbs (ABC).

Figure 7.10 Onion rings with American Breadcrumbs (ABC).

elongated, and porous and provides crispier texture upon frying. Even though the standard crumbs are white, they can be colored to achieve a color range varying from yellow to caramel. JBC is commonly used as a shrimp breader because it offers tender and crunchy bite. This crumb is delicate and requires specific equipment for application to ensure uniform coverage.

JBCs can be formulated to deliver hard crumb, soft crumb, or fresh crumb (30% moisture). Nama panko, which translates to fresh breadcrumbs, is available in freshly frozen, partially dry and dry formulas. Dried nama panko is shelf stable and functions in a similar way to fresh nama panko upon reconstitution. Figure 7.12 to 7.13 show fish sticks and vegetables using Japanese style panko.

Figure 7.11 Shrimps with American Breadcrumbs (ABC).

Figure 7.12 Fish sticks with Japanese style panko.

Figure 7.13 Vegetables with panko breading.

7.2.4 Coating pick up

Batter viscosity plays a major role in determining the way the batter adheres to the substrate. The viscosity of batter is a critical coating characteristic that may affect the quantity and quality of batter pick up, appearance, texture, and the handling properties of the coated product (Mukprasirt, Herald and Flores, 2000). Coating pick up varies according to the type of coating. Generally, pick up rates are 5–7% for a predust, 15–18% for a batter, and 10–12% for a breading. However, the total pick up can vary between 20–60% depending on the fried product.

Total coating pick up is calculated as follows:

$$\% \text{ total pick up} = \frac{(\text{Final weight of coated substrate} - \text{Initial weight of substrate})}{\text{Final weight of coated substrate}} \times 100$$

According to Code of Federal Regulations (9-CFR-381.166) the term 'breaded' is applicable to any meat or poultry product which is coated with a batter and/or breader and amount should not exceed 30% of the weight of the finished breaded product. If the pick-up is in between 30–50%, the product must be classified as a fritter. If the pick-up exceeds 50% it cannot be classified as a meat product. Processors often classified these products as breaded products.

For seafood, raw breaded shrimp product should contain at least 50% shrimp, and the lightly breaded products should contain at least 65% shrimp. In general coated products for foodservice are fried and lightly coated (i.e. coated fish with 20–35% coating) than those are for retail (i.e. coated fish with 40–50% coating). Nevertheless, coated fish products with no standard of identity should contain at least 50% fish flesh.

A pass is defined as a wet step followed by dry step or in other terms, a batter application followed by a breading application. The number of passes required for a substrate depends on the nature of the substrate and also the final characteristics of the fried product.

Single pass application is typically used in poultry or red meat. In this process the substrate is dipped in batter and then coated with crumbs. The pick-up is regulated to be at 30% maximum. Double pass applications achieve approximately 35–45% pick up and are typically used for vegetables and seafoods in which the process is comprised of dipping the substrate in batter, coated with predust, dipped in batter again and coated with breading. Triple pass is done for substrates to provide additional coverage to prevent 'blowouts'. Blowout is defined as the formation of undesirable holes when excess moisture and/or gas escapes through a batter during frying. Blowout is typically caused by having large amounts of water in the coating. When placed in the fryer, the water creates a negative exchange of moisture and fat that takes off the coating. Thus products such as cheese sticks are given triple pass coating to eliminate this risk of blowouts.

7.3 INGREDIENTS IN COATING SYSTEMS

7.3.1 Flour and starches

Flours and starches provide a variety of functionalities in batters and breadings. They help with viscosity control, batter adhesion, moisture control and crispiness upon frying. Wheat flour is the main component of most batters and breadings. The amount of protein in the wheat flour and the availability of protein to bind on the product surface play an important role in the quality of the final product. Flours contribute to viscosity and promote adhesion by forming a glutinous network which is responsible for providing structure. A dry batter mix generally contains between 80–90% flour. Other flours such as rice, corn, soy, or potato may be added to modify texture as desired (Gerdes, 2001). For example rice flour or rice starch provides higher crispiness due to the smaller particle size distribution. Corn flour imparts a specific yellow coloration and improves crispiness. Potato flours typically provide brown hues upon frying. (Mukprasirt et al., 2002; Varela and Fiszman, 2011).

Treated flours make them more suitable for use in batter with improved functionality. Heat treatment of the flour has a positive effect on the viscosity and microbiological stability. Heat treatment either destroys or lowers the amylase activity. Alpha-amylase is the enzyme responsible for the breaking down of the starch that leads to lower viscosity. This also called batter thinning. The improved functionality achieved by the heat treatment ensures that the viscosity of the batter is high which translates to higher batter pick up and cohesiveness. Depending on the amount of protein and the type of treatment, wheat flour also influences the color of the final batter layer. Fully cooked cereals are also used in adhesion batters. In comparison with starch, wheat flour provides a superior adhesion on the product and contributes to a more cohesive structure in the final batter layer.

Modified corn starches are commonly used in coating systems. Other starches include modified wheat, potato, tapioca, and pea starches (Fiszman and Salvador, 2003). Modified starches (i. e., acetylated or hydroxypropylated starch) can add freeze-thaw stability to a par-fried product that is fried and then frozen. Granular cold water swelling starches help to increase the batter viscosity. The amount of modified starch content in formulations may vary from 5–40% depending on the application and the type of modified starch.

Starches with higher amylose content are better film formers and provide stiffness. These starches and their flours provide improved crispiness and extend shelf life under the heat lamp. High amylose starches have also been shown to help reduce fat pick up (Vongsawasdi, 2008).

Starch dextrins which are commonly used in clear coating formulas help retain extra crunchiness in French fries. Often dusting formulas contain 10–12% of dextrin. Excessive usage of dextrin in these formulas creates excessive browning and stickiness in fried foods (Shinsato et al., 1999; Baixauli, 2003).

7.3.2 Hydrocolloids

Hydrocolloids or gums help to suspend flour particles and to provide consistent viscosity in batters during processing. They also function as controlling moisture migration from the substrate to the coating, especially when the food is frozen. Gums work effectively with or without starch in the batter formulation. Hydrocolloids help the coating to adhere to the substrate and evenly spread around the product before it reaches the fryer.

Common hydrocolloids used in batters include xanthan gum, guar gum and methylcellulose. Xanthan gum is cold water soluble and builds batter viscosity which helps in adhesion. Xanthan gum will provide cling to the batter to help it adhere and prevent it from falling off. The viscosity built up by xanthan gum helps to improve the pumping ability of batters when they are transferred using pumps. In general, the particle size of xanthan gum influences the rate of viscosity build up. For example, finer particles around 75 microns produce instant viscosity. Typical usage levels of xanthan gum in batters range from 0.2–0.5% on a dry basis. Levels more than 0.5% dry basis creates chewy texture in fried foods in addition to thicker batter that may be difficult to apply on the substrate.

Guar gum functions as a thickening agent in batters. Guar gum is cold water soluble and finer particle size improves dissolution. In typical batter formulations, guar gum is used at levels less than 0.2%.

Methylcellulose provides adhesion by thermal gelation similar to egg whites and helps to keep the substrate moist. Because of its ability to gel at high temperatures, methylcellulose creates a barrier that prevents moisture from migrating out too rapidly. Therefore, fat migration into the product is reduced during frying with methycellulose barrier. Typical usage levels of methylcellulose vary from 0.10% to 0.40% in liquid systems and up to 0.80% dry basis (Garcia et al., 2002; Garcia et al., 2004; Myers, 1994).

7.3.3 Proteins

Proteins play an important role in batters by providing functional properties. In general, proteins are added at a level of 10% to 15%. Researchers have shown that products with higher protein contents are generally more effective as binding agents. Protein in flours helps hydration and improves water binding ability of flour which leads to higher viscosity of the batter system. Proteins act as emulsifiers by binding to both hydrophilic and hydrophobic areas of the substrate. This helps the substrate retain more moisture which in turn improves the moistness, the texture and the flavor of the fried product.

Upon frying, proteins coagulate or denature and help to stabilize the structure. Amino acids in proteins react with reducing sugars to form desirable colors through Maillard browning reaction giving the fried food its golden brown color and distinct savory flavor. Proteins such as egg and dairy powders are commonly used to improve adhesion. Eggs are used widely, both as a batter ingredient and as pre-dips. The protein in egg whites improves adhesion, while the phospholipids in yolk proteins provide emulsification. Baker and Scott-Kline (1988) showed that a predust composed of 24.5% of dried egg albumin reduced the fat absorption and retained higher percentage of moisture in coated chicken drumsticks.

Milk and whey proteins provide structure and the lactose in milk participates in browning reaction to develop color. Researchers have shown that 3% whey protein concentrate in a batter formula developed darker color and helped to reduce oil absorption significantly in chicken nuggets. Whey proteins concentrates either in predust or in batters also have been shown to increase the moisture retention under the heat lamp.

7.3.4 Chemical Leavening

Leavening acids react with bicarbonates and produce carbon dioxide gas when they come into contact with water. The gas forms bubbles that are trapped in the batter or dough and then expand during frying delivering expanded texture.

Neutralization reaction is as follows:

$$HA + NaHCO_3 \rightarrow NaA + H_2O + CO_2$$

Leavening acids differ depending on their level of reactivity and the temperatures at which they react. Reactivity depends mostly on solubility, which in turn depends on chemical composition, particle size, and special treatments such as coating. The leavening acids are categorized as slow, moderate or fast. The gas release characteristics of the leavening system affect the texture. Timing of the gas release affects coating adhesion.

Generally, baking soda and various phosphates are used as leavening agents in tempura batters whereas no leavening agents are used in adhesion batters. Table 7.3 contains

Table 7.3 Neutralizing values.

Leavening acid	NV*
Ascorbic acid	52
Sorbic acid	70
Sodium acid pyrophosphate	72
Lactic acid	93
Sodium aluminum sulfate	100
Malic acid	110
Adipic acid	115
Propionic acid	115
Fumaric acid	145
Citric acid	159

NV* Neutralizing value for sodium carbonate

the neutralizing values of various leavening acids. Neutralizing value is defined as the amount of bicarbonate that can be neutralized by 100 parts of the selected leavening acid.

Neutralizing value = amount of bicarbonate × 100 / amount of leavening acid.

7.4 PRINCIPLES OF DEEP FAT FRYING

Deep fat or immersion frying is one of the common food processing systems used worldwide. In deep fat frying, food is immersed in edible oil at a temperature above the boiling point of water (the temperature inside the food). The oil temperature may range from 170° to 190°C (Farkas et al., 1996). Frying helps to set the coating and provides additional flavor and texture to the coated product. The two major phenomena taking place during deep fat frying are heat and mass transfers that influence the texture, appearance and the nutritional properties of the fried food.

7.4.1 Heat transfer

In general, foods that are coated for deep fat frying have high moisture contents. For example, meat and fish products contain around 75% moisture, whereas, vegetables can carry up to 90% moisture. The heat from the oil transfers to the surface of the coated product when the coated product is introduced into hot oil. Absorption of heat by the water at the surface of the product brings about moisture evaporation.

Farkas et al. (1996) explained the heat transfer process that occurs during frying in four steps.

Initial heating. During this period, the temperature of the food surface increases to the temperature of the hot oil. This phase is generally considered as non-boiling and little moisture loss takes place.

Surface boiling. During this phase, the free water on the surface of the product starts evaporating and heat from the oil starts changing the surface of the product and begins the formation of the crust.

Falling rate. This phase is relatively longer than the other phases and many changes take place. The core temperature increases and moisture migrates from the core to the exterior of the food by evaporation. Simultaneously oil penetrates into the food. Other changes that take place during this process include protein denaturation, starch gelatinization and crust formation.

Bubble end-point. The final phase is called the bubble end point in which no heat transfer or boiling takes place. Thus, the rate of moisture loss decreases as well.

According to Aguilera and Hernández (2000), the heat transfer rate is the leading factor affecting continuous evaporation of water vapor at the food-oil interface, whereas diffusion of water from the core towards the exterior of the food is the leading process when evaporation decreases.

7.4.2 Mass transfer

Moisture evaporation from the food product creates a vacuum inside the food product. To fill up the vacuum, oil moves into the food product simultaneously. This phenomenon is characterized as mass transfer during frying. Mehta and Swinburn (2001) reported that fat uptake is predominantly determined by the moisture content of the food.

Coatings act as barriers for mass transfer during frying. Rapid heat transfer during frying creates a crust on the surface of the coating. This crust formation or the hardening of the surface locks moisture inside the coated product thus keeping the product moist while keeping the outer layer dry and crunchy. Faster setting of coating helps to reduce fat uptake and retain more moisture inside the coated product making it moister and less greasy. The crunchy texture influences the sensory properties of the product making it more appealing and desirable.

The greater part of oil uptake (about 80%) takes place after removing the fried foods from the fryer. Several researchers (Bouchon et al., 2003; Mellema, 2003) demonstrated that fat is hardly absorbed during frying but mostly absorbed through the cooling effect. During deep fat frying, the water vapors are released from the inside of the fried foods by small voids while the crust is formed. The microstructure of crust also plays an important role in determining oil uptake by fried food product. When the fried food is removed from the hot oil, air inside the voids cools down and develops a vacuum. This allows oil to penetrate from the surface. Therefore, it is necessary to allow the fat to drip off under the warm conditions to minimize the fat absorption after frying. Mass transfer influences the nutritional properties of the fried food. The fat content of the finished product may vary depending on the amount of moisture retained during frying and the amount of oil absorbed during or after frying. Higher fat absorption leads to poor nutritional value of the finished product.

7.5 FRYING OILS

Frying is one of the oldest cooking methods used in preparing foods. Frying oils play a great role in determining the flavor, appearance, nutritional value, quality, storage stability, handling and cost of the fried product. The oils and fats used in frying vary from fully refined products to specialty products designed for frying. However, commonly used frying oils include corn, peanut, soybean, cottonseed, or blends of these oils.

Prolonged heating of a frying medium will change its composition and finally leads to the thermal and oxidative degradation of the oil or fat. The degradation process varies depending upon temperature of frying and the duration of the deep fat frying process. Slight initial oxidation in frying oils is important to develop the sensory characteristic of fried foods. As oils breakdown further due to oxidation, hydrolysis, and polymerization, compounds are formed that can cause off flavors and may even be toxic if formed in high concentrations (Warner, 2007). Therefore, the frying process must be optimized to obtain desired sensory characteristics and to maintain conditions for the oil to last longer as much as possible. The main considerations are both to maintain high quality fried breaded foods and to minimize unnecessary cost by frequent oil replacement. In general, the temperature range for frying is 325° to 375°F. Most foods cook rapidly in the 325° to 375°F range and develop crispy texture, flavor and golden color. When the products are fried in the normal temperature range they absorb up to 25% oil. When higher temperatures of 375° to 400°F are used, the outer layer is cooked faster than the interior of the product and leads to thinner crust, which facilitates less oil absorption. Lower temperatures require longer frying time and results in less flavor development, lighter color, and increased oil absorption. In general the ratio of the deep fat fried food to the frying oil should be at or below 1:10, in order to avoid temperature fluctuations.

When selecting frying oils, one should consider the optimization of the process with regards to sensory and nutritional properties of the finished products as well as the cost of frying oil. During deep fat frying, a substantial portion of frying oil is absorbed by the food. Thus, the fatty acid composition of the deep fat fried food is determined considerably by the fatty acid composition of the deep frying medium. From the nutritionist point of view, the monounsaturated oleic acid, which can be found in olive and rapeseed oils, and polyunsaturated (linoleic or linolenic) fatty acids are preferred. Monounsaturated fatty acids have sufficient temperature stability. However, higher amounts of polyunsaturated fatty acids are prone to faster oxidative degradation and will affect the quality and the stability of the fried product. It is known that certain amounts of polyunsaturated fatty acids contribute to favorable flavors. However, if more than 3% of linolenic acid is present in frying oil, this could add fishy flavor to the fried product instead of the desirable flavor.

Saturated long chain fatty acids are more stable than the corresponding unsaturated fatty acids in terms of temperature and oxidation stability. Therefore, animal fats with a higher proportion of saturated fats are less prone to thermal or oxidative degradation. However, deep fat frying with animal fats is not desired since a high portion of saturated fatty acids can end up in the finished food product. Thus, deep frying with vegetable oils is preferred instead of the more heat stable animal fats. It may also be necessary to add stabilizing agents to maintain the quality of the vegetable oil to reduce the oxidative degradation. Oxidation of oils leads to rancidity of the product on storage. Use of packaging material with high oxygen and moisture barrier properties can significantly reduce oil degradation and increase the shelf life of packaged fried food.

Most vegetable oils are not stable when exposed to high temperatures. Therefore, vegetable oils had been partially hydrogenated to provide the stability. Partial hydrogenation of vegetable oils has been the major source of dietary trans fatty acids. Since the current law by FDA 21 CFR 101.9(c), (c)(2)(ii), (c)(2)(iii), and (c)(2)(iv), does not allow trans fats in foods, the manufacturers are looking into stable but non-hydrogenated oils to be used in frying. One alternative is the low-linolenic soybean oil, made from a specially bred soybean. This oil is less likely to go through oxidation degradation, and therefore does not need to be hydrogenated. Some other alternatives include palm oil which is naturally stable or blends of palm oil with soybean oil which are stable in frying application.

7.6 DESIGNING THE TEXTURE OF BREADED AND BATTERED FOODS

Texture is a property of food that is perceived by the mouth, tongue, palate, teeth or fingers. Consumers relate texture to food acceptability and quality. There are different descriptors of textural quality: crispiness (potato chip, chicken nugget), crunchiness (celery, corn flakes), chewiness (cookies, beef jerky), and creaminess (mayo, sauces and dressings, cake frosting). Each textural property can be further classified according to different levels or intensity.

There are two ways of evaluating texture – through the use of sensory panelists who will describe textural attributes, differentiate the degrees of textural differences (e.g., crispy versus very crispy; crispy versus crunchy) and rate their acceptability. Another way is through the use of an instrument to measure textural properties such as the texture analyzer. A texture analyzer is equipment that measures the force necessary to cut, shear, chew, compress or stretch food. The texture analysis protocol is done in such a way as to simulate ingestion of the food (i.e., chewing, cutting with the teeth, etc.) or as close to it as possible so the two methods can be well correlated. Food texture is one of the important key attributes in the acceptability of breaded and battered foods.

Breaded and battered food products have been popular ever since their introduction in the mid-1960s. From that time on, there has been a lot of attention devoted to development and improvement of breaded and battered foods as evidenced by numerous product introductions, publications on the subject, new and improved ingredients as well as equipment used to efficiently process or prepare them. There have also been developments in the improvement of the equipment used to objectively measure textural properties of breaded and battered foods.

Searching the topic of battered and breaded food products using Google came up with 1,520,000 results, while breaded food products resulted in 7,800,000 hits and battered food products, 19,000,000 hits, supporting the importance and huge interests on this type of food products both academically and commercially. Searching further using Google Scholar returned 5,110 hits which included publications and patents on the subject and 59 legal opinions (i.e., patent infringement cases) and journals. A search on battered and breaded food texture resulted in 197,000 hits.

Designing the texture of breaded and battered foods takes into consideration several important factors known to affect the final food quality. Important parameters include ingredients selection, employment of appropriate processes, packaging and storage or product handling after preparation or cooking. Texture design also takes into account the physicochemical effect of the process on the ingredients to achieve the desired food microstructure and hence product texture. For the selection of the right ingredients, reliance on structure-function relationship will help in zeroing in on the best ingredients rather than employment of trial and error method to achieve targeted food texture. There are basically two types of desired texture when it comes to breaded and battered foods – crispy and crunchy. Also considered in product formulation is the stability of food texture, that is, how long this texture would last when exposed to various conditions before consumption. Since most fried, baked or microwaved breaded and battered foods are not consumed immediately after cooking, it is desired that their texture will not change for a period of a few seconds up to several minutes after preparation. This is especially true for battered and breaded food products served to customers in a restaurant or food service setting.

7.6.1 Crispy versus crunchy versus 'crinchy'

Texture is a very important quality attribute for battered and breaded products. Humans measure texture using their senses which include touch (texture), eyes (appearance) and ears (sound). The textural profile of a crispy product is different from a crunchy product. This difference can be measured using sensory evaluation or a texture analyzer and a knife probe. One model of texture analyzer typically used is TA-XT2i Texture Analyzer by Stable Micro Systems. Figure 7.14 shows the typical probes that can be used to differentiate a crispy versus a crunchy texture. The textural profile of a crispy versus a crunchy battered and breaded chicken using a metal knife as a probe is shown in Figure 7.15. The peak force required for the crunchy product is 1400 grams occurring at about 4.5 seconds while the crispy product required only about 900 grams of force which occurred much earlier at 2.8 seconds. The area under the curve (g-sec) for the crunchy product is almost twice that for the crispy product. Thus there is significant difference in the peak force and the work or energy required to cut through the same thickness of crispy as compared to a crunchy battered and breaded chicken.

Table 7.4 summarizes selected work on battered/breaded and other related food products including variables tested and the method used to assess textural properties.

In addition to the actual force needed to cut through the product, Corn Products/National Starch Food Innovation considers the sounds made when a consumer bites into the product (Gelski, 2012). The crunchy or crispy sounds can influence product consumer acceptance. Crunchy products tend to be louder, have a longer duration of sound (as chewing) and a lower pitch, whereas crispy products tend to be quieter, dissolve more quickly and have a slightly higher pitch (Gelski, 2012). 'Crinchy' according to Corn Products/National Starch's Texicon, is in between crunchy and crispy sounds and may well describe the textural properties of some battered and breaded products especially those with baked cereal breadings.

7.6.2 Adhesion in textural design

Adhesion of the coating to the meats or vegetables is one of the important steps in a well designed batter and breading formulation. Adhesion is the binding of the coating to the food substrate it coats which occurs through chemical and physical means. Poor adhesion will result in poor texture of the crust after frying. Early research in the 1960s

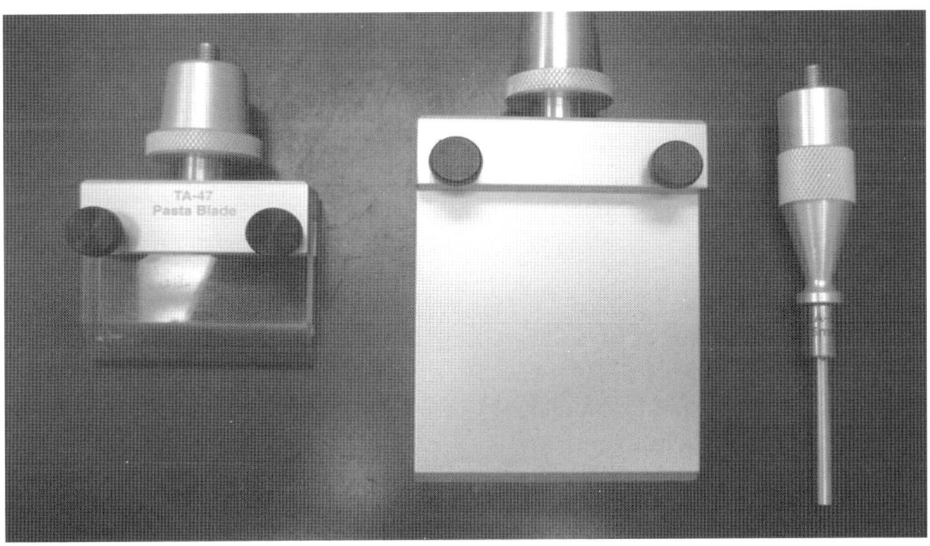

Figure 7.14 From left to right acrylic blade, metal blade and punch probe. These are the probes for TA.XT used to measure the texture profile of battered and breaded food products.

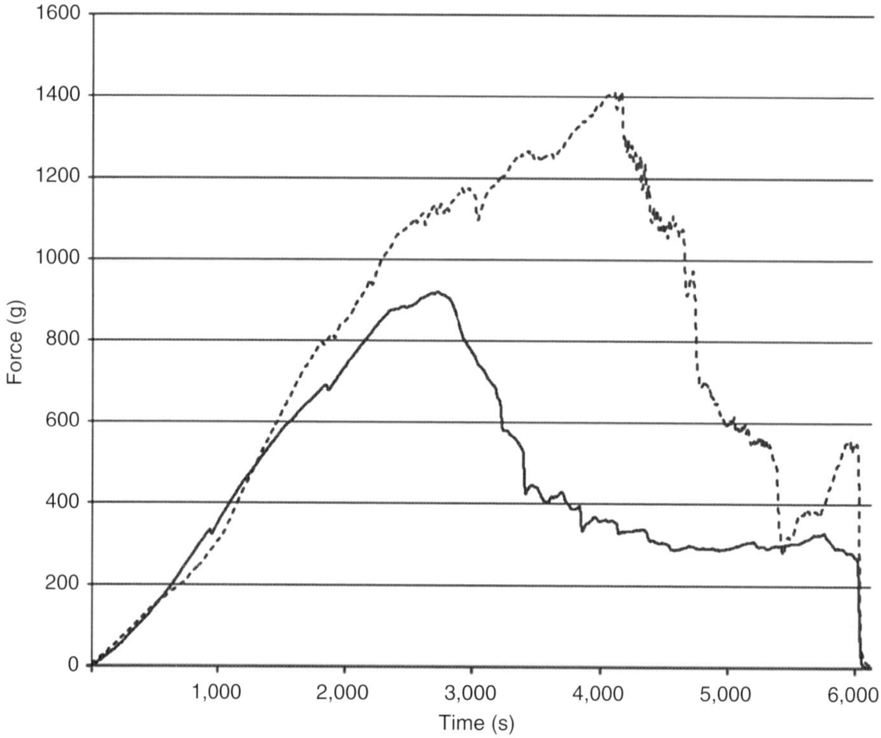

Figure 7.15 Comparison of texture profile of crunchy (dash line) with crispy (solid live) battered and breaded chicken strips.

focused on pre-cooking methods to improve adhesion. After this period, research scientists worked on selection of the right ingredients or blending the right ingredients to achieve batter properties that will enhance binding adhesion. In the early 1980s, when the demand for battered and breaded food products was increasing, several studies dealt with the effects of proteins, gums, phosphate predips and freezing of substrate on adhesion. Suderman et al. (1981) investigated the effects of proteins (whey, soy, non fat dry milk, egg albumen and gelatin) and gums (sodium carboxymethyl cellulose or CMC, guar, tragacanth and xanthan) on adhesion of coating to poultry skin. They found that gelatin and egg albumen significantly improved adhesion while among the gums, only CMC effectively improved adhesion. Proteins were better than gums in improving adhesion. In another study, Suderman and Cunningham (1981) found that freezing poultry meat prior to breading application affected coating adhesion. Frozen storage and the use of phosphate predips as they affect coating adhesion was studied by Corey et al. (1987). Both adhesion and coating were not affected when breaded fish portions was dipped into phosphate predip solution then subjected to three freeze-thaw cycles during a three-month storage period.

In 1992, Hsia et al. studied the effect of three hydrocolloids on the rheological properties and adhesion characteristics of flour-based batters on chicken nuggets. Flour-based batters with 30% solids and containing guar, xanthan, or CMC at various concentrations (0.25, 0.5 and 1.0%) were used in the preparation of chicken nuggets. Percent coating

Table 7.4 Summary of selected research work on battered/breaded and related food products and assessment of textural quality.

Food product	Variables tested	Findings	Textural assessment	Author
Pork nuggets	Low fat pork nuggets with gums and modified starch	Sodium alginate with calcium-lactate or corn starch increased yield and tenderness	Instron shear measurement	Berry, B.W. (1994)
Chicken drumstick	Rice flour-based batter, yellow corn flour, oxidized cornstarch, methylcellulose, xanthan	Microstructural analysis showed that batter formulated with 50:50 mixture of rice and corn flours adhered better to drumstick. Rice flour showed potential alternative to wheat flour.	Laser scanning confocal microscopy TA.XT2 texture analuzer	Mukprasirt et al. (1999)
Battered squid rings	Corn flour, salt & leavening	Leavening affected texture of squid rings	TA.XT2 texture analyzer	Salvador, A. et al. (2002)
Breaded fried chicken nuggets	Correlate objective measurements (ultrasonic parameters and mechanical properties) with sensory crispness	Ultrasonic velocity had high correlation with sensory crispness	Trained sensory panel Ultrasonic nondestructive evaluation system developed at Virginia Tech	Antonova, I. et al. (2003)
Breaded fried chicken nuggets	Effect of edible coating and pressure frying using nitrogen on quality	Edible coatings and pressure has a significant effect of quality attributes of breaded fried chicken nuggets	Ultrasonic nondestructive evaluation system developed at Virginia Tech Kramer shear press apparatus	Ballard, S.B. & Mallikarjunan, P. (2006)
Bread crust	Effect of water content and water activity on the glass transition of model bread crust	Water content of the crust was found to be affect transition point as measured by PTA and NMR; both water content and water activity had an effect on perceived crispness	Phase transition analysis (PTA) Nuclear magnetic resonance (NMR)	Van Nieuwenhuizen, N.H. et al. (2008)
Horse mackerel mince	Effect of ingredients and flour/water ratio, gluten, hydroxypropylmethylcellulose, cornstarch, crisping agent.	The S- and SH-patterm crusts exhibited a firm and brittle texture, the W- and WC-pattern crusts exhibited a crispy texture (S- and SH-pattern exhibited Bingham-pseudoplastic, W and WC-pattern exhibited psuedoplatics fluidity)	Sun Rheometer (R-150, Sun Scientific Co., LTD, Tokyo)	Chen, H., Kang, H. & Chen, S. (2008)

(*Continued*)

Table 7.4 *(Cont'd)*

Food product	Variables tested	Findings	Textural assessment	Author
Crusted foods – chicken nuggets	Method development in crispness assessment – effect of cooking method	Use of blade probe (not sharp) proved to be an effective technique; differences seen in deep fried versus oven cooked nuggets	TA.XT plus	Varela, P. et al. (2008)
Battered and breaded chicken patties	Effect of whey protein isolate as a postbreading dip	Use of whey protein concentrate did not affect the flavor but affected color, hardness and crunchiness but resulted to reduced fat content	Descriptive sensory analysis Penetrometry tests using TA.XT2i	Mah, E. & Brannan R.G. (2009)
Chicken breast	Effect of microwave drying and different types of flours (chickpea, rice or corn)	Type of frying affected the microstructure of fried batter depending on the type of flour used	TA.XT2i SEM	Barutcu, I. et al. (2009)

pickup increased significantly with increasing hydrocolloid concentration. Addition of xanthan gum at 0.5% and 1.0% also increased the percentage of cooked yield by 16% and 30%, respectively. According to Hsia et al. (1992), addition of hydrocolloids that will increase apparent viscosity and adhesion will improve product yield but the right gums must be selected (i.e., those gums which exhibit lower time-dependency), to ensure uniform batter pickup independent of mixing time.

Salvador et al. (2002) worked on the effect of corn flour, salt and leavening on the texture of fried, battered squid rings. Addition of salt significantly reduced the batter viscosity compared but batter pickup was not significantly affected. Overall, the leavening agent has the greatest effect on the final texture of the coatings as measured by the TA-XT2 Texture Analyzer. The leavening agent contributed to the crispness and tenderness of the fried outer crust.

In 2009, Barutcu et al. conducted a study to determine the effect of various types of flours on the microstructure of battered products during conventional and microwave frying. They used chickpea, rice or soy flours and replaced 30% of the corn and wheat flour mix. It was postulated that flours with different starch and protein composition will influence the microstructures of the batter coatings and their combination may provide desirable effects. Protein and starch content are the main contributors to flour functionality in batter. The development of the microstructure of batter coatings is brought about by the denaturation of the proteins and gelatinization of the starch during frying. The batter microstructure after frying will affect its texture, fat absorption and moisture loss. These in turn will have implication on the sensory and textural properties of the product as well as its nutritional value. The texture of samples prepared using deep frying and microwave frying were compared. The appearance of the inner and outer structure of microwave fried and conventionally deep fried batter was different, with the microwave fried samples containing vented holes on the inner surface. Microwave fried samples have lower hardness values as measured by the Texture Analyzer TA-XT2i than the conventionally deep fried samples depending on the flour used. Batter formulation containing chickpea has the lowest hardness value followed by soy flour.

Baur et al. applied for a patent in 2001 on formulated battered and battered/breaded foods with enhanced textural characteristics brought about by using cereal-based batter containing sodium caseinate with no egg albumen added. The addition of 0.3% sodium caseinate produced a crispy battered fish as compared to a slightly tough battered fish when sodium caseinate was used in combination with 0.3% egg white. The control sample was described as slightly crispy.

7.6.3 Advances in coating systems

Fried foods are typically associated with an unhealthy image due to their high fat content. Nevertheless fried foods provide added sensory appeal to many consumers. Consumers demand healthier, fried breaded foods with sensory properties similar to those of traditional fried breaded products. Formulators are trying to develop healthier alternatives by introducing specialty coatings with reduced oil absorption, either by using specialty frying oil or altering the processes to minimize the moisture loss and thereby minimizing the fat absorption. Other requirements for healthier coating systems may include natural, organic, reduced calorie, use of whole grains, gluten-free and allergen-free options.

7.6.3.1 Coatings to reduce fat uptake

Coatings for fried products that help reduce fat uptake are an interesting and challenging task among formulators. Greasy battered and breaded product is not typically desirable. Moisture evaporation from the substrate during frying creates capillary pores in the substrate. To fill up the capillary pores with newly created vacuum, oil penetrates into the substrate through these pores. Thus, microstructure of the crust plays a major role in determining oil uptake (Pinthus et al., 1995). Studies have shown that there is a direct relationship between the moisture loss and the amount of fat absorption during frying (Rice and Gamble, 1989). Fat uptake is also influenced by the surface area to mass ratio of the fried food. For example, the fat content of French fries is from 6–13%; onion rings, 10–15%; donuts, 17–23%; and potato chips 28–38%.

Several studies have shown that the coatings can be manipulated to influence the mass transfer during frying and thereby reduce moisture loss and subsequently, fat absorption. Mellema (2003) showed that hydrophilic biopolymers can be used as water binders in a coating to reduce water loss from the coat. Polysaccharide coatings provide firm gel strength through thermal gelation or cross-linking. The resulting high gel strength leads to less evaporation damage and hence to a lower water diffusivity.

Methylcellulose is one of the commonly used hydrocolloids in the food industry and has been shown to be an effective edible film former and resistant to oil and oxygen transmission. The thermal gelling properties of methylcellulose help to reduce the final oil uptake of fried foods by forming gelatinous coatings upon frying. Salvador and co-workers (2007) showed that 0.5–1% levels of methyl cellulose in batters can reduce the oil uptake of fried food products.

Williams and Mittal (1999) reported that coatings with methylcellulose reduced the fat content of potato spheres by 34.5% compared to control. This result was attributed to the presence of methylcellulose acting as a fat barrier upon thermal gelation.

According to García et al. (2004), an edible methylcellulose coating, showing thermal gelation properties, was applied to reduce oil uptake during frying of a dough system formulated with wheat flour. The coating formulation included 1% methylcellulose and 0.75% sorbitol as plasticizer. Dough samples were dipped in the coating suspensions and immediately fried. For coated dough discs, the oil uptake reduction was 30% compared to uncoated discs. These results demonstrated the effectiveness of methylcellulose as an oil barrier.

Sahin and co-workers (2005) studied the effect of different gums on oil absorption reduction in fried chicken nuggets. The batter dry mix consisted of 48.75% wheat flour, 48.75% corn flour, 1% gum, 1% salt and 0.5% leavening agent. The water to solid ratio was at 3:5. The batter pick up by the gums are given in Table 7.5. At 1% addition gum Arabic did not increase the batter pick up significantly compared to the control. Upon frying, addition of xanthan and HPMC significantly reduce the oil content of fried chicken.

Konjac flour (H&A Canada, Inc. or TIC Gums) is another versatile ingredient that contributes to a unique texture of battered and breaded foods. Figure 7.16 shows the texture profiles of battered and breaded chicken strips with added konjac flour or selected modified starch in a control batter. The texture profile shows a crunchy texture when 10%

Table 7.5 Effect of gum types on coating pick up of deep fat fried chicken nuggets during frying. (Source: Sahin et al. 2005.)

Formula	% coating pick up
Control	29
Guar gum	47
Xanthan gum	49
HPMC	47
Gum Arabic	25

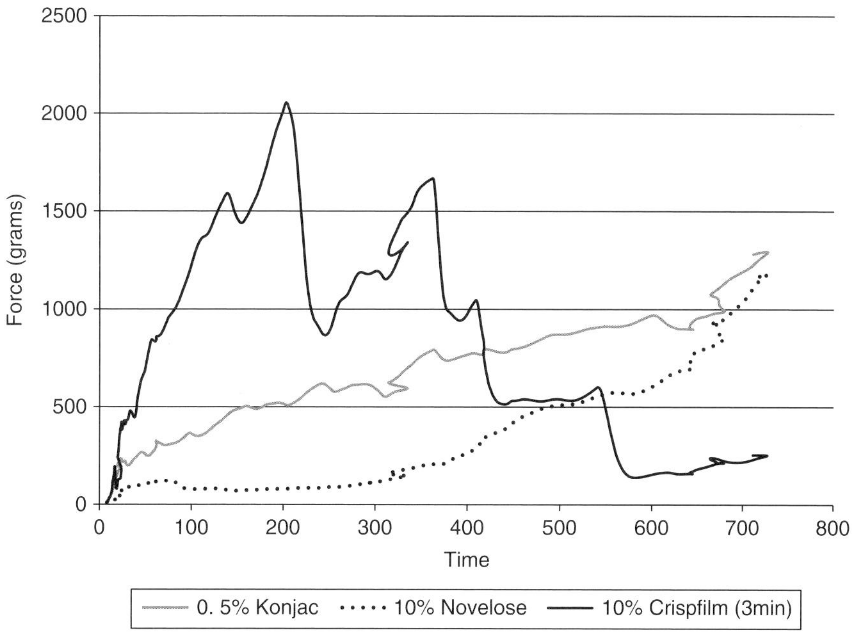

Figure 7.16 Texture profiles of battered and breaded chicken strips with the addition of konjac flour or selected modified starch in the control batter.

CrispFilm (Ingredion) is added to a control batter. This texture is characterized by high peak force that occurred early on including a few peak forces while the knife probe was cutting through the product. The addition of konjac flour or Novelose (Ingredion) resulted in crispy battered and breaded chicken strips. This is shown by a smoother curve as the knife cuts through the product (Figure 7.16). The peak force needed is much lower than that required when Crispfilm was used.

There is an additional advantage of using gums such as konjac flour or modified starch. Table 7.6 shows the increase in batter pick up before and after frying as compared with a control batter. There was a significant increase in batter pick up before and after frying when konjac flour was added in the batter. There was also a significant increase in the batter pick up before and after frying when methylcellulose was added to a control predust.

Table 7.6 Effect of starch or gum on batter pick up before and after frying.

Batter composition	Predust composition	% batter pick up before frying	% pick up after frying
Control	Control	62.81	28.38
Control with 0.5% Konjac flour	Control	115.86	74.77
Control with 8.0% CrispFilm	Control	73.37	39.77
Control	Control with 5% methylcellulose	71.21	38.01
Control with 0.5% Konjac flour	Control with 5% methylcellulose	86.33	50.12
Control with 8.0% CrispFilm	Control with 5% methylcellulose	68.22	33.8

Control (batter): soft wheat flour, corn flour, modified starch, salt, spices, sodium phosphate, sodium bicarbonate
Control (predust): hard wheat flour, salt, modified starch, spices, wheat gluten, egg white solids, sodium phosphate, sodium bicarbonate

Table 7.7 American breadcrumb formula with resistant dextrin.

Ingredient	percent
Wheat flour	46.9
Active dry yeast	1.4
Salt	0.9
Resistant dextrin	11.2
Vegetable oil	2.2
Sucrose	2.1
Wheat gluten	4.4
Water	30.9
Total	100.0

Polysaccharides with film forming properties, for example, resistant dextrin and high amylose starch, have also shown reduced fat diffusivity when applied in coating systems. Resistant dextrin derived from dextrinization of either corn or wheat starch has shown to reduce fat uptake in fried fish and chicken nuggets by 12–35% depending on the usage level in the coating system. For example, in a coating system for chicken nuggets which contained 33% of resistant dextrin resulted in 35% reduction in fat uptake, whereas 12.5% reduction in fat uptake was observed when 25% of breader was replaced with resistant dextrin in a commercial coating system (consisting of a predust, batter and a breader). Coatings with resistant dextrin provided improved crunchiness and color in fried chicken nuggets and extended the time under heat lamp. However, higher levels of resistant dextrin in a breading lead to excessive browning (Perera, 2011).

American breadcrumb containing 11% resistant dextrin was developed for fish nuggets using the formula given in Table 7.7. Fish nuggets were coated with the breading with resistant dextrin and the control was coated with similar type of breading without resistant dextrin. The result showed that the fish nuggets coated with breading containing resistant dextrin had 28% lower oil uptake and was crunchier than the control (Perera, 2011).

Whey protein isolate was shown to have positive effects on fat absorption reduction when applied in coatings. Mah and Brannan (2009) conducted a study with whey protein isolate (0–10%w/w) in a post-breading dip for chicken patties coated either with

Japanese breadcrumbs or crackermeal breadcrumbs. They found that the patties coated with Japanese breadcrumbs dipped in 10% whey protein concentration provided the least amount of fat absorption upon frying.

7.6.3.2 Gluten-free coatings

According to Packaged Facts, the market for gluten-free foods and beverages has continued to grow even faster than anticipated, reaching $4.2 billion in 2012, for a compound annual growth rate of 28% over the 2008–2012 period. A consumer survey by Packaged Facts in August showed that 18% of adults are buying or consuming foods products labeled as gluten-free, up from 15% in October 2010. The major reason for the increased sales of gluten free products is that the consumers believe that gluten-free products are generally healthier than products with gluten (Packaged Facts, 2012).

In addition to the gluten free ingredients such as potato, rice, corn and tapioca, pulses (peas, lentils and beans) are gaining momentum as a valuable source of nutrients in the diet, due to their high protein, starch, fiber, vitamin and mineral content. The proteins in peas, beans and lentils do not contain glutens that cause celiac sprue. These nutritional and health attributes have positioned pulses for use in the rapidly growing markets of nutritionally enhanced and gluten-free foods.

Ingredient manufacturers and government agencies have explored the use of pulses and their fractions in coating systems for fried foods. Fractions of flour, protein, starch and fiber from yellow pea have been extensively studied in batters and breading to replace traditionally used wheat flour, corn flour and their starches. The results have shown that these pulse fractions could provide functional and nutritional benefits in the coating systems for fried products.

Pea starch and pea protein were used in a gluten-free crumb breading formula for breaded fish (Table 7.8). The control formula consisted of wheat flour and corn starch. Findings showed that fried breaded fish with pea protein and pea starch coating was significantly crispier and more golden in color compared to the coating systems which contained wheat flour and corn starch.

Gluten-free breading is an emerging market. The ingredients used for gluten-free breading include mixtures of starches and flours from grains such as rice, oats, amaranth, flax and millet and proteins from non-allergen sources such as pulses.

Table 7.8 Crumb breading with pea starch.

Ingredient	Percent
Pea starch	42.00
Pea protein	7.30
Vegetable oil	3.12
Sucrose	2.08
Active dry yeast	1.46
Salt	0.83
Xanthan gum	0.71
Water	42.50
Total	100.00

Another beneficial aspect of pulse-based ingredients is that they are well suited for clean labeling. Clean label may refer to having natural, organic, minimally processed ingredients or eliminating additives or chemically modified ingredients, all of which have negative consumer perception. Pea fractions such as flours and fibers can be used to increase batter viscosity and have the potential to replace modified starches or gums used as thickeners in the batter. Pea fiber can function as a thickener and absorbs water up to about nine times its weight. Unlike native starches, pea fiber does not lose its viscosity upon heating. Thus, pea fiber can be used to replace some modified starches that are commonly applied in batter formulation to achieve clean labeling.

7.6.3.3 Whole grain breading

Whole grains contain all the essential parts and naturally occurring nutrients of the entire grain seed. Whole grains are ideal for use in breaded products. Whole grain breading provides added texture, color, flavor and nutrition to the fried foods. However, whole grain breading may be better suited for the baked breaded products than those which are designed to be fried since frying can easily destroy micronutrients compared to baking. Furthermore, frying can adversely influence the color of whole grain coated product.

7.6.4 Methods for reducing fat uptake

Use of specialty coating systems to reduce oil uptake was discussed in a previous section. In addition to coatings, manufacturers are also looking into processing methods to manipulate the oil absorption on frying.

Kock and co-workers (1995) reported that the method of freezing fries had an important impact on the fat content and other qualities after they were deep fried. In practice, French fries are frozen either by rapid freezing (cryogenic freezing) or by a slow process of mechanical freezing. Kock et al. (1995) compared the quality of fries that had been frozen by these two different methods. Water lost was minimal during cryogenic freezing compared with the water lost caused by the mechanical freezing. Thus, cryogenically frozen fries had significantly higher moisture content (73%) than the mechanically frozen fries (67%). This can be explained by the structural changes taking place during freezing. Cryogenic freezing involves rapid freezing which does minimal damage to the surrounding tissues and produces minute ice crystals whereas mechanical freezing is a slow process and creates large ice crystals that damage the surrounding tissues. During frying, some of the moisture is removed, and more fat is absorbed into the holes left by the ice crystals. Big ice crystals are replaced by more fat. Thus, cryogenically frozen fries have better texture and better organoleptic qualities with lower fat content.

Acknowledgment

We thank Saulo Embuscado in the preparation of some figures.

References

Aguilera, J.M. and Hernández, H.G. (2000) Oil absorption during frying of frozen parfried potatoes. *Journal of Food Science*. 65, 476–479.

Antonova, I., Mallikarjunan, P. and Duncan, S. (2003) Correlating objective measurements of crispness in breaded fried chicken nuggets with sensory crispness. *Journal of Food Science*. 68, 1308–1315.

Baker, R.C. and Scott-Kline, D. (1988) Development of a high protein coating batter using egg albumen. *Poultry Science* 67, 557–560.

Ballard S.B. and Mallikarjunan, P. (2006) The effect of edible coatings and pressure frying using nitrogen gas on the quality of breaded fried chicken nuggets. *Journal of Food Science*. 71, S259–S264.

Baur, J., Darley, K., Janda, J., Martin, J., Bernacchi, D., and Donhowe, I. (2001) Battered and battered/breaded foods with enhanced textural characteristics. US Patent 6,288,179 B1.

Barutcu, I., Sahin, S., and Sumnu, G. (2009) Effect of microwave frying and different flour types addition on the microwave of batter coatings. *Journal of Food Engineering*. 95, 684–692.

Baixauli, R., Sanz, T., Salvador, A. and Fiszman, S.M. (2003). Effect of the addition of dextrin or dried egg on the rheological and textural properties of batters for fried food. *Food Hydrocolloids*. 17, 305–310.

Berry, B.W. (1994) Properties of low-fat, nonbreaded pork nuggets with added gums and modified starches. *Journal of Food Science*. 59, 742–746.

Bouchon, P., Williams, R. and Mittal, G. (1999) Low-fat fried foods with edible coatings: modeling and simulation. *Journal of Food Science*. 64, 317–322.

Chen, H., Kang, H., and Chen, S. (2008) The effects of ingredients and water content on the rheological properties of batters and physical properties of crusts in fried foods. *Journal of Food Engineering*. 88, 45–54.

Corey, M., Gerdes, D., and Grodner, R. (1987) Influence of frozen storage and phosphate predips on coating adhesion in breaded fish portions. *Journal of Food Science*. 52, 297–299.

Farkas, B.E., Singh, R.P. and Rumsey, T.R. (1996) Modeling heat and mass transfer in immersion frying, I. Model development. *Journal of Food Engineering*. 29, 211–215.

Fiszman, S.M. and Salvador, A. (2003) Recent developments in coating batters. *Trends in Food science & Technology*, 14, 399–407.

Garcia, M.A., Ferrero, C., Bertola, N., Martino, M., and Zaritzky, N. (2002) Edible coatings from cellulose derivatives to reduce oil uptake in fried products. *Innovative Food Science and Emerging Technologies*. 3, 391–397.

García, M.A., Ferrero, C., Campana, A., Bertola, N., Martino, M. and Zaritzky, N. (2004) Methylcellulose coatings applied to reduce oil uptake in fried products. *Food Science and Technology International*. 10, 339–342.

Gelski, J. (2012) *Crispy to crunchy: Formulate texture to achieve a desired sound in baked snacks*. FoodBusinessNews.net. [Online], Available: http://www.foodbusinessnews.net/News/News%20Home/Features/2012/3/Crispy%20to%20crunchy.aspx [27 March 2012].

Gerdes, S. (2001) Batter and breading liven taste. *Food Product Design*, 12, 129–131.

Hsia, H., Smith, D. and Steffe, J. (1992) Rheological properties and adhesion characteristics of flour-based batters for chicken nuggets as affected by three hydrocolloids. *Journal of Food Science*. 57, 16–18, 24.

Kock, S.D., Minnar, A., Berry, D., and Taylor, J.R.N. (1995) The effect of freezing rate on the quality of cellular and non-cellular par-cooked starchy convenience foods. *Lebensmittelwissenschaft und Technologie*. 28, 87–91.

Loewe, R. (1990) Ingredient selection for batter systems. In *Batters and Breadings in Food Processing* (eds K. Kulp and R. Loewe), pp. 11–28. AAAC: St. Paul, MN.

Mah, E. and Brannan, R.G. (2009) Reduction of oil absorption in deep fried battered and breaded chicken patties using whey protein isolate as a postbreading dip: Effect on flavor color and texture. *Journal of Food Science*. 74, S9–S16.

Mehta, U. and Swinburn, B. (2001) A review of factors affecting fat absorption in hot chips. *Critical Reviews in Food Science and Nutrition*. 41, 133–154.

Mellema, M. (2003) Mechanism and reduction of fat uptake in deep-fat fried foods. *Trends in Food Science & Technology*. 14, 364–373.

Mukprasirt, A., Herald, T.J., and Flores, R. (2000) Rheological characterization of rice flour-based batters. *Journal of Food Science*, 65(7), 1194–1199.

Mukprasirt, A., Herald, T., Boyle, D. and Raisch, K. (2000) Adhesion of rice flour-based batter to chicken drumstick evaluated by laser scanning confocal microscopy and texture analysis. *Poultry Science.* 79, 1356–1363.

Mukprasirt, A., Herald, T.J., and Flores, R.A. (2000) Rheological characterization of rice flour-based batters. *Journalf Food Sci.* 65, 1194–1199.

Myers, M.A. (1994). Functionality of hydrocolloides in batercoating systems, batters and breadings in food systems. *American Association of Cereal Chemists*, 117–141.

Packaged Facts (2012) Gluten free foods and beverages in the US. Fourth edn. Oct 17th , Pub ID: LA4880536.

Perera, C. (2011) Formulating batters and breading for reduced fat absorption in fried foods. *R & D Seminar Chicago, IL* August 2–3.

Pinthus, E.J., Weinberg, P., and Saguy, I.S. (1995) Oil uptake in deep fat frying as affected by porosity. *Journal of Food Science.* 60, 767–769.

Rice, P. and Gamble, M. (1989) Technical note: modeling moisture loss during potato slice frying. *International Journal of Food Science Technology.* 24,183–187.

Sahin, S., Sumnu, G. and Altunakar, B. (2005) Effect of batters containing different gum types on the quality of deep fried chicken nuggets. *Journal of Science, Food and Agriculture.* 85, 2375–2379.

Salvador, A., Sanz, T., and Fiszman S. (2002) Effect of corn flour, salt, and leavening on the texture of fried, battered squid rings. *Journal of Food Science.* 67, 730–733.

Salvador, A., Sanz, T. and Fiszman, S. (2007) The performance of methylcellulose in coating batters of fried foods. *Journal of Food Hydrocolloids.* 5, 15–19.

Shinsato, E., Hippleheuser, A.L., and Van Beirendonck, K. (1999) Products for batter and coating systems. *The World of Ingredients*, (January/February), 38–42.

Singh, R.P. (1995) Heat and mass transfer in foods during frying. *Food Technology.* 49(4), 134–137.

Suderman, D., Wiker, J., and Cunnigham, F. (1981) Factors affecting adhesion of coating to poultry skin: Effects of various protein and gum sources in the coating composition. *Journal of Food Science.* 46, 1010–1011.

Suderman, D. and Cunningham, F. (1981) Effect of freezing broiler drumstick on breading adhesion. *Journal of Food Science.* 46, 1953–1955.

Van Nieuwenhuizen, N.H., Primo-Martin, C., Meinders, M., Tromp, R., Hamer, R. and van Vliet, T. (2008) Water content or water activity: What rules crispy behavior in bread crust. *Journal of Agriculture and Food Chemistry.* 56, 6432–6438.

Varela, P., Salvador, A., and Fiszman, S. (2008) Methodological developments in crispness assessment: Effects of cooking method on the crispness of crusted foods. *Lebensmittel Wissenschaft Technologie.* 41, 1252–1259.

Varela, P. and Fiszman, S. (2011) Hydrocolloides in fried foods: A review. *Food Hydrocolloides.* 25, 1801–1812.

Vongsawasdi, P., Nopharatana, M., Srisuwatchree, W., Pasukcharoenying, S. and Wongkitcharoen, N. (2008) Using modified starch to decrease the oil absorption in fried battered chicken. *Associated Journal of the Food Ag-Industry.* 3, 174–183.

Warner, K.A. (2007) Frying oil deterioration. In: *Food Lipids: Chemistry, Nutrition and Biotechnology* (ed. C. Akoh), pp. 71–82. Taylor and Francis Company, Boca Raton, FL.

Williams, R., and Mittal, G.S. (1999) Low-fat fried foods with edible coatings: Modeling and simulation. *Journal of Food Science.* 64, 317–322.

8 Multi-textured foods

Arun Kilara[1] and Tapashi Sengupta*[2]

[1] *Nutri + Food Business Consultants, Chapel Hill, North Carolina, USA*
[2] *PepsiCo Advanced Research, Plano, Texas, USA*

8.1 INTRODUCTION

It is said that variety is the spice of life. When our ancestors discovered fire and began using it to cook food, new textures and flavors began to be introduced to humanity. When hunter gatherers started putting down roots, domesticating animals and growing crops, further advances were made in creating new textures and flavors. Now it is a routine part of every cuisine across the globe to introduce variety into the human diet through textures and flavors. Creation of textures involves various techniques such as drying, freezing, baking, leavening, emulsification, aeration, boiling, frying, roasting, braising and so on, singly or in combination with other techniques. Other common cooking techniques such as slicing, dicing, chopping, grinding and other particle size reduction technologies do also result in altering texture of foods. Some methods to change the texture also render the food safe for consumption by destroying anti-nutrients inherently present in the raw food. Cooking of food by any technique can also render the food microbiologically safe, provided that post cooking contamination is prevented.

The reasons why we consume foods vary. The primary function is to provide nourishment to our bodies through the consumption of water, proteins, fats, carbohydrates, vitamins and minerals. This allows our bodies to perform well, fight diseases and frees humans to pursue happiness. Eating, in the early evolution of humans, was a means to avoid starvation as famines were endemic in our history. With the advancement of civilization humans have moved away from consuming foods for nourishment to consuming foods for pleasure. In fact obesity is becoming a global epidemic and incidences of diseases related to obesity are increasing at an alarming rate. So food consumption is being examined to determine how it can play a role in reversing the trends in the incidence of diseases.

Food texture was defined by Bourne (2002) as 'The textural properties of a food are that group of physical characteristics that arise from the structural elements of the food, that are primarily sensed by the feeling of touch, are related to the deformation, disintegration and flow of the food under a force, and are measured objectively by functions of mass, time

* Tapashi Sengupta is an employee of PepsiCo.Inc. The views expressed in this chapter are those of the authors and do not necessarily reflect the position or policy of PepsiCo.Inc.

Food Texture Design and Optimization, First Edition. Edited by Yadunandan Lal Dar and Joseph M. Light.

and distance'. Brennan (1989) defines it as 'the attribute of a substance resulting from a combination of physical properties and perceived by the senses of touch (including kinesthesis and mouthfeel), sight and hearing.' Food texture is defined as 'all the rheological and structural (geometric and surface) attributes of the product perceptible by means of mechanical, tactile, and where appropriate, visual and auditory receptors' (Lawless and Heymann, 1998).

Texture is an overall impression through the information gathered by several of the senses. Sight, sound, tactile sensation (mechanical, thermal and pain) at first bite and the processing in the mouth (chewing etc.) all play an essential role in the brain's synthesis of texture (Wilkinson et al., 2000). The orbitofrontal cortex and an area of convergence of somatosensory inputs are involved in texture perception (Rolls, 2005). Single neurons in this area have been shown to alter their signal response due to changes in texture of the food (Rolls, 2005).

When a particular food is consumed to satiation (the feeling of fullness), the pleasantness of that food and the desire to eat it diminishes more than foods that have not been consumed. This response is termed sensory specific satiety (Guinard and Brun, 1998). These authors conclude that texture specific satiety may be an important contributor to satiety. One way to delay sensory specific satiety is to change the foods being consumed. Such a change will delay satiety and may also lead to an over consumption of calories. Manipulation of flavor intensity does not affect sensory specific satiety (Havermans et al., 2009).

A recent excellent review (Foegeding et al., 2011) covers the complex relationship between food texture and its measurement by instrumental and sensory methods. Importantly, the review explains the meaning, significance and applications of these measures in practical terms. Another recent review (Chen, 2007) is also recommended for developing an understanding of surface texture. The various descriptors used in texture evaluation by sensory scientists have been described (Anon, undated).

Freshly prepared foods may have more than one texture attribute in a single food. Such foods are called multi-textured foods. Consider a humble Club Sandwich which consists of toasted bread (crunchy) and contains slices of meat (soft, chewy) with lettuce and bacon (crisp) and mayonnaise (emulsion) is one example of a multi-textured food. However, the Club Sandwich is always freshly prepared and cannot be manufactured on an industrial scale because upon storage the desirable textural qualities are lost within a short period of time (less than a day). Why can't we keep fresh multi-textured foods stable over a long time period? The challenges of commercially manufacturing multi-textured foods are to keep the freshly created textures over a long period of time, retaining flavor characteristics and prevention of chemical and microbial degradation of the foods. The reasons why maintenance of freshly created textures is difficult are due to moisture and oil (small solute) migration. Many flavor compounds are oil soluble and migration of this solvent leads to contamination of flavor of the substrate into which the oil migrates, thereby decreasing the desirability of the food. Moisture migration similarly detracts from desirable textures by altering their sensory properties; for example, a crunchy food becomes soggy and loses its consumer appeal.

This moisture migration challenge is best understood by viewing the water activity 'map', originally pioneered by (Labuza et al., 1970). The 'water activity map' shown in Figure 8.1 can be used to predict what types of reactions (i.e., chemical, biochemical, physical, and microbial growth and toxin production) will occur in foods. Only a relative reaction rate is plotted versus a_w, as the actual rate is based on a kinetic phenomenon dependent on the specific food system and reaction being investigated. A nice classification

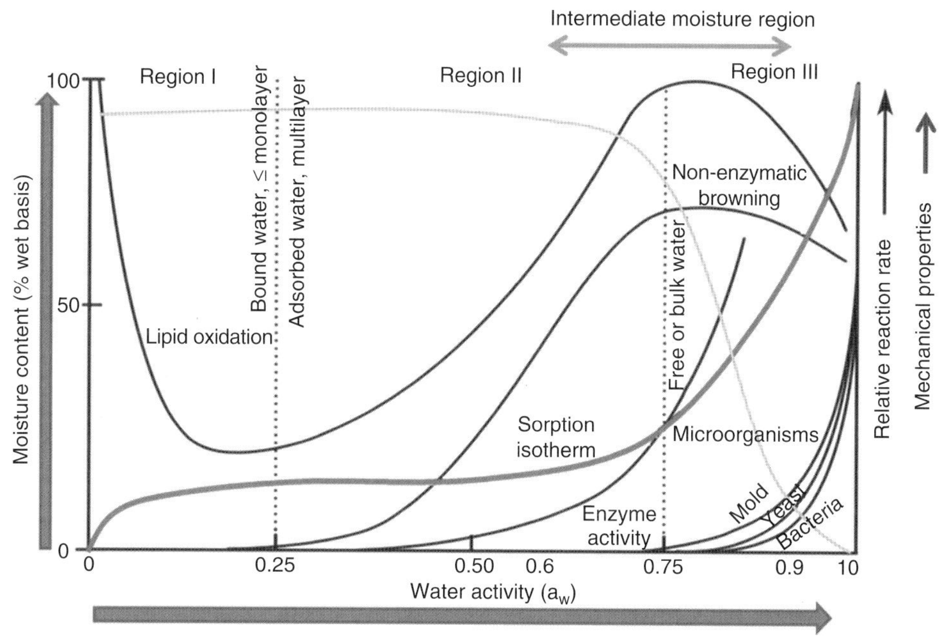

Figure 8.1 Properties of water in the three isotherm regions or zones. (Source: Schmidt 2004. Reproduced with permission of Elsevier. Adapted and reproduced from Labuza et al. 1970. Reproduced with permission of IFT.) "Water activity-stability Map" illustrating the general relationship between the occurrence of various reactions important in foods, as a function of water activity, superimposed on a sorption isotherm. M is for mold, Y is for yeast, and B is for bacteria. The isotherm is plotted as moisture content (left y-axis) versus water activity, whereas all other curves are plotted as relative reaction rate (right y-axis) versus water activity.
a_w ~ 0.2 to 0.3 below this rate of lipid oxidation increases; a_w ~ 0.35 to 0.4 onset of stickiness of powders; a_w ~ 0.4 to 0.5 onset of loss of crispness on gain of moisture; a_w ~ 0.5 onset of hardening (e.g. raisins) on loss of moisture; a_w ~ 0.6 onset of microbial growth; a_w ~ 0.65 to 0.85 maximum reaction rate in amorphous systems as function of a_w (e.g. enzymes, non- enzymatic browning [NEB], lipid oxidation); a_w ~ 0.85 onset of growth of bacterial pathogens.

of water in the three regions and their roles related to Figure 8.1 has been tabulated by Damodaran et al. (Damodaran et al., 2008) in Table 8.1. The impact of water activity on foods in terms of texture, stability and processing aspects are summarized in Table 8.2 (Schmidt, 2004). The combined water activity-stability map in Figure 8.1 and Table 8.2 provide a comprehensive way of viewing the multiple roles of water in foods, from relative stability levels of foods to food processing techniques to textural attributes in exemplified food products.

Matter is classified as solid, liquid or a gas and combinations of these states result in food texture in Figure 8.3 (Roos et al., 1991). Such a classification is limiting as a descriptor for multi-textured foods. A better way to visualize multi-textured foods is to use textural properties or a 'texture map' as a basis for classifying multi-textured foods (Figure 8.2). These types of foods have also been called multi-domain foods (Labuza and Hyman, 1998). Very often, multiple textures can also be found in food products made up of a single component or multiple components in one single domain. A few examples of

Table 8.1 Properties of water in the three isotherm regions or zones. (Source: Adapted from Damodaran et al. 2008. Reproduced with permission of CRC Press, Taylor and Francis Group.)

Property	Zone or Region I a_w 0 to 0.25	Zone or Region II a_w 0.25 to 0.75	Zone or Region III a_w 0.75 to 1.0
Association with solids	Strongest	Less strong	Least (bulk or free water)
Types of associations	Water-interactions with charged groups	Water-dipole and water-water via hydrogen bonding	Mostly water-water via hydrogen bonding
Water's role	Forms monolayer[a]	Acts as a plasticizer; causes swelling	Solvent
Removable?	Difficult	Less difficult to remove	Easy to remove
Freezable?	Unfreezable	Unfreezable	Freezable
Molecular mobility	10^{-2} – 10^{-8} sec.; no solvent power	10^{-9} – 10^{-11} sec; slight-moderate solvent power	10^{-11}–10^{-12} sec; normal solvent power
Participates in reactions	Not very available, except for lipid oxidation	More available	Very available
Microbial growth	No microbial growth	Microbial growth begins at >0.60 a_w (mold)	~0.86 a_w (yeast) and ~0.90 a_w (bacteria)

[a] The BET monolayer is located at the boundary between Regions I and II. The BET monolayer value (g water/g solid) can be thought of as corresponding approximately to the amount of water needed to form a monolayer over only the readily accessible, highly polar groups of the dry matter.

Table 8.2 Impact of water activity and moisture content on relative stability level, processing, preservation and packaging technologies, textural attributes in food products corresponding to each of regions I, II, and III in Figure 8.1. (Source: Adapted from Schmidt 2004. Reproduced with permission of Elsevier.)

Characteristic	Region I	Region II	Region III
Water Activity	0.0 to 0.25	0.25 to 0.75	0.75 to 1.0
Moisture Content (%wb)[a]	Low (0 to 2.6%)	Intermediate (2.6 to 18%)	High (18 to 100%)
Relative Stability Level	High	Intermediate	Low to high, depending on technologies employed
Example Processing, Preservation, and Packaging Technologies	Dehydration, Drying, Extrusion	Concentration, Chemical methods	Refrigeration, Freezing, Modified or Controlled atmosphere packaging, Pasteurization, Canning, Aseptic processing, Fermentation, Chemical methods
Textural Attributes	Dry, Hard, Crisp, Shrunken	Chewy, Firm, Flexible	Soft, Juicy, Moist, Swollen
Example Food Products[b]	Snack foods, Crisp cookies, some Breakfast cereals	Some Breakfast cereals, Granola bars, Soft cookies, Raisins, some Candy bars, IMF Pet foods	Fresh fruits and vegetables, Fresh meat, Poultry, and Fish, Refrigerated dairy products, Canned products

[a] Moisture content values estimated from the composite food isotherm (Schmidt 2004) for 0.25 and 0.75 a_w values.

Table 8.3 State of matter and example foods.

	SOLID	LIQUID	GAS
SOLID	Cream centered cookie	Chocolate cherry cordial	Carbonated candy
LIQUID	Gels (gelatin or pectin)	Mayonnaise	*
GAS	Bread & leavened foods	Carbonated soft drinks	*

*No examples

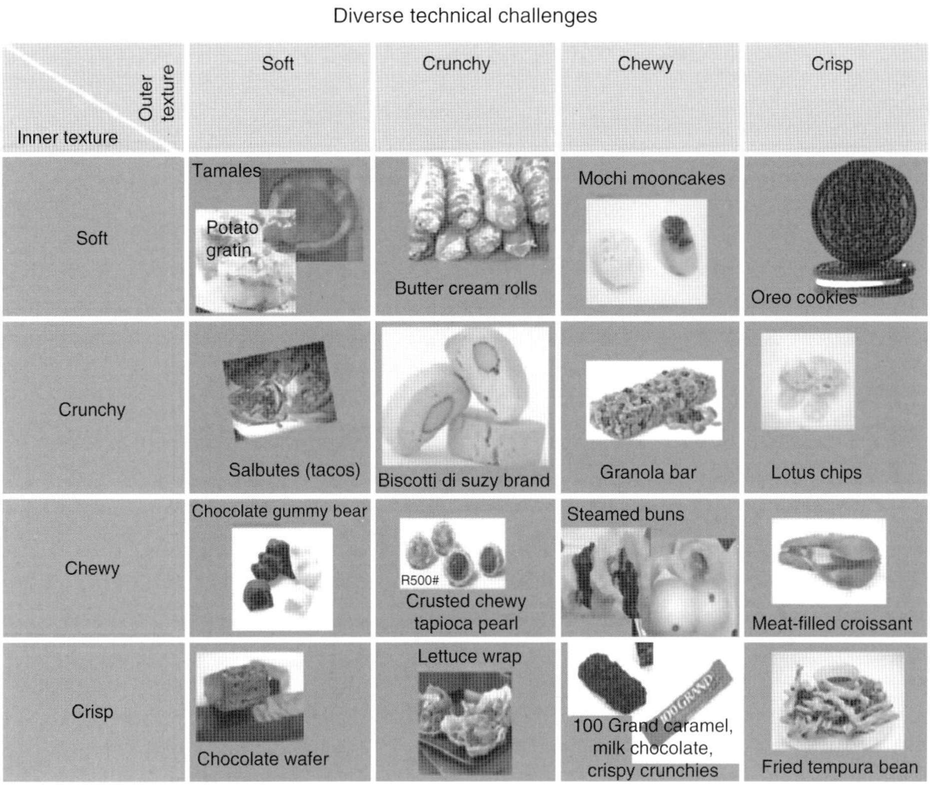

Figure 8.2 Representative examples of multi-textured foods.

single component multi-textured foods are baked bread or French fries; here, baked bread can have a crispy-crunchy outer crust and a soft-chewy crumb, while French fries have a crispy outer layer followed by a chewy inner core. More commonly, multi-textured foods have different domains or compartments made up of different components and having unique texture and moisture content, arranged in one or more geometrical fashion – multiple layers may be layered in a planar or concentric manner or some domains may be dispersed in a continuous matrix. Bread with embedded raisins is an example of the latter, while a layered cake with crispy and soft layers represents the former category. More often than not, the difference in water activities of the domains determines the unique multi-texture and sensory experience for such foods. Some common examples of multi-textured products in market are raisin-cookies, pizza crust with sauce, cereals with raisins, milk with cereal, ice cream in a crispy cone, stuffed crispy/short pie or empanada crumb with soft filling, puffed or baked pastry with fruit or meat filling, crispy/crunchy/chewy Indian samosa dough shell with soft/chewy vegetable filling, crispy sandwich wafers with soft chocolate inside, hard chocolates with liquid centers, hot pockets, crunchy cereal breakfast bars with soft fruit layer inside, chocolate covered nuts and many others like this.

The multi-texture property in foods is most often a direct outcome of differences in water activities or free moisture in different food domains. To overcome this challenge, most commercial products in the market have been prepared with high fat levels to reduce a_w and soften the texture at high solids level, high levels of sugar (mono and disaccharides) and salt, small molecule soluble solutes like polyols (e.g. Glycerol,

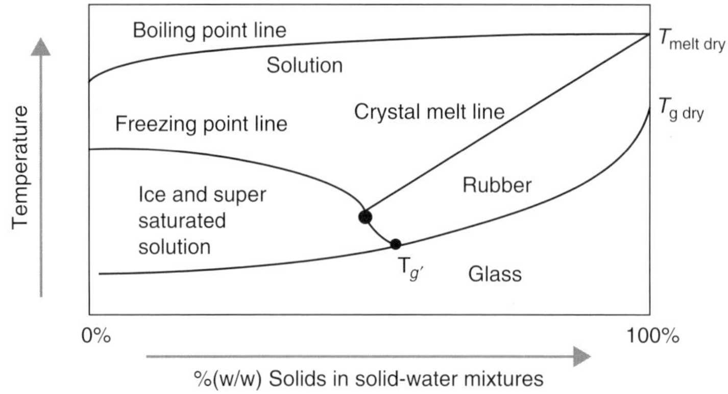

Figure 8.3 Hypothetical State Diagram. (Source: Reprinted from Food Technology, 45, Roos, Y. and Karel, M., Applying state diagrams in food processing and development, 66–71. Copyright (1991), with permission from IFT.)

sorbitol), humectants, and so on to increase the solute concentration and reduce a_w. The water activity of most of these commercial products is below 0.6, or 0.8 with some preservatives (if needed) to inhibit mold and yeast growth. Many of these foods are also refrigerated or frozen to overcome microbial stability issues and slow down the kinetics of moisture migration, like hot pockets and so on. It is even a greater challenge to envision a multi-textured food product that is shelf stable in the intermediate moisture range, and yet not being prepared with high levels of sugar, fat, preservatives, humectants like polyols. Historically, long term shelf stability of foods had been obtained by freezing or dehydration. In both cases, food solids are concentrated either by separating water into another phase (ice) or by removing nearly all of the water (Roos, 1995). Successfully processed and stored frozen and dehydrated materials are extremely stable (Chirife and Buera, 1996). Stability of the materials is based on at least a partially amorphous (glassy) state of the solids. Also, food stability at the same a_w may vary from one food to another from hours to years. For example, a food powder with sucrose as a main component becomes sticky and may crystallize at 0.3 a_w within hours, while dairy powders with high lactose content remains stable for years at the same a_w at typical storage temperatures.

In this chapter moisture migration, its kinetics and effects on physical state of foods, methods of measuring moisture transfer and strategies to delay such undesirable changes are considered.

8.2 GENERAL PROBLEM, DEFINITION OF MULTI-TEXTURE, ROLE OF WATER ACTIVITY, GLASS TRANSITION AND THEIR EFFECTS ON TEXTURE

8.2.1 Water activity

Moisture is a major component in most common foods. A reduction in moisture content leads to preservation of that food. More important than the total moisture content is the concept of water activity or free moisture in foods. Most foods are multiphasic and multicomponent systems. When a multiphase system is in equilibrium, the chemical potential μ of individual components is the same in all phases, obeying the laws of thermodynamics. Chemical potential of component 1 is defined by the equation:

$$\mu_1 (p, T, n_1, n_2, \ldots\ldots\ldots) = \left(\frac{\partial G}{\partial n_1} \right) p, T, n_2, \ldots\ldots\ldots \tag{8.1}$$

Where, $\partial G/\partial n_1$ represents the partial molar free energy of component 1, at constant temperature, pressure and composition of other components. The chemical potential of any component i or water in a solution is given by:

$$\mu_w = \mu_w^{\,0} + RT \ln a_w = \frac{p}{p_o} \tag{8.2}$$

Where μ_w is the chemical potential of water in the solution with other components, $\mu_w^{\,0}$ is the chemical potential of pure water in a standard state, R is the universal gas constant (8.314 $JK^{-1} mol^{-1}$), T is the temperature in K and a_w is the water activity of the solution. Activity is a thermodynamic concept, introduced by Lewis and Randall (Lewis et al., 1961) as 'Activity is, at a given temperature, the ratio of the fugacity, f, of a substance and its fugacity, f^0, in some state which for convenience, has been chosen as a standard state,' expressed as:

$$a = \left(\frac{f}{f^0} \right) T \tag{8.3}$$

Fugacity is a measure of the escaping tendency of a substance and can be replaced by vapor pressure, provided the vapor behaves as an ideal gas; hence, the definition of a_w as 'Water activity is, at a given temperature, the ratio of its fugacity, f_w, in a system and the fugacity, $f_w^{\,0}$ of pure liquid at the same temperature'.

$$a_w = \left(\frac{f_w}{f_w^{\,0}} \right) T \tag{8.4}$$

However, to a good first approximation, fugacity can be replaced by equilibrium vapor pressure or, by equilibrium partial vapor pressure in all systems of interest. The error associated with this approximation under normal conditions is less than 0.1% and hence this substitution is acceptable in food science measurements.

For an ideal solution in contact with an ideal gas, Raoult's law states that the partial pressure of a component *i* (y_i P) in the gaseous phase is equal to the product of the liquid phase mole fraction of the component *i* (x_i) and its saturation vapor pressure (P_i, ^{sat}) at temperature T,

$$y_i\, P = x_i\, P_i^{sat} \tag{8.5}$$

Where y_i is the mole fraction of component i in the gas phase and P is the total pressure of the gas phase at temperature T. The water activity of component i (or, water in this case) becomes equal to mole fraction of water in ideal solution, assuming the gas phase to be ideal, and a_w can vary from 0 to 1.

$$a_w = X_w = \frac{m_w}{m_w + \sum m_{s,i}} = \frac{y_i P}{P_w^{sat}} \tag{8.6}$$

Where, m_w is the molar concentration of water and $m_{s,i}$ is the mole fraction of each of the solutes in solution. Most foods exhibit marked non-ideality, and calculation of a_w from composition is difficult for real food systems (Walstra, 2003). However, since at equilibrium, the chemical potentials of water in the liquid and gaseous phases must be equal in a closed chamber, the water activity of a food can be measured by bringing it to equilibrium with the air above it in the enclosed chamber. Then, under conditions of constant temperature and pressure and at equilibrium (assumptions made for a_w measurements), the a_w values of the aqueous phase of a food ($a_{w,f}$) and of the air ($a_{w,v}$) are equal and can be estimated from the ratio of the partial vapor pressure of water above the food (p) to the saturated vapor pressure of pure water (P_w^{sat}) at the same temperature (Walstra, 2003):

$$a_w = \frac{p}{p_w^{sat}} \tag{8.7}$$

Alternately, a_w can be expressed as follows:

$$a_w = \gamma_w . x_w \tag{8.8}$$

Where γ_w is the activity coefficient of water in the test material providing a measure of non-ideality and x_w is the mole fraction of water in the liquid. Therefore, water activity is a useful measure to indicate the amount of water that is available to participate in chemical and biological reactions and to support microbial growth. Concentration of solutes and substances that 'bind' water are critical factors for water activity. As solute concentration increases water activity decreases. Another important relationship involving water activity is the following:

$$-\ln a_w = \frac{\Pi M_w}{RT\rho_w} \tag{8.9}$$

Where Π is the osmotic pressure, M_w is the molecular weight or molar mass and ρ_w is the density of the solute. The relative humidity of air in equilibrium with a sample is called the Equilibrium Relative Humidity (ERH).

$$\text{ERH} = a_w \times 100 \tag{8.10}$$

Figure 8.4 is an indicative water activity isotherm displaying the hysteresis often encountered depending on whether the water is being added to the dry material or removed (drying) from the wet material. This hysteresis is due to non-reversible structural changes and non-equilibrium effects. There are many empirical equations (and tables) that attempt to describe this behavior but, although indicative, none predict with sufficient accuracy and the water activity isotherm should be experimentally determined for each material. A composite food isotherm is shown in Figure 8.5. It may be interesting to note that although bread and cheese have very high water activities as fruits, vegetables, fish, meat and yogurt, the moisture contents of the two groups

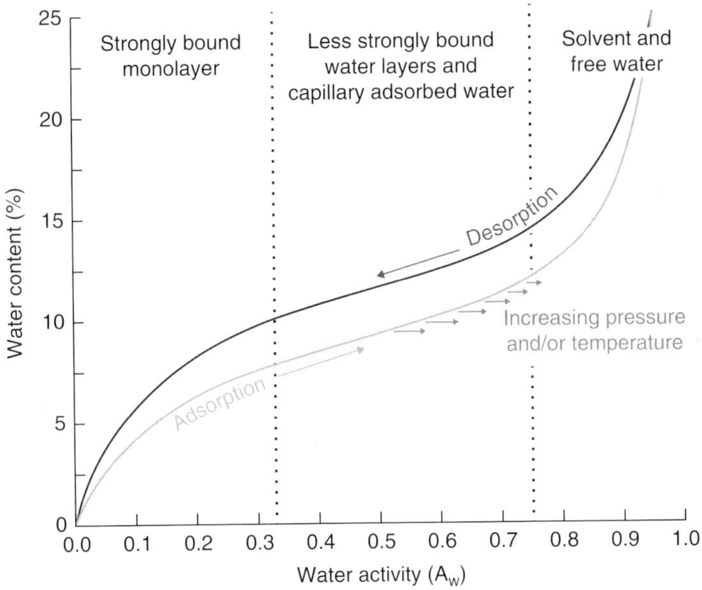

Figure 8.4 Water activity isotherm.

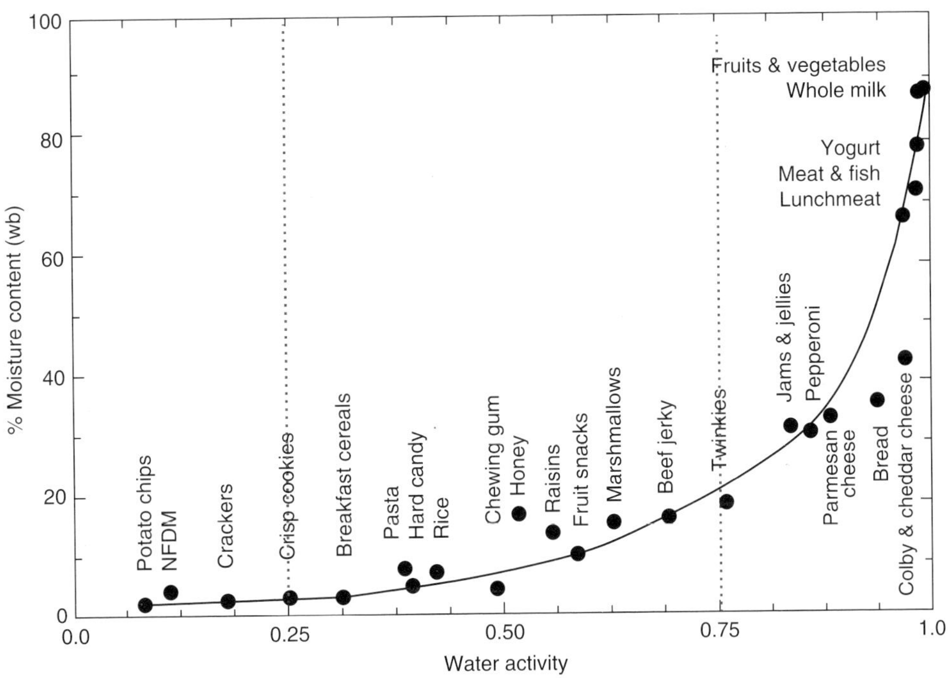

Figure 8.5 Composite food isotherm obtained by plotting moisture content vs. water activity for a variety of food materials from literature sources. (Source: Schmidt 2004. Reproduced with permission of Elsevier.)

are dramatically different indicating that there is a lot of bound moisture in the latter compared to the former.

The water activity (a_w) usually increases with increase in temperature and is pressure dependent (Bell and Labuza, 2000; Kapsalis, 1987). For many food systems, at constant moisture content, water activity increases as temperature increases (Figure 8.6a) (Schmidt, 2004). However, the opposite trend occurs for several small molecular weight solutes, such as crystaline sugars and salts; at constant moisture content, as temperature increases, water activity decreases (Figure 8.6b) (Kapsalis, 1987; Mathlouthi and Roge, 2003; Schmidt, 2004). A combination of the temperature–a_w shifts illustrated in Figures 8.6a and b is observed for foods containing a large amount of solutes/fillers

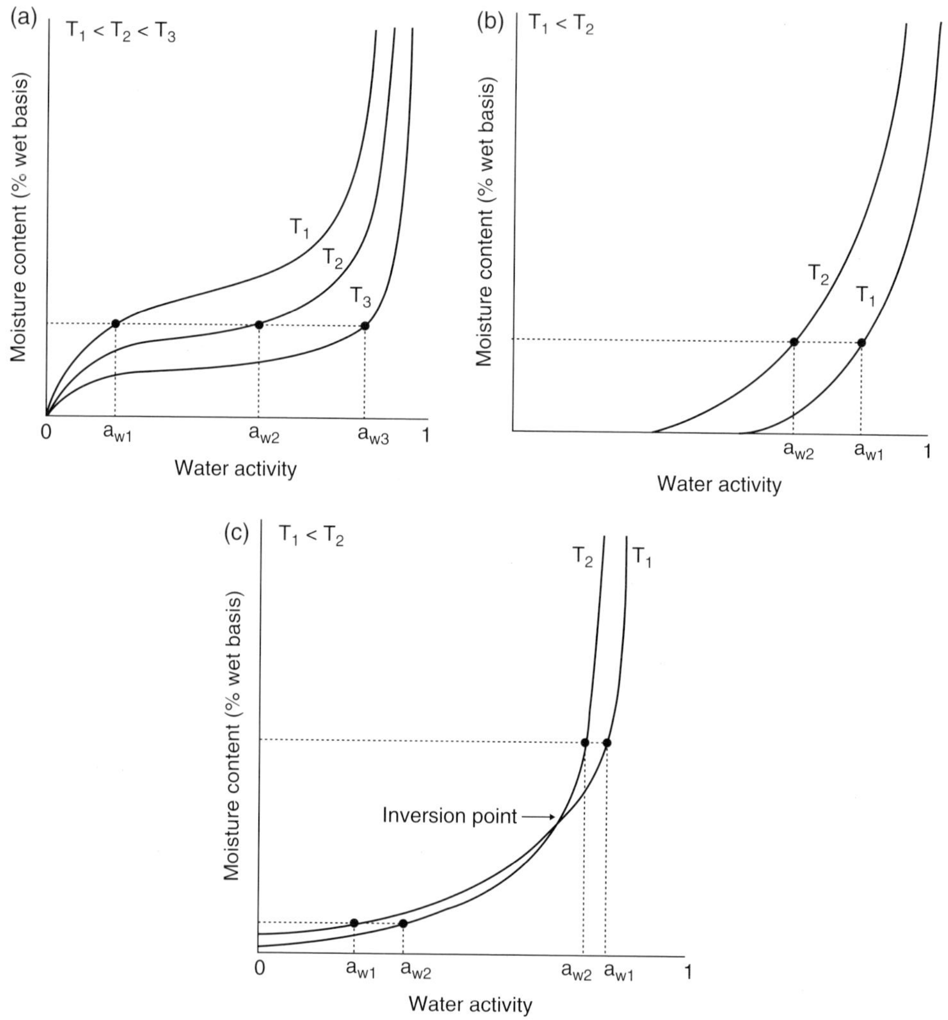

Figure 8.6 Effect of temperature on a_w of (a) a complex food system, (b) a small molecular weight solute such as fructose, and (c) foods containing large amounts of fillers such as raisins. (Source: Reprinted from Water Activity: Theory and Application to Foods, Kapsalis, J.G., Influence of hysteresis and temperature on moisture sorption isotherms, 173–213. Copyright (1987), with permission from Marcel Dekker; Mathlouthi and Roge 2003. Reproduced with permission of Elsevier; Schmidt 2004. Reproduced with permission of Elsevier.)

(e.g., dried fruits) (Figure 8.6c) (Schmidt, 2004). For small temperature increases ($T_1 \rightarrow T_2$) at low a_w, Clausius-Clapeyron is often used:

$$\ln\left(\frac{a_{w1}}{a_{w2}}\right) = \frac{\Delta H_{st}}{R}\left(\frac{1}{T_1} - \frac{1}{T_2}\right) \tag{8.11}$$

Where ΔH is the net isoteric heat of sorption at the moisture content (cal/mole), R is the Universal gas constant (1.987 cal/mole K) and T is the temperature in Kelvin. ΔH_{st} is defined as the difference between the total molar enthalpy change and the molar enthalpy of vaporization of pure water. The assumptions underlying Equation 8.11 are that moisture content is constant and ΔH_{st} is constant over a_w and temperature ranges considered (Kapsalis, 1987; Rahman, 1995) for each system under study. Such changes in water activity induced by temperature may cause water migration between food components and further affect texture and shelf life.

The multi-ingredient nature of food and its processing (for example, cooking) commonly results in a range of water activities being present. Foods containing macroscopic or microstructural aqueous pools of differing water activity will be prone to time and temperature dependent water migration from areas with high a_w to those with low a_w, a useful property used in the salting of fish and cheese but in other cases may have disastrous organoleptic consequences. Such changes in water activity may cause water migration between food components. As the humidity of the air is typically 50–80% (a_w = 0.5–0.8), foods with lower a_w will tend to gain water whilst those with higher a_w will tend to lose water.

Control of water activity (rather than water content) is very important in the food industry as low water activity prevents microbial growth (increasing shelf life) (Table 8.4), causes large changes in textural characteristics such as crispness and crunchiness (for example, the sound produced by 'crunching' breakfast cereal disappearing above about a_w = 0.65) and changes the rate of chemical reactions (increasing hydrophobe lipophilic reactions but reducing hydrophile aqueous-diffusion-limited reactions). The balance between these factors is such that there is an optimum water activity for dehydrated foods, which is usually equated with a monolayer coverage of water and a_w of about 0.2–0.3 (Leake, 2006). Highly perishable foodstuffs have a_w > 0.95 (equivalent to about

Table 8.4 Effect of water activity on the type and growth of microorganisms.

Water activity	Type and extent of microbial growth
0.95–1.0	Most bacteria (Pseudomonas, Escherichia, Bacillus, Clostridium Perfrigens, C. Botulinum E, G) some yeasts
0.95–0.91	most bacteria (salmonella, Listeria Monocytogenes, Bacillus cereus, Clostridium Botilinum A, B)
0.88–0.91	most yeast, Staphylococcus Aeureus (aerobic)
0.80–0.87	most mildew, Staphylococcus Aeureus (aerobic), most Saccharomyces
0.75–0.80	halophilic bacteria
0.65–0.70	Xerophilic molds (Aspergillus chevalieri, Saccharomyces bisporus)
0.61–0.65	osmophilic yeast, few mildew
< 0.6	No microbial growth

43 % w/w sucrose), Growth of most bacteria is inhibited below about $a_w = 0.91$ (equivalent to about 57 % w/w sucrose); similarly most yeasts cease growing below $a_w = 0.87$ (equivalent to about 65 % w/w sucrose) and most molds cease growing below $a_w > 0.80$ (equivalent to about 73 % w/w sucrose). The absolute limit of microbial growth is about $a_w = 0.6$. As the solute concentration required to produce $a_w < 0.96$ is high (typically > 1 molal), the solutes (and surface interactions at low water content) will control the structuring of the water within the range where a_w knowledge is usefully applied. Changes in the natural clustering of water due to low concentrations of solutes will only occur at $a_w > 0.98$. Although low density water (ES) will possess less a_w than collapsed water clustering (CS) and the consequences are very important in biological systems, such changes in the absolute value of a_w are small.

Hygroscopic products may absorb water in different ways: sorption and chemical reaction, sorption and formation of a hydrate, water binding by surface energy, diffusion of water molecules in the material structure, capillary condensation formation of a solution and so on. Therefore, water molecules present in a product are not all retained by the product with the same strength. In fact, only part of the total moisture content can be exchanged between the product and its surroundings under normal conditions.

Moisture content includes both an immobilized part (e.g. water of hydration) and an active part. This explains the lack of accuracy and reliability of most methods for measuring %H_2O. The quantity of water found in a product sample depends on both the method of measurement and on the way it is applied. The active part of moisture content – or free water – is usually measured in terms of its vapor pressure, expressed as percent relative humidity generated in equilibrium with the product sample in a closed system at constant temperature.

Water activity (or ERH) indicates the degree of freedom of the water absorbed in a material and shows dimensions, structure, cohesion, agglomeration properties as well as electrical and chemical properties better than moisture content does. The effect of this water on physical properties, such as differences in water-vapor pressure, and not moisture content govern the interchange of water between a product and its surroundings (e.g. air, packaging, other products etc.). Therefore, the stability of the moisture content of a product is determined by any difference between ERH and %relative humidity (%RH) in the surrounding air or by any difference in a_w or ERH with other products.

8.2.2 Moisture sorption isotherms

At equilibrium, the relationship between %H_2O and a_w of a hygroscopic material can be graphically represented by a curve called a sorption isotherm (Figure 8.4). For each value of a_w, a sorption isotherm shows the corresponding value of H_2O at a given constant temperature. Due to the complexity of sorption phenomena, sorption isotherms cannot be calculated in advance and have to be experimentally determined for each single product. The practical usefulness of sorption isotherms is severely limited by several factors:

- sorption isotherms are valid only for a single specific product
- sorption isotherms are affected by any variation in product composition (e.g. products of natural origin)

- sorption isotherms show a difference between adsorption and desorption curves (sorption hysteresis)
- sorption isotherms are experimentally determined; unfortunately most methods used for measuring %H_2O are neither very accurate or reliable.

From the above it should be clear that moisture content very often cannot be used to accurately determine a_w and vice-versa. The practical usefulness of sorption isotherms is limited to products whose previous moisture history is known and which have a defined composition and a stable temperature. The slope of most sorption isotherms is such that a small variation in %H_2O corresponds to a large variation in a_w. Due to this fact, moisture content cannot be used to make a rough estimate of a_w. In practice, moisture content and water activity should be considered as two independent parameters (Figure 8.4). Water adsorption by foods may be looked upon as a continuous process where water molecules progressively and, to some extent, reversibly combine with the food solids via chemisorption, physisorption, and multilayer condensation.

The isotherm shapes are usually classified by the BDDT classification (Brunauer et al., 1940) (Figure 8.7). This system distinguishes five different types of sorption isotherms (type I – V) which was later extended by an additional type VI. An excellent review of the different types of isotherms and the models used for fitting moisture content and a_w data of foods is given by Basu et al. (2006). Type 1 is the well-known Langmuir isotherm, obtained assuming monomolecular adsorption of gas by the porous solids in a finite volume of voids. Type 2 is the sigmoid isotherm obtained for soluble products, which exhibits asymptotic trend as water activity approaches 1. Type 3, known as the Flory-Higgins isotherm, accounts for a solvent or plasticizer such as glycerol above the glass transition temperature. Type 4 isotherm describes adsorption by a swellable hydrophilic solid until a maximum of hydration sites are reached. Type 5 is the BET multilayer adsorption isotherm, observed for adsorption of water vapor on charcoal; it is related to types 2 and 3 isotherms. However, type 6 isotherm has no relevance for vapor sorption at ambient conditions and type 1 isotherm is rarely observed unless the material is strongly microporous or chemisorption occurs. The two isotherms most commonly found in food products are types 2 and 4. (Mathlouthi et al., 2003). One set, consisting of type 2 and 4, shows a significant uptake at low partial pressures followed by small adsorption at intermediate vapor concentration and again a high uptake at elevated partial pressures. The other set comprises type 3 and 5 isotherms with a characteristic low uptake at low concentration and a strong increase in adsorption at higher vapor concentration. These differences in the adsorption and desorption behavior are due to differences in the sorption enthalpy of solvents and is illustrated in Figure 8.7.

If the vapor molecule has a strong interaction with the solid surface (equivalent to a high heat of sorption) compared to the interaction between two vapor molecules (equivalent to a high heat of condensation) the vapor adsorbs on the surface as a monolayer (in the ideal case) on a perfectly homogeneous surface. After completion of the monolayer, multilayer formation ensues as a condensation of vapor molecules on the initial monolayer. This is typical behavior for molecules with low polarity such as alkanes or nitrogen, which enables their use in the determination of the specific surface area according to the BET theory. This is illustrated in type 2 and 4 isotherms. The opposite case is when the vapor interaction with the surface is energetically similar or

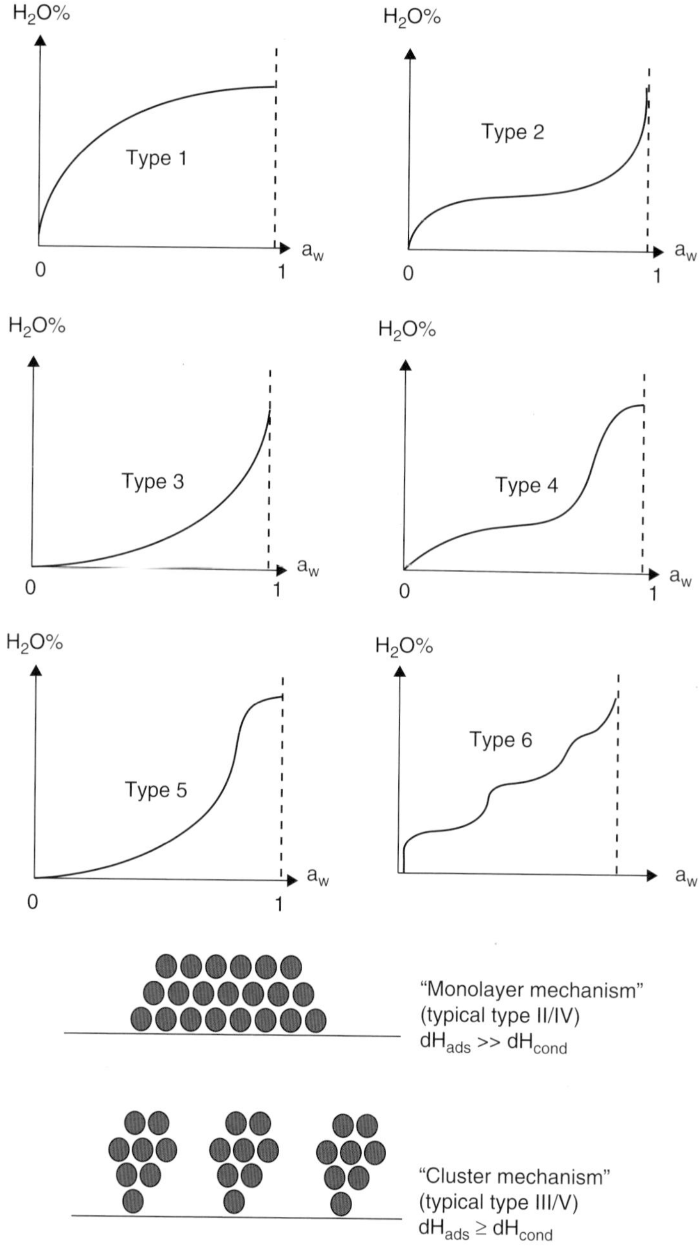

Figure 8.7 Types of adsorption isotherms according to BDDT classification and sorption mechanism of solvent on surfaces. (Source: Levoguer et al. 2006 and Theilman & Burnett 2006.)

just slightly higher than the interaction with another vapor molecule (that is, the heat of sorption is similar to the heat of condensation). This causes only very few molecules to adsorb at low concentration, followed by condensation at higher partial pressures on the few initially adsorbed molecules. Such effect can be described as cluster formation since there is no continuous monolayer formation occurring. This is illustrated in type

3 and 5 isotherms. Some of the most important equations used to explain the isotherms described above are given below.

8.2.2.1 Langmuir equation

On the basis of monomolecular layers with identical, independent sorption sites, Langmuir (Langmuir, 1918) proposed the following physical adsorption model for type 1 isotherm:

$$a_w\left(\frac{1}{M}-\frac{1}{M_0}\right)=\frac{1}{CM_0} \tag{8.12}$$

where, M_0 is monolayer sorbate constant and C is a constant.

8.2.2.2 Brunauer-Emmett-Teller (BET) equation

The BET isotherm equation (Brauner et al., 1938) is one of the most widely used models and yields good fit for a variety of foods over the region 0.05 < aw < 0.45. (Chirife et al., 1978). It provides an estimate of monolayer value of moisture adsorbed on the surface. The BET equation is given below:

$$\frac{M}{M_0}=\frac{a_w}{(1-a_w)\left[1+(C-1)a_w\right]} \tag{8.13}$$

Where, M is the equilibrium moisture content (kg water/kg dry matter), M_0 is the monolayer moisture content on the internal surface (kg water/kg dry matter), and C is a dimensionless parameter related to heat of sorption of monolayer region. The BET equation has these inherent assumptions: (a) the rate of condensation on the first layer is equal to the rate of evaporation from the second layer, (b) binding energy of all adsorbate molecules on the first layer is same, and (c) binding energy of the other layers is equal to that of regular heat of condensation. The assumptions of uniform adsorbent surface and absence of lateral interactions between adsorbed molecules are over simplified and incorrect in view of the heterogeneous food surface interactions.

8.2.2.3 GAB equation

The Guggenheim, Anderson and de Boer equation was derived independently as a refinement to the Langmuir and BET theories, and is widely applicable to food systems over a wider a_w range from 0 to 0.9 (van den Berg, C., 1984). The GAB model underestimates the water content at high water activities ($a_w > 0.93$). However, it has the advantages of containing only three parameters, each parameter has a physical meaning connected to sorption processes, and the equation is able to describe some temperature effect on isotherms through Arrhenius type of equations.

$$\frac{M}{M_0}=\frac{CKa_w}{(1-Ka_w)\left[1-Ka_w+CKa_w\right]} \tag{8.14}$$

$$\frac{M}{M_0} = \frac{(C-1)Ka_w}{[1-Ka_w+CKa_w]} + \frac{Ka_w}{(1-Ka_w)} \tag{8.15}$$

Where, C is a dimensionless GAB parameter related to heat of sorption of monolayer region and K is a dimensionless GAB parameter related to heat of sorption of multilayer region. Equation (8.14) has been divided into two additive terms in Equation 8.15, the first term describing the mono-molecular layer expression in Langmuir's adsorption isotherm and the second term describing the multilayer adsorption corresponding to Raoult's law (van den Berg, 1984). Parameters K and C can be again expressed by Arrhenius type equations:

$$C = C' \exp\left(-\frac{\Delta H_c}{RT}\right) \tag{8.16}$$

$$K = K' \exp\left(-\frac{\Delta H_k}{RT}\right) \tag{8.17}$$

Where, $\Delta H_c = \Delta H_m - \Delta H_q$; $\Delta H_k = \Delta H_l - \Delta H_q$ (kJmol^{-1}); ΔH_l is the heat of condensation of pure water (kJmol^{-1}); ΔH_m is the total heat of sorption of mono layer (kJmol^{-1}); ΔH_q is the total heat of sorption of multilayer covering the monolayer (kJmol^{-1}); and C′, K′ are the constants of entropic character. Table 8.5 illustrates the typical GAB parameters for food matrices (Basu et al., 2006).

Many other models had been developed to relate moisture content of foods to water activity, such as the Modified Oswin equation, Modified Halsey equation, Modified Henderson equation, Chung-Pfost equation, Ferro-Fontan equation, Peleg Model, Timmermann GAB model, Viollaz GAB Model, Lewicki model and Harkins Jura model and so on (Basu et al., 2006). An example of different foods fitted with different sorption isotherms is given in Table 8.5 (Basu et al., 2006).

Table 8.5 Recent works on sorption isotherms of food materials. (Source: Basu et al. 2006. Reproduced with permission of Taylor and Francis Group.)

Food	Temperature, °C	a_w	Best fitted model
Berries	4–27	0.11–0.87	GAB
Mushroom	4–27,	0.12–0.75,	GAB,
	30–70	0.12–0.75	Chung-Pfost
Cocoa beans	25–35	0.08–0.94	BET (a_w<0.5), Harkins-Jura (a_w> 0.5)
Cured beef and pork	10–49	0.1–0.94	Peleg, GAB
Chicken meat	4–30	0.25–0.94	Ferro-Fontan, GAB
Hazelnuts	25	0.11–0.9	GAB
Gelatin gel	20	0.75–0.98	Ferro-Fontan
Crystaline lactose	12–40	0.11–0.98	GAB

8.2.3 Reactions driven by a_w

The concept of 'free' as opposed to the total, including 'bound' water, has gained wide acceptance in the food processing industry. Water activity exerts a decisive influence on such phenomena as change in color, taste and aroma, food poisoning and spoilage (shelf life), loss of vitamins and so on (total moisture content has very little to do with this).

Water activity in foods can be controlled by using various additives (e.g. salts, sugars etc.), by using satisfactory packaging materials, by maintaining favorable maturation and storage conditions. Water activity measurements are increasingly frequent in food research and development as well as in production quality control. On-line measurements are possible to a certain extent: water activity is defined at static equilibrium, a condition which is not always met in the process line itself. Water activity indicates the amount of water in the total water content which is available to micro-organisms. Each species of micro-organism (bacteria, yeast and mold) has its own minimum a_w value below which growth is no longer possible. This limit corresponds to the 'suction power' of the various organisms, that is, to the osmotic pressure they create and which must be higher than in the aqueous phase of food, to absorb from the food the water needed for metabolic activity and growth (Table 8.4). By measuring a_w value of foodstuffs it is possible to determine which micro-organisms will not be able to develop on them. The US Food and Drug Administration has adopted the concept of a_w for establishing moisture limits beyond which certain types of food are considered susceptible to invasion by mold and bacteria. Under this concept, foods with diverse water binding capabilities can be considered as a group (this would not be possible with a $\%H_2O$ concept).

Water activity control is an important factor for the chemical stability of foods. Most foodstuffs contain carbohydrates and proteins and are therefore subject to non-enzymatic browning reactions (Maillard reaction). The Maillard reaction gets stronger at increasing a_w values and reaches its peak at $a_w = 0.6$ to 0.7; with further increase of a_w this reaction gets rapidly weaker.

The spontaneous auto-catalytic breaking of the molecular chains of fats is strongly influenced by a_w. This kind of food spoilage increases at high a_w values. Even at low a_w values, foods with a fat content acquire a rancid taste after being stored for some time. Foods in which the action of glyceride creates fatty acids with short molecular chains are particularly affected by this kind of spoilage: they produce a strong and disagreeable smell. The conservation of foodstuffs is influenced by numerous changes occasioned by oxidation: color change of carotene, oxidation of myoglobin in meat, oxidation of proteins and vitamins and so on. The oxidation of fats and other food components decreases sharply at a_w values below 0.2. Most enzymatic reactions are slowed down at a_w values below 0.8. Some of these reactions occur even at very low a_w values. However, as many foodstuffs are thermally treated during their processing, enzymatic spoilage is usually of very little importance.

With a few exceptions, the molecular activity of water-vapor (i.e. pressure) in hygroscopic materials increases with temperature at roughly the same rate as the increase in pressure of saturated water-vapor. In fact water activity shows only limited temperature dependence.

Increasing the temperature of air in a dryer neither decreases its water-vapor pressure nor increases its 'hygroscopicity'. Warm air increases the water-vapor pressure generated by the product to be dried by increasing its temperature. The drying enthalpy and therefore the energy necessary to dry a product is strongly influenced by water activity. For pure water (i.e. $a_w = 1.0$) the evaporation enthalpy amounts to 2300–2500 kJ/kg. This explains why the last part of the moisture content is so expensive to extract when drying.

The speed at which water can be removed by evaporation from a product is directly proportional to the active surface of the product and to the difference in water-vapor pressure between the product and its surroundings. Differences in moisture content have nothing to do with this. Calculating and controlling drying process requires the knowledge of the relationship between a_w and %H_2O (sorption isotherms). This relationship is especially hard to determine in the case of products of natural origin (variable product composition). In practice it is necessary to consider that sorption isotherms correspond to conditions of static equilibrium. Therefore, differences with theoretical calculations based on sorption isotherms will occur.

Optimising a drying process can only be done by experimentation. As already shown, moisture content measurement (air or product) does not give an accurate image of the drying process itself. Measurement of the temperature drop in the drying air-flow (psychometric effect) gives results which are very dependent on the conditions at the inlet. Besides this, such a measurement neither takes into account the sorption properties of the product to be dried nor the effect of temperature on the water-vapor generated by the product. The dynamic moisture and temperature equilibrium which exists in a dryer cannot be considered as strictly equivalent to water activity. Nevertheless, measuring both the relative humidity and temperature of the product (e.g. %RH and °C at the outlet of a fluid bed dryer) gives the closest image of the drying process. Under consideration of temperature the relative humidity established by dynamic equilibrium in the dryer can be compared to the ERH (a_w) which the product should have after drying. In this way the nature of the product and in particular the effect of temperature on its water-vapor pressure can be taken into account automatically.

Drying a product to the desired moisture content can be a costly exercise. From the point of view of energy conservation it is worth asking if water activity rather than moisture content should be considered in the definition of the product moisture condition after drying. Whenever product quality and stability are the main factors to determine what a dryer should do, the answer should be positive. Significant energy savings as well as better quality control will result from such a decision.

8.2.4 Effect of water on glass transition of food materials

Several factors can influence the physical state of foods, which in turn affects their processability, shelf-life and physical properties upon storage, such as crystalization, stickiness and texture. Multi-domain regions in solid multi-textured foods can exist in either a thermodynamically stable crystaline state or in a metastable amorphous state, which is very sensitive to changes in temperature and moisture content. Usually the amorphous state is a result of rapid removal of water by drying or freezing (White and Cakebread, 1966; Alexander and King, 1985; Levine and Slade, 1988). Amorphous materials can exist in either a very viscous glass or a more mobile liquid like-rubbery state and can move between the two states at a critical temperature, called the glass transition

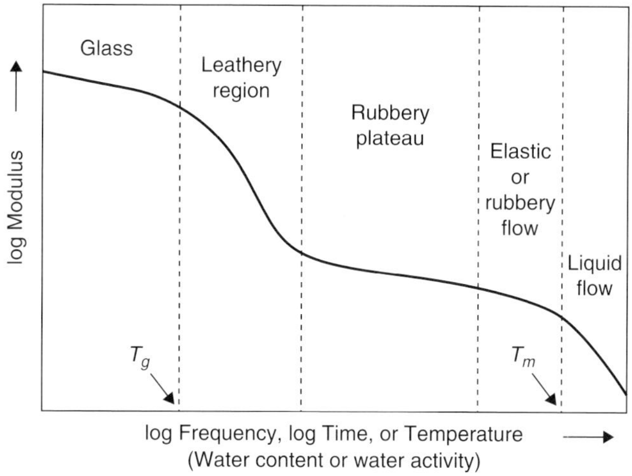

Figure 8.8 Modulus of amorphous materials vs. time or temperature, showing five different viscoelastic regions. (Source: Slade and Levine 1995. Reproduced with permission of Elsevier.)

temperature (Tg), which is specific for each material and is a second-order transition. When such a transition from the glassy state to the rubbery state or vice versa occurs, changes in thermodynamic properties, molecular mobility, dielectric constant and mechanical properties like tensile strength, Young's modulus, complex modulus, elastic or loss modulus and so on, occur (Sperling, 1986). Figure 8.8 demonstrates an example where the modulus changes as a function of time or temperature as the amorphous food matrix changes from one state to another (Levine and Slade, 1990). Also, when a glassy state is heated over the glass transition temperature, molecules become mobile and gain translational mobility. Hence, glass transition is related to changes in food structure, microstructure, crystalization, rates of diffusion-controlled reactions, and possibly, stabilization of microbial cells and spores (Roos, 1995; Chirife and Buera, 1996; Le, Meste et al., 2002). Glass transition may occur at various temperatures depending on the water content (Roos, 1993a, 2002).

In polymers, molecular weight, composition and type and concentration of plasticizers are the main factors affecting Tg (Fox and Flory, 1950; Gordon and Taylor, 1952). The glass transition in food materials can be affected by temperature, moisture and other plasticizers. The amount of plasticizer (e.g. water), along with the temperature, determines what state the region is in. This can best be seen by a state diagram as shown in (Figure 8.3). The state diagram defines the moisture content and temperature region at which a food domain is glassy, rubbery, crystaline, frozen, and so on. State diagrams are particularly useful in the characterization of the physical state of sugars and the water content dependence of transition temperatures (Roos, 1995). Some of the common plasticizers present in foods are water, polyols, sugars (mono and disaccharides), surfactants, oils, and a host of other small molecular solutes. Thus, if a multi-domain region gains or loses moisture, it may move between the rubbery and glassy states. As moisture increases, the rate of crystalization increases, thus limiting the shelf-life of foods (Duckworth, 1981). The glass transition temperature of food components also governs physical properties like diffusivity which may also affect rates of deteriorative reactions (Vrentas and Duda, 1978; Simatos and Karel, 1988). Therefore, the physical state of an amorphous matrix in frozen and dehydrated foods may greatly influence their stability, and affect both physical/chemical changes during food processing and storage (Levine and Slade,1988; Roos and Karel, 1991a, b).

Water is the main solvent of non-fat food solids and acts as the main plasticizer of amorphous food components, such as carbohydrate and proteins biopolymers. As a plasticizer and a solvent, water may dramatically affect rates of mechanical and diffusional properties of foods (Slade and Levine, 1995). Whereas water activity is an equilibrium thermodynamic property, glass transition is a relaxation process accompanying the transformation of amorphous solids to a more liquid-like supercooled state (Roos, 2002). The amount of water effective as plasticizer, corresponds to the total amount of water, independent of a_w, and takes into account any water that is dissolved or miscible with substances in the matrix (Roos, 1995). The glass transition is a primary determinant of the dehydration characteristics and the stability of high carbohydrate containing foods (Roos, 2004).

High molecular weight biopolymers, such as carbohydrates and proteins, have high-temperature anhydrous glass transitions that are often higher than their decomposition temperatures and cannot be measured experimentally, unless water is added to lower their transitions. These ingredients seldom exhibit problems in dehydration or flow in powder form. Low molecular weight carbohydrates such as mono and disaccharides, have comparably low-temperature glass transitions (Roos, 1993b). Hence, foods containing high concentration of these are extremely difficult or impossible to dehydrate or freeze in typical food-freezing processes (Roos, 1995, 2004). Glass transitions of amorphous pure sugars can be measured easily by differential scanning calorimetry and dynamic mechanical analysis. However, complex fat-containing foods or frozen food systems, containing partially amorphous materials like starches and proteins in dispersed forms with fats and so on, cannot be measured simply as the glass transitions may be weak in these systems and may interfere with other transformations such as crystalization and melting of lipid components (Roos, 1995, 2002).

The glassy state of a material is a non-equilibrium state and time dependent, and a material can have a number of glassy states depending upon how the glass has been formed and how these states were frozen in time. Hence, a glass may have a relatively large or low ‘free volume’, and differences in these states reflect the molecular relaxations associated with the glass transition and gained translational mobility. A rapidly formed glass has a large free volume and gives an exotherm in a DSC (Differential Scanning Calorimeter) scan as the extra energy released in heating over the glass transition (accompanying relaxation to the lower free volume state) (Roos, 2002). A very slowly formed dense glass gives an endotherm in the DSC scan as the molecules absorb energy to transition to the more expanded liquid state or rubbery state. Some transformations may occur in the glassy state over storage time at a particular temperature, known as physical aging (Roudaut et al., 2004). As a result of water plasticization, the free volume of the material and molecular mobility increases and glass transition is lowered corresponding to the increase in water content. Typically amorphous materials undergo a glass transition at 100°–150°C below its equilibrium melting temperature (Sperling, 1992; Roos, 1993b).

8.2.5 Glass transition, water activity, relaxation times, and mechanical properties

The effect of glass transition on the mechanical properties of food systems, such as caking, stickiness, collapse of structure and crispiness of low-moisture foods has also been demonstrated (Slade and Levine, 1991; Roos, 1995; Roos et al., 1998; Le Meste et al., 2002). A typical master curve of polymers is based on collecting modulus data at different temperatures by some method, for example, stress relaxation method, which measures the modulus as a function of time after deformation. These measurements of time-dependence of modulus are equivalent to frequency dependent modulus data at a particular temperature. The same curve can be shifted to another temperature, and many such curves can be obtained as a function of temperature to predict the time- and temperature-dependent behavior of amorphous, viscoelastic materials. A typical master curve for a viscoelastic amorphous material is shown in Figure 8.8. Thus, when various mechanical properties of food materials are measured as a function of frequency, time, temperature, water content, or water activity they should produce a similar modulus curve reflecting the physical state of foods. A typical value of Young's modulus of glassy polymer is $3x10^9$ Pa, which applies to a large number of amorphous polymers (Sperling, 1986) and probably to amorphous foods. If the amorphous materials have partial crystallinity, the crystaline regions decrease the depression of the modulus within the leathery region and extend the rubbery plateau region until the liquid flow occurs above the melting temperature. Crosslinking of polymers also extends the rubbery plateau region as shown in Figure 8.9. This is observed in bread dough due to the presence of partially crystaline starch fractions and the crosslinking or thermosetting protein fractions due to disulphide bonds.

Over the glass transition, the glassy state of amorphous materials transform to more liquid-like materials with enhanced flow properties, which indicates a decrease in the relaxation times of mechanical changes, that can be related to the viscosity of the materials (Williams et al., 1955; Sperling, 1992). One of the most common models that is used to relate the temperature of amorphous materials above the glass transition to

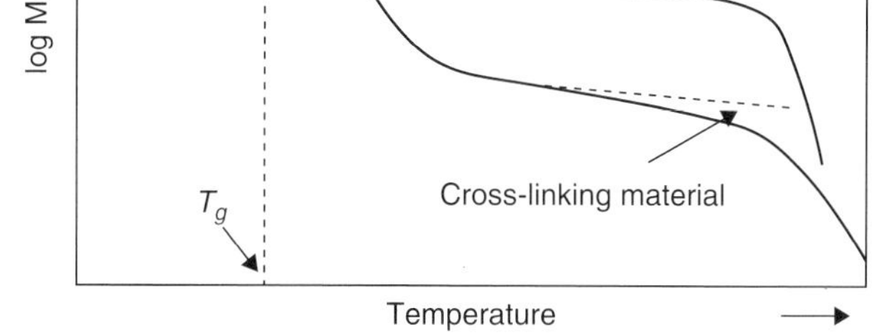

Figure 8.9 Modulus of partially crystaline and cross-linking materials vs. temperature compared to a completely amorphous viscoelastic material. (Source: Roos 1995. Reproduced with permission of Elsevier.)

relaxation times is the Williams-Landel-Ferry (WLF) models, given by the equation below (Williams et al., 1955):

$$\ln a_T = \ln \frac{\tau}{\tau_s} = \ln \frac{\eta}{\eta_s} = -\frac{C_1(T - T_s)}{C_2 + (T - T_s)} \tag{8.18}$$

Where, a_T is the ratio of relaxation times at an observation, T, and reference temperature, T_s, and η, η_s are corresponding viscosities, C_1 and C_2 are constants. This equation can be used to predict viscosity or changes in rates of diffusion controlled reactions in some cases, above the Tg (Nelson and Labuza, 1994; Miao and Roos, 2004). Plasticization by water or temperature may change relaxation times differently and sometimes have independent effects on reaction kinetics (Bell et al., 1998; Lievonen and Roos, 2002; Miao and Roos, 2004). Fermi's model, as suggested by Peleg (1992, 1994) can be used to describe changes in relaxation times in amorphous systems as a function of a_w, or water content, as shown in Figure 8.10. The schematic diagram can be used to explain stability at low water contents for systems existing in glassy states, since the glass transition at low water contents occurs at temperatures higher than the storage temperature. The critical a_w can be defined as that corresponding to water plasticization, where the glass transition is depressed to the observation or storage temperature. Peleg (1993, 1994a,b) also pointed out that the mechanical changes in amorphous foods such as stickiness, caking, collapse modulus or stiffness are characterized by a modulus curve, showing downward concavity, when plotted against a_w, temperature or water content. The changes in mechanical property can be expressed by the equations given below:

$$\frac{Y}{Y_s} = \frac{1}{\left\{1 + \exp\left(\frac{T - T_c(W)}{a(W)}\right)\right\}} \tag{8.19}$$

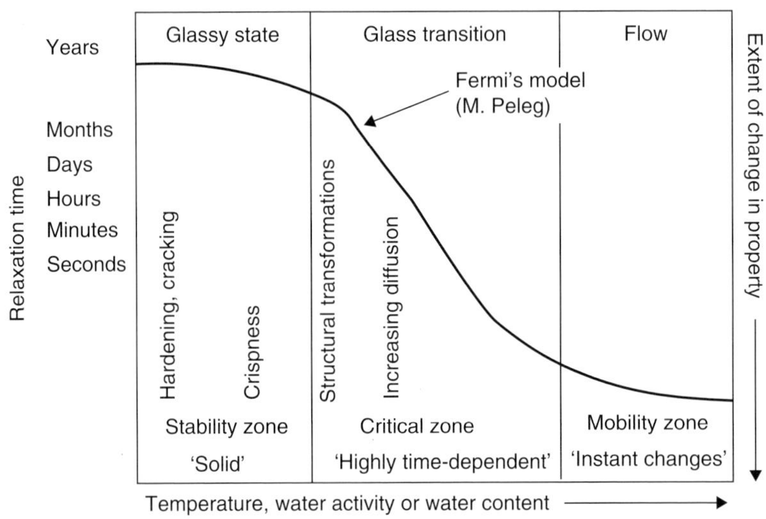

Figure 8.10 Fermi Model – effects of temperature, water activity and water content on relaxation times of amorphous foods. (Source: Canovas et al. 2007. Reproduced with permission of John Wiley & Sons.)

Where Y is the stiffness or texture parameter (e.g. Modulus), Y_s is the stiffness or texture parameter at a reference state (e.g. Glassy state or dry state), T_c (W) is the temperature that characterizes the transition region where $Y = Y_s/2$, a′(W) is a constant that indicates the steepness of the stiffness curve around T_c [(W) refers to the constant water content] and it is assumed that the change in stiffness occurs at a constant water content and can be used to model data as a function of temperature. Stiffness as a function of water activity can also be described by similar equation, where a_{wc} (T) is the water activity characterizing the transition region and a‴ (T) is a constant where T refers to the constant temperature.

$$\frac{Y}{Y_s} = \frac{1}{\left\{1 + \exp\left(\frac{a_w - a_{wc}(T)}{a'''(T)}\right)\right\}} \tag{8.20}$$

In various studies, crispness of low-moisture foods has been related to the water content or relative humidity and water plasticization of structure (Quast and Karel, 1972; Zabik et al., 1979; Katz and Labuza, 1981; Sauvageot and Blond, 1991). The crispiness of foods like potato chips and so on in literature (Bruns and Bourne, 1975) has been defined as the linear slope of the initial portion of the force deformation curve measured in a texture analyzer. Katz and Labuza suggested that the crispy snack foods have a critical water content, where the addition of water enables mobility of macromolecules and consumer perceives the resultant loss of crispiness. The initial slope of the force deformation curve was found to be constant at low water contents, and it was assumed that at such low water activities, the water content was too low to affect mechanical properties. Above a critical water content, a significant decrease of the initial slope occurred and the products lost their crispiness. The steady state water activity at which the crispiness is lost is specific to each product, and is referred to as the critical water activity at that particular temperature. Studies have shown that a change in crispiness occurs between water activity values of 0.35 and 0.5 (Katz and Labuza, 1981; Hsieh et al., 1990). Figure 8.11 shows such a curve representing crispiness of amorphous crispy foods vs. water activity. Sauvageot and Blond (1991) studied the crispiness of breakfast cereals as a function of water activity and using sensory techniques found a good correlation between the results of the sensory methods and an instrumental penetration test. Peleg's Equation 8.20 was also fitted to the sensory data as was shown in Figure 8.11 (Peleg, 1994c). Also, Table 8.6 illustrates the fit of Fermi equation to brittle cereal foods (Harris and Peleg, 1996). Similar analysis can be extended to numerous other textures for multi-textured foods and the critical water content at which the desired texture is lost may also be identified by both instrumental and sensory means providing valuable information about their textural and shelf stability.

8.2.6 Multi-textured foods

These types of foods have also been called multi-domain foods (Labuza and Hyman, 1998). In addition to examples shown in Figure 8.1 other food examples include ice cream in a cone, cereals with semi-moist raisins, pastries with fruit fillings and so on. Such food systems are dynamic and moisture migration occurs continuously until equilibrium is reached. This means that some components lose moisture at the expense

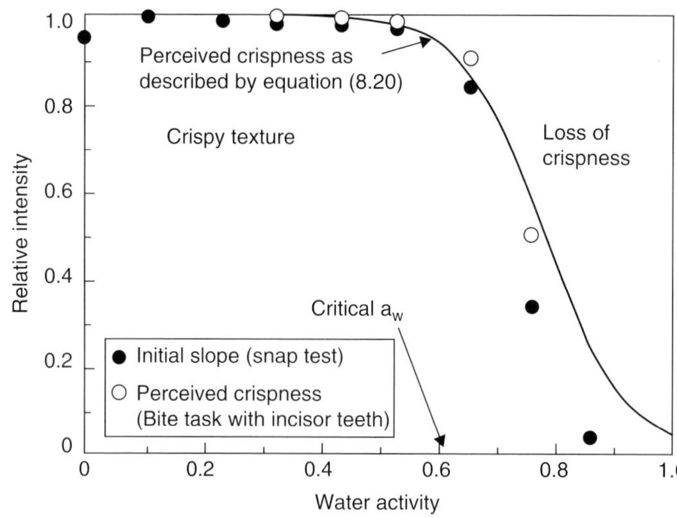

Figure 8.11 Application of Fermi Model to Instrumental and Perceived Crispiness of Crispy foods. (Source: Roos 1995. Reproduced with permission of Elsevier.)

Table 8.6 Application of Fermi equation to brittle cereal foods. (Source: Harris and Peleg 1996. Reproduced with permission of American Association of Cereal Chemists.)

Effect of Water Activity (a_w) on Two Stiffness Parameters of Selected Brittle Cereal Foods[a]

	Force at 20% Deformation					**Force at 30% Deformation**				
Material	**F_o (N)**	**k (N)**	**a_{wc} (–)**	**b (–)**	**r^2 (–)**	**F_o (N)**	**k (N)**	**a_{wc} (–)**	**b (–)**	**r^2 (–)**
Cheese balls	0.35	11	0.62	0.087	0.998	0	14	0.64	0.090	0.985
Cheese puffs	0.65	19	0.58	0.110	0.985	1	23	0.63	0.110	0.992
Crutons										
French bread	4.9	3.2	0.78	0.040	0.990	4.1	4.6	0.79	0.050	0.976
Pumpernickel	5.2	0	0.84	0.005	0.980	5.0	0	0.84	0.005	0.955

[a] Based on the modified Fermi equation: $F(a_w) = F_o + k(a_w)/\{1{=}\exp[(a_w - a_{wc})/b]\}$ where $F(a_w)$ is the magnitude of the mechanical parameter; F_0 is its magnitude in the dry state; k is a constant (roughly the slope of the linear region); a_{wc} roughly marks the infection point of $F(a_w)$ b is a constant that accounts for the steepness of the relationships at a_{wc}.

of other components. The main driving forces for the amount and rates of moisture migration are thermodynamic (difference in chemical potential of water or difference in water activity) and kinetic (diffusion rates or mobility of water in foods). Strategies that can address such migration include barrier films between domains with a differential in a_w, altering diffusivity of water through the domains and changing viscosity of the domains to slow down the kinetics of moisture migration. Other strategies include equalizing the a_w of the different domains or altering the a_w of the domains to have the minimum differential in a_w possible to eliminate completely or reduce the thermodynamic driving force for moisture migration.

When dry (7% moisture) and wet (27% moisture) barley grains were intimately mixed, a majority of the moisture equilibrium was reached in three days (Henderson, 1987).

Equilibration at 10°C was slower than at 20°C. The moisture differential between wet and dry grains was only 20%. In foods systems such differentials can be much greater and equilibration may be reached in a few hours.

Bread staling is a phenomenon that has been intensively investigated. There are two types of bread staling a) crust staling and b) crumb staling. Crust staling is the result of moisture transfer from the crumb to the crust (Lin and Lineback, 1990) which results in a soft leathery texture. Among the several theories of crumb staling one postulate is that moisture transfer from the gluten network to starch results in crumb staling (Kulp and Ponte, 1981) and a comprehensive review of bread staling is recommended (Gray and Bemiller, 2003). Short proved crispy rolls showed lower rate of crust water uptake while longer proved rolls showed faster uptake during cooling down (Primo-Martin et al., 2010). Sensory analysis revealed higher crispness for short proved crispy rolls after the same storage time at low %RH. They hypothesize that shorter proved crispy rolls with finer crumb morphology, more closed structure, and smaller gas cells with less gas cell interconnections and a thicker crust have a significant positive effect on water uptake kinetics and crispness retention.

Equilibrium and dynamic water sorption properties of sponge cakes with varying porosity (86–52%) and fat content (0–0.30 g/g dry basis {d.b.}) were determined using a water vapor sorption microbalance (Roca et al., 2007). Contrary to porosity, addition of fat decreased equilibrium moisture contents. The effective moisture diffusivity (D_{eff}) was identified from a numerical solution of Fick's second law, taking into account an external mass transfer coefficient and the swelling of the solid matrix. D_{eff} increased from 1.61 to $8.67 \times 10^{-10}\,m^2/s$ with moisture content, reached a threshold at moisture content 0.15 g/g d.b. and then decreased until water saturation. D_{eff} decreased from 8.67 to $2.97 \times 10^{-10}\,m^2/s$ with decreasing porosity. This effect was attributed to a change of water diffusion mechanism, from predominant vapor to liquid. D_{eff} decreased from 8.67 to $2.12 \times 10^{-10}\,m^2/s$ with increasing fat content. Addition of fat had an effect on the water diffusion in two ways, decreasing porosity (sagging of the foam) and increasing tortuosity.

Moisture and salt migration are responsible for the softening of cheese, called cold melt, found in cheese-in-sausage products (Mandl et al., 2009). By characterizing the rate of moisture and salt migration in model cheese-in-sausage systems, diffusion coefficients for water and salt in cheese were obtained. As expected, the rate of moisture migration was greater when the initial driving force (difference in water activities of cheese and sausage) was greatest. These conditions also led to most rapid softening of the cheese. An edible barrier was found to reduce but not completely inhibit moisture migration and softening, whereas formulation to match water and salt activity of sausage and cheese virtually eliminated moisture migration and softening.

The effects of milk protein type, formulation and storage were investigated in model bar systems to gain improved insight into deteriorative, textural changes (Hogan et al., 2012). Texture analysis, water activity measurements, confocal laser scanning microscopy and Fourier transform infrared (FT-IR) spectroscopy were used to monitor interactive physical and chemical changes in bars during storage. Hardness of bars was dependent on protein type, concentration and co-solvents. Microstructural changes in bars during elevated temperature storage showed swelling and fusion of powder particles. FT-IR spectra suggest a molecular contribution to hardening due to solvent-induced plasticization and reconfirmation of protein secondary structures. Hardness development in high-protein bars appears to result from non-equilibrium changes due to the hydration behavior of individual components and the competition for available moisture. Minimization of water activity differences between ingredients provides a means of controlling hardness. The work provides improved understanding of textural changes in highly concentrated food systems.

Limitation of moisture transfer in composite food products can be achieved by two main techniques: reduction of a_w difference between components and use of hydrophobic barrier at the interface of the system. Both techniques were tested in a multi-domain model food associating a dry cereal-based component to a wet component using a model based on Fick's second law (Bourlieu et al., 2008). Required input modeling parameters were moisture equilibrium and transport properties of the components. The two stabilization techniques permitted a significant extension of the period of acceptability of the dry component: from 7 min to more than 40 h using solutes addition and up to 12 days using a 300 μm hydrophobic barrier. A further extension of this period of acceptability was achieved by combining the two techniques. However, this had a detrimental effect on the relative efficiency of the barrier due to increased internal resistance of the wet component.

There are five major mechanisms for water-solid interactions in solid food materials: adsorption of water vapor on to the surface, crystal hydrate formation, deliquescence, capillary condensation, and absorption of water vapor into the bulk structure. Solid food materials can generally be divided into two categories: crystaline solids and amorphous solids. Amorphous solids have a glassy state and a temperature above which this amorphous solid changes state from a glass to a rubber. This characteristic temperature is known as the glass transition temperature. Water in food systems acts as a plasticizer and depending upon the temperature acts as an agent that can alter texture of the food.

The above discussion is by no means an exhaustive survey of the effects of water, water activity and its role in texture modification of multi-textured foods. Some strategies to delay texture degradation have been briefly mentioned for some foods. The kinetics of moisture migration is discussed next.

8.3 KINETICS OF MOISTURE MIGRATION

Accurate prediction and control of moisture transfer inside composite food products is still a major challenge in food preservation. There is an abundance of edible barrier performance data from numerous research labs today, using a wide range of biopolymers to optimize moisture migration in these systems. However, due to non-linear food-matrix specific water sorption isotherms, moisture dependent diffusivities, and variations of physical state with changes in moisture or a_w, modeling transport phenomena through edible barriers is complex. Water vapor permeability (WVP), although a property of the material can hardly be considered as an inherent property of films and only gives a relative indication of the barrier efficacy. Most of the models used to predict moisture migration have been validated under a narrow range of conditions. However, literature in this area also cites a few predictive models based on Fick's Second Law, developed to assess shelf-life extension of food products including barriers. Some of these models, assuming complex and realistic hypothesis, have been validated in various model foods.

Several reviews focusing specifically on edible moisture barriers (Bourlieu et al., 2009; Debeaufort et al., 2000a; Koelsch, 1994) and/or lipid-based edible films (Baldwin et al., 1997; Callegarin et al., 1997; Greener and Fennema, 1992; Hernandez, 1994; Morillon et al., 2002; Shellhammer and Krochta, 1997a) and chocolate or chocolate based barriers in confectionary coatings (Ghosh et al., 2002; Ghosh, 2003; Ghosh et al., 2004; Ghosh et al., 2005) have been published. The requirement of complying with regulatory, nutritional, organoleptic, and targeted functional properties to develop effective moisture barriers was underlined in these reviews.

Several mathematical models have been used to predict moisture transfer in foods. Leveraging the diffusion equations derived for different geometries by Crank (1975), several models have been developed to predict changes in moisture content during drying (Biquet and Guilbert, 1986), sorption experiments (Biquet and Labuza, 1988), or soaking (Hsu, 1983a,b; Sayar et al., 2001) or in composite food systems (Guillard et al., 2003a; Guillard et al., 2003b; Guillard et al., 2003c; Guillard et al., 2003d; Bourlieu et al., 2008; Bourlieu et al., 2009; Roca et al., 2007). Other authors used a numerical procedure for solving differential equations of Fick's 2nd law and simulating moisture distribution in food using finite element or finite difference schemes (Hong et al., 1986; Williams and Mittal, 1999; Lomauro and Bakshi, 1985; Lomauro et al., 1985; Sapru and Labuza,1996). Two simplified analytical solutions suitable for implementation without special purpose software had been found by Jens Risbo (2003). The first analytical solution assumes that the moisture exchange of one component is rate limiting. The second analytical solution describes the situation where all components are characterized by the same moisture uptake dynamics (response function or sorption isotherms). The calculations in this chapter show that the headspace water activity is distinctly different from the water activities of the components at any time indicating that no single component sorption process is rate limiting. The analytical solutions prove that if moisture sorption properties of individual food components under constant humidity conditions are known, then these data can be utilized in simple expressions for moisture content vs. time in each component, with the possibility of making shelf life calculations on any combinations of the food components.

The most simplistic assessment of moisture barrier efficacy is directly related to film/barrier Water Vapor Permeability (WVP) measurements and to the analysis of the material physical properties, such as effective diffusivity of moisture in the matrix, moisture sorption properties, and changes in glass transition due to changes in moisture content.

8.3.1 Effective diffusivity and solubility coefficients of water in a matrix

Moisture migration occurs in composite food systems where components of high and low water activity are adjacent. Transfer between regions in a multi-domain system is mainly by diffusion. Water diffuses from the wet into the dry component, leading to irreversible texture loss. Moisture transfer compromises the quality, stability and safety of the product and limit its shelf life (Kester and Fennema, 1986). The transport and equilibrium properties of water for each compartment are important factors needed for modeling and predicting moisture transfer between components and as a consequence shelf life of the product. For multi-domain systems, we are not only concerned with diffusion of water from the domain of higher a_w or chemical potential to the domain of lower a_w or chemical potential, but we also need to determine the rates of diffusion within each domain to understand the rate limiting step to moisture exchange. The two primary factors controlling the rate of self-diffusion of water in each domain are the local viscosity of the adsorbed aqueous phase of the system if diffusion occurs in the liquid state, and the porous structure of the system for gaseous diffusion. Higher viscosity leads to lower molecular movement, and this is also directly related to the glassy or rubbery states of the amorphous food matrix. The amount of water, the amorphous state of the food, the temperature, and the molecular weight of the ingredients all affect the local viscosity (Labuza and Hyman, 1998). Diffusion of water in the glassy or rubbery state of the amorphous matrix can be modeled by the Boltzmann equation (Labuza and Hyman, 1998):

$$\chi^2 = \frac{k_b T^{\tau}}{3\pi\eta D} \tag{8.21}$$

Where χ is vibrational space or diffusion length in time τ; k_b is Boltzmann constant, T is temperature (K), D is the local effective diffusion coefficient, τ is the characteristic time for molecular size movement, and η is the local viscosity. Comparing the glassy state of a domain (η = 10 ^{12}Pa.s) with that of rubbery state (η = 10 ^{6}Pa.s) of the same, keeping all other variables constant, a water molecule takes about 1×10^{6} times longer to vibrate in the glassy phase than in the rubbery phase. An exception to this is the situation in which the pore structure (diffusion space) is large enough, in which case it may have no effect. Ingredients, like long chain polymers can affect the local viscosity and thus mobility of water, as well as intermolecular association or interactions with ingredients.

The diffusion of a small molecule such as water through a food domain is controlled by the molecule's size, pore size, the presence of other molecules in the vapor phase that water molecules may collide with and the surrounding geometry, that is, tortuosity of the path. Mathematical models for given geometries can be found in Geankopolis (1972) and in Crank (1975). Moisture transport is more complex and difficult to analyze in porous matrices, such as sponge-cake, usually encountered in bakery industry. In such porous products, different mechanisms of moisture transport occurred on a microstructure level, such as liquid movement as a result of capillary action, liquid diffusion as a result of concentration gradients, vapor diffusion in air-filled pores as a result of vapor pressure gradients, and liquid diffusion at the surface of pores (Gekas, 1992). The relative contribution of each mechanism listed above to overall internal moisture transfer is difficult to determine. This results in an apparent or effective moisture diffusivity (D_{eff}), which is a function of moisture content for moisture sensitive products.

In a porous domain, molecules migrate more quickly in the liquid state because of suction due to capillary action, together with liquid diffusion. In the absence of pores or in low porosity matrix, liquid diffusion is the main method of moisture transport. The smaller the pore size in the matrix of the food domain, slower is the moisture vapor migration due to hindered diffusion. If the pore diameter in the range of 2–50 nm and length is comparatively much larger, then we may also encounter Knudsen diffusion at low pressures and the vapor phase diffusivity can be further reduced. In Knudsen diffusion, the mean free path of the diffusing gas molecules is smaller than the pore size and the gas molecules collide with the pore walls more frequently than with each other. In addition, membranes, fat or sugar or starch crystals, anisotropic materials like clay or other fillers, and lipids contribute as barriers to moisture migration in a multi-domain system, by increasing the diffusion path length as the water molecules need to bypass them and move in a more tortuous path.

MRI can be used to visualize the actual rates of diffusion in foods during storage and to estimate the diffusivities. Ruan and Litchfield (1992) used MRI to follow moisture mobility and distribution in corn kernels. Schmidt and Lai (1991) reviewed the applications of NMR and MRI to study water relations in foods. Umbach et al. (1992) calculated water self-diffusion coefficients in starch-gluten-water systems by NMR. Heil et al. (1993) studied water migration in baking biscuits by MRI. From these studies, MRI and NMR seem to be useful methods in gaining further understanding of the rates and paths of moisture migration. Theoretically, self-diffusion coefficients for water from NMR data are higher than the effective diffusion coefficients calculated from drying data. The differences between the macroscopic and microscopic data are presumed to be because of changes in sample dimension during the drying experiments as well as due to the impermeable barriers within the matrix (le Meste et al., 1995).

The effective diffusivity of water through any domain of the food matrix or a barrier film can be measured experimentally from the sorption kinetics, as measured by DVS (Dynamic Vapor Sorption by SMS Instruments, UK) and using appropriate equations for specific geometries. The method used to calculate the diffusion constants for thin films utilizes diffusion equations first employed by Crank and Park (1968). A thin film sample is suspended in the DVS pan and the sorption kinetics for a series of steps in humidity is recorded in the usual manner. For a single step in humidity from 0% RH to a specific %RH, with a double-sided thin film of thickness d, the initial kinetics of sorption into the bulk may be described by the following equation (Levoguer et al., 2006):

$$\frac{M_t}{M_\infty} = \frac{4}{d}\sqrt{\left(\frac{Dt}{\pi}\right)} \tag{8.22}$$

Where, M_t is the amount adsorbed at time t, M_∞ is the amount adsorbed at thermodynamic equilibrium, and D is the effective diffusion constant. This equation is generally valid for values of M_t/M_∞ <0.4 or the initial linear part of the sorption curve for a monomolecular layer of solute adsorption, where adsorbed molecules are still in the form of a monolayer with no lateral intermolecular interactions and where a plot of Mt/M_∞ against $t^{1/2}$ /d should be linear. The effective diffusion coefficient D may then be calculated from the slope of this line. This can be repeated for a number of steps in %RH increase from 0% RH and a plot of effective diffusivity vs. a_w can be generated for the film. The same technique can be used to measure an effective diffusivity of moisture in a bread crumb/crust, powders, flakes, or a gelled type of filling (for filled multi-textured snacks) of known geometry and dimensions. An example of diffusion of moisture into a 15 mg sample of polyimide film 7.5 microns thick, studied in a DVS-1 instrument is described below. The sample was exposed to increasing steps in humidity from 0% RH to the desired humidity and back down to 0% RH as shown in

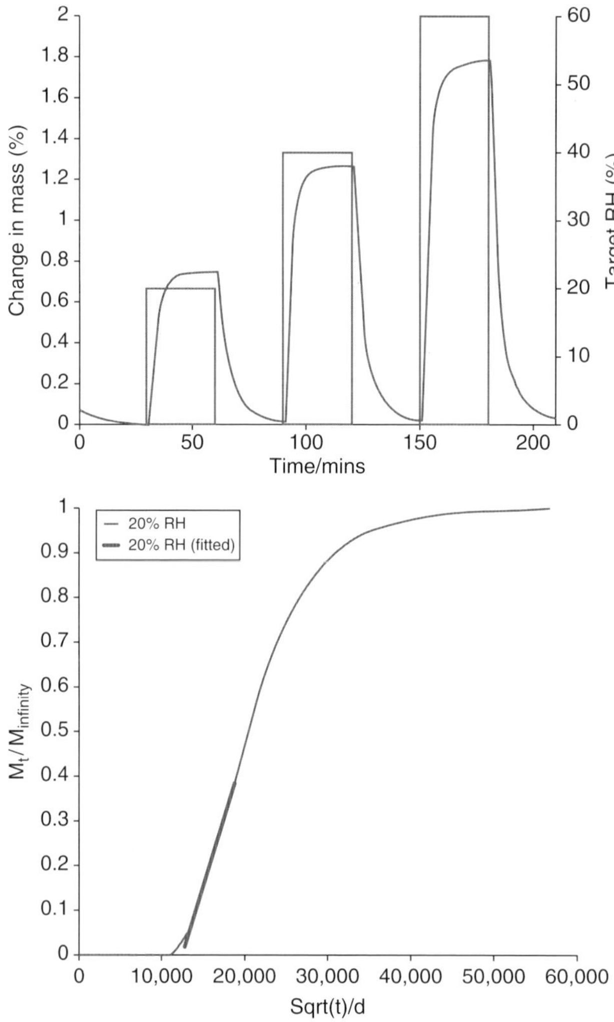

Figure 8.12 Sorption and desorption kinetics on a 7.5 μm polyimide film and calculation of effective diffusivity of the film. (Source: Levoguer et al. 2006. Reproduced with permission.)

Figure 8.12, such that both sorption and desorption steps were measured for each discrete humidity step. The effective moisture diffusion coefficient was then calculated from the initial slope of each step using equation given above and the values are listed in Table 8.7 for both sorption and desorption steps. It may be noted that the effective diffusion coefficient through the film during the sorption step is always smaller than the desorption step.

The shapes of sorption curves are very important for understanding the mechanism of solute migration into the barrier material. Alfrey et al. (1966) developed a classification system for diffusion of vapor through films based on the relative rates of solute mobility and polymer relaxation. The different diffusion mechanisms are strongly dependent on the glass transition temperatures of the polymers and are classified as Case I (Fickian), Case II and Anomalous diffusion. Case I diffusion is involved when

Table 8.7 Diffusion coefficients through polyamide films at different %RH. calculated from initial slopes. (Source: Levoguer et. al. 2006.)

Previous RH (%)	Target RH (%)	Diffusion Coeff. (cm^2/s)	R-squared (%)
0.0	20.0	7.63E-10	99.55
20.0	0.0	4.38E-10	99.58
0.0	40.0	9.04E-10	99.52
40.0	0.0	6.05E-10	99.59
0.0	60.0	9.30E-10	99.54
60.0	0.0	6.55E-10	99.57

solute diffusion rate is smaller than the polymer segment relaxation, characteristics of systems above their glass transition and in rubbery state. Sorption equilibrium is rapidly established under these conditions and time-dependent boundary conditions are independent of swelling kinetics (Crank, 1975). The shape of the sorption curve over time is described by Kt^n, and $n = 1/2$ in Case I diffusion, which implies that the slope of sorption vs. square root of time is linear. In Case II diffusion, diffusion is very rapid compared to polymer relaxation rates, implying that the system is below glass transition and in glassy state and therefore the polymer mobility is not rapid enough. This is the characteristic case of diffusion through glassy polymers and the boundary exists between the outer swollen shell and the inner glassy core of the polymer as the solute front advances into the core. The core diminishes in size until the polymer is at equilibrium. The sorption curve for this case is also represented by by Kt^n, and $n = 1$ in Case II diffusion. Anomalous diffusion occurs when the solute migration rate and the polymer relaxation rate are comparable. The systems lies between Case I and Case II diffusion conditions.

Recently, Karbowiak et al. 2007 have evaluated the transport properties in foods by means of reference molecules in combination with molecular investigation of physico-chemical interactions with the diffusing substances. Translational diffusion of a reference molecule such as fluorescein, determined by the fluorescence recovery after photobleaching (FRAP) method, displays a threshold of a critical water content inducing an increase of the molecular mobility, and demonstrates that multiple populations of a single molecular species can be involved in different diffusion kinetics. Further investigations at a molecular scale through high resolution solid state nuclear magnetic resonance (NMR) enables the understanding of the interactions involved in such a system to go deeper, in particular on the identification of the possible binding sites of the diffusing solute on the polymer and on the overall effect of interactions on the polymer organization.

The solubility of water in a matrix can be obtained from the sorption curve as well as in the DVS instrument, at different relative humidities. Sorption curves for any food material can also be obtained manually by equilibrating the food material in sealed desiccators containing supersaturated salt solutions at different relative humidities at a particular temperature, and measuring the weight gain at equilibrium under each condition. Typical salt solutions that are used for generating specific relative humidities are listed in an article by Debeaufort et al., 1994.

8.3.2 Moisture migration through barrier films with constant water vapor permeability

Barrier properties of dense, non-porous film and coating materials are generally described by a water vapor permeability coefficient, assuming that migration of moisture is purely diffusion driven by a chemical potential difference. Under this condition, the permeability of a solute (such as water molecule) through a film depends on three factors: i) the nature of the film material, ii) the nature of the solute molecule, iii) the interactions between the solute molecule and the film matrix. Hence, the transport process can be divided into three stages: adsorption of the solute penetrating from the high concentration side of the film, diffusion of the migrating solute through the film under concentration gradient, and desorption of the solute on the low concentration side (Rogers et al., 1962). Gas or vapor transport through a film can only occur by two mechanisms – capillary diffusion and activated diffusion. The term 'activated diffusion' has been used extensively in diffusion and permeability literature since 1930s (Chao and Rizvi, 1988; Doty, 1946; Barrer, 1938), and does not replace the term 'facilitated diffusion'. Capillary diffusion dominates in materials with pores or imperfections, while activated diffusion involves solubility of the gas in the film, diffusion through the film and then release at the opposite side of the film. In the absence of pores or imperfections, activated diffusion is the main mechanism, while if the gas is insoluble in the film then the main mechanism for penetration into the film is capillary flow. Activated diffusion applies to diffusion processes which are temperature dependent and possess some activation energy.

Fick's first law (Eq. 8.23) states that the permeate flux, J, is proportional to the concentration gradient in the film, with permeate diffusivity (D) as the constant for proportionality. The permeate diffusivity is a function of temperature. It is also observed that the effective diffusivity of moisture depends on the moisture content in the film. The negative sign indicates that the migration occurs in the direction of lower concentration (Donohowe and Fennema, 1994).

$$J = -D\frac{dc}{dx} \tag{8.23}$$

If Henry's law of solubility is obeyed, then the concentration (C) of permeate in the film is equal to the product of the solubility coefficient (S) and the permeate partial pressure in the adjacent air (P):

$$c = SP \tag{8.24}$$

Henry's law states that at a constant temperature, the amount of a given gas that dissolves in a given type and volume of liquid is directly proportional to the partial pressure of that gas in equilibrium with that liquid.

Combining Equations 8.23 and 8.24, and replacing dC by S. dP, we obtain,

$$J = -DS\frac{dP}{dx} \tag{8.25}$$

Where, dP is the difference in partial pressure of the permeate across the film. In thin, non-porous, non-swelling, highly hydrophobic film, WVP through a barrier is determined by combining Fick's First Law of diffusion (Eq. 8.23) and Henry's Law of Solubility (Eq. 8.24) as expressed in Eq. 8.25 (Park, 1995; ASTM, 1980). In this case, WVP is the product of the solubility and the diffusivity coefficients as shown in Eq. 8.26 below:

$$WVP = DS = -\frac{Jdx}{dP} \tag{8.26}$$

Numerous assumptions are involved in arriving at the above equation: (1) the film is assumed to be homogenous which is often not the case for hydrocolloid or biopolymer films due to varying water concentrations in different parts of the film, (2) film solubility and diffusivity are assumed to be constant, while it is very well known that both depend strongly on a_w and moisture content, and (3) it is assumed that the equilibrium is established at the film surface and the mass transfer resistance at the film surface is negligible. WVP is the result of two parameters: thermodynamic parameter S where the solubility depends on the compatibility between the solute molecule at equilibrium and the material the solute is migrating through; kinetic parameter D indicating water mobility in the material and highly influenced by the structural and morphological characteristics of the material.

Water vapor permeability is the time rate of water vapor transmission through unit area of flat material of unit thickness induced by unit vapor pressure difference between two specific surfaces, under specified temperature and humidity conditions, and is a property of the material. The terms 'permeability' and 'permeability coefficient' are used interchangeably (ASTM, 1990) for homogenous films and have the units g/s.m.Pa. However, 'permeability coefficient' should be used only when it is constant, irrespective of the pressure gradient and under these conditions, both Fick's law and Henry's laws are obeyed. When permeability of a film differs with %RH (for example, from 60% to 95%RH), then 'permeability' is a more suitable term to use (Donohowe and Fennema, 1994) than permeability coefficient. For the same reason, the partial pressure gradient should be reported. 'Permeability' measurement takes into account barrier thickness and it is assumed that a plot of permeation vs. thickness will yield a value of 1, or permeability is a function of thickness^{-x}, where x = 1. Usually, x varies between 0.8 and 1.2 for most polymers (Salame and Steingiser, 1977). Because of this inconsistency, film thickness should always be indicated. 'Permeance' is defined as the permeability coefficient devoid of the term for film thickness, and is not an inherent property of the film (ASTM, 1990). It is used when the film is heterogenous and /or of unknown thickness. Resistance to diffusion is the inverse of permeance.

Permeability is dependent on temperature following the Arrhenius relationship:

$$P = P_0 \exp\left(-\frac{E_p}{RT}\right) \tag{8.27}$$

Where P is permeability or permeance of edible films, P_0 is a constant, R is universal gas constant, and E_p is the activation energy barrier for permeation. The activation energy calculation is only valid for films yielding a linear relationship between log permeability and temperature $(K)^{-1}$, which implies that the permeant-barrier interactions change linearly with temperature. The condition is violated if there are structural changes in the film, such as crystalization or polymorphic alterations, with temperature.

The Water Vapor Permeability (WVP) of a film is a steady state property which describes the rate at which water passes through a unit area of the film of certain thickness, subjected to a given RH difference on the two sides of the film. The definition implies that the measurements are only valid when steady state is reached, which is implied by the condition that the amount of moisture reaching the film at one surface = amount of water migrating out of the film at the other surface. This is obtained when the breakthrough condition is reached, during which the film absorbs moisture to reach the steady state, and the time required to reach this condition depends on the film and its pre-equilibration conditions. Figure 8.13 shows one of the setups (aluminum cups with lids containing gaskets for tight seal) that is used routinely according to ASTM E96/E96M-10 method and is available commercially to perform such measurements. The cup must be made up of a non-corroding water-impenetrable material and the mouth of the dish should be at least 4.65 in^2, and the desiccant or water area should not be less than the mouth area. The test dish should be filled with distilled water to a level $\frac{3}{4} \pm \frac{1}{4}$ in. (19 ± 6 mm) from the specimen. The air space thus allowed has a small vapor resistance. According to ASTM E96/E96M-10 specification, the desiccant method should use 0%RH–50±2% RH condition, while the water method should use 100%RH–50±2% RH condition. The slope of each line yields the WVTR (water vapor transmission rate in $g/s\,m^2$) when divided by the active surface area of the film. The lower the slope, the better the moisture barrier property of the film. The slope of the line sometimes changes with time, as the film physical properties (glass transition etc.) change with increasing

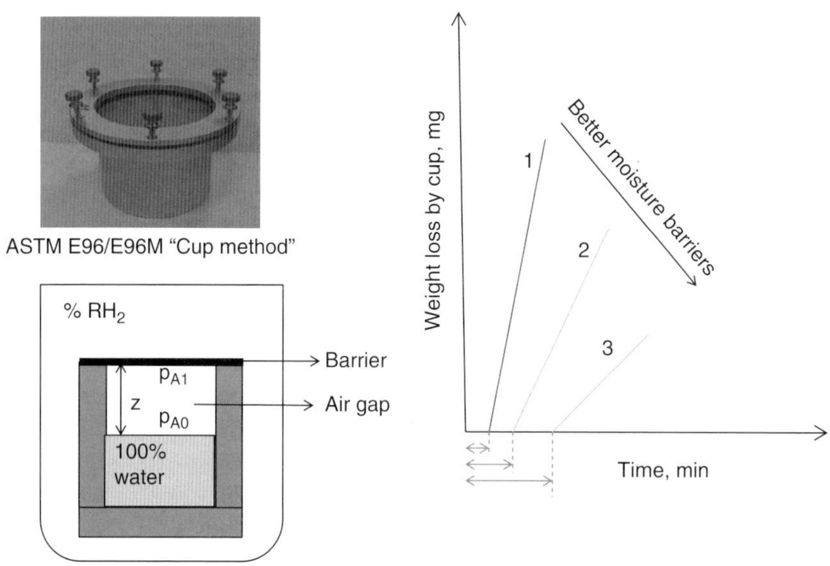

Figure 8.13 Water vapor permeability of film measuring setup (ASTM E96/E96M-10) and schematic of data obtained.

a_w and moisture content, and the effective diffusivity of moisture through the film matrix changes as well. The WVP is calculated from experimental data using the following equations:

$$WVTR = \frac{Weight\ of\ moisture\ lost}{time.\ Area\ of\ film} \quad \text{in gm/s m}^2 \tag{8.28}$$

$$WVP = \frac{WVTR.t}{P\left(RH_1 - RH_2\right)} \quad \text{in gm/s m Pa} \tag{8.29}$$

$$\text{Permeance} = \frac{\text{WVTR}}{\Delta\text{p}} = \frac{\text{WVTR}}{\text{P}\left(\text{RH}_1 - \text{RH}_2\right)} \quad \text{in gm/s m}^2\ \text{Pa} \tag{8.30}$$

$$\text{Average permeability} = \text{Permeance} \times \text{thickness} \quad \text{in gm/s m Pa} \tag{8.31a}$$

$$\text{Average permeability} = \text{Permeance}_1 \times \text{thickness}_1 + \text{Permeance}_2 \times \text{thickness}_2 + \ldots\ldots\ldots \tag{8.31b}$$

$$\frac{\Delta x_{total}}{slope_{total}} = \frac{\Delta x_1}{slope_1} + \frac{\Delta x_2}{slope_2} + \ldots\ldots\ \text{in ms/gm} \tag{8.31c}$$

Equations 8.31b and 8.31c demonstrate that if the individual permeance or thickness or slope of a composite laminated barrier or film is known, then the average permeability can be calculated easily and vice versa. Water vapor permeability of edible films, and flexible non-edible films is normally measured using a 100–0% RH gradient. However, several studies by (Greener and Fennema, 1989a; Kester and Fennema, 1989a, and Kemper and Fennema, 1984b) have revealed that for lipid-hydrocolloid bilayer films at 97–65% RH gradient is the worse test condition. Therefore, it will be prudent to conduct permeability studies with edible barriers under conditions, which is more realistic of the specific food product with specific a_w conditions. Permeability of edible films to lipids can be measured using techniques such as colorimetric assays, isotopes, or incorporation of lipid probes (Nelson and Fennema, 1991; Tabouret, 1987).

8.3.2.1 Corrections for resistance due to still air and specimen surface

The ASTM E96-80 method relies upon the assumption that there is negligible resistance to mass transfer in the gas phases on both sides of the film surface. This is true for the mass transfer resistance above the cup, if convection is provided by a fan at 500 ft/min, as recommended by ASTM (1980). However, the small distance between the mounted film and the saturated salt solution, or desiccant in cup, does not account for the desired level of convection and therefore, $P_{A0} \neq P_{A1}$. Due to this, large errors can occur in permeability calculations with highly permeable hydrophilic films, when the gas film provides the highest resistance to diffusion. Schwartzberg (1986) developed a model to correct for water diffusion through a stagnant gas layer to calculate the mass transfer through the stagnant air layer in the test cup, with a few assumptions, as given below:

$$WVTR = K_w(p_{A0} - p_{A1}) = K_w(p_{A0} - P^* a_w) \tag{8.32}$$

Where, WVTR is the steady state transmission rate or water per unit area of film, K_w is the mass transfer coefficient = D/RTz, D is the diffusivity of water through air, R is the universal gas constant, z is the depth of the stagnant air layer, P* is the vapor pressure of pure water, p_{A0} is the vapor pressure at the solution surface inside the cup, p_{A1} is the vapor pressure at the inner film surface, and a_w is the water activity in equilibrium with the respective film moisture content at surface 1.

$$p_{A1} = p_{A0} - WVTR\frac{RTz}{D} \tag{8.33}$$

Where D is the diffusivity of water vapor through air at the experimental temperature, T, and R is the universal gas constant. WVTR is obtained experimentally using Equation 8.28. The water vapor partial pressure at the inner surface of the film, p_{A1} can be calculated using Equation 8.33, and then WVP can be calculated using Equation 8.29. McHugh et al. (1993) suggested the use of the WVP Correction Method to account for the water vapor partial pressure gradient in the stagnant air layer within the test cups. The correction method involved a classical calculation for diffusion of water vapor through air (Bird et al., 1960) to calculate the mass transfer through the stagnant layer in the cup, knowing the experimental WVTR via Equation 8.28, and obtaining the resultant equation below (Krochta, 1992) with P is the total pressure:

$$WVTR = PD \ln\left[\frac{\dfrac{(P - p_{A1})}{(P - p_{A0})}}{RTz}\right] \tag{8.34}$$

Knowing the corrected WVTR from Equation 8.34, the WVP can be calculated from Equation 8.29. Equation 8.34 is the popular corrected WVTR equation used in most experiments, since the WVP correction method takes into account the convective flux induced by diffusion within the cup, contrary to the Schwartzberg method.

ASTM E96/E96M-10 specifies that if the material is highly permeable, these corrections are more significant. With known thickness of the still air layer in the cup, the corresponding vapor resistance can be calculated using the following Equation **8.31a** for permeability:

$$\delta_a = \frac{2.306\times10^{-5} P_0}{R_V\, T\, P}\left(\frac{T}{273.15}\right)^{1,81} \tag{8.35}$$

where: δ_a = permeability of still air, kg·m^{-1}·s^{-1}·Pa^{-1},
T = temperature, K,
P = ambient pressure, Pa,

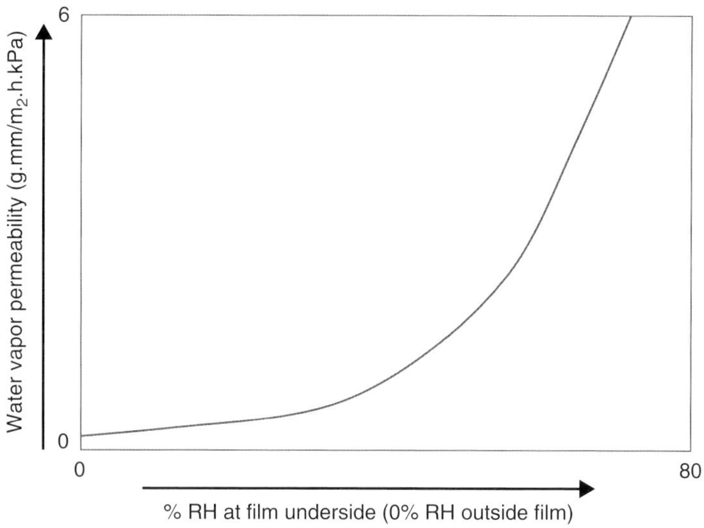

Figure 8.14 A schematic of the water vapor permeability of 62% WPI/ 38% glycerin films vs. % RH. (Source: McHugh and Krochta 1994.)

P_o = standard atmospheric pressure, that is, 101325 Pa, and
R_v = ideal gas constant for water, that is, 461.5 $J \cdot K^{-1} \cdot kg^{-1}$.

Knowing the permeability of still air and thickness of the film, permeance of still air can be calculated from Eq.8.31a. The average permeability of the setup is obtained experimentally, and hence from Eqs. 8.31a and 8.32b, the permeability of the film can be calculated as well. For cup methods that follow this standard, the total surface resistance (Hansen and Lund **(5)**) should be ≈4.3 $10^7 Pa \cdot s \cdot m^2 \cdot kg^{-1}$.

An excellent review on the permeability properties of edible films, instruments and methods used to measure them, mechanisms and factors affecting permeability is covered by McHugh and Krochta (1994). The water vapor permeability is not constant but is strongly dependent upon %RH as shown in Figure 8.14. The structure of the polymer chains, effect of chain packing, presence of bulky side groups are all important in determining the WVP of a polymer film, polymer crystallinity, glass transition temperatures, glassy vs, rubbery states of polymers, orientation of polymers, crosslinking of polymer chains have significant effects on WVP of polymer films.

8.3.3 Moisture migration through barriers including interactions of barriers with moisture

With regard to moisture permeation through hydrophilic, composite, and moderately hydrophobic films, D and S coefficients vary due to plasticization/swelling as a result of moisture absorption. Hydrophilic materials such as proteins and polysaccharides absorb moisture much more at the high a_w range, than hydrophobic materials. Water sorption in hydrophilic polymers is usually a non-ideal process leading to plasticization and/or clustering phenomena and, as a result, to complex type II or sigmoidal isotherms (Despond et al., 2005). A single value of WVP (often measured with a 0–100% RH difference),

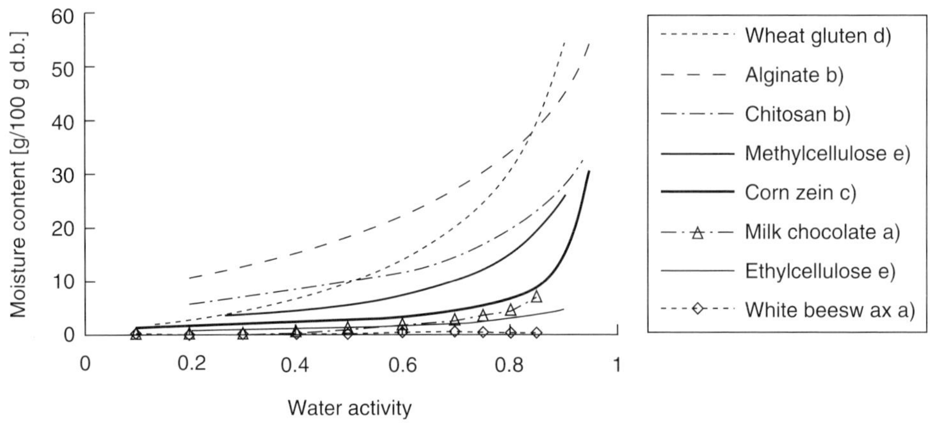

Figure 8.15 Examples of Guggenheim-Anderson-De-Boer (GAB) fitted moisture sorption isotherms of hydrophilic and hydrophobic barrier films at 20°C reported in literature. (Source: Bourlieu et al. 2009. Reproduced with permission of Taylor and Francis Ltd.)

therefore, cannot be considered as an inherent property of the film to reflect its water barrier efficiency over a broad range of testing conditions. On the contrary, WVP is then considered as an inherent property of the couple 'penetrant-barrier material' in a given testing configuration (Pasquat, 1986). Buonocore et al. (2005) presented the following classification of hydrophilic components based on increasing moisture sorption isotherm over the full aw range (25°C): chitosan (MW ~ 310 KDa, deacetylation degree >75%) < corn zein (glycerol 20% w/w) < sodium caseinate (glycerol ~40% w/w, cross-linked with transglutaminase) < sodium alginate (cross-linked with $CaCl_2$), as shown in Figure 8.15.

Buenocore et al. (2005) recently proposed a model that describes the dependence of WVP to water activity in hydrophilic film, as shown in Equation 8.36. This model integrated the non-linear sorption of water of hydrophilic polymer-based films (alginate, casein, chitosan, and zein) and moisture concentration dependence of their effective diffusivity as shown in Figures 8.15 and 8.16, respectively. The model was validated over a 0.3–0.8 a_w range at 20 °C. In particular, the moisture sorption isotherms of these polymers were described using the Flory equation and variations of moisture effective diffusivity was considered as an exponential function of the moisture content of the polymer.

$$\begin{aligned} WVP_{(aw1,aw2)} &= \frac{1}{p_0 \cdot (aw_1 - aw_2)} \cdot \int_{C_w(aw_2)}^{C_w(aw_1)} D \cdot dC_w \\ &= \frac{1}{p_0 \cdot (aw_1 - aw_2)} \cdot \int_{C_w(aw_2)}^{C_w(aw_1)} (D_0.\exp(\alpha.C_w) \cdot dC_w \end{aligned} \quad (8.36)$$

Where WVP(a_{w1}, a_{w2}) is the Water Vapor Permeability coefficient for a water activity difference between the upstream and downstream side of the film equal to a_{w1} and a_{w2} respectively, p_0 is water vapor pressure, D is the water diffusion in the polymer at a

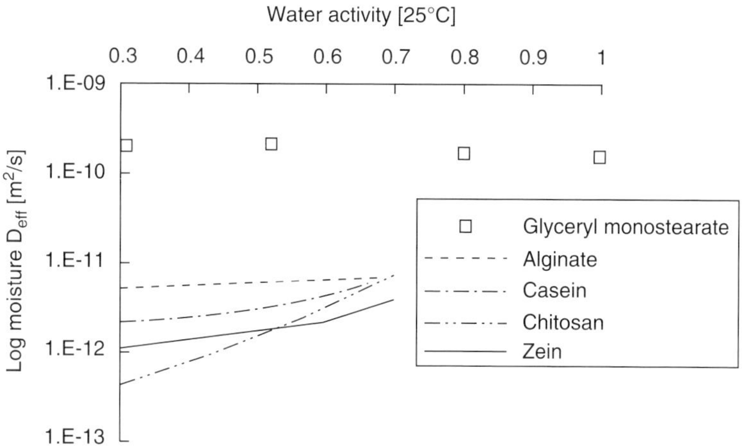

Figure 8.16 Moisture effective diffusivity (experimental data or model) as a function of water activity at 25°C. (Source: Buonocore et al. 2005. Reproduced with permission of John Wiley & Sons; Higuchi and Aguiar 1959. Reproduced with permission of John Wiley & Sons.)

given moisture content, D_0 is the water diffusion at zero moisture concentration, α is a constant that accounts for the ability of water to plasticize the polymeric matrix, C_w is the polymer moisture content.

For hydrophobic substances, the affinity and interaction with water is limited and hence their permeability should be less affected by the %RH conditions. Exceptions to this were found for fat materials, including acetyl-stearyl glycerols, milk chocolate, glyceryl monostearate, beeswax, and blends of these last two components (Buonocore et al., 2005; Higuchi and Aguiar, 1959; Landmann et al., 1960). This is attributed to the sorption of water by hydrophilic groups (ester, hydroxyl, carboxyl groups) of the hydrophobic substances, which results in a less dense structure and modifies the film permeability (Morillon et al., 2002).

Variation in WVP coefficients have been reported as a function of barrier film thickness. More specifically, hydrophilic films exhibit a positive slope relationship between thickness and WVP. This relationship has even been reported to be exponential for some materials (McHugh et al., 1993). Several explanations have been forwarded to explain the effect of thickness on WVP of barriers: i) the barrier film thickness can influence the barrier film structure and its homogeneity; ii) air gap between the solution and films lead to equilibrium moisture relationships at film/air interface differing from test cup solution equilibrium conditions, increase in thickness lowers the effect of such limit layers, iii) increased thickness lead to a higher quantity of hydrophilic component which may interact with water molecules causing swelling and apparent thickness effect (Mac Hugh et al.,1993). These observations seem to confirm that there exists a critical thickness, below which WVP increases due to the presence of defects and less favorable structure in thin barrier films. Above the critical thickness, which depends on the film composition and mode of application, similar homogeneity can be expected and anomalous behavior can only be explained by increased non-linear phenomena (swelling and sorption) due to the increased amount of hydrophilic groups that may interact with water.

8.3.4 Moisture migration in low, intermediate and high moisture foods

Low moisture foods are generally characterized as those having a_w lower than 0.6, intermediate moisture foods have a_w ranging between 0.6 and 0.9 (Table 8.8), and high moisture foods have a_w higher than 0.9. Generally, moisture transfer during storage of composite model foods has been less extensively analyzed and modeled than mass transfer phenomena during food processes such as baking, drying, frying, or rehydration. Karathanos, Kostaropoulos, and Saravacos (1995) discussed water diffusion in a dough/raisin system; (Guillard, Broyart, Bonazzi, Guilbert, and Gontard, 2003a, 2003b, 2003c, 2004) validated a predictive model on several composite foods composed of high water activity agar gels and cereal based-components, possibly separated by barrier films; Roca, Guillard, Guilbert, and Gontard (2006) successfully used the same model to determine the effect of porosity and fat content on the moisture transfer in a composite product (sponge-cake/high aw gel). Due to the lack of suitable analytical or simplified models, the shelf life of composite products is still often determined by the usual storage tests, which have to be repeated when any parameter of the product manufacturing process is modified. The mathematical models which are more complex and take into account the variation of sorption, diffusivity, WVP vs. a_w for all domains in the composite food system often take longer to validate due to the sheer enormity of the data required to generate and fit the model using some of these experimental parameters. Hence, the validation of such complex models for each specific composite product systems takes time and defeats the purpose of predictive utility of such models for shelf life of these systems. Therefore, easily validated and yet versatile, predictive models to assess moisture transfer in a wide range of food products are in demand and can benefit most of the sectors of the food industry and help meet the challenges associated with marketing innovative, long shelf stable and nutritionally acceptable products.

Reduction of moisture transfer in composite food products can be achieved by mainly two main techniques: reduction of a_w difference between components and use

Table 8.8 Examples of traditional intermediate-moisture foods. (Source: Taoukis and Richardson 2007. Reproduced with permission of John Wiley & Sons.)

Food	Water activity (a_w)	Principal Humectants[a]
Dried foods		
raisins	0.51–0.62	a, b (naturally occurring)
prunes, figs	0.65–0.83	a,b (naturally occurring)
Sugared fruits	0.57–0.79	a,b
Jams, jellies, marmalade	0.82–0.84	a,b
Fruit fillings of pastries	0.65–0.71	a,b
Dry salami	0.82–0.85	c,d
Country ham	0.88–0.90	a,c,f
Dried or salted fish	0.74–0.82	c,e,f
Honey	0.58–0.68	a,b
chocolate	<0.55	a
Parmesan	0.66–0.74	c,e

[a] Sucrose, a; other sugars, b; NaCl, c; food acids, d; amino acids, e; humectants f

of edible moisture barrier at the interface of the system. Both techniques were tested in a multi-domain model food associating a dry cereal-based component to a wet component using a model based on Fick's second law (Bourlieu et al., 2008). The input modeling parameters were moisture equilibrium and transport properties of the components. The two stabilization techniques permitted a significant extension of the period of acceptability of the dry component: from 7 min to more than 40 h using solutes addition and up to 12 days using a 300 μm hydrophobic barrier. A further extension of this period of acceptability was achieved by combining the two techniques.

Using the DVS balance, kinetics data were obtained over a wide range of a_w with low levels of a_w variation corresponding to transient-state of moisture content in the product. The dry biscuit D_{eff} values were obtained from the DVS balance and the transient-state moisture content during sorption are plotted on Figure 8.17. The dry biscuit D_{eff} demonstrated a strong moisture concentration dependence with three main regions of variation: (i) a steady state around 4.00 $10^{-10}\,m^2/s$ for moisture content ranging from 3 to 15 g/100 g (d.b.) corresponding to low and intermediate a_w and the stable initial porosity of the material; (ii) an increase from 4.00 10^{-10} to 12.20 $10^{-10}\,m^2/s$ between 15 and 21 g/100 g (d.b.), corresponding to 0.75 a_w at which dissolution of sugar can occur; (iii) a decrease between 21 and 30 g/100 g (d.b.) which can be linked to the closure of porosity starting after 0.75 a_w and subsequent collapse of the cereal matrix. At higher moisture content, both liquid diffusion through the solid matrix and vapor diffusion through the pores are taking place. Typical values are 10^{-9} and $10^{-5}\,m^2/s$ for liquid and vapor-phase diffusion, respectively (Saravacos and Maroulis, 2001). The effective moisture diffusivity through the hydrophobic lipid barrier was found around $0.05 \times 10^{-10}\,m^2/s$, and it remained almost constant. This low D_{eff} value is in line with its hydrophobic partially crystaline nature (Solid Fat Content at 20 ^{0}C ~88%). The moisture effective diffusivity identified on the 0.75 a_w gel dynamic desorption data demonstrate that D_{eff} increases with increase in moisture content. The plots of the diffusivity of the biscuit, agar gels and the lipid barrier are shown in Figure 8.17, taken from Bourlieu et al., 2008. The material properties of all these phases are listed in Table 8.9 (Bourlieu et al., 2008) and the moisture migration experimental and model predictions covered in details by Bourlieu et al. (2008). The experimental and predictive errors are most prominent for Δa_w of 0.8 (a_w range of 0.11–0.99) for the wet components both with and without the barrier; the experimental and predictive errors are also quite large for Δa_w of 0.8 (a_w range of 0.11–0.99) for dry component without the barrier. The experimental and predictive errors are quite small for all other a_w conditions, indicating that the model is capable of reasonable prediction for Δa_w of 0.5–0.6. These data provide a very good confirmation of the importance of both edible barriers and reduction in differential in a_w to retard moisture migration in low to high moisture food domains in composite food products.

8.3.5 Preservation issues with intermediate and high moisture foods

Low moisture foods are generally characterized as those having a_w lower than 0.6, intermediate moisture foods have a_w ranging between 0.6 and 0.9, and high moisture foods have a_w higher than 0.9. Table 8.2 demonstrates the category of foods in each of these regions. Table 8.4 and Figure 8.1 show that below 0.6 a_w, there are no issues with

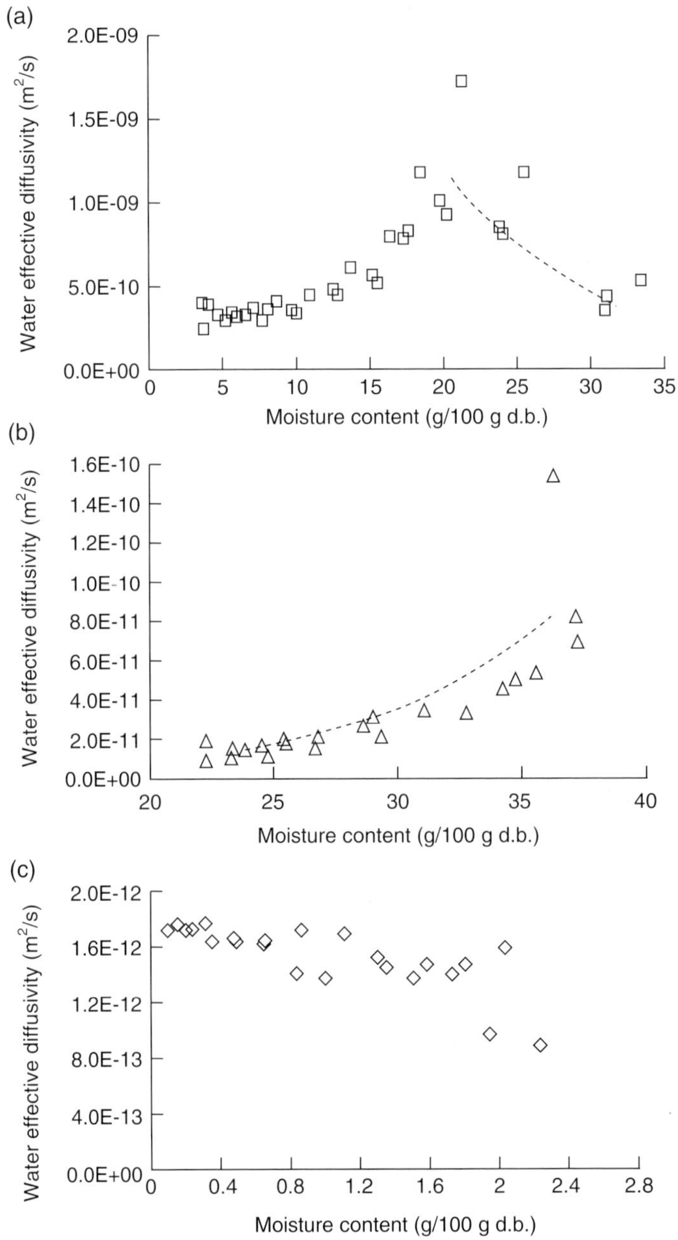

Figure 8.17 Moisture diffusivity in dry biscuit, agar gels of different aw, and the lipid barrier as a function of moisture content. (a) Moisture diffusivity in the dry biscuit at 20 °C identified from DVS balance sorption kinetics data: experimental data (□, 3 Replicates), fitting (- - -, RMSE = 4.99 x $10^{-10}$$m^2/s$) of the moisture effective diffusivity dependence with moisture content using the empirical form $D_{eff} = D_o.e^{(-k.X)}$ with D_{eff} the moisture effective diffusivity (m^2/s), D_o the pre-exponential factor (m^2/s), k a constant, X the moisture content (g/g d.b.). (b) Moisture diffusivity in the 0.75 a_w gel at 20 °C identified from DVS balance sorption experiments: experimental data (Δ, 3 Replicates), fitting (- - -, RMSE = 1.49 × 10^{-11} m^2/s) of the moisture effective diffusivity dependence with moisture content using the empirical form $D_{eff}= D_o.e^{(-K.X)}$ with D_{eff} the moistture effective diffusivity (m^2/s), D_og d.b.). (c) Moisture diffusivity in the Wax/ACETEM barrier at 20 °C identified from DVS balance sorption experiments (2 replicates). (Source: Bourlieu et al. 2008. Reproduced with permission of Elsevier.)

Table 8.9 Material characteristics of the model food components used for modeling purpose in the intermediate and low aw systems at 20°C. (Source: Bourlieu et al. 2008. Reproduced with permission of Elsevier.)

	Initial moisture content, (g/100g d.b.)	GAB parameters			Effective diffusivity (10^{-11} m²/s)
		X_w	C_g	K	
Biscuit	3.0	0.0545	8.6747	0.9365	10
0.75 a_w Gel	38.4	0.0877	10.5374	1.0440	8.81
0.70 a_w Gel	29.2	0.0877	10.5374	1.0440	2.43
0.60 a_w Gel	21.7	0.0877	11.4021	1.0385	0.15
0.50 a_w Gel	13.3	0.0809	11.5317	0.9497	0.10
0.64 a_w Jam	23.8	0.0845	4.0727	0.9826	0.15
Wax/ACETEM barrier	0.3	0.0085	1.2351	0.7893	0.05

pathogen but spoilage due to molds and some osmophilic yeasts, below 0.86–0.88, yeasts and molds are the main issues for spoilage while some Staph aureus may grow between 0.86 and 0.9. Above a_w of 0.9, most microbes will grow. Since intermediate moisture foods have a_w ranging from 0.6–0.9 and 10–50% water by weight, additional hurdles besides a_w, such as pH and temperature, are usually used for these foods. By bringing down a_w below 0.86, lowers the risk of pathogens and bringing down the pH below 4.6 (close to the pKa of organic acids used for preservation), provides additional stability to the growth of molds and yeasts. Lowering of a_w is often combined with addition of chemical preservatives (nitrites, sorbates, sulphites, benzoates, antimicrobials of natural origin, smoke components from wood), reduction of pH (which usually inhibits or decreases bacterial growth, enhances the action of preservatives by keeping the acids mostly in their active undissociated states, and increases the minimum a_w values for bacterial growth). Other intermediate-moisture products receive a thermal treatment or pasteurization during the manufacturing process that inactivates heat-sensitive microorganisms and in some other cases, hot filling in containers further improve microbial stability (Leistner and Gould, 2002). Most intermediate-moisture foods are designed or desired to be storable for several months at room temperature, even in tropical climates, and eaten without rehydration. They should be wet enough to be ready-to-eat without giving a sensation of dryness when eaten, but dried enough to be shelf stable (Karel, 1973,1976; Jayaraman, 1995). However, most intermediate-moisture foods are lowered in a_w with high amounts of solutes (sugar or salt) and are either too sweet or salty and have both undesirable nutritional and sensory characteristics. High moisture foods with a_w above 0.9 are prone to the prolific growth of all microorganisms and hence stability at room temperature for this category is obtained by applying carefully designed hurdle technology, involving interaction of a_w with pH, preservatives and heat treatment. This represents a rational application of the combined approach making the products storable without refrigeration (Alzamora et al., 1995; Leistner and Gould, 2002).

Jorge Welti-Chanes et al. (2007) listed the following ways that a_w can be influenced during the preservation of intermediate-moisture foods: Water can be removed by dehydration, evaporation, or concentration processes. Additional solutes can be incorporated as dry or wet infusion, that is, added as dry or pre-wet manner. When dry solute incorporation is done, the process is called 'osmotic dehydration' as the solutes will prefer-

Table 8.10 Minimal a_w for growth of Staph. Aereus in media where a_w is adjusted with different solutes. (Source: Chirife 1994. Reproduced with permission of Elsevier.)

Solute	Minimal a_w
Ethanol	0.975
Porpylene glycol	0.96
Polyethylene glycol 200	0.93
Glycerol	0.89
NaCl	0.86
Sucrose	0.867

entially absorb and bind water making it less available for a_w. This allows the inclusion of desired amount of antimicrobial and antibrowning agents, solutes for improving sensory and nutritional quality. For porous foods, moist infusion can be done under vacuum. The internal gas or liquid occluded in the pores is exchanged for an external liquid phase (of controlled composition) due to pressure changes. Combination of methods 1 and 2 – in this case the food pieces are infused with the solutes and additives, and then partially dried.

Several mathematical models exist to predict the most stable a_w value for intermediate and high moisture foods, which can be used to reformulate a food product for quality and stability improvement. Studies by van den Berg and Bruin (1981), Leung (1986), Rizvi (1986), and others (Welti-Chanes et al. 2007) show several reviews about procedures used to evaluate or predict a_w. Several theoretical and empirical models suggested for the calculation of a_w in semi-moist and moist foods are given by van den Berg and Bruin (1981), van den Berg (1986), Chirife (1995). It should be noted that different solutes have specific a_w for arresting the growth of each microbe, as shown in Table 8.10. Compatible solutes (which do not appreciably interfere with the metabolic and reproductive functions of the cell) are usually nonionic solutes as many enzymes lose their activity in high salt concentration. Amino acids (proline, α-keto glutarate, γ-aminobutyric acid, glutamic acid) appear to be common solutes in bacteria, while polyols (mannitol, arabitol, sorbitol, glycerol, erythritol etc.) are compatible solutes for many fungi. Although many bacteria are more tolerant to glycerol and less to NaCl, Staph aereus is affected adversely by glycerol and ethanol. Hence, care should be taken in choosing solutes to bring down water activity, keeping in mind the target organisms and other hurdle technologies associated with formulation and processing.

8.4 METHODS OF PREVENTING OR DELAYING MIGRATION (FILM BARRIERS, REDUCTION OF GRADIENTS)

8.4.1 Reduction of a_w gradients or water activity equalization

Water activity reduction is usually aimed to bring the different domains to an equilibrium value, which will prevent any further moisture migration during storage of the composite food. Ingredients, capillary size and distribution, and surface interactions are three factors contributing to the water activity of each food domain in a multi-domain system.

The first factor is the colligative property of water activity, the reduction of which can be achieved by the use of humectants such as salts and sugars or small water soluble molecules in solution and increasing the number of solute molecules. However, there is only a restricted number of safe, effective, and palatable a_w lowering agents. The biggest challenge in adding humectants is that the concentration of humectants that can realistically be added is limited by sensory and nutritional considerations. Glycerol, the most effective a_w reducing agent, tends to provide a syrupy, sweet and medicinal taste to savory foods, which is undesirable. High salt and sugar are not nutritionally favorable. Other solutes like sorbitol or lactose have side effects – the former has laxative properties and the other is not suitable for consumption by lactose-intolerant consumers. Also, by simply reducing the level of water in foods or dehydrating them, the water activity can be reduced at the expense of loss of creamy or soft texture. Multi-textured foods are all about the sensory appeal of multiple textures that will trigger the pleasant sensory feeling when they are eaten. If the texture differences are minimized or driven towards unfavorable directions (like making a creamy product tough and chewy by dehydration) then the product is no longer viable for commercial purpose. Hence, it will be highly beneficial to the food industry to come up with more innovations in identifying water activity reducing small molecules that are bland to taste and have no physiological or nutritional deleterious effects. Historically, small solutes have been used to reduce the water activity by binding water to themselves. Some of the common solutes that are widely used for this purpose are sugars (sucrose, glucose, fructose, and lactose), salt (sodium chloride, potassium chloride, and calcium chloride), and polyols (propylene glycol, glycerol, sorbitol and other sugar alcohols, amino acids, organic molecules) as shown in Table 8.11. Potassium lactate, trehalose, oligosaccharides, and small molecular weight carbohydrates, surfactants, with high water binding properties are also used to reduce the water activity of high moisture domains.

To predict the effect on change of a_w by adding an ingredient, S, to the food, Raoult's law (Eq. 8.8) can be utilized. Raoult's law is fairly good at predicting the a_w of a multi-domain component at high water activities or low solute content since it applies to ideal dilute solutions only. Raoult' s law can be modified to accommodate deviations from

Table 8.11 Common solutes for reducing a_w of foods. (Source: Adapted from Galmarini et al. 2008; Labuza and Hyman 1998. Reproduced with permission of Elsevier.)

Solute	Maximum solubility in water (%)	a_w of saturated solution
Lactose	20	0.97
Glucose	47	0.92
Fructose	75	0.63
Sucrose	69	0.86
Trehalose*	40.6% (20 °C), 46.8% (30 °C), 52.4% (40 °c)	0.953, 0.943, 0.922
KCl	27	0.86
NaCl	26	0.75
Glycerol	100	0.00
Sorbitol	70	0.79

ideality as a function of concentration [28, 30] by using the Norrish Equation, which accounts for multiple solutes and the concentration dependence of γs:

$$\ln a_w = \ln X_w + \sum_{i=1}^{j} \frac{K_i X_i^2}{X_i^2}(1 - X_w^2) \tag{8.37}$$

Where, x_i is mole fraction of each solute i , K_i is correction factor derived from the activity coefficient for each solute, and j represents the number of solutes where

$$\gamma = K_i X_i^2 \qquad \text{Eq. 8.38}$$

The second factor that can reduce water activity is the capillary effect. The change in degree of intermolecular hydrogen bonding between water molecules due to surface curvature effect leads to a difference in vapor pressure of water above a curved liquid meniscus versus that of an infinite plane of pure water. Near a concave meniscus, more water molecules are near to each other and interact, compared to a planar surface and hence cannot escape from the liquid to the air phase. If a food matrix has a large number of nano pores, then a_w will be lowered appreciably. The water activity in a capillary structure can be related to the meniscus radius as follows:

$$\ln a_w = \frac{\Delta P v_m}{RT} \qquad \text{Eq. 8.39}$$

$\Delta P = \Delta\rho gh = 2\gamma \mathrm{Cos}\theta / r$, where ΔP is pressure surface force or Poynting pressure, Δρ is density of water minus density of air, g is gravitational constant, h is height of meniscus, γ is surface tension=71.9 dynes/cm for pure water at 25 °C; r is capillary radius, v_m is liquid molar volume, and θ is wetting angle(=0 if perfectly wet) (Labuza and Hyman, 1998). Therefore a smaller capillary will have a higher suction pressure, which will reduce the vapor pressure above it and thereby a_w. If pore size is in the range of 10–300 μm, the lowering of a_w is very small in the range of 0.989–0.999. But for very small pore sizes (0.01–0.001 μm), a_w can be lowered above the meniscus to values of 0.28 (Capillary effect can occur over the entire range of moisture sorption isotherm in foods, with small pores filling first and then the larger pores).

The third factor that affects a_w is the interaction of water with other chemical groups of molecules through dipole-dipole forces, ionic bonds, van der Waals forces and hydrogen bonding. These water molecules require that extra energy beyond the heat of vaporization available to transfer one molecule from the liquid to the vapor state.

8.4.2 Edible film barriers

Due to non-linear water sorption isotherms, water-dependent diffusivities, and variations of physical state, modeling transport phenomena through edible barriers is complex. Water vapor permeability can hardly be considered as an inherent property of films and only gives a relative indication of the barrier efficacy. However, a general classification of polysaccharides and protein-based films cannot be established due to: i) the influence of formulation (degree of polymerization of the macromolecules, presence of lateral groups, addition of plasticizer, addition of non-lipid components), ii) the influence of film forming conditions (pH, thermal treatment, or any cross-linking treatment) and iii) the shift in the relative positions of moisture sorption isotherms over the full range of a_w.

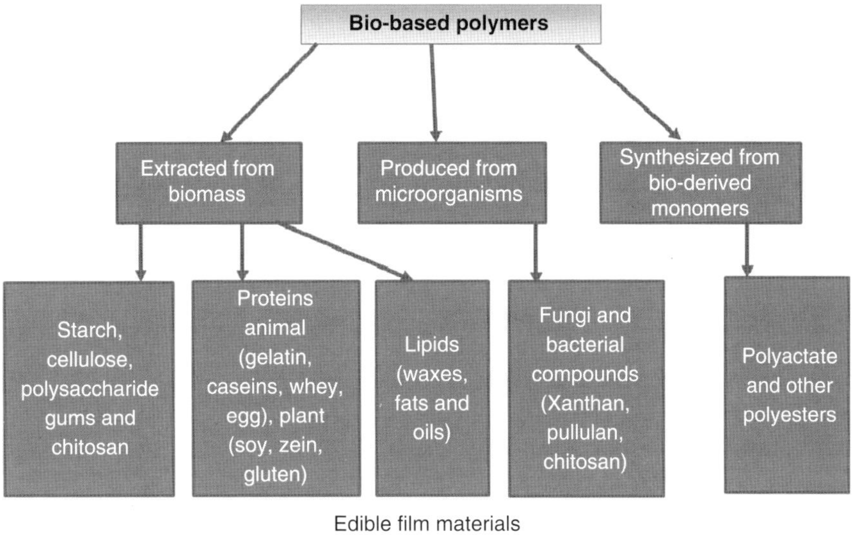

Figure 8.18 Different types of biomaterials used in edible films and packaging. (Source: Debeaufort and Voilley 2009. Reproduced with permission of Springer-Verlag.)

Edible protective films or coatings have been generally defined as thin layers of material which are eaten by the consumer and provide a barrier to mass transfer (moisture, oxygen, and solute movement) in the food itself or between the food and its environment. Films are distinguished from coatings, since they are formed as stand-alone sheets of material, whereas coatings are directly formed on the product (Guilbert, 1986). Several comprehensive reviews on edible barrier films and coatings have been provided by Han et al., 2005; Perez-Gago et al., 2005; Debeaufort et al., 2009; Rhim et al., 2005; Shellhammer et al., 1997a; Sothornvit et al., 2005; Krochta et al., 1994; Gennadios, A. 2002. Some commonly used biomaterials used for edible films and biodegradable packaging are illustrated in Figure 8.18. Some commonly used edible films are of vegetable or seaweed origin – such as rice papers used in Asian spring rolls, seaweed based Nori films used in Sushi, grape leaves used in Turkish Dolma, cabbage and lettuce leaves used for similar stuffed foods preparation. Recently, vegetable and fruit based Oregami films are also available to replace seaweed Nori films for consumers that are allergic to seaweeds.

Moisture barriers have been used for centuries to extend the shelf-life of fruits and vegetables and one of the common agents used for this purpose is bee's wax. Edible coatings are used to retard moisture transfer between components in composite foods. Often these coatings are hydrophobic and repel water; oils are hydrophobic materials that can serve this function well. Pure proteins and carbohydrates are hydrophilic and thus are poor coating materials to prevent moisture migration. However, proteins and carbohydrates in conjunction with hydrophobic materials can provide better protection because of the alteration of mechanical properties of hydrophobic materials. Such combinations provide better surface coverage and flexibility than hydrophobic materials alone.

An ideal film would be cost advantageous, safe, made with food grade materials, easy to manufacture, provide effective deterrence against moisture, flavor and oxygen migration, should be biochemically, physico-chemically and microbially stable, and would be neutral flavor-wise and sensorially acceptable. Migration of moisture, oxygen and flavor can occur from either the exodus from one food component to another or the ingression of water, air and flavor from the food component adjacent to the one being protected. Neither case is desirable. There is no material that is capable of forming an ideal film. Therefore combinations of materials are often used to form edible films. The three major classes of macronutrients, carbohydrates, proteins and lipids have been used successfully in conjunction with other materials to form edible films and barriers.

The methods of manufacturing edible films are varied. The film forming material is dissolved in solvent and other materials such as antimicrobial agents, color, flavor, plasticizers are added and the solution may be heated to facilitate molecular changes. Adjustments in pH can also facilitate film formation. The solution is then cast and solvent evaporated under controlled temperature and relative humidity conditions. As moisture evaporates, a film is formed as a free standing sheet. Alternately, the solution can be used to dip the food substrate being protected, or the solution can be sprayed, brushed or panned to evaporate the solvent (Donohowe and Fennema, 1993). Edible films can be obtained by extrusion, co-extrusion, spreading, casting, roll coating, drum coating, pan coating, or laminating techniques. Edible coatings, on the other hand, are mainly applied using spraying, drum coating, spray-fluidization, pan coating, or falling film techniques. Although plastics can effectively protect a food product from its surrounding environment, they cannot be used to reduce mass transfer within a composite food. For instance, pizza crusts rapidly lose their crispiness because of moisture transfer from tomato paste. In heterogeneous composite foods, moisture migration always occurs from the wet compartment to the dry one, either as liquid or as vapor, even when the wet compartment contains solid water, as is the case with frozen products. To prevent such transfers, lipid based coatings can be applied. Table 8.12 lists a range of edible materials used as barriers along with their water vapor permeability under suitable conditions.

8.4.2.1 Carbohydrate based films

Generally polysaccharides like starch, gums, cellulose and so on have molecular structures conducive to film formation. However, since carbohydrates interact readily with water they may not be ideal film formers for preventing moisture migration. Carbohydrates are used in conjunction with proteins or lipids to obtain the desired properties.

8.4.2.1.1 Cellulose and its derivatives

Cellulose is the most abundant material in nature. It is a polymer of glucose linked through β-1,4 linkages. It has a highly crystaline structure and in plants other substances like lignin etc. coexist with cellulose. Due to its crystallinity cellulose *per se* is not used for film formation because it is virtually impossible to solvate it in any medium. Cellulose is converted to carboxymethyl-cellulose (CMC), hydroxypropyl cellulose (HPC), hydroxypropylmethyl cellulose (HPMC) and methyl-cellulose (MC) derivatives through chemical reactions. These derivatives are hydrophilic and are capable of holding large amounts of water and form gels.

Methylcellulose films are rough, clear and brittle posing problems in practical applications of the films. To improve the properties of methylcellulose films plasticizers are used. Common plasticizers include water, polyhydric alcohols (propylene glycol,

Table 8.12 Comparison of moisture WVP (20 °C) ranges for edible coatings in comparison with most commonly used synthetic packaging films. (Source: Adapted from Saravacos and Maroulis 2001. Reproduced with permission of CRC Press; Bourlieu et. al 2009. Reproduced with permission of Taylor & Francis Ltd.)

Film or coating	WVP[10^{-11} g.(m.s.Pa)$^{-1}$]	RH difference [%]	Thickness [mm]	Temperature [°C]
Synthetic films				
Aluminium	0.0005	0–98	(–)	38
HPDE	0.002	0–100	0.019	27.6
LPDE	0.014		0.010	27.6
PP	0.010		0.025	25
PVC	0.041		0.012	27.6
Lipid films	0.03–1.0	0–100		20
Waxes				
Candelilla wax	0.014	0–100	0.14	24.9
Paraffin	0.023		0.66	25
Beeswax	0.103		0.14	25.9
Carnauba wax	0.114		O.130	27.5
Fatty acids				
Capric acid	0.38	12–56		
Palmitic acid	0.65			
Stearic acid	0.22			
Triglycerides and derivatives				
Tripalmitin	0.225	0–100	0.130	27.5
Triolein	12.100	22.84		25
Anhydrous milkfat	1.028	0–100	0.130	24.9
Hydrogenated peanut oil	3.863		3.39	25
Glyceryl monostearate	0.77		1.026	24.3
Acetomonopalmitine	11.35		0.200	20
Tempered cocoa butler	26.5–61.2		1.59–2.92	26.7
Milk chocolate	88.99		0.68	20
Protein films	1.0–100.0	0–100		20
Wheat Gluten	12.97	0–100	0.053	20
Soya	281.18	50–66	0.072	25
Corn zein	11.6	0–85	0.12–0.33	21
Polysaccharide	1.0–10.0	0–100		20
Cellulose derivatives	9.2–11.0	0–85	0.04–0.07	21
Starch	25–78	11–100	0.005–0180	30

glycerol or polyethylene glycol) which provide improved tensile strength and modulus of elasticity (Grover, 1993). Plasticized films are flexible and can be used to provide a barrier to lipids and flavor compounds. Both MC and HPMC are stable over a wide range of pH but are affected by salt concentrations depending on the type of salt and the concentration or ionic strength. Plasticized films of MC and HPMC are used with extruded seafood and potatoes to reduce oil pick-up during deep fat frying. HPC films are flexible, lack tackiness and exhibit good heat sealability and can be used to modify other film forming materials like shellac, ethyl-cellulose, CMC and starches to improve flexibility, toughness and prevent cracking (Zecher and Gerrish, 1997).

Rarely is CMC used for free or unsupported film formation but it can form clear films by evaporation of solvent. Formed films are strong, flexible and withstand oils, fats, grease. Inclusion of 10–30% glycerol in film formation increases elongation and folding endurance. Water insoluble film can be formed by CMC by treating a formed film with trivalent cations which replace sodium (Feddersen and Thorp, 1993). This ion exchange insolubilizes the CMC film.

Moisture sorption isotherms and water barrier properties of films made from hydroxypropyl methylcellulose (HPMC) containing surfactant mixtures of sorbitan monostearate (SPAN 60) and sucrose palmitate (sucrose ester P-1570) were evaluated at 10° C (Villalobos et al., 2006). The effect of hydrocolloid/surfactant ratio (H/S) (0.5, 1.0 and 1.5) and the hydrophilic/lipophilic balance of the mixture (HLB: 4.7, 6.0 and 8.0) was analyzed. GAB and BET sorption models were tested to fit the experimental data. The equilibrium moisture content of films increased dramatically above aw=0.6. Films with greater H/S ratio have greater water binding capacity. For a specific hydrocolloid/surfactant ratio, equilibrium moisture content of the coatings decreased as the HLB of the surfactant mixture increased. Equilibrium moisture contents of the composite films could be predicted from the corresponding values of pure compounds and the respective component mass fractions by a linear model, in the water activity range of 0.11–0.75. At the high relative humidity at which water permeability of the films was evaluated, the water vapor barrier was effective when the films reached a critical amount of surfactants (H/S=0.5). The equilibrium moisture content of HPMC films depends on the percentage and chemical nature of the surfactant compounds incorporated into the hydrocolloid matrix. The addition of surfactants decreased the equilibrium moisture content of the films; the higher the polarity of the surfactant mixture the greater the decrease. This can be attributed to hydrogen bond interactions between hydrocolloid and polar groups of surfactant, thereby reducing the number of polar groups available to interact with water molecules. At the high relative humidity at which the water transmission rate of the films was measured; water permeance is controlled by a critical ratio of surfactant without significant effect of the equilibrium moisture content probably due to the high level of plasticization of the hydrocolloid matrix.

8.4.2.1.2 Starch based films

Starch is abundant in nature and can be obtained from grains, root tubers and tree trunks. A majority of the starch is derived from corn but rice, wheat, sorghum, potato, sago and tapioca starches are also available as food ingredients. Starch is made up of amylose and amylopectin which are glucose polymers with α, 1–4 linkages between the glucose units. Amylopectin is a highly branched polymer with branch points occurring with α, 1-6 linkages. Starch exists as granules and the relative proportions of amylose and amylopectin vary with the source of the starch. Amylopectin is responsible for the crystallinity of the starch granule and as the content of amylopectin increases so does crystallinity (Zobel and Stephen, 1995). Starch granules in aqueous suspension heated to 90° C or above undergo gelatinization. Gelatinization is an irreversible process in which hydrogen bonds are ruptured, starch granules take up a lot of water leading to swelling and melting of the crystaline structure leading to solubility of the granules. These changes can be detected by differential scanning calorimetry (DSC). Gelatinization temperatures vary based on the source of starch because of the differences in the amounts of amylose and amylopectin in each source. Upon cooling gelatinized starch, a phenomenon called retrogradation occurs. Retrogradation is the aggregation and ordering of linear starch chain segments and can occur in the hydrated swollen granule state or in solution. Retrogradation results in textural changes described as hardening and staling and in solution cloudiness develops followed by precipitation.

Starch is modified to obtain different functionalities (Wurzberg, 1995). Among these modifications are acid conversion (viscosity lowering), oxidation (stabilizing, adhesion, gelling and clarity), dextrinization (binding coating, encapsulation and solubility), crosslinking (thickening, stabilization, suspension and texturization), esterification (stabilization, thickening and clarification), etherification (stabilization and low temperature storage) and dual modification (combination of properties).

Starch and its derivatives have to be solubilized prior to film formation and the concentration of starch has to >4% to achieve gelling and uniform spreadability (Liu, 2005). On the other hand, high concentrations (>30%) of starch are too viscous to form uniform films. Solubilizing starch necessitates heating to achieve gelatinization or the use of solvents such as butanol. Self-supporting films are produced by casting the starch solution on a rotating drum or a moving belt and then gently removing the cast film. Sometimes a release agent (such as a surfactant) may facilitate the ease of removal of the film. Drying the cast film is essential and is best achieved under very gentle conditions in order to maintain film integrity.

Water transport in edible films of starch based products is a complex phenomenon due to the strong interaction of sorbed water molecules with polymeric structure of starch. Water sorption isotherms on high amylose starch based films are highly non-linear in the range studied (5–45 °C). Permeability is usually affected by a number of parameters such as temperature, film thickness and plasticizer content. This knowledge is required to simulate and predict transport patterns. Bertuzzi et al. (2007) studied water vapor permeability (WVP) of high amylose corn starch (HACS) films following a standard procedure. They observed that WVP increases as temperature increases between 5 and 40 °C and results obtained can be correlated with an Arrhenius-type expression. Activation energy of WVP for the starch film investigated is 5.61 kJ/mol and values of WVP ranged between 2×10^{-10} and 1×10^{-9} g m^{-1} s^{-1} Pa^{-1}. A direct relation between WVP with plasticizer content and film thickness was reported. These observations could be explained in terms of variations in the number of potential intermolecular interactions in the film structure.

8.4.2.1.3 Non-starch, non-cellulosic carbohydrate films

This category encompasses gums/stabilizers/hydrocolloids and materials such as chitin and its derivatives. Algin is derived from the brown seaweeds and there are several sources of this hydrocolloid. Some common brown seaweed types are *Macrocystis pyrifera*, *Laminaria hyperborea*, *Laminaria digitata*, *Laminaria japonica*, *Ascophyllum nodosum*, *Eclonia maxima* and *Lessonia nigrescens*. Depending on the type of seaweed the algin content (measured as alginic acid) varies from 20–41% and their gel strengths at isoconcentrations vary as well (Clare, 1995). Algin is made up of D-mannuronic and L-guluronic acid molecules occurring in blocks referred to as M-block and G-block, respectively. They can also occur as repeating blocks of M and G blocks referred to a MG-block. The abundance of these blocks determines the functionality of alginate for example, when the G-block content is high the gel strength is high, high M-block content produce medium gel strength and greater resistance to syneresis.

Soluble alginate films are obtained by water removal or exchange with a water miscible solvent such as acetone or isopropanol. Insoluble alginate films can also be made by converting soluble alginate to an insoluble form by the addition of appropriate cations. Alginate films have very good wet and dry strength properties (Clare, 1995). When completely dried, alginate films are brittle but are softer and flexible at higher moisture content. Alternately, hygroscopic materials such as glycerol can be added to increase flexibility. Alginate films are often reversible; dry soluble films can be rehydrated and dissolved while insoluble alginate films are rendered soluble by changing the pH > 7 by or contacting with dilute salt solutions. These films have great oil and grease barrier properties. In terms of practical applications fresh meat and poultry can be sprayed or dipped in an alginate solution, followed by spraying or dipping in a calcium chloride bath to form a film. This protects the meat and poultry from dehydration and handling damage (Gennadios et al., 1997).

Chitin is a polysaccharide derived from the exoskeletons of shellfish, insects and mushrooms. Chitin is a linear polymer of N-acetyl-D-glucosamine and D-glucosamine in which > 70% of the polymer consists in the acetylated form. Chitosan is a related polymer that contains the same two monomers as chitin in which the D-glucosamine is predominant. Both chitin and chitosan contain reactive hydroxyl and amine groups (Winterowd and Sanford, 1995). An extensive review of the use of chitosan in food preservation (No et al., 2007) lists numerous instances where chitosan films have been used to preserve fruits and vegetables.

The effects of acid (acetic, formic, lactic, propionic) concentrations, plasticizer concentrations, and storage time (up to 9 wk.) on mechanical properties, water vapor permeability, and oxygen permeability of solution-cast chitosan films were determined (Caner et al., 1998). Measured water vapor permeability coefficients ranged from 5.35–13.20 $\times 10^{-1}$ g/m·day·atm. Oxygen permeation coefficients ranged from 0.08 to 31.67 $\times 10^{-3}$ cc O_2/ m·day·atm. Neither property was affected by storage time. Tensile strength (6.85–31.88 MPa) also was not time dependent, but elongation (14–70%) decreased with storage time. Lactic acid solutions produced the lowest oxygen permeability values, formic acid the highest. Films formed with 7.5% lactic acid solutions had uniquely high values for elongation at break.

Oxygen permeability coefficients (OP), water vapor permeability coefficients (WVP), ethylene permeability coefficients (EP), tensile strength (TS) and percent elongation (%E) at break values were determined for chitosan films plasticized with glycerin at two concentrations (0.25 and 0.50 mL/g chitosan) (Butler et al., 1996). Film samples were tested after 0, 2, 4, 8 and 12 wk. of storage. After an initial drop in permeability during the first 2 wk. of storage, mean OP (4.6×10^{-5} cc/m.day.atm) and mean EP (2.3×10^{-4} cc/m.day.atm) remained constant while mean WVP (2.2×10^{-1} g/m.day.atm) decreased with respect to storage time. TS values (15–30 MPa) decreased and %E values (25%–45%) increased with respect to storage time. The stability of OP and EP values with storage was not expected, while the change in mechanical properties was as expected.

In a study (Wiles et al., 2000) measured the water vapor transmission rates (WVTR) and moisture sorption of chitosan films over a range of water vapor pressures at 25 °C. Films of a constant thickness were made using chitosan with 3 levels of deacetylation. Films were tested at test relative humidity (RH) difference between RH values of 84%, 75%, 69%, 53%, 43%, 33%, 23%, 11%, and 0 at 25°C using ASTM F1249-90 or ASTM E 96-80. The equilibrium moisture content in the films ranged from 3.7% to 31.8% (dry basis) corresponding to 11% to 84% RH. WVTRs of films increased with increase in water vapor pressure. The mean WVTR ranged from 6.7 to 1146 (g/m^2/d) over the range of water vapor pressure from 2.68 (11% RH) to 19.9 mmHg (84% RH). The percentage of deacetylation of chitosan films and the viscosity of the cast solution did not have an effect on the WVTR properties of chitosan films.

Edible chitosan coatings showed anti-*Listeria monocytogenes* effect evaluated by numeration and epifluorescence methods, imparting a strong localized functional effect at the food surface by active packaging (Coma et al., 2002). The use of film-forming solution in culture liquid medium showed a known flocculant phenomenon combined with bactericidal activity, keeping 20% of the initial microbial charge as viable cells in flocculant, which could develop subsequently. However, chitosan film showed 100% of *L. monocytogenes* inhibition for at least 8 d, completed by bactericidal activity measured by epifluorescence assays. A decrease in anti-bactericidal effect with time was obtained, most probably due to a decreasing availability of amino-groups of chitosan. Latter results were validated on Emmental cheese samples using *L. innocua* as model strain because of its non-pathogenicity.

Plastic is readily available and inexpensive, so it is becoming the main material for packaging. Unfortunately plastics do not biodegrade and, if reduced in small pieces, contaminate soil and waterways. In a recent publication (Mura et al., 2011) studied natural films composed of chitosan (CH), methylcellulose (MC) and silica (SiO2) nanoparticles (NPs) as new packaging materials. The effect of the incorporation of NPs into the polymeric film matrix was evaluated. An excellent improvement of the mechanical properties was obtained for nanostructured films with a composition of CH : MC 50:50 and NPs 1% w/v that make these materials able to replace plastics and derivatives, reducing environmental pollution.

Several microbially derived polysaccharides have also been studied for their ability to form edible films. Edible films made with gellan (deacylated) gum to carry L-(+)-ascorbic acid (AA) were formulated either in mixture with gellan acylated form or by addition of calcium ion (Leon et al., 2008). AA-stability and the subsequent non-enzymatic browning (NEB) development were evaluated during storage at 33.3%, 57.7% or 75.2% relative humidity and 25 °C. Presence of acyl-side chains in the gellan backbone led to a less rigid macromolecular network of the polymer mixture, which allowed lowering glycerol content, obtaining higher AA retention and lower NEB development with adequate mechanical characteristics of films. Otherwise, additional calcium presence in deacylated gellan film slowed down the browning development at the highest relative humidity of storage (75.2%). In general, it was observed an accelerating effect of local moisture mainly on non-enzymic browning reactions, as well as the compromise of the film water content in plasticization of the polymeric networks when stored at 33.3% or 57.7% relative humidity.

Edible films based on gellan were developed (Yang and Paulson, 2000). Of the plasticizers tested, glycerol was found to be the most suitable with respect to mechanical properties and transparency. The mechanical properties (tensile and puncture), water vapor permeability (WVP) and glass transition temperature (Tg) were examined for gellan films as a function of glycerol concentration. The lowest effective glycerol concentration was ~60% (film dry weight basis); below this concentration, the films tended to be brittle and difficult to handle, whereas films with more than ~75% glycerol tended to be sticky. Addition of glycerol to gellan films increased extensibility (tensile elongation and puncture deformation) moderately, but decreased tensile strength, elastic modulus and Tg, and increased WVP of the films. Increasing a_w caused marked decreases in tensile strength and elastic modulus, but decreased tensile elongation only slightly. In general, tensile strength and elastic modulus appeared to be more sensitive to changes in glycerol content and aw than puncture strength.

The physicochemical properties of Ca^{2+}-treated gellan films plasticized with glycerol were investigated as a function of $CaCl_2$ concentration (0–20%, w/w) in an aqueous soaking solution (Yang et al., 2010). Films were examined based on their mechanical properties, water vapor permeability (WVP), swelling index (SI), thickness and opacity. The SI was lower for Ca^{2+}-treated films relative to a control, however, above 13% (w/w) $CaCl_2$ no differences in SI were found. At 13% (w/w) $CaCl_2$, tensile and puncture strengths reached a maximum. Tensile elongation, puncture deformation, film thickness and WVP were lower for $CaCl_2$-treated films than untreated, however, all increased with $CaCl_2$ concentration. Opacity of gellan films increased with the Ca^{2+}-treatment relative to the untreated film, however declined as $CaCl_2$ concentration increased. In general, Ca^{2+}-treated gellan films were stronger, acted as better water vapor barriers, swelled less when in contact with water, and became less pliable and transparent.

8.4.2.1.4 Lipid based films

The most commonly used hydrophobic lipid based film-forming barrier materials are listed in Table 8.13. The moisture barrier efficiency decreases sue to the presence of polar side groups (esters, carboxylic acids etc.), decreasing hydrocarbon chain length, and number of unsaturation or acetylation. For chemically similar molecules, an increasing chain length increases non-polar nature of the molecule and reduces water-solubility of the film. Fresh fruits and vegetables are usually coated with wax coatings to control

Table 8.13 Types of hydrophobic lipid type coatings used in edible films. (Source: Adapted from Debeaufort and Voilley 2009. Reproduced with permission of Springer-Verlag; Adapted from Rhim and Shellhammer 2005. Reproduced with permission of Elsevier.)

Lipid type	Examples
Waxes	Insect and animal waxes: Beeswax, spermaceti, wool grease, lanolin Vegetable waxes: Carnauba wax, candelilla wax, ouricouri wax, sugar cane wax, jojoba oil, baybenywax, Japan wax, rice bran oil Non-natural mineral waxes: Ozocerite, montan wax, paraffin waxes, microcrystalline waxes Synthetic wax: Polyethylene, Synthetic esters, Synthetic amides, Carbowax
Lacs/ Lacquers	Shellac, natural resins
Oils, fats, shortening, margarine	Animal and vegetable native oils and fats (peanut, coconut, palm, palm kernel oils, milk butters, lard, tallow, etc.) Fractionated, concentrated and/or reconstituted oils and fats (fatty acids, mono-, di-, and triglycerides,) Hydrogenated and/or transesterified oils (margarine, shortenings, etc.)
Fatty acids/ alcohols	Fatty sucrose esters, fatty alcohols, fatty acids
Acetylated glycerides	Lecithins, mono- and diglycerides, mono- and diglyceride esters, acetylated monoglyceride
Cocoa-based compounds and their derivatives	Cocoa, cocoa butter substitutes, etc.
Essential oils and liquorices	Camphor, mint, and citrus fruit essential oils, Liquorices

desiccation and both moisture and oxygen transport through the skin. Cocoa butter and cocoa-based films are widely used in confectionery and biscuit industries. Lipid based films are also used in pizza and dough products to prevent moisture migration from the sauce to the dough and in crispy and dry cereal-based products to prevent loss of flavor and water from fat or aqueous stuffing and filling. Most food-grade lacs and varnishes are applied in the pharmaceutical industry, as a protective layer for active substances and on candies in the food industry to improve surface appearance (color, sheen, and gloss) and reduce surface stickiness.

Composite films have been covered extensively by Wu et al. 2002. Composite edible films are prepared by combining pure lipids with hydrocolloids, such as proteins, starches or celluloses, and their derivatives, either by incorporating lipids into the hydrocolloid film-forming solution (in the form of an emulsion) or by depositing lipid layers onto the surface of the pre-formed hydrocolloid film to obtain a bilayer (Kamper et al., 1985; Krochta et al.,1997; Morillon et al., 2002). Bi-layer film systems can again be formed by the 'coating technique' or the 'emulsion technique' (Perez-Gago et al., 2005). The coating technique is a two-step process that involves casting a lipid layer, either molten or from solvent, onto a previously formed edible film. The 'emulsion technique' is a one-step coating process that involves emulsifying the lipid into the film-forming solution prior to film casting, with the bi-layer film resulting from destabilization of the emulsion and phase separation during drying. A schematic of lipid film formation is depicted in Figure 8.19 and the WVP of several lipid films and coatings has been tabulated in Table 8.14 (Morillon et al., 2002). The composite films take advantage of the functional properties of each component of the film to provide both barrier and mechanical properties; the resulting water barrier efficiency of bilayered films is often of the same order of magnitude as that of pure lipid and is much higher than that of emulsion-based films (Debeaufort et al., 2009). Carbohydrates and polysaccharides are often combined with stearic or palmitic acids, beeswax, acetylated monoglycerides, and lecithins (Wu et al., 2002). Permeability of lipid materials having

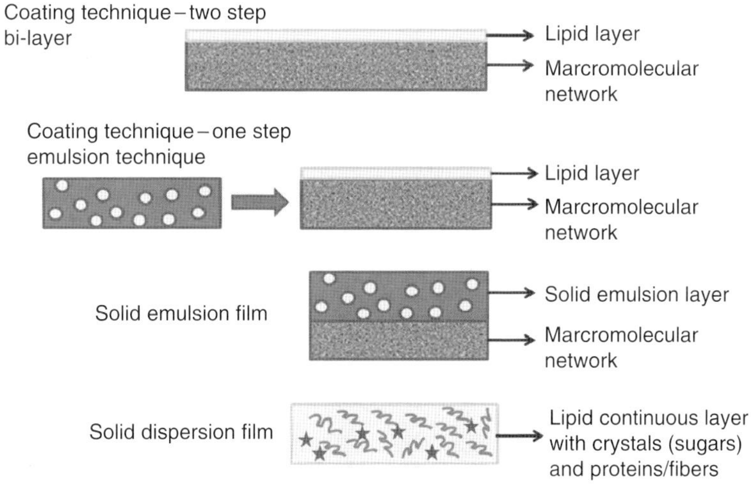

Figure 8.19 Different structures of lipid composite films. (Source: Debeaufort and Voilley 2009. Reproduced with permission of Springer-Verlag.)

Table 8.14 Types of hydrophobic lipid type coatings used in edible films. (Source: Adapted from Morillon et al. 2002. Reproduced with permission of Taylor and Francis.)

Films	Temp (°C)	RH gradient (%)	WVP ($\times 10^{-11}$ g m/m^{2}s Pa)
Monolayer films			
Paraffin wax	25	0/100	0.02
Candelilla wax	25	0/100	0.02
Carnauba wax + glycerol monostearate	25	22/100	35
Microcrystalline wax	25	0/100	0.03
Beeswax	25	0/100	0.06
Capric acid	23	12/56	0.38
Myristic acid	23	12–56	3.47
Palmitic acid	23	12–56	0.65
Stearic acid	23	12–56	0.22
Shellac	30	0/100	0.42–1.03
Hydrogenated cottonseed oil	26.7	0/100	0.13
Hydrogenated palm oil	25	0/85	227
Hydrogenated peanut oil	25	0/100	390
Native peanut oil	25	22/44	13.8
Bilayer films			
MC/paraffin wax	25	22/84	0.2–0.4
MC/paraffin oil	25	22/84	2.4
MC/beeswax	25	0/100	0.058
MC/carnauba wax	25	0/100	0.033
MC/candelilla wax	25	0/100	0.018
MC/triolein	25	22/84	7.6
MC/hydrogenated palm oil	25	22/84	4.9
HPMC/stearic acid	27	0/97	0.12
Emulsion films			
MC + PEG400 + behenic acid	23	12–56	7.7
MC + triolein	25	22/84	14.4
MC + hydrogenated palm oil	25	22/84	13.3
MC + PEG400 + myristic acid	23	12–56	3.5
Wheat gluten + oleic acid	30	0/100	7.9
Wheat gluten + soy lecithin	30	0/100	10.5
Wheat gluten + paraffin wax	25	22/84	1.7
Wheat gluten + paraffin oil	25	22/84	5.1
Wheat gluten + triolein	25	22/84	9.7
wheat gluten + hydrogenated palm oil	25	22/84	7.4
Na-caseinate + acetylated monoglyceride	25	0/100	18.3–42.5
Na-caseinate + lauric acid	25	0/92	11
Na-caseinate + beeswax	25	0/100	11.1–42.5
Whey protein isolate + palmitic acid	25	0/90	22.2
Whey protein isolate + stearyl alcohol	25	0/86	53.6
Whey protein isolate + beeswax	25	0/90	23.9–47.8

From Morillon *et al.* (2002)

a high solid fat content is usually much lower than that of liquid lipids. However, most solid lipid-based films are brittle when used alone; so they often are combined with hydrocolloids to form bilayer or emulsion films. Their structure and barrier properties depend on the preparation technique.

References

Alexander, K. and King, C.J. (1985) Factors governing surface morphology of spray dried amorphous substances, *Drying Technology*, 3, 321.

Alfrey F., Gurnee, E.F. and Lloyd, W.G. (1966) Classification system for diffusion of vapor through films, *Journal of Polymer Science*, 12, 249.

Alzamora, S.M., Cerrutti, P., and Lopez-Malo, A. (1995) Minimally processed fruits by combined methods. In: *Food Preservation by Moisture Control (Fundamentals and Applications)* eds. G. Barbosa-Canovas and J. Welti-Chanes, pp. 463–492, Lancaster, PA: Technomic.

Anonymous, (undated). The Texture Revolution Lexicon. Texture Innovation Center, TIC Gums, White Marsh, MD, Internet available: www.ticgums/lexicon.

Baldwin, E.A., Nisperos, M.O., Hagenmaier, R.D. and Baker, R.A. (1997) Use of lipids in coatings for food products, *Food Technology*, **51**, 56–63.

Barrer, R. M. (1938) Permeation diffusion and solution of gases in organic polymers, Trans. *Faraday Society*, 35, 628–643.

Basu, S., Shivhare, U.S. and Majumdar, A.S. (2006) Models for sorption isotherms for foods: a review, *Drying Technology*, 24(8),917–930.

Bell, L.N. and Labuza, T.P. (2000) *Moisture sorption: practical aspects of isotherm measurement and use*, 2nd American Association of Cereal Chemists, St. Paul, MN.

Bell, L.N., Touma, D.E., White, K.I., and Chen, Y.H. (1998) Glycine loss and Maillard browning as related to the glass transition in a model food system, *Journal of Food Science* 63, 625–628.

Bertuzzi, M.A., Castro Vidaurre, E.F., Armada, M. and Gotfredi, J.C. (2007) Water vapor permeability of starch based films, *Journal of Food Engineering*, 80, 972–978.

Bird, R.B., Stewart, W.E., and Lightfoot, E.N., eds. (1960) *Transport Phenomena*, New York: John Wiley and Sons.

Bourlieu, C., Guillard, V., Powell, H., Vall`es-P`amies, B., Guilbert, S. and Gontard, N. (2006) Performance of lipid-based edible moisture barriers in food products with intermediate water activity, European Journal of Lipid Science Technology 108,1007–1020.

Bourlieu, C., Guillard, V., Powell, H., Valles-Pamies, B., Gulibert, S., and Gontard, N. (2008) Modeling and control of moisture transfers in high, intermediate and low a_w composite foods, *Food Chemistry*, 106, 1350–1358.

Bourlieu, C., Guillard, V., Valles-Pamies, B., Guilbert, S., and Gontard, N. (2009) Edible moisture barriers: How to assess of their potential and limits in food products shelf life extension? Critical Review of Food Science Nutrition, 49, 474–499.

Bourne, M.C. (2002) *Food Texture and Viscosity Concept and Measurement*, 2nd Ed. Academic Press, New York, NY.

Brennan, J.G. (1989) Texture perception and measurement, In: *Sensory Analysis of Foods*, Ed. Piggott, J.R., pp. 69–101, Elsevier, London.

Brunauer, S., Emmet, P.H., and Teller, E. (1938) Adsorption of gases in multimolecular layers, *Journal of the American Chemical Society* 60, 309–319.

Brunauer, S., Deming, L.S., and Teller, E. (1940) On a theory of Van der Waals adsorption of gases, *Journal of the American Chemical Society* 62, 1723–1732.

Bruns A.J. and Bourne, M.C. (1975) Effects of sample dimensions on the snapping force of crisp foods, *Journal of Texture Studies*, 6, 445–458.

Buonocore, G.G., Conte, A. and Del Nobile, M.A. (2005) Use of a mathematical model to describe the barrier properties of edible films. *Journal of Food Science*, **70**, E142–E147.

Butler, B. L., Vergano, P. J., Testin, R. F., Bunn, J. M. and Wiles, J. L. 1996, Mechanical and Barrier Properties of Edible Chitosan Films as affected by Composition and Storage, J. Food Sci. 61:5 953–956.

Callegarin, F., Gallo, J.A.Q., Debeaufort, F. and Voilley, A. (1997) Lipids and biopackaging, Journal of American Oil Chemistry Society, 74, 1183–1192.

Caner, C., Vergano, P. J. and Wiles, J. L. 1998, Chitosan Film Mechanical and Permeation Properties as Affected by Acid, Plasticizer, and Storage, J. Food Sci., 63:6 1049-1052.

Canovas, G.V.B., Fontana, A.J. Jr., Schmidt, S.J. and Labuza, T.D. (2007) *Water Activity in Foods, Fundamentals and Applications*, Blackwell Publishing, UK.

Chirife, J., and Iglesias, H.A. (1978) Equations for fitting water sorption isotherms of foods: Part 1. A review, *Journal of Food Technology* 13, 159–174.

Chirife, J. (1994) Specific solute effects with special reference to Staphylococcus Aureus, *Journal of Food Engineering*, 22, 409–419.

Chirife, J. (1995) An update on water activity measurements and prediction in intermediate and high moisture foods: The role of some non-equilibrium situations. In: *Food Preservation by Moisture Control (Fundamentals and Applications)*, eds. G. Barbosa-Canovas and J. Welti-Chanes, pp. 169–189, Lancaster, PA: Technomic.

Chirife, J. and Buera, M.P. (1996) Water activity, water glass dynamics, and the control of microbial growth in foods, *Critical Reviews in Food Science and Nutrition*, 36, 465–513.

Chao, R.R. and Rizvi, S.H. (1988) *Oxygen and Water Vapor Transport Properties through Polymeric Films in Food and Packaging Interactions*, J.H. Hotchkiss, ed., Washington DC, American Chemical Society.

Clare, K. (1995) Algin. In: *Industrial Gums*, Ed. Whistler, R.L. and BeMiller, J.N., pp. 105–143, Academic Press, New York, NY.

Coma, V., Martial-Gros, A., Garreau, S., Copinet, A., Salin, F. and Deschamps, A. 2002, Edible Antimicrobial Films Based on Chitosan Matrix, 67:3 1162–1169.

Crank J. and Park, G.S. (1968) *Diffusion in Polymers*, Academic Press, New York.

Crank, J. (1975) *Mathematics of Diffusion* (2nd ed.),Clavedon Press, Oxford, UK.

Damodaran, S., Parkin, K.L. and Fennema, O.R. (2008) *Fennema's Food Chemistry* 4thedition, pp. 67–70, CRC Press, Taylor and Francis Group, Boca Raton Florida.

Debeaufort, F., Voilley, A. and Meares, P. (1994) Water vapor permeability and diffusivity through methylcellulose edible films, *Journal of Membrane Science*, 91, 125–133.

Debeaufort, F., Quezada-Gallo, J.A. and Voilley, A. (2000a), Chapter 2. Edible barriers: A solution to control water migration in foods, **In**: *Food Packaging: Testing Methods and Applications*, pp. 9–16. Amer. Chemical. Soc., Washington.

Debeaufort, F. and Voilley, A. (2009) Lipid-based edible films and coatings. In: *Edible Films and Coatings for Food Applications*, Eds. Embuscado, M.E. and Huber, K.C., Ch. 5, Springer, Dordrecht Heidelberg London New York.

Despond, S., Espuche, E., Cartier, N. and Domard, A. (2005) Hydration mechanism of polysaccharides: A comparative study. *J. Polym. Sci. Pt. B-Polym. Phys.* **43**, 48–58.

Donohowe, I.G. and Fennema, O.R. (1993) The effects of plasticizers on crystallinity, permeability, and mechanical properties of methylcellulose films, *Journal Food Process Preservation*, 17, 247–257.

Donohowe, I.G. and Fennema, O.R. (1994) Edible films and coatings: characteristics, formation, definitions and testing methods, In: *Edible Coatings and Films to Improve Food Quality*, Chapter 1, pp. 1–24, Technomic Publishing Company, PA.

Doty, P.M. (1946) On diffusion of vapors through polymers, *Journal of Chemical Physics*, 4(4), 244–251.

Duckworth, R.B. (1981) Solute mobility in relation to water content and water activity. In: *Water Activity Influences on Food Quality* (Rockland, L.B. and Stewart, G.F., eds), pp. 295–317, Academic Press, NY.

Feddersen, R.I. and Thorp, S.N. (1997) Sodium carboxymethylcellulose. In: *Industrial Gums*. 3rd ed. Ed. Whistler, R.L. and BeMiller, J.N., pp. 537–578, Academic Press, New York, NY.

Foegeding, E.A., Daubert, C.R., Drake, M.A., Essick, G., Trulsson, M., Vineyard, C.J. and Van de Veld, F. (2011) A comprehensive approach to understanding textural properties of semi and soft-solid foods. *Journal of Texture Studies* 42, 103–129.

Fox, T.G. and Flory, P.J. (1950) Second-order transition temperatures and related-properties -of polystyrene. I. Influence of molecular weight. *Journal of Applied Physics*, 21, 581.

Galmarini, M.V., Chirife, J., Zamora, M.C., and Pe´rez, A. (2008) Determination and correlation of the water activity of unsaturated, supersaturated and saturated trehalose solutions, *LWT*, 41, 628–631.

Geankoplis, C.J. (1972) *Mass Transport Phenomena*, Holt, Reinhart, and Winston, New York, NY, USA.

Gekas, V. (1992) *Transport Phenomena of Foods and Biological Materials*, 237 p. 1st ed. CRC Press, London.

Gennadios, A., Hanna, M. A. and Kurth, L. B. (1997). Application of edible coatings on meats, poultry and seafoods: a review. *Lebensm. Wiss. u. Technol.* 30(4), 337–350.

Gennadios, A. (2002) *Protein-Based Films and Coatings*, CRC Press, Boca Raton, FL.

Gerrish, T. (1997) Cellulose derivatives , In: *Thickening and Stability of Low Moisture and Intermediate Moisture Foods, Gelling Agents for Food Technology.* 24, 543–550.

Ghosh, V., Ziegler, G.R. and Anantheswaran, R.C. (2002) Fat, moisture, and ethanol migration through chocolates and confectionary coatings. Critical Review of Food Science and Nutrition. **42**,583–626.

Ghosh, V. (2003) Moisture Migration through Chocolate-Flavored Coatings. Ph.D. Diss., Pennsylvania State University, USA.

Ghosh, V., Duda, J.L., Ziegler, G.R. and Anantheswaran, R.C. (2004) Diffusion of moisture through chocolate-flavored confectionery coatings. *Food Bioproduction Process.* 82, 35–43.

Ghosh, V., Ziegler, G.R. and Anantheswaran, R.C. 2005, Moisture migration through chocolate-flavored confectionery coatings. Journal of Food Engineering. 66,177–186.

Gordon, M. and Taylor, J.S. (1952) Ideal copolymers and the second-order transitions of synthetic rubbers. I. Non-crystalline polymers. *Journal of Applied Chemistry*. 2, 493.

Gray, J.N. and Bemiller, J.N. (2003) Bread staling: Molecular basis and control. Complete *Review of Food Science and Food Safety*, 2, 1–21.

Greener, I.K. and Fennema, O. (1992) Lipid-based edible films and coatings. *Lipid Technology*. March–April: 34–38.

Greener, I.K. and Fennema, O. (1989) Barrier properties and surface characteristics of Edible Bilayer Films, *Journal of Food Science*, 54(6), 1393–1399.

Grover, J.A. (1993) Methyl cellulose and its derivatives. In: *Industrial Gums* 3rd edn., Ed. Whistler, R.L. and BeMiller, J.N., pp. 475–504, Academic Press, New York, NY.

Guilbert, S. (1986) Technology and application of edible protective films. In: *Food Packaging and Preservation*, Mathlouthi, M., Ed., pp. 371–394, Elsevier Applied Science, New York.

Guillard, V., Broyart, B., Bonazzi, C., Guilbert, S., and Gontard, N. (2003a) Moisture diffusivity in sponge cake as related to porous structure evaluation and moisture content. *Journal of Food Science*, 68(2), 555–562.

Guillard, V., Broyart, B., Bonazzi, C., Guilbert, S. and Gontard, N. (2003b) Evolution of moisture distribution during storage in a composite food modeling and simulation. *Journal of Food Science*, 68(3), 958–966.

Guillard, V., Broyart, B., Bonazzi, C., Guilbert, S. and Gontard, N. (2003c) Modeling of moisture transfer in a composite food: Dynamic water properties in an intermediate aw porous product in contact with high a_w filling. *Chemical Engineering and Resource Design*, 81(A9), 1090–1098.

Guillard, V., Broyart, B., Bonazzi, C., Guilbert, S. and Gontard, N. (2003d) Preventing moisture transfer in a composite food using edible films: Experimental and mathematical study. *Journal of Food Science*, 68(7), 2267–2277.

Guillard, V., Broyart, B., Guilbert, S., Bonazzi, C. and Gontard, N. (2004) Moisture diffusivity and transfer modeling in dry biscuit. *Journal of Food Engineering*, 64(1), 81–87.

Guinard, J.X. and Brun, P. (1998) Sensory specific satiety: Comparison of taste and texture effects. *Appetite* 31: 141–157.

Han, J.H. and Gennadios, A. (2005) *Edible Films and Coatings: A review*, in Innovations in Packaging, Ch. 15, ed. Han, J.H., Academic Press, NY.

Harris, M. and Peleg, M. (1996) Patterns of textural changes in brittle cellular cereal foods caused by moisture sorption, *Rheology*, 73(2), 225.

Hansen, K.K. and Lund, H.B. (1990) Cup Method for Determination of Water Vapor Transmission Properties of Building Materials. Sources of Uncertainty in the Methods, Proceedings of the 2nd Symposium, Building Physics in the Nordic Countries, Trondheim pp. 291–298.

Havermans, R.C., Geschwind, N., Filla, S., Nederkoorn, C., Jansen, A. (2009) Sensory specific satiety is unaffected by manipulations in flavor intensity. *Physiological Behavior* 97, 327–333.

Heil, J.R., Oezilgen, M., and McCarthy, M.J. (1993) Magnetic resonance imaging analysis of water migration and void formation in baking biscuits in *AICHE Symp. Ser.* 89(297), 39–45.

Henderson, S. (1987) Moisture transfer between wet and dry barley grains. *Journal of Agriculture and Engineering Resources*, 37, 163–170.

Hernandez, E. (1994) Edible coatings from lipids and resins. In: Edible films and coatings to improve food quality. Pp. 279–304. Krochta, J.M., Baldwin, E.A., Nisperos-Carriedo, M.O., eds., Technomic Publishing Company, Lancaster.

Higuchi, T. and Aguiar, A. (1959) A study of permeability to water vapor of fats, waxes and other enteric coating material. Journal American Pharmacology Association. 48, 574–583.

Hogan, S.A., Brendan, V.C., O'Kennedy, B.T. and Kelly, P.M. (2012) Influence of dairy proteins on textural changes in high protein bars. *International Dairy Journal*. 24, 1–8.

Hong, Y.C., Bakshi, A.S., and Labuza, T.P. (1986) Finite element modeling of moisture transfer during storage of mixed multicomponent dried foods. *Journal of Food Science*, 51(3), 554–558.

Hsieh, F., Hu, L., Huff, H.E. and Peng, I.C. (1990) Effects of water activity on textural characteristics of puffed rice cake. *Lebensm. –Wiss. U. Technol.* 23, 471–473.

Jayaraman, K.S. (1995) Critical review on intermediate moisture fruits and vegetables. In: *Food Preservation by Moisture Control (Fundamentals and Applications)*, eds. G.V. Barbosa-Canovas and J. Welti-Chanes, pp. 411–442, Lancaster, PA: Technomic.

Kamper, S.L. and Fennema, O. (1984a) Water vapor permeability of edible bilayer films, *Journal of Food Science*, 49(6), 1478–1481, 1485.

Kamper, S.L., and Fennema, O. (1985) Use of edible film to maintain water vapor gradients in food, *Journal of Food Science*, 50, 382–384.

Kapsalis, J.G. (1987) Influence of hysteresis and temperature on moisture sorption isotherms, In: *Water Activity: Theory and Application to Foods*, L.B. Rockland and L.R. Beuchat, eds., pp. 173–213. Dekker, New York.

Karathanos, V.T., Kostaropoulos, A.E., and Saravacos, G.D. (1995) Diffusion and equilibrium of water in dough raisin mixtures. *Journal of Food Engineering*, 25(1), 113–121.

Karbowiak, T., Gougeon, R.D., Rigolet, S., Delmotte, L., Debeaufort, F., and Voilley, A. (2008) Diffusion of small molecules in edible films: Effect of water and interactions between diffusant and biopolymer, *Food Chemistry* 106, 1340–1349.

Karel, M. (1973) Recent research and development in the field of low and intermediate moisture foods, *CRC Critical Reviews in Food Technology*, 3, 329–373.

Karel, M. (1976) Technology and application of new intermediate moisture foods, in *Intermediate Moisture Foods*, eds. R. Davies, G.G. Birch, and K.J. Parker, pp. 4–31. London: Applied Science Publishers.

Katz, E.E. and Labuza, T.D. (1981) Effect of water activity on the sensory crispiness and mechanical deformation of snack food products, *Journal of Food Science*, 46, 403–409.

Kester, J.J. and Fennema, O. (1986) Edible films and coatings: A review. *Food Technology*, 40(12), 47–49.

Kester, J.J. and Fennema, O. (1989a) An edible film of lipids and cellulose ethers: Barrier properties to moisture vapor transmission and structural evaluations, *Journal of Food Science*, 54, 1383–1389.

Koelsch, C.M. (1994) Edible water vapor barriers: properties and promise, *Trends Food Science and Technology*. 5, 76–81.

Krochta, J.M. (1992) Control of mass transfer in foods with edible coatings and films, advances. In: *Food Engineering*, R.P. Singh and M.A. Wirakartakusumah. Eds., pp. 517–538, Boca Raton, FL: CRC Press, Inc.

Krochta, J.M., Baldwin, E.A. and Nisperos-Carriedo, M. (1994) *Edible Coatings and Films to Improve Food Quality*, Lancaster, PA.

Krochta, J.M. and DeMulder-Johnston, C. (1997) Edible and biodegradable polymer films: challenges and opportunities, *Food Technology*, 51, 61–74.

Kulp, K. and Ponte J.G. Jr. (1981) Staling of white pan bread: Fundamental causes. *CRC Crit. Rev. Food Sci. Nutr.* 15, 1–43.

Labuza, T.P., Tannenbaum, S.R. and Karel, M. (1970) Water content and stability of low moisture and intermediate moisture foods, *Journal of Food Technology*, 24, 543–550.

Labuza, T. P. and Hyman, C. R. 1998 Moisture migration and control in multi-domain foods, Trends in Food Science & Technology 9, 47–55.

Landmann, W., Lovegren, N.V. and Feuge, R.O. (1960) Permeability of some fat products to moisture, *JAOCS*, 1–4.

Langmuir, I. (1918) The adsorption of gases on plane surfaces of glass, mica and platinum, *Journal of the American Chemical Society*, 40, 1361–1402.

Lawless, H.T. and Heymann, H.H. (1998) *Sensory Evaluation of Food: Principles and Practices*, Chapman and Hall, New York, NY.

Le Meste, M. (1995) *Mobility of Small Molecules in Low and Intermediate Moisture Foods in Food Preservation by Moisture Control* (Barbosa-Canavas, G.V. and Welti-Chanes, J., eds.), pp. 209–225, Technomic.

Le Meste, M., Champion, D., Roudaut, G., Blong, G., and Simatos, D. (2002) Glass transition and food technology: A critical appraisal, *Journal of Food Science*, 67, 2444–2458.

Leake, L.I. (2006) Water activity and food quality. *Food Technology*. 60, 62–67.

Leistner, L. and Gould, G.W. (2002) *Hurdle Technologies, Combination Treatments for Food Stability, Safety and Quality*, New York: Kluwer Academic/Plenum Publishers.

Leon P. G., Lamanna, M. E., Gerschenson, L. G. and Rojas, A. M. Influence of composition of edible films based on gellan polymers on l-(+)-ascorbic acid stability, 2008, Food Research International, 41:6 667–676.

Levine, H. and Slade, L. (1988) Principles of 'cryostabilization' technology from structure/property relationships of carbohydrate /water systems-A review. *Cryo-Letters* 9: 21.

Levoguer, C.L. (2006) (Surface Measurement Systems Ltd., UK), Booth, J. (Scientific and Medical Products Ltd., UK), Calculation of Diffusion Constants in Thin Polymer Films using DVS, DVS Application Notes.

Leung, H.K. (1986)Water activity and other colligative properties of foods, In: *Physical and Chemical Properties of Foods*, ed. M.R. Okos, pp. 138–185, St. Joseph, MI: American Society of Agricultural Engineers.

Lievonen, S.M. and Roos, Y.H. (2002) Nonenzymatic browning in amorphous food models: Effects of glass transition and water. *Journal of Food Science* 67, 2100–2106.

Lin, W. and Lineback, D.R. (1990) Changes in carbohydrate fractions in enzyme supplemented bread and the potential relationship to staling. *Starch/ Starke* 42, 385–392.

Liu, Z. (2005) Coatings from starches, Ch.19, pp. 318–337. In: *Innovations in Food Packaging*. Ed. Han, J.H., Academic Press, NY.

Lomauro, C.J., Bakshi, A.S., and Labuza, T.P. (1985) Moisture transfer properties of dry and semimoist foods. *Journal of Food Science*, 50, 397–400.

Lomauro, G.J., and Bakshi, A.S. (1985) Finite element analysis of moisture diffusion in stored foods. *Journal of Food Science*, 50, 392–396.

Mandl, K., Hartel, R.W. and Wendorff, W. (2009) Effects of salt and moisture migration on cheese firmness in cheese-in-sausage products, *Journal of Food Engineering*. 91, 164–172.

Mathlouthi, M. and Roge, B. (2003) Water vapor sorption isotherms and the caking of food powders. *Food Chemistry*. 82, 61–71.

McHugh, T.H., Avena-Bustillos, R., and Krochta, J.M. (1993) Hydrophilic edible films: Modified procedure for water vapor permeability and explanation of thickness effects, *Journal of Food Science*, 58(4), 899–903.

McHugh, T.H., and Krochta, J.M. (1994) Permeability property of edible films in edible films and coatings: Characteristics, formation, definitions and testing methods, Chapter 7, pp. 139–187, In: *Edible Coatings and Films to improve Food Quality*, Technomic Publishing Company, PA.

Miao, S. and Roos, Y.H. (2004) Nonenzymatic browning kinetics of a carbohydrate-based low-moisture food system at temperatures applicable to spray drying, *Journal of Agricultural and Food Chemistry*, 52, 5250–5257.

Morillon, V., Debeaufort, F., Blond, G., Capelle, M. and Voilley, A. (2002) Factors affecting the moisture permeability of lipid-based edible films: A review. Critical Review of Food Science Nutrition. 42, 67–89.

Mura,S., Corrias, F., Stara, G., Piccinini, M., Secchi, N., Marongiu, D., Innocenzi, P., Irudayaraj, J., and Greppi, G. F. 2011, Innovative Composite Films of Chitosan, Methylcellulose, and Nanoparticles, J. Food Sci., 76:7 N55–N60.

Nelson, K. and Fennema, O. (1991) Methylcellulose films to prevent lipid migration in confectionary products, *Journal of Food Science*, 56(2), 504–508.

Nelson, K.A., and Labuza, T.P. (1994) Water activity and food polymer science: Implications of state on Arrhenius and WLF models in predicting shelf life. *Journal of Food Engineering* 22, 271–289.

No, H. K., Meyers, S. P., Prinyawiwatkul, W. and Xu, Z. 2007, Applications of Chitosan for Improvement of Quality and Shelf Life of Foods: A Review, J. Food Sci., 87–100.

Peleg, M. (1992) On the use of WLF model in polymers and foods, *Critical Reviews in Food Science and Nutrition*, 32, 59–66.

Peleg, M. (1993) Mapping the stiffness-temperature-moisture relationship of solid biomaterials at and around their glass transition. *Rheol. Acta*. 32, 575–580.

Peleg, M. (1994a) A model of mechanical changes in biomaterials at and around their glass transition, *Biotechnology Progress* 10, 385–388.

Peleg, M. (1994b) Mathematical characterization and graphical presentation of the stiffness-temperature-moisture relationship of gliadin, *Biotechnology. Prig*. 10, 652–654.

Peleg, M. (1994c) A mathematical model of crunchiness/ crispiness loss in breakfast cereals. *Journal of Texture Studies*, 25, 403–410.

Perez-Gago, M. and Krochta, J.M. (2005) Emulsion and edible bilayer films. In: *Innovations in Food Packaging*, ed. Han, J.H., Ch. 22, Academic Press, NY.

Primo-Martin, C., van Dalen, G., Meanders, M.B.J., Don, A., Hamer, R.H., and van Vliet, T. (2010) Bread crispness and morphology can be controlled by proving conditions, *Food Resources International*. 43, 207–217.

Quast, D.G. and Karel, M. (1972) Effects of environmental factors on the oxidation of potato chips, *Journal of Food Science*, 37, 584–588.

Rahman, S. (1995) *Food Properties Handbook*, CRC Press, Boca Raton, FL.

Rhim, J.W. and Shellhammer, T.H. (2005) Lipid based edible films and coatings, in *Innovations in Packaging*, Ch. 21, Academic Press, NY.

Rizvi, S.S.H. (1986) Thermodynamic properties of foods in dehydration. In: *Engineering Properties of Foods*, eds. M.A. Rao and S.S.H. Rizvi, p. 133, New York: Marcel Dekker.
Roca, E., Broyart, B., Guillard, V., Guilbert, S. and Gontard, N. (2007) Controlling moisture transport in a cereal porous product by modification of structural or formulation parameters, *Food Resources International*. 40, 461–469.
Rogers, C.E., Meyer, J.A., Stannett, V. and Szwarc, M. (1962) PART II. Some factors affecting the permeability constant, In: *Permeability of Plastic Films and Coated Papers to Gases and Vapors*, Barghava, R.L., Rogers, C.E., Meyer, J.A., Stannett, V., Szwarc, M., Eds., Technical Association of the Pulp and Paper Industry, New York.
Rolls, E.T. (2005) Taste olfactory and texture processing in the brain and the control of food intake, *Physiological Behavior* 85, 45–56.
Roos, Y. and Karel, M. (1991) Applying state diagrams in food processing and development, in *Food Technology*. 45, 66–71, 107.
Roos, Y. and Karel, M. (1991b) Plasticizing effect of water on thermal behavior and crystallization of amorphous food models, pp. 38–43, Chapman Hall, New York, NY.
Roos, Y. and Karel, M. (1991c) Phase transitions of mixtures of amorphous polysaccharides and sugars, *Biotechnological Progress*, 7, 49.
Roos, Y.H. (1993b) Melting and glass transitions of low molecular weight carbohydrates, *Carbohydrates Research*, 238, 39–48.
Roos, Y.H. (1995) Characterization of food polymers using state diagrams. In: *Journal of Food Engineering*. 24, 229–360.
Roos, Y.H. (1995) *Phase Transition in Foods*, pp.157–192, Academic Press Inc., CA.
Roos, Y.H., Roininen, K., Jouppila, K., and Tuorila, H. (1998) Glass transition and water plasticization effects on crispiness of a snack food extrudate, *International Journal of Food Properties*, 1, 163–180.
Roos, Y.H. (2002) Thermal analysis, state transitions and food quality, *Journal of Thermal Analysis and Calorimetry*, 71, 197–203.
Roos, Y.H. (2004) Phase and state transitions in dehydration of biomaterials and foods, In: Dehydration of products of biological origin, Ed. A.S. Majumdar, pp. 3–22, Enfield: Science Publishers.
Roy, S., Gennadios, A., Weller, C.L. and Testin, R.F. (2000) Water vapor transport parameters of a cast wheat gluten film, *Industrial Crops and Products*, 11, 43–50.
Ruan, R. and Litchfield, J.B. (1992) Determination of water distribution and mobility inside maize kernels during steeping using magnetic resonance imaging, in *Cereal Chemistry*. 69, 13–17.
Roudaut, G., Simatos, D., Champion, D., Contreras-Lopez, E., Le Meste, M. (2004) Molecular mobility around the glass transition temperature: A mini review, *Innovative Food Science and Emerging Technology*, 5, 127–134.
Salame, M. and Steingiser, S. (1977) Barrier polymers, *Polym. Plast. Technol. Eng.*, 8(2), 155–175.
Sapru, V., and Labuza, T. (1996) Moisture transfer simulation in packaged cereal-fruit systems. *Journal of Food Engineering*, 27, 45–61.
Saravacos, G.D., and Maroulis, Z.B. (2001) Diffusivity and permeability of small solutes in food systems, In: *Transport Properties of Foods* (pp. 237–269), New-York: Marcel Dekker.
Sauvageot, F. and Blond, G. (1991) Effect of water activity on crispiness of breakfast cereals. *Journal of Texture Studies*, 22, 423–442.
Schmidt, S.J. (2004) Water and solids mobility in foods, *Advances in Food Nutrition Resources*, 48, 1–101.
Schmidt, S.J. and Lai, H.M. (1991) *Use of NMR and MRI to Study Water Relation in Foods, Advances in Experimental and Medicine and Biology* Vol. 302 (Levine, H. and Slade, L., eds.), pp. 405–452, Plenum Press.
Schmidt, S.J. (2004) Water and solids mobility in foods, Advances in Food Nutrition Resources, 48, 1–101.
Simatos, D. and Karel, M. (1988) Characterization of the condition of water in foods – Physico-chemical aspects. In: *Food Preservation by Water Activity*, p. 1, C.C. Scow (2nd edn.). Elsevier, Amsterdam.
Shellhammer, T.H. and Krochta, J.M. (1997) Chapter 17, Edible coatings and film barriers. In: *Lipid Technologies and Applications*, pp. 453–479, Gunstone, F.D., Padley, F.B., Eds., Marcel Dekker, New York.
Slade, L. and Levine, H. (1995) Glass transition and water-food structure interactions. In: *Advances in Food Nutrition and Research*, 38, 103269.
Sperling, L.H. (1986) *Introduction to Physical Polymer Science*, John Wiley & Sons, New York, NY, p.594.
Sperling, L.H. 1992, *Introduction to Physical Polymer Science*, p. 594, John Wiley & Sons, New York, NY.
Standard Test Methods for Water Vapor Transmission of Materials, Designation: E96/E96M – 10, ASTM International, PA.

Schwartzberg, H.G. (1986) Modeling of gas vapor transport through hydrophilic films. In: *Food Packaging and Preservation*, pp. 15–135, M. Mathlouthi, ed. New York, Elsevier Science Publishing Co. Inc.
Sothornvit, R. and Krochta, J.M. (2005) Plasticizers in edible films and coatings. In: *Innovations in Packaging*, Ch. 23, Academic Press, NY.
Tabouret, T. (1987) Technical Note: Detection of Fat Migration in a Confectionary Product, *International Journal of Food Science and Technology*. 22, 163–167.
Taoukis, P.S. and Richardson, M. (2007) *Principles of Intermediate-Moisture Foods and Related Technology in Water Activity in Foods- Fundamentals and Applications*, Canovas, G. V.B., Fontana, A.J. Jr., Schmidt, S.J. and Labuza, T.D. edited, Blackwell Publishing, UK.
Thielmann, F. and Burnett, D. (2006) Isotherm Types and Adsorption Mechanisms of Solvents on Pharmaceutical Excipients, Application Notes 26, Stable Micro Systems Ltd.
Umbach, S.L., Davis, E.A., Gordon, J. and Callaghan, P.T. (1992) Water self-diffusion and dielectric properties determined for starch-gluten-water mixtures heated by microwave and by conventional methods, in *Cereal Chemistry*. 69, 637–642.
Velazquez, G., Torres, J.A. and Martin-Polo, M.O. (2001) Temperature effect on the moisture sorption isotherms for methylcellulose and ethylcellulose films. *Journal of Food Engineering*. 48, 91–94.
Van den Berg, C. and Bruin, S. (1981) Water activity and its estimation in food systems: theoretical aspects. In: *Water Activity: Influences in Food Quality*, eds. L.B. Rockland and G.F. Stewart, pp. 1–64, New York: Academic Press.
van den Berg, C. (1984) Desorption of water activity of foods for engineering purposes by means of the GAB model of sorption, pp. 311–321. In: *Engineering and Foods*, Mckenna, B.M., Ed., Elsevier: London, UK.
Van den Berg, C. (1986) Water activity. In: *Concentration and Drying of Foods*, ed. D. McCarthy, pp. 11–61, London: Elsevier Applied Science.
Villalobos, R., Hernandez-Muños, P. and Chiralt, A. (2006) Effect of surfactants on water sorption and barrier properties of hydroxypropyl methyl cellulose films, *Food Hydrocolloids* 20, 502–509.
Vrentas, J.S. and Duda, J.L. (1978) A free-volume interpretation of the influence of the glass transition on diffusion in amorphous polymers. *Journal of Applied Polymers Sciences*. 22, 23–25.
Walstra, P. (2003) *Physical Chemistry of Foods*, pp. 60–85, Dekker, New York, NY.
Welti-Chanes, J., Perez, E., Angel Guerro-Beltran, J.A., Alzamora, S.M., and Vergara-Balderas, F. (2007) *Application of Water Activity Management in the Food Industry, In: Water Activity in Foods: Fundamentals and Applications* eds. G. Barbosa-Canovas, A.J. Fontana Jr., S. Schmidt, and T.P. Labuza, pp. 341–357, Blackwell Publishing and IFT Press, Ames, Iowa.
White, G.W. and Cakebread, S.H. (1966) The glassy state in certain sugar containing food products. *Journal of Food Technology*, 1, 73.
Wiles, J. L., Vergano, P. J., Barron, Bunn, J. M. and Testin, R. M. 2000, Water Vapor Transmission Rates and Sorption Behavior of Chitosan Films, J. Food Sci. 65:7, 1175–1179.
Williams, M.L., Landel, R.F., and Ferry, J.D. (1955) The temperature dependence of relaxation mechanisms in amorphous polymers and other glass forming liquids, *Journal of the American Chemical Society*, 77, 3701–3707.
Williams, R. and Mittal, G.S. (1999) Low-fat fried foods with edible coatings: modeling and simulation, *Journal of Food Science*. 64(2),317–322.
Wilkinson, C., Dijksterhuis, G.B. and Minekus, M. (2000) From food structure to texture. *Trends in Food Science Technology*, 11, 442–450.
Winterowd J.G., Sanford P.A., 1995. Chitin and Chitosan. In:Stephen A.M., Marcel D. (eds). Food Polysaccharides and Their Applications, pp. 441–462. Marcel Dekker, NewYork, NY, USA.
Wu, Y., Weller, C.L., Hamouz, F., Cuppett, S.L. and Schnepf, M. (2002) Development and application of multicomponent edible coatings and films: A review. *Adv. Food Nutr. Res.*, 44, 347–394.
Wurzburg, O.B. (1995) Modified starches. In: *Food Polysaccharides and their Applications*, pp. 67–98, Ed. Stephen, A.M., Marcel Dekker, New York, NY.
Yang, L. and Paulson, A. T., Mechanical and water vapor barrier properties of edible gellan films, 2000, Food Research International, 33:7 563–570.
Yang, L. and Paulson, A. T. and Nickerson, M. T., Mechanical and physical properties of calcium-treated gellan films, 2010, Food Research International, 43:5, 1439–1443.
Zabik, M.E. Fierke, S.G. and Bristol, D.K. (1979) Humidity effects on textural characteristics of sugar-snap cookies, *Cereal Chemistry*. 56, 29–33.
Zobel, H.F. and Stephen, A.M. (1995) Starch structure: Analysis and application. In: *Food Polysaccharides and their Applications*, pp. 19–66, Ed. Stephen, A.M., Marcel Dekker, New York, NY.

9 Textural attributes of wheat and gluten free pasta

Alessandra Marti[1], Maria Ambrogina Pagani[1], and Koushik Seetharaman[2]

[1]Environmental and Nutritional Sciences (DeFENS), Università degli Studi di Milano, Milan, Italy
[2]Department of Food Science and Nutrition, University of Minnesota, St Paul, MN

9.1 DEFINING 'GOOD' TEXTURE FOR PASTA

Dried pasta from durum wheat semolina is a basic food for populations living in some countries of Southern Europe, particularly in Italy. Nevertheless, pasta worldwide popularity is increasing, thanks to several features such as its convenience, long and easy shelf-life, palatability, simplicity of ingredients and process. Moreover, an extensive variety of dishes can be prepared using the many pasta shapes and sizes available on the market. Last but not least, pasta has an excellent nutritional profile, being a good source of slowly digestible polysaccharides (assuring a low glycemic index) and a moderate source of protein and some vitamins. Even if the nutrient balance in dried pasta is far from the optimal distribution, the use of sauces with high nutritional value can compensate for nutritional deficiencies.

Despite the fact that pasta is now re-evaluated from a nutritional point of view, the texture properties of cooked pasta are the primary factor in overall assessment of pasta quality and play a dominant role in influencing consumer acceptance. Good cooking quality is related to high firmness, absence of stickiness and low cooking loss (D'Egidio and Nardi, 1996), characteristics which describe high cooking and overcooking tolerance (D'Egidio et al., 1990). This behavior can be attributed to the specific structural organization of the two main components of pasta: starch and proteins. In particular, good cooking quality is strongly related to a compact structure of the matrix, characterized by swollen and gelatinized starch granules, each of them trapped inside the continuous network formed by coagulated gluten proteins (Figure 9.1) (Resmini and Pagani, 1983; Cunin et al., 1995; Petitot et al., 2009a). The extent of starch gelatinization and protein coagulation, and consequently the overall cooking quality of the final product, are greatly affected not only by the native properties of wheat (especially protein quantity and quality) but also by the processing conditions applied during pasta-making (De Noni and Pagani, 2010).

As regards the raw material, semolina from durum wheat is recognized as the most suitable raw material for dried pasta production due to its unique color, flavor, and, above all, technological functionality of proteins which assure an optimal cooking performance,

Food Texture Design and Optimization, First Edition. Edited by Yadunandan Lal Dar and Joseph M. Light.

(a)

(b)

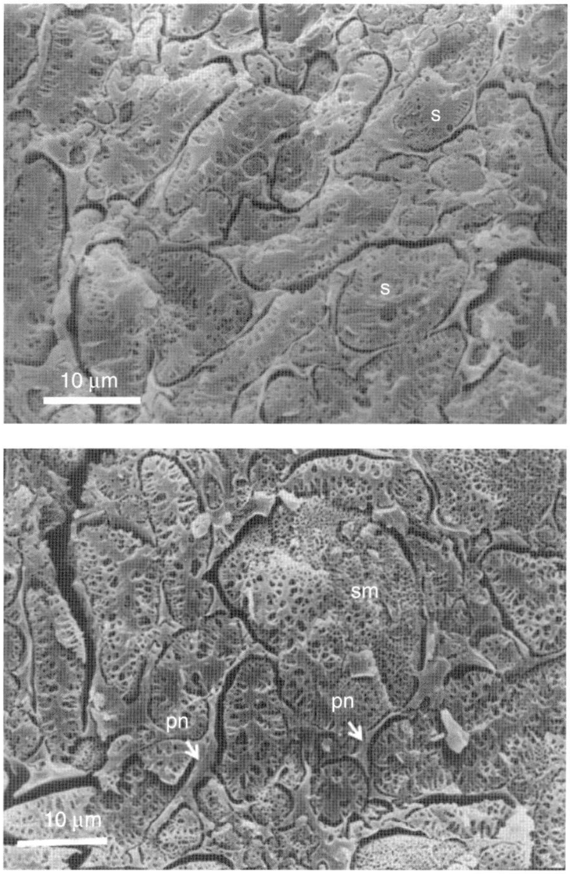

Figure 9.1 SEM images of cooked pasta characterized by good (a) and poor (b) cooking quality. pn=protein network; s=starch granules; sm=swollen starch material.

in terms of low stickiness and good firmness (D'Egidio et al., 1990; Feillet and Dexter, 1996, Petitot et al., 2009a; De Noni and Pagani, 2010). For this reason, in Italy, France and Greece, dry pasta intended for the national market can only be made from durum wheat semolina (Italian law n. 580, 4 July 1967 modified by the presidential decree n. 187, 9 February 2001; French law, 3 July 1934 modified by the law n. 99 e 574, 9 July 1999; Greek law n. 359, 4 May 1993). Beside the conventional pasta product made by durum wheat semolina, since the 1980s, it has been proposed to enrich pasta in some cereals (barley, rye, etc.), pseudocereals (buckwheat, amaranth, quinoa), and legume flours (pea, chickpea, lentils, etc.), as a source of fiber, minerals, antioxidants, and polyphenols (Pagani, 1986; www.barillacfn.com). In addition, in the last few decades, a third group of pasta products, the gluten-free goods, is more and more consumed by the growing number of celiacs.

Taking into account the wide variety in raw materials and processing conditions, a good texture for pasta is very difficult to define and to measured in an objective way. In fact, consumer preference for pasta textural properties varies around the world, although the Italian tradition of 'al dente' eating properties, characterized by high degrees of firmness and elasticity, generally first comes to mind.

Texture is defined as an attribute of a substance resulting from a combination of physical properties and perceived by the senses of touch, sight and hearing (Brennan, 1980). Defining pasta texture is a hard task, since the structure present in the uncooked product greatly affects its texture after cooking. The complexity is also due to the key role played by raw-material characteristics and processing conditions in determining the pasta texture. Last but not least, pasta texture generally refers to a composite of different textural parameters. On the basis of the classification proposed by Szczesniak (1963) for food texture, the textural characteristics of cooked pasta can be distinguished in two main categories: geometrical and mechanical. The geometrical characteristics are related to the physical structure of the material, such as size, shape and orientation of particles. As regards the mechanical properties, such as firmness, cohesiveness, elasticity, stickiness, chewiness, and bulkiness, even today, there is some confusion regarding the definitions of each parameter, such that, in many cases, a number of terms are used to describe the same feature or the same terms are used to described different properties (D'Egidio and Nardi, 1996). The main textural properties for pasta products can be defined as follows (D'Egidio and Nardi, 1996; Marchylo and Dexter, 2004):

1. Firmness or Hardness: the force necessary to penetrate pasta with the teeth; it represents the degree of resistance to the first bite.
2. Cohesiveness or Consistency: the extent to which pasta can be deformed before its breaking.
3. Elasticity or Springiness or Recovery: the rate at which a deformed material returns to its non-deformed state after the deforming force is removed.
4. Adhesiveness or Stickiness: the work necessary to overcome the attractive forces between the surface of pasta and the surface of other materials which come into contact with the food (e.g. tongue, teeth, palate, fingers).
5. Chewiness: the time (or the energy) required to masticate a sample for reducing it to a state ready for swallowing; it is related to firmness, cohesiveness, and elasticity.
6. Bulkiness or Clumpiness: the adhesion rate of cooked pasta strands among them; it is related to stickiness.
7. Fracturability: the force with which the sample crumbles, cracks or shatters; it is a textural characteristic typically for extruded and crunchy products such as puffed snacks and cereals, or dried uncooked pasta.

9.2 MEASURING THE TEXTURE OF PASTA

The quality of pasta is mostly determined by its behavior during cooking. The overall cooking quality of the final product is the result of several and simultaneous phenomena occurring within pasta, such as progressive hydration, protein polymerization, starch gelatinization, and interaction among biopolymers. Microscopic observations have revealed that structural transformations in starch and proteins during cooking are differently enhanced by water availability, and that the kinetics of hydration of each biopolymer have a major role as determinants of the final texture of cooked pasta (Resmini and

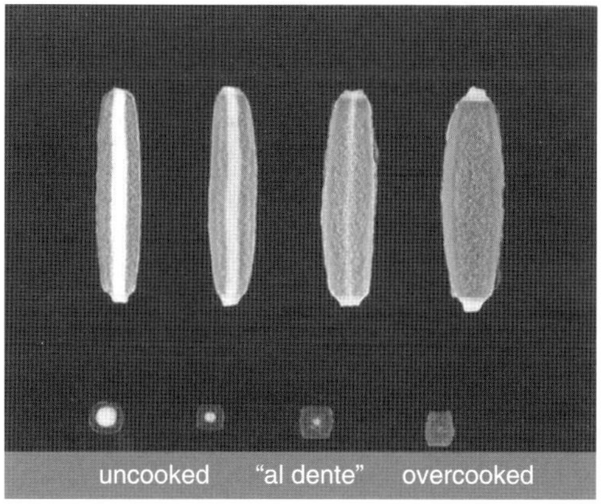

Figure 9.2 Determination of the optimal cooking time by squeezing pasta between two glass plates.

Pagani, 1983). When heat-induced coagulation of gluten proteins in the continuous network prevails, the swollen starch material is trapped within the protein network and the cooked pasta will be firm with no stickiness on the surface (and consequent clumpiness). On the contrary, when the protein network is not strong and elastic enough, the starch swells and gelatinizes before protein coagulation takes place (Resmini and Pagani, 1983). Although interesting information could be obtained by ultrastructural investigations (using light and/or SEM and TEM microscopy), these techniques are time-consuming, labor-intensive, highly expensive, and they require a deep knowledge of macromolecule organization. Moreover, they allow only a qualitative overview and no quantitative evaluation of the involved phenomena (Pagani et al., 1986; Pagani et al., 1989). In other words, microscopy observations are helpful and powerful tools for the interpretation and the explanation of the involved phenomena, but these techniques cannot represent routine approaches for pasta quality evaluation.

Taking into account these considerations, pasta quality is evaluated by means of its textural properties, which are commonly measured by sensory evaluation or by instrumental methods. The differences between the sensorial and instrumental approach will be discussed in the following section.

The measures from each technique are highly affected by pasta intrinsic characteristics, such as shape and dimension. If the samples are to be compared with each other, they should be uniform in shape and dimension in order to provide reproducible results. The cooking procedure also needs to be standardized, since cooking time, cooking temperature, draining time, salt concentration and pH of water, water: pasta ratio highly affects the results (Menger, 1979).

D'Egidio et al. (1990) proposed a cooking procedure that is widely used in Italy as reference. Pasta is cooked in boiling natural water with no salt added, keeping the pasta: water ratio equal to 1:10. Pasta is usually cooked at the optimal cooking time (OCT) and water is drained for one minute. The OCT is evaluated as the required disappearance of dry central core when gently squeezed between two glass plates (D'Egidio et al., 1990) (Figure 9.2). Sometimes the measurements are carried out at a fixed cooking time (12–13 minutes) and in overcooking, evaluated as the time 25%

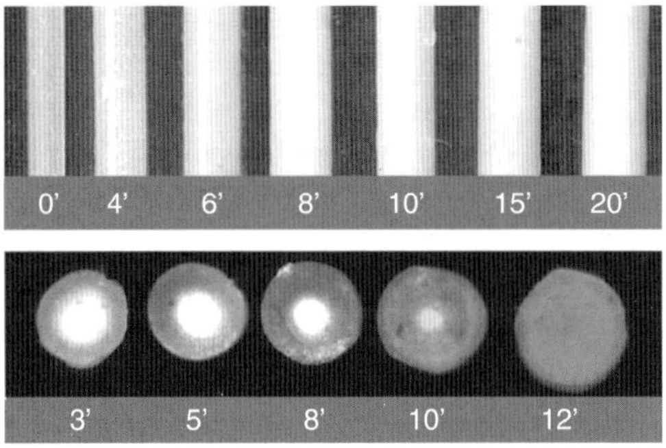

Figure 9.3 Increase in size during pasta cooking (upper panel). Loss of birefringence during cooking (lower panel).

higher than optimum cooking (D'Egidio and Nardi, 1996). In France and other European countries, the cooking characteristics are determined at the normal cooking time and in overcooking. The minimum cooking time is attained when the continuous white line seen at the center of the spaghetti strand disappears; while normal cooking time is the minimum cooking time plus 1 minute or 2–3 minutes, depending on shape and diameter. Overcooking corresponds to the normal cooking time or by doubling the minimum or normal cooking time (Cubadda, 1988) (Figure 9.2).

During cooking, pasta is subjected to physical and chemical changes related to the amount of absorption of water and heating to which it is subjected. The most obvious phenomenon refers to the ability of water to penetrate and diffuse into the matrix, with consequent increase in the weight and size of the product (D'Egidio et al., 1990) (Figure 9.3). The weight gain and the increase in size during cooking are performed with linear kinetics as a function of time, and turn out to be strongly linked to the characteristics of the raw material and to the drying conditions adopted (Ferrari, 2008; Marti et al., 2013).

Advanced approaches, such as Magnetic Resonance Imaging (MRI), have been used for allowing discrimination between free and bound water, and for assessing meanwhile the total amount of water present in pasta (McCarthy et al., 2002; Lai and Hwang, 2004). Recent studies pointed out that gradients of water distribution and mobility in spaghetti play a major role in contributing to the 'al dente' feeling in cooked pasta (Horigane et al., 2009; Bonomi et al., 2012).

De La Pena and Manthey (2011) investigated the influence of cooking surface temperature, cooking water:pasta ratio, and cooking vessel material (glass and stainless) on the final quality of cooked pasta. The best cooking parameter combination was to cook a pasta: water ratio of 13 g : 400 ml at a surface temperature of 400 °C in a stainless steel vessel.

As regards the physicochemical properties of cooking water, although it is traditionally recognized to have an important role in determining the quality of the product, very few and quite old studies have been carried out to ascertain which characteristics of water are related to pasta cooking quality. Alary et al. (1979) stated that the pH value of the cooking water has an important influence on cooking properties and appearance of the cooked pasta. The appearance (surface disintegration and stickiness) of mineral water-cooked spaghetti was greatly improved by cooking at pH 6. Improvement in viscoelastic properties (compressibility, elastic recovery), decrease in cooking losses, increase in luminance and decrease in colorimetric parameter b* (yellowness) were also observed. On the other hand, at low pH (<4) pasta exhibited high cooking loss and an unattractive appearance (Alary et al., 1979). Cooking pasta in distilled or deionized water has been reported to minimize differences among pasta samples whereas preparation in tap water, mineral water, or prepared solutions has been demonstrated to accentuate product differences (D'Egidio et al., 1981; Dexter et al., 1983). Recently, evaluating the effect of cooking water on both textural and cooking properties of spaghetti, Sozer and Kaya (2008) highlighted that a certain amount of salt (2.5 and 5%) in the cooking water improves the textural characteristics of cooked pasta, in terms of hardness and adhesiveness, when compared with samples cooked in deionized water.

Several tests have been proposed and a number of testing instruments have been developed commercially for evaluating the textural properties of pasta. The Instron Universal Tester (Instron Corp., Canton, MA; www.instron.com) and the TA.XT2 texture analyzer (Texture Technologies Corp., Scarsdale, NY; www.texturetechnologies.com) have been the most adopted today. The basic components of these types of instruments include a drive mechanism for deformation and a recording system of force, time and compression rate.

9.3 INSTRUMENTAL METHODS

Instrumental tests are based on measures of compression, adhesion, tension or stretching, shear or cutting and hybrid methods for measuring the rheological properties of the dough and pasta, both before and after cooking.

The rheological features of semolina dough for pasta-making is completely different from the flour dough for bread-making, mainly due to the different amounts of water used in the two processes. Semolina dough characterization is generally obtained on the basis of the alveographic parameters which show a strong relationship with pasta cooking quality (D'Egidio et al., 1990). However, in the alveograph mixing bowl (www.chopin.fr), the sample is prepared adding a water amount such as to obtain a pre-fixed moisture of dough, equal to 43.3% (Faridi and Rasper, 1987). This moisture value is higher than that one used in an actual pasta-making process, where dough moisture range is 30–32% (Dalbon et al., 1996). Recently, Marti et al. (2012a) have developed a new technique by adapting the Glutograph (Brabender GmbH&Co., Germany; www.brabender.com) to measure the changes in elasticity induced by heating on sheeted dough prepared from semolina of six durum wheat varieties with differences in protein quantity and quality. The information obtained by this procedure was related to pasta cooking quality. Stretching and relaxation cycles were applied repeatedly during the test time, while dough temperature was increased until 90 °C at 1.2 °C/min (Figure 9.4a). For each peak recorded, stretching and recovery values were calculated and plotted against sample temperature. Both stretching and recovery curves were integrated and the loss of elasticity was calculated (Figure 9.4b). A number of critical variables were identified that, when controlled, provide consistent results. Results shown as high elasticity loss during heating is related to low firmness of the corresponding cooked pasta. Following testing with larger sample numbers, this technique has the potential to be a valid screening tool for durum wheat quality.

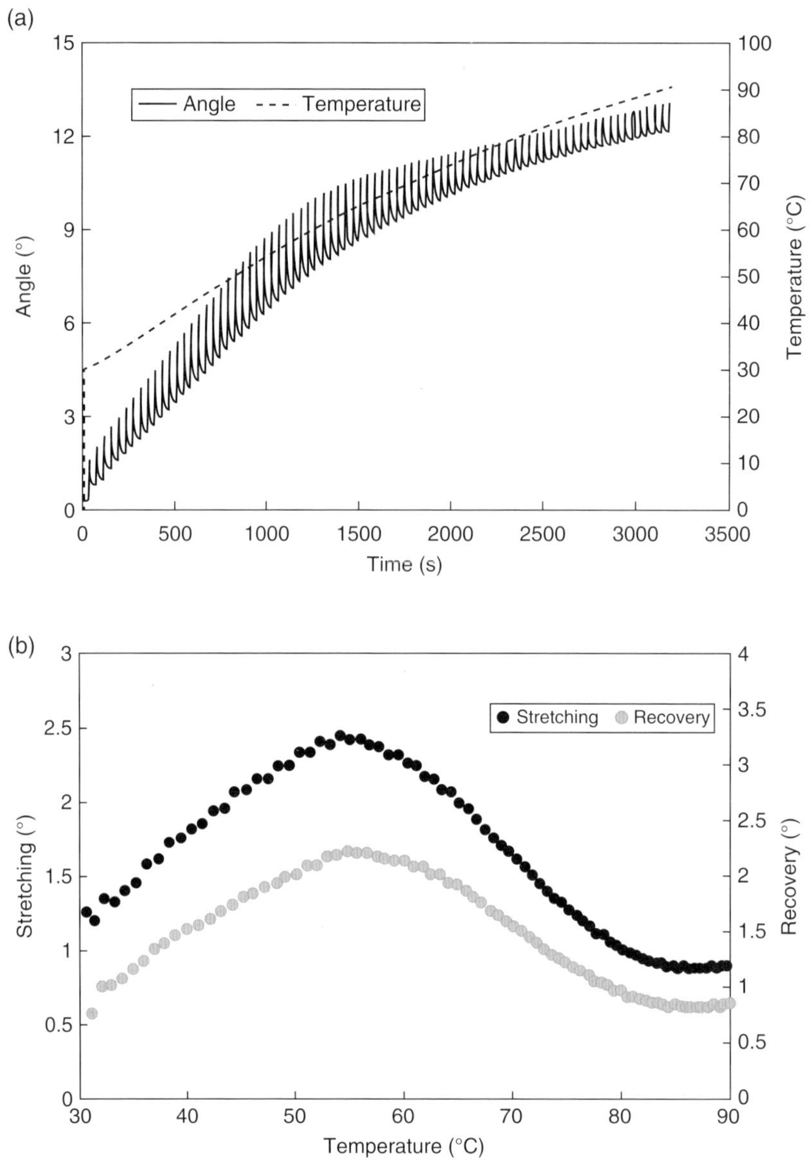

Figure 9.4 Sample curve obtained with the modified-glutograph (a). Changes in stretching and recovery during heating of sheeted dough (b).

As described below, extruded pasta is dried in order to guarantee a long and easy shelf-life. The drying step represents a key point of the pasta-making process: water should be carefully removed from the sample, assuring compactness to the product and avoiding the appearance of 'checking', or fracture lines. Bending tests have been utilized to evaluate mechanical strength of dried pasta products since Binnington et al. (1939) adapted the shortometer (Bailey, 1934) for this application. Winston (1960) designed an instrument called the 'bending stress tester' that applied a continuously increasing force to a strand of dried spaghetti until the fracture occurred. Breaking strength provides an indication of susceptibility of dried pasta to breakage during packaging and distribution but does not relate directly to textural characteristics of pasta during or after cooking. Furthermore, moisture content and deformation rate strongly influence the results (Cole, 1991). Bending tests are unsuitable for cylindrical pasta shapes because fractures occur randomly at numerous points within the sample test piece and contribute to the high variability of breaking strength values (Binnington et al., 1939). Karacsonyi and Borsos (1961) designed an apparatus for application of torque to dry pasta samples and found torsional strength to be more appropriate than bending or breaking strength for detection of irregular checking and axial fissures in dried pasta; however, the relationship of torsional strength to cooking characteristics of the finished product was extremely weak.

Finally, it is the cooking step that gives the product its final structure and consumer acceptability. The first instrument designed for measuring pasta texture was proposed by Matsuo and Irvine (1969), in order to simulate the bite of pasta. Known as the Grain Research Laboratory (GRL) spaghetti tenderness tester, the apparatus applied an increasing force to the cutting edge that moved downward perpendicular to a strand of spaghetti placed in a groove. A tenderness index was derived from the slope of the linear portion of the curve, which was indicative of the time required to cut the strand of spaghetti: the softer the sample, the shorter the time. The instrument had sufficient sensitivity to differentiate spaghetti made from different varieties of durum wheat. Although the results were found to correlate with sensory measurements of firmness (Matsuo and Irvine, 1974), the authors found that small changes in diameter greatly affected the instrumental answer; therefore, only samples with similar diameters could be compared (Matsuo and Irvine, 1969). In the same years, Walsh (1971) measured the firmness of cooked spaghetti using a plastic blade. The test produced a force versus distance recording, where the area under the curve was the amount of work (g*cm) required to shear the cooked spaghetti. The resulting firmness score was highly correlated with sensory panel score. However, the limitation of only measuring the force (or the work) required to cut through spaghetti has been recognized by other authors. Kramer and Hawbecker (1966) suggested that, in order to provide a complete description of textural quality, different parameters need to be considered: 1) deformation, the ability to withstand a certain amount of change in shape before breaking; 2) strength or firmness, a measure of the force required to penetrate or break the sample; 3) uniformity, a measure of the 'internal' characteristics of the bulk of the material; and 4) adhesiveness or stickiness, a measure of the energy required to remove the probe from the sample. In an attempt to obtain a better overall evaluation of cooked spaghetti, in terms of 'chewiness' or 'springiness', the cutting blade was replaced with a blunt-edged blade for compression (Matsuo and Irvine, 1974). Indeed, according to the authors, a product can be firm but yet lack elasticity or springiness. The limitation of the Walsh method (1971) due to the measurement of only one point on an individual strand of spaghetti was discussed by Voisey and Larmond (1973), highlighting the variation within and among spaghetti strands. After comparing several compression, shear, and extrusion type texture probes, it was found that stress and force measured by a multi-blade shear cell were highly correlated with firmness (Voisey, 1970; Voisey and Larmond, 1973).

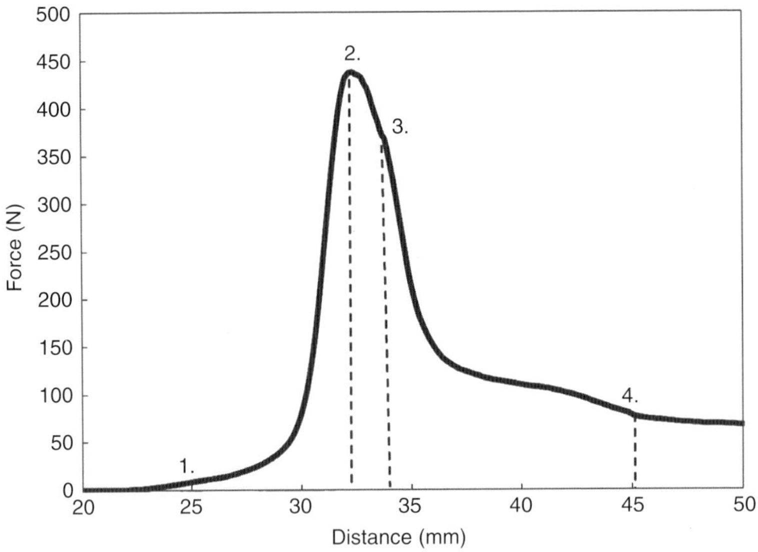

Figure 9.5 Curve obtained by the compression-shear-extrusion test on macaroni using the Kramer cell.

More recently, the American Association of Cereal Chemists (2000) adopted a compression method (AACC Approved Method 16-50.01, 2000; Approved Method 66-50.01, 2000) for determining the texture of cooked pasta, particularly for long shape pasta such as spaghetti and noodles. Five strands of spaghetti are sheared and firmness is expressed as the energy or work (g*cm) to cut one strand of spaghetti. Since variations exist within and among strands, it is important that more than one strand of spaghetti is measured. For short goods pasta products, a single-blade test is not applicable due to the physical constraints of aligning irregular pasta shapes with the straight blade. Thus, a multi-blade shear cell such as the Kramer cell is more appropriate for measuring cooked firmness of short-shape pasta with irregular or particular shapes (Manthey and Dick, 2012). Despite that, very few studies have been done on the use of the Kramer test on short-shaped pasta (Sahlstroem et al., 1993; Tudorica et al., 2002; Raina et al., 2005a; Raina et al., 2005b; Marti et al., 2010). A typical curve obtained by using a Kramer cell on six pieces of macaroni prepared from durum wheat semolina is shown in Figure 9.5. In order to provide a complete description of texture quality, four parameters have to be considered (Table 9.1):

1. Compression Energy (N*mm): the area under the first portion of the curve (from point 1 to point 2 in Figure 9.5); it represents the energy required to compress the cooked pasta.
2. Firmness (N): the maximum force registered by the instrument (point 2 in the chart).
3. Shear Force (N): the force developed when the blades cut the sample; high values indicate high resistance to shearing.
4. Extrusion Energy (N*mm): the area under the portion of the curve between point 3 and point 4 in the chart; it represents the energy required to chew cooked pasta; high values indicate high springiness.

Also in this test, changes in diameter greatly affected the results, as reported in Table 9.2.

Table 9.1 Indices obtained during compression-shear-extrusion test with the Kramer cell.

Point	Parameter	Definition	Index
From point 1 to point 2	Compression Energy (N*mm)	Energy required to compress cooked pasta	High values indicate high consistency
Point 2	Firmness (N)	Maximum force	High values indicate high consistency
Point 3	Shear Force (N)	Force developed when the blades cut the sample	High values indicate high resistance to shearing
From point 3 to point 4	Shear-Extrusion Energy (N*mm)	Energy required to chew cooked pasta	High values indicate high springiness

Table 9.2 Effect of size on texture of pasta produced from the same semolina.

	Pasta 1	Pasta 2
External Diameter (mm)	7	15
Optimal Cooking Time (min)	12	12
Compression Energy (N*mm)	824	1281
Firmness (N)	442	500
Shear Force (N)	186	288
Shear-Extrusion Energy (N*mm)	1226	3478

As an alternative to compression tests, tensile strength tests are commonly used, particularly for assessing the texture of cooked noodles. This family of tests have been widely used for fresh pasta (Alamprese et al., 2005), but rarely for dried pasta (Ferrari, 2008) probably because of the difficulties in gripping the ends of the pasta strands and ensuring breakage occurs along the extended region of the sample. Maximum stress values obtained by tension tests give an indication of the resistance of cooked pasta to break and the distance to break values indicate its extensibility (Smewing, 1997). However, Voisey and Larmond (1973) reported that correlation between tensile stress and measurements was not so high like the correlation among sensory attributes and the instrumental shear readings obtained from a multi-blade shear compression cell. This can be attributed to the shearing and compression forces that take place during mastication rather than the application of tensile forces (Marchylo and Dexter, 2004).

Pasta elasticity is an important parameter assessing the final texture of cooked pasta. Microscopic observations have revealed that an elastic network is able to trap starch material within the protein matrix, resulting in a firm cooked pasta (Resmini and Pagani, 1983). Several researchers have developed methods for measuring the elasticity of pasta. Currently, there is great interest in using fundamental rheology for evaluating the mechanical properties of pasta. Usually, the properties of viscoelastic materials are measured by creep and recovery, stress relaxation or dynamic oscillatory tests. Although results are very sensitive to the chemical composition and physical structure (Steffe, 1996) and a strong correlation was found between spaghetti firmness measured by the

Instron Universal Tester and the storage modulus and dynamic viscosity obtained by dynamic rheometry (Edwards et al., 1993), the technique requires sophisticated equipment and considerable experience.

The viscoelastograph (Chopin, Tripette et Renaud, Villeneuve La Garem, France) has also been used to measure compressibility and recovery of cooked pasta (Autran et al., 1986; Delcour et al., 2000). It was developed by the Institute National de la Recherche Agronomique (INRA) in Montpellier, France (Feillet et al., 1977) to assess the viscoelastic properties of cereal products. With this instrument, the sample is compressed between two plates with a constant load applied perpendicularly and then removed. Changes in thickness during and after loading, as well as the deformation and the capacity to return to the initial form, are recorded. Compressibility, consistency and relative recovery are obtained from the curve and an index of viscoelasticity can be calculated from compressibility and relative recovery values (D'Egidio and Nardi, 1996). Based on findings by D'Egidio et al. (1993) the viscoelastograph is considered a suitable technique to predict sensory firmness of pasta, particularly when pasta is dried at high temperatures.

As regards the measurement of surface stickiness of cooked spaghetti, Voisey et al. (1978) used a test cell containing a grooved lower platen to compress to a fixed compression force ten strands of cooked spaghetti simultaneously. After allowing the spaghetti to relax, the plates were pulled apart and the maximum tensile force required to separate the probe from the strand surface was used as the index of stickiness. However, the maximum adhesive force as measured by this approach was unrelated to sensory assessment of adhesiveness. By contrast, Dexter et al. (1983) modified the GRL compression tester by the incorporation of several features into its design to measure the stickiness of cooked pasta, such as a rectangular plunger for compressing the specimen. However, instrumental stickiness values were poorly correlated with other cooking characteristics, including cooking loss, cooked weight, and instrumental measurements of tenderness, compressibility and recovery (Dexter et al., 1983). Kim et al. (1989) adapted this technique for use with the Instron but encountered some difficulties with the measurements which they attributed to surface water released from the spaghetti during compression. According to the authors, a low stickiness score could mean either a water-logged (over-cooked) spaghetti with water released during compression or a firm spaghetti that was not sticky. It is possible that the use of a lower compression force may have overcome this problem (Marchylo and Dexter, 2004).

Several researchers have found that the amount of rinsed material collected from the surface of pasta is a good indicator of pasta stickiness. D'Egidio et al. (1976) have developed a chemical method for quantitative determination of total organic matter (TOM) rinsed from the surface of cooked spaghetti. It was accepted as a standard method by the International Association for Cereal Science and Technology (ICC, 1992). TOM values reflected the quantity of starch released from the protein matrix of cooked pasta and subsequently adhere to the product surface. Since it was highly correlated with sensory (D'Egidio et al., 1982) and instrumental (Dexter et al., 1985) values of firmness, stickiness and resilience of cooked spaghetti, TOM values were proposed as suitable indices for prediction of sensory assessments of pasta, particularly when a low-temperature drying cycle is applied (D'Egidio et al., 1993).

Even if several tests have been proposed for measuring the texture of cooked pasta, none of them allows all the features of cooked pasta to be evaluated at the same time. For this reason, sensorial approaches are widely used for assessing pasta quality.

9.4 SENSORY METHOD

Sensory evaluation is considered to be the most reliable method for measuring the textural properties of cooked pasta since panelists have the ability to measure the overall textural characteristics of cooked pasta at the same time. The International Standards Organization (ISO 1985) has developed a standard method (TC 34 SC4 7304) using trained assessors for evaluating the firmness and surface condition (stickiness) of cooked spaghetti. A series of reference photographs are used to estimate surface condition. The general appearance, degree of swelling and stickiness are taken into account for assessment of the overall rating. Scores for firmness and surface condition are assigned using nine point rating scales. In Italy, the quality of pasta cooked in water without salt is assessed by using the scoring method proposed by D'Egidio and Nardi (1996), for the evaluation of stickiness, bulkiness and firmness. Stickiness is described as the material adhering to surface of cooked pasta evaluated by visual inspection with the aid of standard reference samples and by handling. Bulkiness, which is related to stickiness, measures the adhesion degree of pasta strands to each other evaluated both visually and manually. Firmness relates to the resistance of cooked pasta to chewing. Each parameter was scored on a 10 to 100 scale, where 10 = absent for stickiness and bulkiness, and 100 = very good for firmness. An overall quality score is determined by calculating the mean of each of the three scores. An overall score greater than 70 indicates good quality while a score below 50 indicates poor-quality pasta (D'Egidio and Nardi, 1996).

9.5 INSTRUMENTAL VS SENSORY TEXTURE

The preference of adopting instrumental methods instead of sensory analysis for measuring the texture of the product remains an open debate still today. Sensory evaluation is often criticized as being subjective techniques. Actually, two distinct types of sensory test exist: consumer-oriented and product-oriented (D'Egidio and Nardi, 1996; Marchylo and Dexter, 2004). The former involves a large number of consumers for determining product acceptability or degree of liking, based on subjective criteria. In this case, the sensorially evaluated characteristics of pasta may be highly different according to the consumer eating habits. In other words, the evaluation scale is highly subjective, because it is based on personal preference (Cole, 1991; D'Egidio and Nardi, 1996). This problem is solved using a product-oriented test that involves the use of selected and trained panelists under controlled testing conditions to evaluate the quality of a product. For this reason, sensory techniques are time-consuming, expensive and may exhibit poor reproducibility if proper procedures are not followed. In addition, the use of sensory panels to evaluate a large number of samples or to evaluate samples when sample size is limited is not feasible (Marchylo and Dexter, 2004). As regards the instrumental methods, they are often criticized as being able to measure only a limited number of characteristics. Moreover, objective methods very often have been employed without a valid preliminary comparison with the sensory evaluation, on a limited series of samples, or on samples very different in quality (D'Egidio and Nardi, 1996). Least but not last, it is often difficult to relate the instrumental measurements to the sensory properties of the product, since textural differences among samples are too small to be detected by sensory perception (Manthey and Dick, 2012). Bourne (1982) stated that it cannot be assumed that one instrumental method can measure all of the textural properties of a food. Recently, evaluating the association between sensory and instrumental measurements on commercial pasta, Martinez et al. (2007) found a good correlation between firmness from the sensory evaluation and its respective texture measurement (Pearson correlation coefficient 0.92; $P=0.02$). While, no correlation between stickiness, elasticity, springiness, and chewiness from the sensory evaluation and instrumental measurement was found.

9.6 FACTORS INFLUENCING THE TEXTURAL PROPERTIES OF PASTA

The main ingredient of premium quality pasta is 100% durum wheat semolina. Good quality pasta can also be made with durum wheat flour or a blend of semolina and durum wheat flour (particle size $<200\,\mu m$), as assessed by Italian, French, and Greek Laws. Common wheat flour can also be used for preparing dried pasta, but the corresponding product will be generally inferior in appearance (color) and cooking quality compared to pasta made from durum wheat (D'Egidio et al., 1990; Pollini, 1990).

To improve the quality of low protein raw materials, different methods have been explored; Dexter et al. (1981) applied elevated temperatures during the drying stage to improve the quality of low protein pasta. More recently, Majzoobi et al. (2011) showed that it is possible to improve the textural properties of spaghetti made with low protein wheat flour using gluten powder.

Recently, the production of dried pasta from a mixture of semolina and other flours such as wheat flour, whole wheat flour, bran-semolina, corn, rice, oat, legumes flours, and so on, is more and more increasing. The main reasons for such replacements are the lower availability of semolina in some parts of the world, the production of new types of pasta using other flours, the elimination of gluten for special diet products, and the increase of the nutritional benefits of enriched-pasta. However, partial or total substitution of semolina with other flours has adverse effects on the quality of the final product. In particular, the introduction of fiber greatly affected the structure of pasta and consequently its cooking quality and textural properties as measured by instrumental or sensorial analysis. Indeed, high fiber content interrupts the continuity of the pasta and gluten structure upon overcooking, both in gluten (Tudorica et al., 2002; Brennan et al., 2003) and non-gluten (Marti et al., 2010; Cabrera-Chavez et al., 2012) pasta.

As regards the addition of legume flours, the enrichment in non-gluten protein and the lower water uptake of cooked pasta fortified with legume flours may be responsible for higher hardness. However, the resulting protein network can be weaker than the one developed in durum wheat pasta, due to dilution of the gluten proteins (Petitot et al., 2010). The increase in drying temperature strengthened legume pasta structure as shown by the decrease in cooking loss and the increase in pasta hardness, cohesiveness and breaking energy evaluated by mechanical test and higher hardness, elasticity and lower fracturability determined by sensorial analysis (Petitot et al., 2010).

Processing conditions play a key role in determining the cooking quality of semolina pasta, not only in the case of enriched-pasta. As well-known, the pasta-making process comprises subsequent steps, represented by mixing of semolina and water, forming (by extrusion or lamination), drying and, finally, packaging of the finished product. In the first step of the conventional pasta-making, semolina and water are blended together in proportions suitable for creating a mass to final moisture content of 30–35%. At this low moisture level, gluten development is only partial (Dexter and Matsuo, 1979; Pagani et al., 1989) and other semolina components may affect the physical properties of pasta (De Noni and Pagani, 2010). For this reason, water quantity has to be adjusted according to ingredients features. For example, large amounts of damaged starch in semolina require a high level of water (Manthey and Schorno, 2002). Moreover, as damaged starch increases, there is greater loss of solids during cooking as the starch is quickly gelatinized (Matsuo and Dexter, 1980) and a greater tendency to surface stickiness (Grant et al., 1993). Semolina particle size distribution should be as narrow as possible to minimize uneven hydration, and generally between 250 e 350 μm (Dalbon et al., 1996). The presence of such a defect results in the incomplete and non-homogeneous hydration of semolina particles during mixing and, consequently, in an incomplete and non-homogeneous structuring of protein network, giving origin to a more fragile product, both before and after cooking, due to the inclusion of white specks or streaks.

Mixing and extrusion chambers operate under vacuum to minimize oxidation of the carotenoid or xanthophyll pigments. Oxidation of these pigments is known to decrease the yellow color of finished pasta (Feillet and Dexter, 1996). In addition, the vacuum prevents air bubble formation, which can give rise to defects and an unsightly appearance in the dried product.

After hydration, semolina mixture is formed into pasta by extrusion or lamination. It is generally accepted that pasta obtained by sheeting has a better cooking quality than extruded products. The gluten network should be more developed by sheeting than by extrusion (Pagani et al., 1989) (Figure 9.6). In the extrusion process commonly used for semolina pasta production, the dough is submitted to significant shear stresses, which can weaken the poor quality protein structure (Pagani et al., 1989), but, at the same time, to pressure (9–10 MPa), responsible for giving the product a high compactness, and allowing it to withstand cooking (Petitot et al., 2009b). On the other hand, the sheeting process by rolls allows the protein to align in a continuous network whose goodness strongly depends on protein quantity and quality (Figure 9.6). In a recent work, Marti et al. (2011a) reported low breaking strength for dry pasta prepared from composite flours and by sheeting, usually responsible for a more porous structure, and related to high cooking losses during cooking. On the other hand, the roll-sheeting step that allowed a better alignment of the gluten strands (Pagani et al., 1989), accounting for low adhesiveness during cooking (Marti et al., 2011a).

During the extrusion process, dough temperature should not increase above about 50 °C to prevent swelling of starch granules and coagulation of gluten network: both of these phenomena would adversely affect the cooking properties of the finished product (Abecassis et al., 1994).

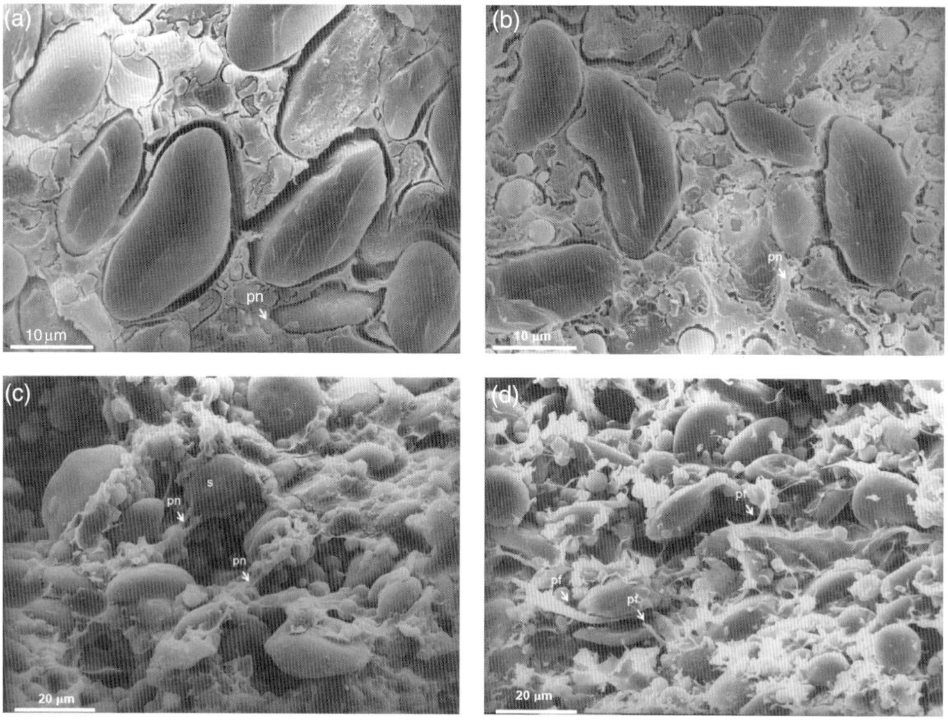

Figure 9.6 SEM images of spaghetti before cooking. Roll-sheeted pasta (a, c); Extruded pasta (b, d); Cryo-fracturing (a, b); Dry-fracturing (c, d). pf = protein fibrils; pn = protein network; s = starch granules.

The type and condition of the die will affect extrusion pressure, the dimensions of the pasta and the condition of the surface (Maldari and Maldari, 1993; Lucisano et al., 2008). The advantages of using Teflon die are longer die lifetime, smoother pasta surface, better appearance and good cooking quality, due to a more continuous protein network surrounding starch granules within the matrix, and also covering the pasta surface, compared to pasta extruded through bronze dies (Donnelly, 1982). Despite that, the use of bronze dies produces a product with a rough surface which helps the sauce to stick to the cooked pasta (Dalbon et al., 1996).

During drying, the product has to be dried down to a moisture of around 12.5%, in order to allow a long shelf life (generally three years). Drying conditions (temperature, humidity and airflow) greatly affect pasta cooking quality. If pasta dries too quickly, the surface will harden, resulting in 'checking', or fracture lines in the product, resulting in a weak structure which may rise to breakage of the product during packaging and handling, negatively affecting pasta quality (Feillet and Dexter, 1996). The introduction of High Temperature (> 60 °C; HT) in the drying process, improved the cooking quality of pasta, in terms of low cooking loss, low stickiness, high firmness and better tolerance to overcooking (D'Egidio et al., 1990). As was explained in the review of Resmini and Pagani (1983), the improvement of cooking performances is related to the changes induced by HT treatments to starch and, above all, protein: the latter, in appropriate conditions of temperature (from 65–70 °C) and regardless of the conditions of humidity, coagulate and create, already in the raw product, a network very similar to that which is formed during cooking. This macromolecular organization is effective in counteracting an excessive swelling of starch granules and, therefore, a high viscosity and the absence of firmness in the cooked pasta (D'Egidio and Nardi, 1991). Although raw materials with low protein quantity and quality required a HT drying cycle to ensure acceptable cooking behavior, a drying cycle is often wrongly assessed as much more effective the higher the temperatures in it. On the contrary, the production of dried pasta of good quality is the result of correct combination of time, temperature and humidity of the product. It should be reminded that the onset of stresses and fractures may occur because the water profile inside pasta can cause a different local dough shrinkage (De Cindio et al., 2003). The best results were obtained by increasing the temperature only when the moisture of the product is below 16–18% (Resmini and Pagani, 1983): HT drying applied to a product with a high moisture content may lead to an excessive swelling of starch granules with the consequent rupture of the protein network and the deterioration of the cooking quality of pasta (Resmini et al., 1988).

The HT-drying cycles, now so widespread because of the many advantages already mentioned, can induce significant thermal damage, mainly due to the Maillard reaction (or non-enzymatic browning). The reaction not only affects some sensory characteristics of dried pasta (such as color and flavor) but it also causes a decrease of the content of bioavailable lysine, up to 40% of the lysine originally present in the semolina (Pagani et al., 1992).

Among the new methods of drying the pasta, recently De Pilli et al. (2009) studied the effect of drying by the use of microwaves on pasta cooking quality. This study showed that microwave drying is related to absence of fissures, higher consistency and lower degree of gelatinization in the dried pasta compared to the product dried in a conventional manner (HT).

9.7 GLUTEN FREE PASTA

In the last few decades, beside the conventional pasta-product made by durum wheat semolina, pasta from gluten-free (GF) cereals (rice, corn, sorghum, oat) and pseudcereals (buckwheat, amaranth, quinoa, etc.) is more and more consumed not only by the growing number of celiacs but also by other consumers who wish to exclude gluten-based products from their diet for health reasons (Marti and Pagani, 2013). Moreover, celiac disease can occur at any age (Cureton and Fasano, 2009), therefore, the production of good quality GF products for people with a tradition of wheat-based products is required. In fact, pasta prepared only from non-gluten flour frequently exhibits inferior quality in comparison with semolina pasta. Because of the lack of a protein regular network in rice pasta samples, starch polymers are less efficaciously entrapped in the matrix, resulting in a sticky product with low firmness, low tolerance to overcooking, and high cooking loss (Resmini et al., 1979; Marti et al., 2012b). The addition of texturing ingredients can be a good solution for improving cooking quality. Hydrocolloids or gums are commonly used due to their ability to make a gel, thus assuring the formation of a three-dimensional network (BeMiller, 2008). A wide range of hydrocolloids have been proposed (such as arabic gum, xanthan-gum, locust bean gum, carboxymethylcellulose) in order to provide high viscosity at room temperature, to improve the firmness, and to give body and mouth-feel to pasta (Huang et al., 2001; Singh et al., 2004; Chillo et al., 2007). In addition, because of their ability to bind water, gums can increase the rehydration rate of pasta and give positive effects for handling and processing (Sozer, 2009). Regarding emulsifiers, they act as a lubricant in the extrusion process and provide a firmer consistency and a less sticky surface, as they control starch swelling and leaching phenomena during cooking (Lai, 2001; Kaur et al., 2005; Charutigon et al., 2008). Despite the several wellknown positive effects of the addition of emulsifiers and hydrocolloids (Huang et al., 2001; Lai, 2001; Singh et al., 2004; Kaur et al., 2005; Chillo et al., 2007; BeMiller, 2008; Charutigon et al., 2008; Sozer, 2009) the consumers often associate their presence in GF pasta to a 'no-natural' or 'artificial' food. Starting from this finding, the use of proteins as structuring building ingredients could be an interesting approach for producing GF pasta, also for its positive nutritional effects (Thompson, 2009). On this subject, egg proteins are well known for their gelling, foaming and emulsifying characteristics, in addition to their high nutritional quality (Plancken van der et al., 2007). Whey and caroubin proteins form a protein network that holds starch during cooking (Feillet and Roulland, 1998). Recently, the addition of egg white powder (6%) and emulsifier (1.2%) improved the texture as well as the cooking quality of GF pasta prepared from buckwheat, amaranth and quinoa flour blend, that reached sensorial evaluation comparable to wheat pasta (Schoenlechner et al., 2011).

While gluten proteins have a key role in conventional durum wheat pasta properties, starch assumes a structuring role in GF pasta-making, related to the tendency of its macromolecules to reassociate and interact after heating treatment in the presence of water (retrogradation phenomenon) (Colonna and Buleon, 1992). Marti et al. (2012b) reported the cooking characteristics of GF pasta prepared according to the two technologies currently used in the GF sector (Figure 9.7). In one case, heat-treated flour in which starch is already partially gelatinized is extruded in the conventional pasta-making plant used for semolina pasta (Figure 9.7a). In the second approach (extrusion-cooking

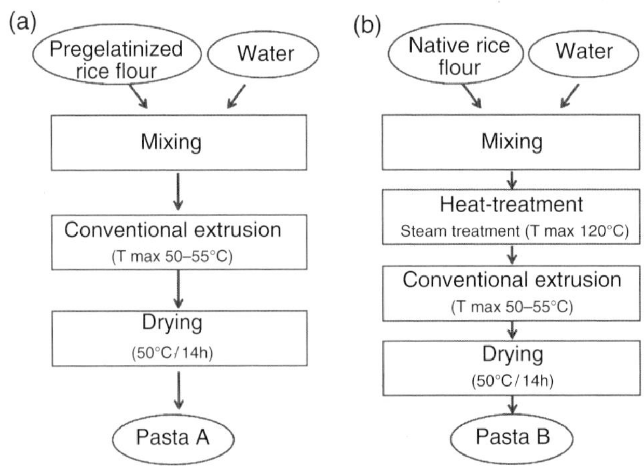

Figure 9.7 Conventional extrusion (a) and extrusion-cooking process (b) for gluten-free pasta-making.

Table 9.3 Cooking quality of experimental rice pasta.

	Pasta A	Pasta B	Commercial semolina pasta
Cooking loss (g/100g)	10.3±0.7	9.8±0.2	3.5±0.3
Water absorption (g/100g)	78.1±3.6	90.7±4.2	98.7±1.5
Compression Energy (N*mm)	552±58	328±7	824±106
Firmness (N)	310±35	191±7	442±9
Shear Force (N)	293±21	150±5	186±5

process), starch gelatinization directly occurs in the extruder by using steam (Figure 9.7b). Both pasta samples showed high cooking loss – even four times higher than those of pasta from durum wheat semolina – and poor texture, characterized by lower firmness than semolina pasta (Table 9.3). Rice parboiling can represent a pre-processing suitable by itself to induce a new macromolecular structure, able to assure good texture after cooking behavior in pasta products. The combination of parboiling process and extrusion-cooking gave a product with low cooking loss, but an extreme firmness (Marti et al., 2010). The strong interactions of amylopectin and/or amylose promoted by the extrusion-cooking process suggested that in the product the amylopectin matrix is likely a combination of amylose and amylopectin chains. The average molecular weight of amylose and amylopectin, as well as their molecular organization within the granule, affected starch functionality and, consequently, cooking performances (Marti et al., 2011b; 2011c).

9.8 CONCLUSIONS AND FUTURE TRENDS

The protein content and gluten quality are recognized as being responsible for the cooking quality of pasta with good textural characteristics, before and after cooking. A high protein content and a 'strong' gluten (in terms of its viscoelasticity) are required to process semolina into a suitable final pasta product with optimal cooking performances. Clear-cut relationships between protein content/composition and pasta cooking quality in terms of elasticity, extensibility, and resistance to overcooking have been proposed. Although up to 80% of dry matter in semolina is starch, this component has received limited research attention up-to-now in comparison with protein fraction. Some evidence points out that starch in pasta cannot be considered an inert filler, but it contributes substantially to pasta structure and quality. The role of starch and more specifically of its interaction with other semolina components in pasta cooking quality has been also investigated indirectly by analyzing the pasta cooking water and rinsing water. Lipids could also affect the cooking quality of this product. In particular, polar lipids may promote both hydrophobic and hydrophilic interactions between gluten and starch, improving the cooking quality by enhancing the stability of both protein network and starchy material. A better understanding of the interactions among macromolecules will be useful for improving the texture of the product, by the optimization of the processing conditions applied to each pasta-making step.

As regards gluten-free pasta, despite the great efforts made in the last few decades to produce GF pasta that is similar in quality to durum wheat products, the GF pasta currently on the market is still far from what the consumer is looking for. Moreover, little information is available regarding starch arrangements that could assure good quality cooking and eating. Understanding the relationship between starch structure and processing conditions will help the industry re-formulate and develop products with the desired texture as well as improved nutritional and digestive properties.

References

AACC, American Association of Cereal Chemists (2000) Approved Methods of the AACC, St Paul, MN, USA.

Abecassis, J., Abbou, R., Chaurand, M., Morel, M.H., and Vernoux, P. (1994) Influence of extrusion conditions on extrusion speed, temperature, and pressure in the extruder on pasta quality. *Cereal Chemistry* 71, 247–253.

Alamprese, C., Iametti, S., Rossi, M., and Bergonzi, D. (2005) Role of pasteurisation heat treatments on rheological and protein structural characteristics of fresh egg pasta. *European Food Research and Technology* 221, 759–767.

Alary, R., Abecassis, J., Kobrehel, K., and Feillet, P. (1979) Effects of cooking water, particularly its pH, on the characteristics of cooked spaghetti. *Bulletin des Anciens Eleves de l'Ecole Francaise de Meunerie* 293, 255–262.

Autran, J.C., Abecassis, J., and Feillet, P. (1986) Statistical evaluation of different technological and biochemical tests for quality assessment in durum wheats. *Cereal Chemistry* 63, 390–394.

Bailey, C.H. (1934) An automatic shortometer. *Cereal Chemistry* 11, 160–163.

BeMillers, J.N. (2008) Hydrocolloids. In: *Gluten-free Cereal Products and Beverages*. (eds. Arendt E.K. and Dal Bello, F.), pp. 203–216, Academic Press, London, UK.

Binnington, D.S., Johannson, H., and Geddes, W.F. (1939) Quantitative methods for evaluating the quality of macaroni products. *Cereal Chemistry* 16, 149–167.

Bonomi, F., D'Egidio, M.G., Iametti, S., Marengo, M., Marti, A., Pagani, M.A., and Ragg, E. (2012) Structure–quality relationship in commercial pasta: a molecular glimpse. *Food Chemistry* 135, 348–355.

Bourne, M.C. (eds.) (1982) *Food Texture and Viscosity*, Academic Press Inc., New York, NY, USA.

Brennan, J.G. (1980) Food texture measurement. In: *Developments in Food Analysis Techniques*. (eds. King R.D.), 2nd edn., pp. 1–78. Applied Science Publishers, London, UK.

Brennan, C.S., Kuri, V., and Tudorica, C.M. (2003) Inulin-enriched pasta: Effects on textural properties and starch degradation. *Food Chemistry* 86, 189–193.

Cabrera-Chavez, F., Calderon de la Barca, A.M., Islas-Rubio, A.R., Marti, A., Marengo, M., Pagani, M.A., Bonomi, F., and Iametti, S. (2012) Molecular rearrangements in extrusion processes for the production of amaranth-enriched, gluten-free rice pasta. *LWT-Journal of Food Science and Technology* 47, 421–426.

Charutigon, C., Jitpupakdree, J., Namsree, P., and Rungsardthong, V. (2008) Effects of processing conditions and the use of modified starch and monoglyceride on some properties of extruded rice vermicelli. *Food Science and Technology* 41, 642–651.

Chillo, S., Laverse, J., Falcone, P.M., and Del Nobile, M.A. (2007) Effect of carboxymethylcellulose and pregelatinized corn starch on the quality of amaranthus spaghetti. *Journal of Food Engineering* 83, 492–500.

Cole, M.E. (1991) Review: prediction and measurement of pasta quality. *International Journal of Food Science and Technology* 26, 133–151.

Colonna, P. and Buleon, A. (1992) New insights of starch structure and properties. Proceedings of Cereal chemistry and technology: along past and a bright future. In: *Proceedings of 9th International Cereal and Bread Congress*, pp. 25–42. Paris, FR.

Cubadda, R. (1988) Evaluation of durum wheat, semolina and pasta in Europe. In: *Durum Wheat: Chemistry and Technology*. (eds. Fabriani, G. and Lintas, C.) pp. 217–228. AACC International, St. Paul, MN, USA.

Cunin, C., Handschin, S., Walther, P., and Escher, F. (1995) Structural changes of starch during cooking of durum wheat pasta. *Lebensmittel-Wissenschaft und Technologie* 28, 323–328.

Cureton, P. and Fasano, A. (2009) The increasing incidence of celiac disease and the range of gluten-free products in the marketplace. In: *Gluten-free Food Science and Technology*. (eds. Gallagher, E.), pp. 1–15, Blackwell Publishing Ltd, London, UK.

D'Egidio, M.G., Sgrulletta, D., Mariani, B.M., Galterio, G., DeStefanis, E., and Fortini, S. (1976) Method for the measurement of stickiness and quality of pasta products. *Tecnica Moliroria* 27, 89–93.

D'Egidio, M.G., DeStefanis, E., Fortini, S., Galterio, G., Nardi, S., and Sgrulletta, D. (1981) Influence of the type of cooking water used on the quality of pasta. *Tecnica Molitoria* 32, 505–511.

D'Egidio, M.G., DeStefanis, E., Fortini, S., Galterio, G., Nardi, S., and Sgrulletta, D. (1982) Standardization of cooking quality analysis in macaroni and pasta products *Cereal Foods World* 27, 367–368.

D'Egidio, M.G., Mariani, B.M., Nardi, S., Novaro, P., and Cubadda, R. (1990) Chemical and technological variables and their relationships: A predictive equation for pasta cooking quality. *Cereal Chemistry* 67, 275–281.

D'Egidio, M.G. and Nardi, S. (1991) Effects of high temperature drying system on pasta quality of durum wheat cultivars. *Tecnica Molitoria* 12, 429–434.

D'Egidio, M.G., Mariani, B.M., Nardi, S., and Novaro, P. (1993) Viscoelastograph measures and total organic matter test: suitability in evaluating textural characteristics of cooked pasta. *Cereal Chemistry* 70, 67–72.

D'Egidio, M.G. and Nardi, S. (1996) Textural measurements of cooked pasta. In: *Pasta and Noodle Technology*. (eds. Kruger, J.E., Matsuo, R.B., and Dick, J.W.), pp. 133–156. AACC International, St. Paul, MN, USA.

Dalbon, G., Grivon, D., and Pagani, M.A. (1996) Continuous manufacturing process. In: *Pasta and Noodle Technology*. (eds Kruger, J.E., Matsuo, R.B., and Dick J.W.), pp. 13–58. AACC International, St. Paul, MN, USA.

De Cindio, B., Gabriele, D., Migliori, M., and Pollini, C.M. (2003) Modelling of drying for high quality pasta production. In: *Proceedings of 3rd International Symposium of Rheology and Food Structure. Eurorheo 2003-01*, pp. 635–636, Zurich, CH.

De La Pena, E. and Manthey, F.A. (2011) Parameters that affect pasta coking. *Cereal Foods World* 56, A37.

De Noni, I. and Pagani, M.A. (2010) Cooking properties and heat damage of dried pasta as influenced by raw material characteristics and processing conditions. *Critical Reviews in Food Science and Nutrition* 50, 465–472.

De Pilli, T., Giuliani, R., Derossi, A., and Severini, C. (2009) Study of cooking quality of spaghetti dried through microwaves and comparison with hot air dried pasta. *Journal of Food Engineering* 95, 453–459.

Delcour, J.A., Vansteelandt, J., Hythier, M.C., Abecassis, J., Sindic, M., and Deroanne, C. (2000) Fractionation and reconstitution experiments provide insight into the role of gluten and starch interactions in pasta quality. *Journal of Agriculture and Food Chemistry* 48, 3767–3773.

Dexter, J.E. and Matsuo, R.R. (1979) Effect of water content on changes in semolina proteins during dough mixing. *Cereal Chemistry* 56, 15–19.

Dexter, J.E., Matsuo, R.R., and Morgan, B.C. (1981) High temperature drying: Effect on spaghetti properties. *Journal of Food Science* 46, 1741–1746.

Dexter, J.E., Matsuo, R.R., and Morgan, B.C. (1983) Spaghetti stickiness: some factors influencing stickiness and relationship to other cooking quality characteristics. *Journal of Food Science* 48, 1545–1559.

Dexter, J.E., Matsuo, R.R., and MacGregor, A.W. (1985) Relationship of instrumental assessment of spaghetti cooking quality to the type and the amount of material rinsed from cooked spaghetti. *Journal of Cereal Science* 3, 39–53.

Donnelly, B.J. (1982) Teflon and non-teflon lined dies: effect on spaghetti quality. *Journal of Food Science* 47, 1055–1058, 1069.

Edwards, N.M., Izydorczyk, M.S., Dexter, J.E., Biliaderis, C.G. (1993) Cooked pasta texture: comparison of dynamic viscoelastic properties to instrumental assessment of firmness. *Cereal Chemistry* 72, 564–567.

Faridi, H. and Rasper, V. (1987) The Alveograph Handbook. AACC International, St. Paul, MN, USA.

Feillet, P., Abecassis, J., and Alary, R. (1977) Description d'un nouvel appareil pour mesurer les propriétés viscoelastiques des produits céréaliers. *Bulletin des anciens élèves de l'ENSMIC* 273, 97–101.

Feillet, P. and Dexter, J.E. (1996) Quality requirements of durum wheat for semolina milling and pasta production. In: *Pasta and Noodle Technology*. (eds. Kruger, J.E., Matsuo, R.R., Dick, J.W.), pp. 95–131. AACC International, St Paul, MN, USA.

Feillet, P. and Roulland, T.M. (1998) Caroubin: a gluten-like protein isolated from carob bean germ. *Cereal Chemistry* 75, 488–492.

Ferrari, I. (2008) Influenza della materia prima e delle condizioni di essiccamento sulla qualità in cottura della pasta. Thesis in Food Science, University of Milan, IT.

Grant, L.A., Dick, J.W., and Shelton, D.R. (1993) Effects of drying temperature, starch damage, sprouting, and additives on spaghetti quality characteristics. *Cereal Chemistry* 70, 676–684.

Horigane, A.K., Kawabuchi, M., Uchijima, S., and Yoshida, M. (2009) Effects of seasonings on physical properties and MRI T2 map of cooked spaghetti. *Food Research International* 42, 41–50.

Huang, J.C., Knight, S., and Goad, C. (2001) Model prediction for sensory attributes of non-gluten pasta. *Journal of Food Quality* 24, 495–511.

ICC, International Association for Cereal Science and Technology (1992) Standard method of ICC, Wien, Austria.

ISO, International Standards Organization (1985) International Standard TC34 SC4 7304 Durum wheat semolinas and alimentary pasta – Estimation of cooking quality of spaghetti by sensory analysis, Geneva Switzerland, 04–15.

Karacsonyi, L.P. and Borsos, A.C. (1961) An apparatus for measuring the torsional strength of macaroni. *Cereal Chemistry* 38, 14–21.

Kaur, L., Singh, J., and Singh, N. (2005) Effect of glycerol monostearate on the physico-chemical, thermal, rheological and noodle making properties of corn and potato starches. *Food Hydrocolloids* 19, 839–849.

Kim, H.I., Seib, P.A., Posner, E., Deyoe, C.W., and Yang, H.C. (1989) Improving the colour and cooking quality of spaghetti from Kansas hard winter wheat. *Cereal Foods World* 34, 216–223.

Kramer, A. and Hawbecker, J.V. (1966) Measuring and recording rheological properties of gels. *Food Technology* 20, 209.

Lai, H. and Hwang, S. (2004) Water status of cooked white salted noodles evaluated by MRI. *Food Research International* 37, 957–966.

Lai, H.M. (2001) Effects of hydrothermal treatment on the physicochemical properties of pregelatinized rice flour. *Food Chemistry* 72, 455–463.

Lucisano, M., Pagani, M.A, Mariotti, M., and Locatelli, D.P. (2008) Influence of die material on pasta characteristics. *Food Research International* 41, 646–652.

Majzoobi M., Ostovan, R., and Farahnaky, A. (2011) Effect of gluten powder on wheat flour spaghetti. *Journal of Texture Studies* 42, 468–477.

Maldari, M. and Maldari, C. (1993) Design and performance of pasta dies. *Cereal Foods World* 38, 807–810.

Manthey, F.A., and Schorno, A.L. (2002) Physical and cooking quality of spaghetti made from whole wheat durum. *Cereal Chemistry* 79, 504–510.

Manthey, F. A. and Dick, T. (2012) Assessment of probe type for measuring pasta texture. *Cereal Foods Worlds* 57, 56–62.

Marchylo, B.A. and Dexter J.E. (2004) Improving the texture of pasta. In: *Texture in Food. Solid Foods*. (eds Kilcast, D.), pp. 465–500. Woodhead Publishing Ltd., Cambridge, UK.

Marti, A., Seetharaman, K., and Pagani, M.A. (2010) Rice-based pasta: a comparison between conventional pasta-making and extrusion- cooking. *Journal of Cereal Science* 52, 404–409.

Marti, A., Fongaro, L., Rossi, M., Pagani, M.A. (2011a) Quality characteristics of pasta enriched with buck-wheat flour. *International Journal of Food Science and Technology* 46, 2393–2400.

Marti, A., Pagani, M.A., and Seetharaman, K. (2011b) Characterizing starch structure in a gluten-free pasta by using iodine vapor as a tool. *Starch/Starke* 63, 241–244.

Marti, A., Pagani, M.A., and Seetharaman, K. (2011c) Understanding starch organisation in gluten-free pasta from rice flour. *Carbohydrate Polymers* 84, 1069–1084.

Marti, A., D'Egidio, M.G., Dreisoerner, J., Seetharaman, K., and Pagani, M.A. (2012a) New procedure for evaluating pasta-making aptitude of durum wheat semolina. Cereal Foods World 57, A60.

Marti, A., Caramanico, R., Bottega, G., and Pagani, M.A. (2012b) Cooking behavior of rice pasta: effect of thermal treatments and extrusion conditions. *LWT-Journal of Food Science and Technology* 54, 229–235.

Marti, A., Seetharaman, K., Pagani, M.A. (2013) Rheological approaches suitable for investigating starch and protein properties related to cooking quality of durum wheat pasta. *Journal of Food Quality* 36, 133–138.

Marti, A. and Pagani, M.A. (2013) What can play the role of gluten in gluten free pasta? Trends in Food Science and Technology 31, 63–71.

Martinez, C.S., Ribotta, P.D., Leon, A.E., and Anon, M.C. (2007) Physical, sensory and chemical evaluation of cooked spaghetti. *Journal of Texture Studies* 38, 666–683.

Matsuo, R.R. and Irvine, G.N. (1969) Spaghetti tenderness testing apparatus. *Cereal Chemistry* 46, 1–6.

Matsuo, R.R. and Irvine, G.N. (1974) Relationship between the GRL spaghetti tenderness tester and sensory testing of cooked spaghetti. *Canadian Institute of Food Science and Technology Journal* 7, 155–156.

Matsuo, R.R. and Dexter, J.E. (1980) Comparison of experimentally milled durum wheat semolina to semolina produced by some Canadian commercial mills. *Cereal Chemistry* 57, 117–122.

McCarthy, K.L., Gonzalez, J.J., and McCarthy, M.J. (2002) Change in moisture distribution in lasagna pasta post cooking. *Journal of Food Science* 67, 1785–1789.

Menger, A. (1979) Crucial points of view concerning the execution of pasta cooking tests and their evaluation. In: *Comptes Rendus ICC. Proceeding of International Symposium: Matieres premieres et pates alimentaires.* (eds Fabriani, G. and Lintas, C.) pp. 53–60. Istituto Nazionale della Nutrizione, Rome, IT.

Pagani, M.A. (1986) Pasta products from non-conventional raw materials. In: *Pasta and Extruded Products.* (eds Mercier, C. and Cantarelli, C.), pp. 52–68. Elsevier Applied Science.

Pagani, M.A., Gallant, D.J., Bouchet, B., and Resmini, P. (1986) Ultrastructure of cooked spaghetti. *Food Microstructure* 5, 111–129.

Pagani, M.A., Resmini, P., and Dalbon, G. (1989) Influence of the extrusion process on characteristics and structure of pasta. *Food Microstructure* 8, 173–182.

Pagani, M.A., Resmini, P., and Pellegrino, L. (1992) Technological factors affecting the Maillard reaction in pasta processing. *Tecnica Molitoria* 43, 577–592.

Petitot, M., Brossard, V., Barron, C., Larré, C., Morel, M. H., and Micard, V. (2009a) Modification of pasta structure induced by high drying temperatures. Effects on the in vitro digestibility of protein and starch fractions and the potential allergenicity of protein hydrolysates. *Food Chemistry* 116, 401–412.

Petitot, M., Abecassis, J., and Micard, V. (2009b) Structuring of pasta components during processing: impact on starch and protein digestibility and allergenicity. *Trends in Food Science & Technology* 20, 521–532.

Petitot, M., Boyer, L., Minier, C., and Micard, V. (2010) Fortification of pasta with split pea and faba bean flours: Pasta processing and quality evaluation. *Food Research International* 43, 634–641.

Plancken Van der, I., Love Van, A., and Hendrickz, M.E. (2007) Foaming properties of egg white proteins affected by heat or high pressure treatment. *Journal of Food Engineering* 78, 1410–1426.

Pollini, C.M. (1990) Production of pasta from soft wheat. *Tecnica Molitoria* 41, 257–264.

Raina, C.S., Singh, S., Bawa, A.S., and Saxena, D.C. (2005a) Effect of vital gluten and gum arabic on the textural properties of pasta made from pre-felatinised broken rice flour. *Food Science and Technology International* 11, 433–442.

Raina, C.S., Singh, S., Bawa, A.S., and Saxena, D.C. (2005b) Textural characteristics of pasta made from rice flour supplemented with proteins and hydrocolloids. *Journal of Texture Studies* 36, 402–420.

Resmini, P. and Pagani, M.A. (1983) Ultrastructure studies of pasta: a review. *Food Microstructure* 2, 1–12.

Resmini, P., Lenner, A., Debernardi, G., and Gerardo, D. (1979) Technology and ultrastructure of 'pasta' (noodles) produced with rice flour. Centro Ricerche Società Braibanti, Milan, IT.

Resmini, P. and Pagani, M.A. (1983) Ultrastructure studies of pasta: a review. *Food Microstructure* 2, 1–12.

Resmini, P., Pagani, M.A., and Dalbon, G. (1988). The influence of raw material characteristics and of pasta-processing conditions on quality of pasta products. *Tecnica Molitoria* 39, 425–436, 442.

Sahlstroem, S., Mosleth, E., Baevre, A.B., Gloria, H., and Fayard, G. (1993) Influence of starch, gluten proteins and extraction rate on bread and pasta quality. *Carbohydrate Polymers* 21, 169–175.

Schoenlechner, R., Drausinger, J., Ottenschlaeger, V., Jurackova, K., and Berghofer, E. (2011) Functional properties of gluten-free pasta produced from amaranth, quinoa and buckwheat. *Plant for Foods and Human Nutrition* 65, 339–349.

Singh, S., Raina, C.S., Bawa, A.S., and Saxena, D.C. (2004) Sweet potato-based pasta product: optimization of ingredient levels using response surface methodology. *International Journal of Food Science and Technology* 39, 191–200.

Smewing, J. (1997) Analyzing the texture of pasta for quality control. *Cereal Foods World* 42, 8–12.

Sozer, N. and Kaya, A. (2008) The effect of cooking water composition on textural and cooking properties of spaghetti. *International Journal of Food Properties* 11, 351–362.

Sozer, N. (2009) Reological properties of rice pasta dough supplemented with proteins and gums. *Food Hydrocolloids* 23, 849–855.

Steffe, J.F. (eds.) (1996) *Rheological Methods in Food Engineering.* Freeman Press, East Lansing, MI, USA.

Szczesniak, A.S. (1963) Objective measurements of food texture. *Journal of Food Science* 28, 410–420.

Thompson, T. (2009) The nutritional quality of gluten-free foods. In: *Gluten-free Food Science and Technology.* (eds. Gallagher, E.), pp. 42–51, Blackwell Publishing Ltd., London, UK.

Tudorica, C.M., Kuri, V., Brennan, C.S. (2002) Nutritional and physicochemical characteristics of dietary fiber enriched pasta. *Journal of Agricultural and Food Chemistry* 50, 347–356.
Voisey, P.W. (1970) Test cells for objective textural measurements. *Canadian Institute of Food Thechnology Journal* 3, 93–102.
Voisey, P.W. and Larmond, E. (1973) Exploratory evaluation of instrumental techniques for measuring some textural characteristics of cooked spaghetti. *Cereal Science Today* 18, 127–133, 142–143.
Voisey, P.W., Wasik, R.J., and Lougheed, T.C. (1978) Measuring the texture of cooked spaghetti. 2. Exploratory work on instrumental assessment of stickiness and its relationship to microstructure. *Canadian Institute of Food Science and Technology Journal* 11, 180–188.
Winston, J.J. (1960) Evaluating the macaroni strength of a spaghetti product by the use of a bending stress tester. *The Macaroni Journal* 42, 142–143.
Walsh, D.R. (1971) Measuring spaghetti firmness. *Cereal Science Today* 16, 202–205.
www.barillacfn.com
www.brabender.com
www.chopin.fr
www.instron.com
www.texturetechnologies.com

10 Addressing texture challenges in baked goods with fiber

Rajen S. Mehta

SunOpta Ingredients Group, Chelmsford, Massachusetts, USA

10.1 INTRODUCTION

The beneficial effects of dietary fiber (DF) have been known since ancient times. Hippocrates is known to have recommended the positive effects of whole-grain bread in the bowel. Burkitt, Trowell and coworkers developed the hypothesis that increased DF consumption would lead to reductions in diverticular disease, diabetes, cancer, heart disease, and a range of other diseases (Mehta, 2009a; DeVries, 2011). The health benefits of DF are now well accepted by consumers (Mehta, 2005) and health professionals. A recent NIH-AARP diet and health study (Park et al., 2011) confirmed previous learnings about the health benefits of DF. It was determined that DF intake was associated with a significantly lowered risk of total death in both men and women (multivariate relative risk comparing the highest with the lowest quintile was 0.78). DF intake also lowered the risk of death from cardiovascular, infectious, and respiratory diseases by 24% to 56% in men and by 34% to 59% in women. In another recent analysis (Krishnamurthy et al., 2012) of data from 14,543 participants in the National Health and Nutrition Examination Survey III it was demonstrated that for each 10-g/day increase in DF intake, the odds of elevated serum C-reactive protein levels were decreased by 38% and consequently high DF intake is associated with lower risk of inflammation and mortality in those with kidney disease. Other authors have reviewed the impact of DF on health and disease (IOM, 2005; Jones, 2004 and Mehta, 2005) and have concluded that the lack of DF can increase the risk of a variety of diseases.

The definitions of DF have evolved over the years (DeVries, 2011; Mehta, 2009a). The current definition that is now accepted extensively worldwide is the Codex DF definition. CODEX defines DF as carbohydrate polymers with ten or more monomeric units, which are not hydrolyzed by the endogenous enzymes in the small intestine of humans and belong to the following categories: (1) Edible Carbohydrate polymers naturally occurring in the food as consumed, (2) Carbohydrate polymers, which have been obtained from food raw material by physical, enzymatic or chemical means and which have been shown to have a physiological effect of benefit to health as demonstrated by generally accepted scientific evidence to competent authorities, and (3) Synthetic

Food Texture Design and Optimization, First Edition. Edited by Yadunandan Lal Dar and Joseph M. Light.

carbohydrate polymers which have been shown to have a physiological effect of benefit to health as demonstrated by generally accepted scientific evidence to competent authorities. There are two footnotes to this definition. (a) When derived from a plant origin, DF may include fractions of lignin and/or other compounds associated with polysaccharides in the plant cell walls. These compounds also may be measured by certain analytical method(s) for DF. However, such compounds are not included in the definition of DF if extracted and re-introduced into a food. (b) Decision on whether to include carbohydrates of three to nine monomeric units should be left up to national authorities. Most countries either have or are in the process of including carbohydrates of three to nine monomeric units as DF; the exceptions are Chile for health claims and South Africa. Many other countries are in the process of deciding (DeVries, 2011).

Since 1981, the analytical methods and the definitions of DF have almost become synonymous and the methods have become the defacto definitions (DeVries, 2011). The current newly developed method for DF is AOAC 2009.01/AACC 32-45 (AOAC, 2010; AACC, 2010). This AOAC 2009.01 method is based on the principles of AOAC methods 985.29, 991.43, 2001.03, and 2002.02 (McCleary et al., 2010). Older methods including the commonly used AOAC 985.29 (AACC 32-05) and AOAC 991.43 (AACC 32-07) are outlined in recent publications (Mehta, 2009a; DeVries, 2011). Methods for resistant maltodextrins (AOAC 2001.03/AACC 32-41), galactooligosachharides (AOAC 2001.02/ AACC 32-33) and polydextrose (AOAC 2000.11/AACC 32-28) have also been described. The advantages of the new AOAC 2009.01 method has been described by DeVries (2011): Small amounts of resistant starches and other DFs that were degraded/lost in older methods are now recovered in AOAC 2009.01 (see Table 10.1). Resistant starch digestion matches ileostomy results. DF reported by AOAC 2009.01 includes water and alcohol soluble resistant oligosaccharides not included in earlier methods such as AOAC 991.43/985.29 and reports all soluble DF with three or larger carbohydrate units and more of the different types of resistant starch. Lastly, it reduces the need to run multiple DF methods such as for resistant maltodextrins (AOAC 2001.03), fructans (AOAC 997.08), polydextrose (AOAC 2000.11) and galactooligosaccharides (AOAC 2001.02) in addition to the Total DF methods such as AOAC 991.43 and AOAC 985.29.

Table 10.1 Efficacy of analysis of fibers by AOAC methods 991.43 and 985.29.

Fiber type	**Efficacy of analysis by AOAC 991.43 and 985.29**
Cellulose, hemicellulose, lignin	Completely analyzed
β-glucan	Completely analyzed
Galactomannan	Completely analyzed
Arabinoxylan	Completely analyzed
Pectin	Mostly analyzed
Arabinogalactin	Mostly analyzed
Resistant starches	Mostly analyzed
Fructans (Inulin, Fructo-oligosaccharides	Partially analyzed
Polydextrose	Partially analyzed
Resistan maltodextrins	Partially analyzed
Galacto-oligosaccharides	Not analyzed
Raffinose/stachyose	Not analyzed

Adapted from DeVries (2011)

Table 10.2 Types of fiber. (Source: Adapted from DeVries 2011.)

Category	Fiber
Insoluble	Cellulose
	Hemicellulose
	Lignin (bound to cellulose)
	Cutin
	Suberin
	Chitin
	Chitosan
	Arabinoxylan
Soluble (non viscous)	Inulin
	Fructooligosaccharides
	Galactooligosaccharides
	Raffinose
	Stachyose
	Resistant maltodextrins
	Isomaltooligosaccharides
Soluble (viscous)	Pectins
	β-glucan
	Galactomannan
	Arabinogalactan
	Gums
Resistant starch	RS-1
	RS-2
	RS-3
	RS-4
	RS-5
Polydextrose	

Adapted from DeVries (2011)

The many different types of DF and have been summarized by various authors (Nelson, 2001; Mehta, 2009a; DeVries, 2011) and have been listed in Table 10.2. One novel and unique DF that has recently been identified is the RS5 type of resistant starch (Jane, 2011; Hasjim et al., 2010, Miller et al., 2011). This is a starch-lipid complex whereas RS1 through RS4 are respectively physically modified starch, granular starches, retrograded starches, and chemically modified starches (DeVries, 2011; Mehta, 2009a). The RS4 resistant starches are chemically modified and thus the most stable to processing; for example 78 to 89% TDF was retained after extrusion (Miller et al., 2011). In contrast, although RS3 survives typical baking conditions, they may not withstand the high temperature, pressure, and shear of the extrusion process.

10.2 FIBER'S ROLE IN FOODS

DF is added to almost every type of food. In 2011, 223 new food products were introduced into the US where DF was added as an ingredient (Mintel, 2012). The new food products represented 46 subcategories including almost all types of foods and beverages (Table 10.3). The bakery, snack, and cereal category accounts for almost two-thirds of the new products introduced with added DF. Consequently, the focus of this chapter will be primarily on this category though other foods and beverages will also be discussed.

Table 10.3 New product introductions containing fiber (USA 2011). (Source: Adapted from Nelson 2001, Mehta 2009a and DeVries 2011.)

Category	Percent of new product introductions
Bakery, Snack & Cereal (63%)	
Bread & Bread Products	18.8
Snack/Cereal/Energy Bars	13.5
Cold Cereals	11.7
Cakes, Pastries & Sweet Goods	6.7
Savoury Biscuits/Crackers	3.1
Sweet Biscuits/Cookies	2.7
Baking Ingredients & Mixes	1.3
Wheat & Other Grain-Based Snacks	1.3
Bean-Based Snacks	0.9
Corn-Based Snacks	0.9
Potato Snacks	0.9
Pastry Dishes	0.4
Shelf-Stable Desserts	0.4
Snack Mixes	0.4
Meals & Meats (12%)	
Sandwiches/Wraps	3.6
Meat Substitutes	3.1
Pizzas	2.7
Meal Kits	0.9
Prepared Meals	0.9
Hors d'oeuvres/Canapes	0.4
Meat Products	0.4
Beverages (8%)	
Meal Replacements & Other Drinks	6.3
Beverage Mixes	0.4
Coffee	0.4
Soy Based Drinks	0.4
Tea	0.4
Dairy, Sauces & Condiments (7%)	
Dairy-Based Frozen Products	1.3
Dips	1.3
Other Frozen Desserts	1.3
Processed Cheese	0.9
Water-Based Frozen Desserts	0.4
Cooking Sauces	0.4
Dressings & Vinegar	0.4
Baby Food (5%)	
Baby Formula (0-6 months)	2.7
Growing Up Milk	0.9
Baby Cereals	0.4
Baby Formula (6-12 months)	0.4
Baby Snacks	0.4
Pasta & Noodles (3%)	
Pasta	2.2
Instant Noodle, Pasta & Rice	0.9
Miscellaneous (2%)	
Fruit	0.4
Fruit Snacks	0.4
Chocolate Tablets	0.4
Non-Individually Wrapped Chocolate Pieces	0.4
Other Natural Sweeteners	0.4
Total Sample	100.0

Adapted from Nelson (2001); Mehta (2009a); DeVries (2011)

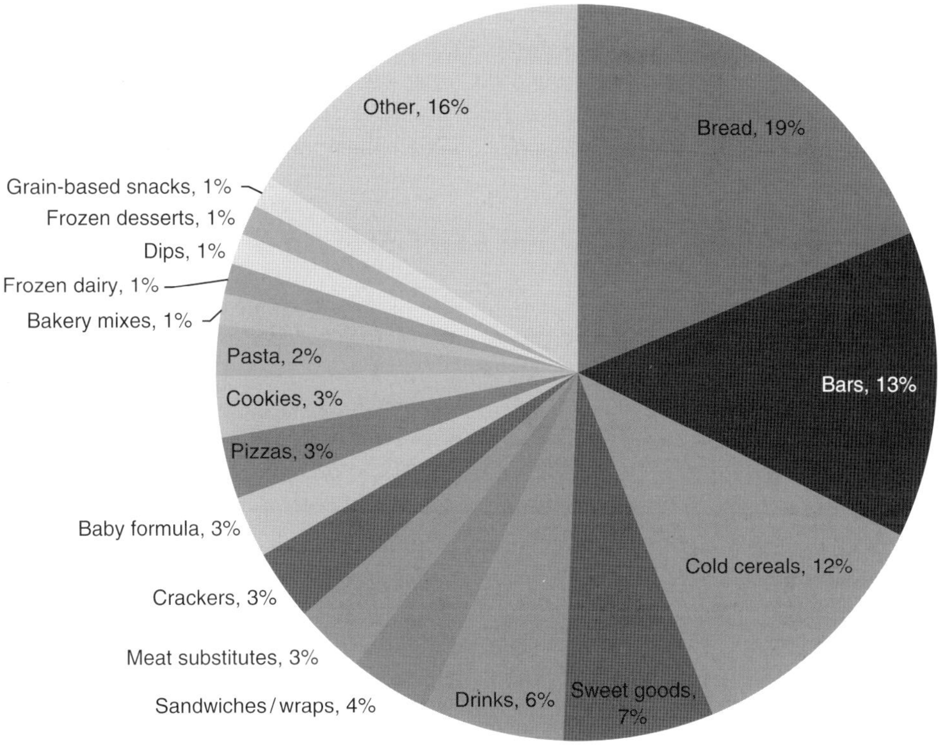

Figure 10.1 New product introductions with fiber as ingredient (USA 2011). (Source: Mintel 2012.)

Table 10.4 Reasons why fiber is added to foods and beverages. (Source: Mintel 2012.)

1	Total dietary fiber (TDF) content
2	Physiological benefits
3	Texture improvement
4	Breakage reduction
5	Shelf-life improvement
6	Cost reduction & productivity improvements
7	Healthy product quality enhancement
8	Miscellaneous benefits

Interestingly, DF was added to most other foods, though to a considerably lesser extent than to the bakery, cereal, and snack category. Also, it is worth noting that unlike a few years ago, DF is now being added more frequently to meats and meals; on the other hand, it appears to be added to fewer new beverages. The top 17 subcategories are shown in Figure 10.1. Bakery, snack, and cereal products represent the top four subcategories and include bread, bars, cold cereals, and sweet goods. The key reasons why DF is added to foods is summarized in Table 10.4. Traditionally, and even today, it is added predominantly to fortify or increase the TDF content. Often it is added at levels to make nutrient content claims; however, it is increasingly added at higher levels to exploit marketing opportunities and for competitive reasons. Physiological benefits vary widely and have also been reviewed extensively in literature (Jones, 2004; Gordon and Goda, 2008; Cho and Almeida, 2012; Cho and Samuel, 2009; van der Kamp et al., 2010). Functional (such as breakage reduction) and texture improvement benefits have been reviewed (Mehta, 2009b; Mehta et al., 2008; Matkovic and Mehta, 2010).

10.3 TOTAL DIETARY FIBER CONTENT

Despite the association of lack of DF to the risk of a wide range of diseases, the intake around the world continues to be far from adequate (Jones, 2004). A recent NHANES data analysis from 1999 to 2006 (Nicklas et al., 2012) found that the mean DF intake in the US was 15.8% and only 9.4% of adults met the adequate intake (AI) guidelines for DF. Most of the DF consumed was grain DF. Higher intake occurs in some north European countries than in southern Europe and US (Jones, 2004). Males generally eat more DF but females generally eat more DF dense diets. While worldwide recommendations on intake vary, populations generally do not meet the guidelines despite compelling epidemiological data that DF reduces the risk of many chronic diseases including obesity, diabetes, cardiovascular disease, and some types of cancer. Based on intake level observed to protect against coronary heart disease, the IOM (2005) has set an AI of 30 to 38 g/day of TDF for men and 21 to 25 g/day TDF for women; younger men and women need the higher TDF. AI of 28 and 29 g have been set for pregnant and lactating women respectively. AI for children as young as one year old have been set at 19 g/day and starting at four years of age, the ranges are between 25 and 38 g/day depending on age and gender. Trumbo (2012) outlined the DF related claims that are allowed in the US. Nutrient content claims based on a recommended Daily Value (DV) of 25 g DF are a good source (2.5 g TDF/serving) and excellent source (5 g TDF/serving). A structure/function claim often used for DF is 'fiber maintains bowel regularity'. The four health claims related to DF are 21 CFR 101.76 and 101.78 for cancer, CFR 101.77 and 101.81 for coronary heart disease and these claims require good source and/or soluble DF levels of 0.6 g or higher per serving.

Typically it is not too difficult to formulate to deliver 5 g TDF for larger serving foods such as bread and 2.5 g for snacks and cookies which have a serving size of only 28 to 30 g. Even some of the 6 g TDF targets needed for nutrient content claims in some countries other than the US are not too difficult to achieve. Although some food manufacturers use multiple DFs to achieve these claims, often they can technically be achieved with only one or two DF sources by careful selection of DFs. To achieve TDF levels higher than that, diverse arrays of DFs need to be carefully balanced to achieve desired textures. Many bakery products intrinsically have 1 g TDF per serving (Mehta, 2009b). Thus an incremental 1.5 to 4 g TDF needs to be added to meet a nutrient content claim. Many wheat flour based whole grain products intrinsically contain 1.5 g TDF per serving. Whole grains vary widely in TDF content (see Figure 10.2) and thus for formulation, typically nutritional analysis programs are used. Comparing the various grains, it is clear that even one full serving of most whole grains will not provide 2.5 g TDF, which is needed for a nutrient claim. Also, it is obvious that the amount of DF that will need to be added for nutrient content claims will be higher for rice based whole grain food

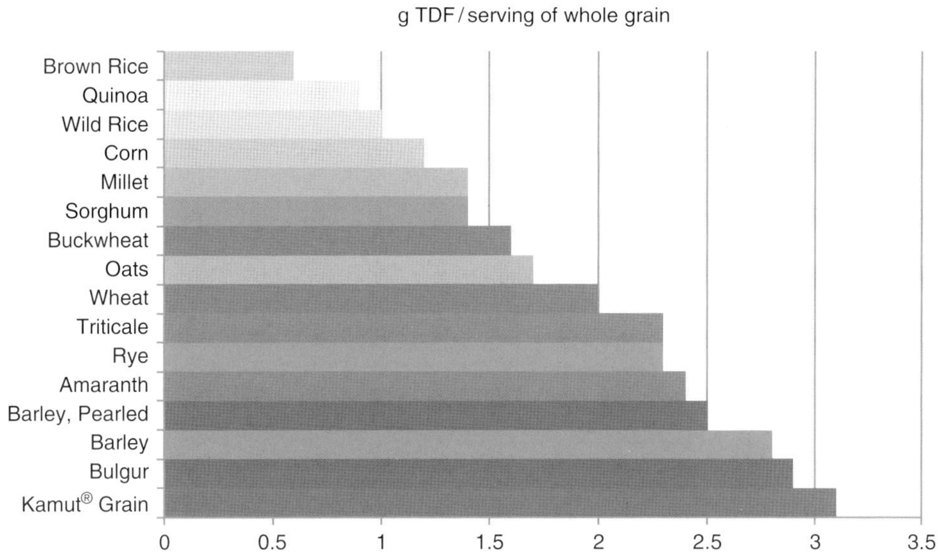

Figure 10.2 Total dietary fiber contribution of different whole grain sources.

products than for wheat, barley, or rye based food products. It is important to deliver a minimum of 2.5 g TDF/serving in whole grain foods due to consumer expectations and association of whole grains with a high TDF content. Usually for the low levels of fortification, the first DF chosen is either for its lowest cost, or in the case of beverages and even bars, it is chosen for ease of use and/or availability during formulation. Often, when optimum DFs are not utilized, in addition to the higher cost, they have a deleterious effect on texture, mouthfeel, and flavor. Some DFs reach tolerance level limits in foods and consequently cause side-effects such as abdominal cramping, gas and bloating if consumed at too high a level. Thus it is best to have a balance of DFs to achieve optimum product quality. Insoluble DFs typically do not cause any gastric distress but have to be chosen for optimum textural benefits. Soluble DFs typically are easier to use because they usually have less of a deleterious textural effect regardless of the type of soluble DF used; however, some of them can cause gastric discomfort if consumed at too high a level beyond their tolerance level.

Bakery products with differing amounts of TDF were compared (Mehta et al., 2008 and Mehta, 2009b). In breads, only one source of insoluble DF was used to achieve 2.5 g TDF/50 g serving. For 11 g TDF/serving, two insoluble DFs, one soluble DF, one resistant starch and one source of bran were used for a total of five sources. It was shown that using only one of those multiple sources to achieve this high TDF level had a deleterious effect on mixing properties, volume, crumb structure, and other loaf characteristics. For bagels, three sources of DFs were used to achieve 5 g/55 g serving. This is often the case in commercial formulas when targeting an excellent source of TDF level to allow for optimum texture. For 15 g TDF/serving in bagels, five diverse DF sources in addition to gums as texturizers needed to be used. For tortilla chips, since the serving size was small, to achieve 9 g TDF/28 g, four DF sources and multiple gums had to be used. For chocolate milk, a ratio of 60% insoluble and 40% soluble DF needed to be used to achieve 4 g TDF/240 ml serving; this achieves a balance of insoluble and soluble DF. In thin beverages, finely ground insoluble DF needs to be suspended by combination of gums such as gellan and xanthan.

10.4 TEXTURE AND FUNCTIONAL BENEFITS

When choosing a DF during formulation, there are five key drivers (Figure 10.3). First, the target needs to be clearly visualized and defined. Is the goal to have a specific nutrient content claim with its associated TDF target, or is it to achieve a functional benefit? That is the first key determinant of DF choice. If the DF level is for a low TDF, then it is best to choose a low cost DF source. For functional benefits, specific DFs will have to be chosen. Independent of whether the target is specific TDF levels or specific functional effects, the second key consideration is to define the desired texture of the product. Does the texture of the current category need to be maintained or is the goal another specific texture based on consumer and marketing goals? There are also process limitations that dictate DF choice and sometimes are the predominant determinant. Costs or price points are the next key determinant. If cost is not that important, then it widens choices and a variety of DFs off-the-shelf can be used. However, more often, there is a restriction and it is not unusual to have a cost reduction goal. At that point, the lowest cost DF needs to be used at the highest levels, and then using multiple DFs allows the achievement of multiple objectives. However, often goals can be more challenging. For example, the goal may be to provide a satiety benefit, or is targeted towards reducing caloric value or to maximize the TDF level. For these targets, multiple DFs chosen from different categories need to be utilized. For specific health or physiological benefits, such as prebiotic effects, natural, organic, or for gluten-free diets, the DF choice will be more focused.

When formulating for optimum texture, one has multiple macromolecules (Figure 10.4) to choose from. Traditionally, and even today, most researchers think of DF as just a filler. The role of different types of starches, such as modified starches, on texture of foods is well known. Similarly, the relationship between properties of fats, proteins, water, and adjunct components such as emulsifiers and enzymes is well understood. Research on finding relationships between DF structure, properties and its

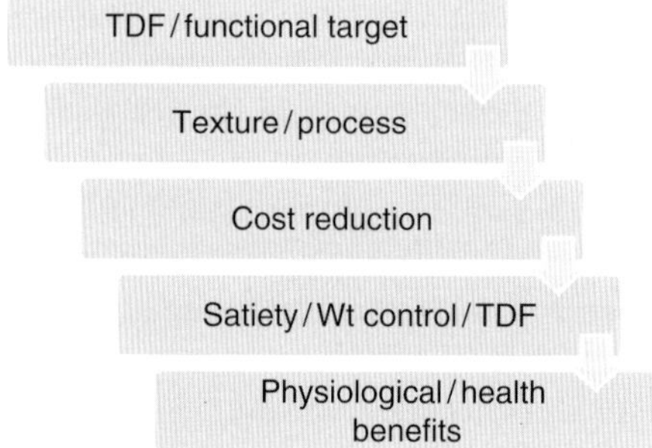

Figure 10.3 Key drivers for fiber choice.

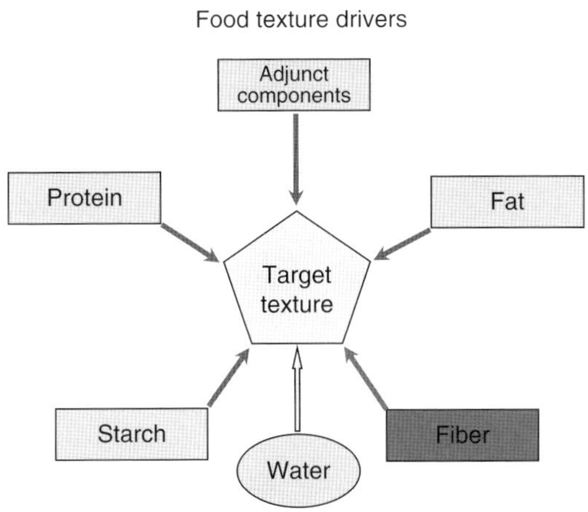

Figure 10.4 Food components that drive texture.

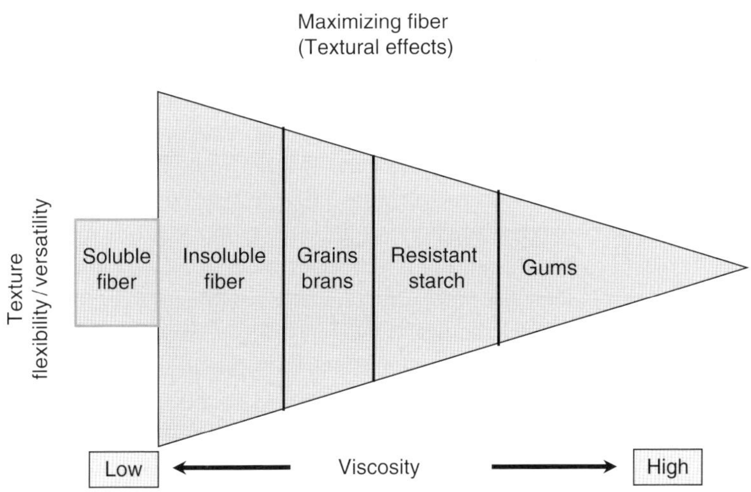

Figure 10.5 Effect of fiber source on textural effects when maximizing TDF content of foods.

functionality and effect on texture is relatively at its infancy. The mechanism of influence of structure and properties on texture is considerably different in each category of macromolecules. For example for starch, in addition to the agricultural crop source of the granules, it is primarily the degree of cross-linking and the chemical used for stabilization that influences texture. For fats, it is the melt-point and the fatty acids including the number and type of double bonds that impacts texture. For proteins it is the molecular weight and other special properties from gelling to coagulating and others that influence texture. There has to be a similar paradigm change in understanding the impact of different DFs on texture.

When formulating, multiple categories of DFs (Figure 10.5) are available. For the bakery, snack, and cereal category, the formulator starts with the flour from wheat, other grain, or starchy sources which form the background matrix. These have low levels of

TDF but typically also the cost is low and thus form the bulk of the formulation. The brans are usually a little higher in TDF because the fraction of the grain they represent typically has a higher ratio of DF compared to the aleurone layer and other fractions of grains. Each bran source from various grains provides its unique texture to foods ranging from crunchy for red wheat bran to crispness or softness for corn. Conventional and innovative commercial ingredient processes such as heat stabilization and grinding also impact absorption patterns in foods during processing (rate and total amount) and thus impact the texture of foods. Insoluble DFs have functional properties that follow definite patterns (Mehta, 2012a). During processing of oat DFs, different amounts of lignin and hemicelluloses are removed by using a combination of steeping time, temperature, pressure and pH. The cellulose content goes up proportionately as the amount of lignin and hemicelluloses is reduced. Microscopic work (unpublished data) has shown that surface properties are affected to a varying extent as a result of this processing as well as due to grinding. As a result a number of properties of the DFs in food products are influenced. The most obvious ones are water and oil holding capacity increases and this can be measured by using standard AACC, or modified, water absorption techniques. These also influence the absorption in flours as measured by Farinographs (Matkovic and Mehta, 2010). In general, as the lignin and hemicellulose content is decreased, in addition to increasing absorption, the DFs are softer, provide a finer mouthfeel, control syneresis more effectively, and add more flexible structure to the finished product in addition to strength thereby becoming more effective at reducing breakage and resulting in more resiliency in finished bakery products (Mehta, 2009b). Careful study of the choice of DFs in various application formulas provided by DF manufacturers demonstrates the use of these properties in choosing DFs from applications ranging from bakery products to meats to beverages and a variety of other food and non-food products. For low viscosity soluble DFs, the main effect on texture appears to be molecular weight related. The softness, body, and to some extent flowability during processing are affected by the molecular weight and sometimes the types of bonds in the soluble DF. A classic example is in cookies where the spread and the shape of the cookie are influenced by the selection of soluble DF. This can become especially important during processing of non-traditional bakery products such as gluten-free (Mehta, 2012b); gluten in traditional bakery products provides a structure that is forgiving. However, in gluten-free products, lower molecular weight soluble DF imparts a certain flowability that is at times important to balance out whichever macromolecules (hydrocolloids and/or proteins) are used to form the basic structure to replace gluten. The properties of soluble DFs also become important in formulating extended shelf-life products to impart microbial stability by lowering water activity and/or by maintaining a stable texture during shelf-life. Resistant starches are relatively easy to work with because they have characteristics that are similar to naturally occurring starch in grains and flours. The key consideration in addition to the grain source of DF is the type of resistant starch; these affect the stability during processing especially at high temperature processing. Viscous soluble DFs add viscosity and slickness. Thus, their choice is extremely important in both traditional products ranging from high moisture systems such as beverages and salad dressings to intermediate moisture systems such as bread and tortillas to low moisture systems such as snacks. In each, the types of properties and structures imparted by the gums have to be selected to target different objectives. For example in beverages,

Table 10.5 The fiber toolbox for texture.

Fiber Source	Function
Flours and whole grains	Backbone matrix
Brans (e.g. wheat)	Crunch
Insoluble fibers (selected for functionality)	Wide texture range and functionality
Soluble fibers low viscosity – high or low molecular weight	Softness, flowability
Resistant starches	Flour-like
Soluble fibers high viscosity – gums or β-glucans	Viscosity, slick

properties such as viscosity, shear-thinning, mouthfeel, ability to suspend particles, stability during storage are critical. In intermediate moisture systems, it varies widely by type of bakery product but can range from shelf-life enhancement to binding characteristics. For low moisture systems, it is typically mouthfeel and bite characteristics that are the most critical.

The DF toolbox (Table 10.5) is a useful starting point to choose DFs during formulation. In addition to the textural effects, it is important to note that in general the cost impact on the formula is least for the DF sources on the top of the table (flours) and gradually increases as the formulator goes lower in the table culminating with the expensive gums. However, gums are usually used at lower levels and thus usually have a lower cost-in-use impact on the overall formula. The exception is in some healthy foods such as gluten-free foods or where β-glucan is added for health claims.

The textural versatility of various DF sources is visually outlined in Figure 10.5 when the goal is to maximize the TDF of foods. Soluble DFs, except for gums and β-glucans, have the lowest viscosity and have the lowest versatility for their impact on texture. Insoluble DFs in general have the largest versatility and thus choosing the optimum insoluble DF is important; raw materials that have lignin, hemicelluloses, and cellulose have the most potential of being processed and ground to different extents to deliver DFs with diverse textural effects. Whole grain sources and brans, as outlined previously, have the next most versatility between grain sources and within grain sources by choosing variety, removing and concentrating fractions, and by grinding. Resistant starch is the next and gums are relatively limited on their impact on texture versatility for very high TDF foods.

Mehta (2009b) has discussed the functionality effects of insoluble DF. Long flexible insoluble DF, when used at 4% (on a flour weight basis) in ice cream cones, reduced breakage by 68% versus control by increasing strength by 27% and flexibility by 34% (Mehta, 2009b). Oil uptake during frying was decreased by up to 14% when long flexible oat DF at 3.6% (on a flour weight basis) was used. Bowl life of cereal (time for which the cereal remains crispy or crunchy in milk) was increased two or more minutes by using long insoluble DFs. Stiff oat DFs allowed the cereal to retain a crispy texture and flexible oat DFs allowed the cereal to retain crunch. Matkovic and Mehta (2010) studied the interactions of insoluble oat DF and other dough constituents to optimize breads with high DF and whole grain content. They found that processing of cellulosic DFs has a large impact on DF functionality likely through differences in surface properties, porosity, water absorption, DF length and shape. Farinographs demonstrated that porous DFs have higher water absorption levels and high affinity for water slowing down the

Table 10.6 Cost reduction with 1% of the optimum insoluble fiber.

Benefit	Examples
Reduce expensive ingredients	Replace expensive soluble with cost-effective insoluble fibers
	17% tomato solids reduction in tomato sauce
	7% chicken reduction in chicken patties
Yield Increase (increase water/oil retention)	2 to 5% increase cheese yield
	0.5 to 4% increase meat yield
Reduce bulk density & Increase volume/fluffiness	Muffins/waffles/pasta
Reduce breakage/cracking	1 to 5% reduces breakage by 10 to 50%
Extend shelf-life	Bakery product returns (texture deterioration)
	Syneresis prevention (freeze-thaw)

distribution of water to the flour component resulting in slower dough hydration and consequently increased time to develop the dough and improved dough stability; oat DFs and other DFs that have been manufactured with varying processing conditions can be chosen to achieve specific desirable dough mixing characteristics. These DFs are typically used at lower levels to achieve functional effects such as reducing breakage and for properties such as maintaining moisture during microwaving or during shelf-life. With most whole grains, dough stability is reduced. Mixing properties of flours for DF-fortified bread with different types of whole grains were effectively optimized by choosing DFs with complimentary mixing characteristics to deliver good quality bread.

One ancillary effect of the textural benefits of insoluble DF is that it provides opportunities for cost reduction (Table 10.6). It can allow reduction of expensive ingredients by replacing expensive soluble DF with more cost-effective DFs that are optimum for a specific food. The longer high absorbing softer DFs provide a fiber-like texture seen in natural tomato pulp and thus allow for a reduction in tomato solids. Similarly, because of their high water and oil retention along with natural texturizing effect, they can increase the yield of meats and dairy products. They reduce the bulk density of many foods and thus allow for calorie reduction and also as an indirect result allow the costs to be reduced since the volume of the food appears to be similar but will require less of the food ingredients by weight. Lastly, one of the realities of the commercial food distribution system is that many products such as bakery products are returned or reworked during manufacturing due to breakage (e.g. snacks) and deterioration in quality on storage, or are delivered on a very frequent basis to ensure that freshness is maintained. Breakage reduction and the extension of shelf-life with DF are impactful and thus reduce reworking during manufacturing of bakery products and allow for a reduction of returns and a reduction in the frequency of store deliveries needed.

10.5 PASTAS, MEATS, BEVERAGES, AND PET FOOD

DF can be used in pasta for many reasons. A heart-healthy claim can be made using a concentrated source of β-glucan from barley or oats. Using 11% of a concentrated barley β-glucan source (containing about 25% β-glucan), excellent pasta was made (unpublished data) to deliver 0.6 g β-glucan/55 g serving of dry pasta and thus enabling a heart healthy claim in the USA. In pasta, typically a long flexible oat DF with water absorption values on the high end of the spectrum delivers the best textured pasta that has significantly reduced cooking time but if overcooked by five minutes, retains firmness. This DF also provides dramatically improved flexibility, while retaining strength, for dry spaghetti as measured by a texture analyzer and thus reduces breakage in packages. The addition of only about 5% provides 5 g TDF/55 g dry pasta thus allowing an excellent source claim and up to 15% can be added to achieve about 9 g TDF/serving. Traditionally lower cost insoluble DFs are added for higher margins sometimes at the expense of a loss in quality because some DFs may have a deleterious effect on cooking characteristics and on texture. Another benefit of adding DF to pasta is that it reduces bulk density of the final pasta; again the long flexible insoluble oat DFs provide the best efficacy. Using 10% of this insoluble DF reduced volume of 25 g of rotini by 33% and increased number of pieces per unit weight by 15%. This density effect is linear in proportion to the amount of DF used up to 10% and can be an effective tool to increase the appearance of the amount of pasta in for example canned pasta or as one of the tools to decrease the caloric content of a meal. Another useful property of insoluble DF is that it allows the replacement of egg-whites thereby eliminating allergens, and during the times of high egg prices, providing a cost reduction opportunity. About 8% to 16% of the long flexible oat DF can replace 4% or more of the egg whites and deliver equivalent cooked firmness or slightly more firm cooked pasta as measured by Kramer shear cell in a texture analyzer (unpublished data). As outlined in Table 10.6, the use of 1% of a flexible long oat DF allows reduction of 11% tomato sauce and thus a cost saving of about 4% in the cost of sauce while maintaining the natural tomato pulpiness. Similarly costs of cheeses and meats can be reduced by replacing expensive dairy and meat solids with DF and water; the eating quality of cheeses and meats is often concurrently improved.

The ability of the highly absorptive DFs to hold water and oil provides several benefits in meat products. It allows greater retention of moisture and fat, prevents purge (with concurrent detraction of appearance of packaged meats), increases juiciness and tenderness and thus improves overall sensory qualities of a variety of meats including processed meats, ground beef, poultry, fish, sausage, and vegetarian meat substitutes. Other benefits include improved machineablity, sliceability, and improved texture in reduced fat products, reduced cost, and stability under heat lamps and on steam tables. The use of 3% of a highly absorptive oat DF increased yield on cooked 80% lean ground beef from 70% for Control to 82% with a concurrent decrease in cost by 14% (unpublished data). The increase in yield was in general proportional to the water absorption values of the insoluble DFs; however the highly absorptive DFs provided the best quality product. These properties have commercial application in a variety of products. For example they can increase the yield and reduce the oily appearance of pizza toppings. They can replace meat or other binders in ground beef fillings targeted for Mexican foods, meatballs, meatloaf, chicken nuggets and sausages for cost reduction by increasing the yield while maintaining or enhancing moistness and juiciness.

In beverages, typically soluble DFs are used. However, there are opportunities to balance out soluble DFs with insoluble to represent a balance of insoluble and soluble recommended by most nutritionists. There are also opportunities to provide unique textures for consumers. It is often a challenge to add insoluble DFs but by careful selection and formulation techniques, beverages with unique textures with good sensory characteristics can be formulated. The pulpy character enhancement properties of insoluble DF provide unique benefits and have been used commercially. One challenge in some categories of beverages is the suspension of insoluble DFs. The use of fine particulate insoluble DFs at 1% addition rate along with 0.03% gellan gum for example allowed the suspension of insoluble DFs for a prolonged time in even low viscosity translucent beverages (unpublished data). In thick beverages such as milk shakes, fruit juices, and breakfast drink mixes, use of insoluble DF also allows opportunities for cost reduction, versus soluble DF, in addition to having the nutritional benefit of a more balanced proportion of soluble to insoluble.

DFs are an important ingredient in pet food formulation. They have traditionally been used to provide a nutritionally balanced formula. An important difference is that in pet foods and in animal foods, the DF level is quantitated using crude fiber (CF) versus TDF in foods for people. What is surprising is that CF content is not directly proportional to TDF. For example highly concentrated oat DFs that have a high TDF (90 to 95%) have CF values ranging from 36% to 70% (unpublished data). Similarly for other commonly used CF sources such as brans, the CF content is usually about one-fifth the TDF content; for brans especially the CF values are increased when the particle size is increased.

Insoluble DFs have traditionally been used for CF and for nutritional benefits in pet foods such as for weight control diets. However, they are now starting to be used for benefits that are not well known. For example, they are used to reduce tartar and plaque build-up. These dental diets improve the dental health in between trips to the vets. Again choice of DFs becomes important especially to achieve efficacy and dental claims related to tarter, plaque, and breath that are critical in the pet food industry. Insoluble DFs can be used to control the texture of dry kibble as well as canned food; a wide range of textures can be achieved by choosing the insoluble DF. This can be especially important for finicky cats that tend not to like food that is too sticky sometimes due to the use of traditional thickeners such as gums; the use of an appropriate insoluble DF improves the overall palatability. The development of differing textured foods can also provide a point of differentiation for the human owners that buy the pet food. DFs can also be used to absorb liquid sprayed-on palatability enhancers in dry pet food.

10.6 HEALTHY FOODS

Increasingly, foods that are healthier are being developed to meet increasing consumer demand. Because they are often based on formulas that are not typical for the type of food being developed, these foods have textural characteristics that are not optimum and do not meet the sensory characteristic expectations of consumers. A variety of commercial gluten-free mixes were bought in grocery stores and were found to have textural issues (see Table 10.7). Choosing the optimum insoluble DF allowed the resolution of these issues. Gluten-free muffins, made from mixes, which were gummy, hard, and dry, were made fluffy, moist and soft by using long flexible oat DFs that had a mixture of lengths and softness. Gluten-free pancake mixes yielded hard and dense pancakes that did not have a round shape and fell apart; they also tended to be mushy in texture and splattered when poured onto a griddle. Again DFs similar to the ones used for muffins alleviated all the issues. The use of general purpose insoluble DFs routinely used in the

Table 10.7 Effect of insoluble fibers on texture of commercial gluten-free mixes.

Commercial Mix	Control	Fiber Effect	Insoluble Fibers Used
Muffins	Gummy, hard, dry	Fluffy, soft, moist	Long, flexible Mixture of length/softness
Pancakes	Hard, dense Not cohesive/round shape, mushy, splatter	Fluffy, soft Holds together/round, crisp surface, no splatter	Long, flexible Mixture of length/softness General purpose
Pizza crust	Doughy, soft	Rigid, crunch	Mixture of length/softness Long, rigid
Cookies	Spread, dome-shaped (thin at edges), sticks to pan	Maintain spread, gradual tapering, no sticking	Long, rigid Long, rigid, very slightly soft

Table 10.8 Using fibers to improve the texture of non-traditional baked goods.

Baked Good	Challenge	Solution	Mechanism
Bread	Not resilient	Long, flexible insoluble fiber	Strength, flexibility
Bread	Gummy	Insoluble fiber mixture - long or short/ flexible or rigid	Open up structure
Frozen dough	Drying/releasing water/oil	High water absorbing insoluble fiber	Hold water
Dough	Stiff dough	Short insoluble fiber	Relaxes dough
Fine bakery	Dense	Soft, fluffy, long insoluble fiber	Opening structure, flexibility, strength
Fine bakery	Powdery	Long flexible insoluble fiber & low viscosity soluble fiber	Provide structure during leavening
Bakery	Dry	High water absorbing insoluble fiber & high viscosity soluble fiber	Hold water and adds slick mouthfeel

bakery industry for fortification also alleviated most of the issues. A soft and doughy pizza crust typical of gluten-free pizzas was made rigid by using long insoluble DFs; rigid DFs added crunch to the crust. Again, mostly rigid insoluble DFs were able to maintain spread and improved the shape of the cookies to achieve more typical shapes that were not thin. Also the unacceptable stickiness of the cookie dough was eliminated. Thus insoluble DFs can be effectively chosen and utilized to alleviate issues in bakery goods. Table 10.8 provides an overview of examples of defects seen in some specific healthier baked goods and provides examples of a variety of soluble, insoluble and resistant starches to minimize those defects.

10.7 APPLICATION OVERVIEW OF FIBERS

Mehta (2009b) summarized example uses of DF in varying applications. Example formulas and recommendations on DF used in almost every conceivable application can be found on most DF company websites or by contacting them. In addition to traditional bakery goods, DFs are used in cereals, bars, meats, pet foods, beverages, dairy products, salad dressings, dips, condiments, fish, analogues, and skin care products. Two examples that provide a vivid overview on the versatility of DF use in two applications are detailed below.

In pizza, insoluble DFs can be used for a variety of reasons: to add structure and rigidity to the crust that can withstand sauce, to add crunchiness to the crust, to reduce the oiliness from cheese or meat, to reduce costs of the meat, cheese and sauce and to add a more pulpy character to the sauce. Soluble DF can be used to manage the water activity of sauce, and the water activity differential between sauce and crust, thereby slightly slowing the migration of water from the sauce to the crust. Gums can be used to add a slick mouthfeel which can be important especially if the crust is lower in fat or dries out during storage. A combination of insoluble DFs, soluble DFs, resistant starches, and gums can be used to maximize TDF beyond nutrient content claims and also to improve textural shelf-life and to increase the amount of water that can be added to the crust or the sauce thereby reducing calories and costs.

In gluten-free tortillas, insoluble DFs can be used for a variety of reasons: to increase crust crunch, reduce stickiness during storage as well as during processing, to reduce gumminess, and last but not the least to reduce denseness and maintain the structure of the tortilla during storage. Because of the demands for extra water from non-traditional ingredients, water activity control with lower molecular weight soluble DFs becomes important. Gums are a key part of the formulation since along with the protein; they provide the basic structure and thus a combination of gums carefully selected for functionality are used at considerably higher levels than in traditional bakery products. Combinations of DFs are used to maximize the TDF and to reduce staling. TDF is very important in gluten-free products because the lack of traditional ingredients typically results in very low TDF content contributed by the baked products. The DFs allow significant improvement in the texture of gluten-free tortillas.

Tapola and Sarkkinen (2009) stated that three forms of oat DF are commonly available for human consumption. These are from oat groats, oat bran, and oat hulls. Oat hulls are very high in TDF (79 to 95 g/100g) and contain mainly insoluble DF, very little soluble DF and no β-glucan. Oat bran has a total β-glucan content of at least 5.5% on a dry weight basis and a TDF of 16%. Commercial oat bran concentrated to as high as 22% β-glucan has been made through dry-milling or wet-milling processes. They list the various commercial oat products available. Bodner and Cho (2009) reviewed oat DF (from hulls) use in a variety of foods. They recommended the use of minimally extracted oat DF with improved water-binding capacity for crackers and cookies and reviewed literature that showed that it was possible to produce high TDF cookies without sacrificing sensory properties. Oat DF was also added to bread and pasta without a deleterious effect. Longer-strand DF reduced susceptibility to breakage in bagel chips. Oat DF with high water-holding capacity was used to decrease cooking losses and purge of various meat products. Oat DF up to 3% levels increased yield for bologna, frankfurters, pork products, and beef patties. Oat DF addition at 1.32% improved body and texture due to increase in viscosity in unsweetened yogurts, but lowered overall scores for body and texture in yogurts sweetened with sucrose. Inman et al. (2012a) also reviewed oat DF use in a variety of foods. They provided example formulas and processes for cookies, crackers, bread and meatballs. Inman et al. (2012b) discussed utilization of sugarcane DF in cookies, muffins, bread, rolls, cheesecake, pancakes, confectionery products, beverages, and eggs. Like other insoluble DFs, it is very stable in food manufacturing and its nutritional value is stable to heat, freeze/thaw and pH. They have provided formulas and processes for many of the products and have demonstrated a cost reduction in many formulas by using the DF.

Fastnaught (2009) extensively reviewed food applications for barley products. The amount of β-glucan and TDF varies considerably in different barley products. Extrusion cooking of barley increases the solubility of the β-glucan and increases resistant starch thereby increasing both the soluble and insoluble DF. The solubility of β-glucan and DF in oven-puffed ready-to-eat cereal increased 23%. The hedonic scores of 8% TDF extruded rice cereals was similar to 1.3% TDF cereal. Expansion ratio was negatively correlated to DF content from barley; however, product with 5% soluble and 10% TDF had the best expansion and was rated highest by a sensory panel. The texture profile analysis of Udon noodles containing 15% ground pearled barley was similar to 100% wheat flour noodles but required a shorter cooking time. Pastas with 5% and 7.1% β-glucan had sensory characteristics similar to control. One percent of an 85% extracted β-glucan barley increased dough development time but improved bread grain and texture while maintaining loaf volume. Higher levels of β-glucan decrease loaf volume; however, sensory tests showed that 3% β-glucan breads have similar ratings as control. Studies have shown that xylanase or extrusion processing of barley flour has overcome some of the loaf reduction effects of barley. Flatbreads and biscuits containing very high levels of barley flour or bran (30 to 100%) have been made. Acceptable muffins, pancakes, biscuits, cookies, and tortillas have been made with barley flour substituted for wheat flour. β-glucan was found to mimic fat in low fat muffins and quick bread. Tortillas with β-glucan were easier to roll, had less breakage, were softer and less dry and chewy. Low-fat sausages with 0.3% β-glucan were liked as well as high-fat sausages. They also described the effects of barley β-glucan use in beverages, yogurt, soups, and corn-flakes.

Takahashi (2009) briefly reviewed food applications of cellulose and its derivatives such as microcrystaline cellulose, methylcellulose, sodium carboxymethylcellulose and hydroxypropylmethyulcellulose and observed that they can be used as emulsifiers, stabilizers, dispersing agents, and thickeners. Cellulose increases cake volume and reduces shrinkage, improves texture of beef burgers and can be added to pasta, imitation cheese and cereal.

References

AACC International (2000) *Approved Methods of the American Association of Cereal Chemists*, 10th Ed. AACC International, St. Paul, MN.

AOAC International (2000) *Official Methods of Analysis of AOAC International*, 17th edn. AOAC International, Gaithersburg, MD.

Bodner, J. and Cho, S.S. (2009) Oat fiber from oat hull. In: *Fiber Ingredients Food Applications and Health Benefits*. pp. 249–262. CRC Press, Boca Raton, FL.

Cho, S.S. and Almeida, N. (eds.) (2012) *Dietary Fiber and Health*. CRC Press, Boca Raton, FL.

Cho, S.S. and Samuel, P. (eds.) (2009) *Fiber ilgredients Food Applications and Health Benefits*. CRC Press, Boca Raton, FL.

DeVries, J. (2011) Global fiber definition and methods. *USDA Soft Wheat Research Review Conference*, [Online], Available: http://www.ars.usda.gov/SP2UserFiles/Place/36070500/RR2011/Presentations/Dietary%20Fiber-Jonathan%20DeVries-2011RR.pdf [18 Nov 2012].

Gordon, D.T. and Goda, T. (eds.) (2008) *Dietary Fiber: An international perspective for harmonization of health benefits and energy value*. American Association of Cereal Chemists, St. Paul, MN.

Hasjim, J., Lee, S., Hendrich, S., Setiawan, S., Ai, Y., and Jane, J. (2010) Characterization of a novel resistant-starch and its effects on postprandial plasma-glucose and insulin responses. *Cereal Chemistry*, 87, 257–262.

Inman, D., Lee, A.W., and Cho, S.S. (2012a) Oat fiber. In: *Dietary Fiber and Health*. (eds. S. S. Cho and N. Almeida) pp.329–341. CRC Press, Boca Raton, FL.

Inman, D., Lee, A.W., and Cho, S.S. (2012b). Sugarcane fiber. In: *Dietary Fiber and Health*. (eds. S.S. Cho and N. Almeida) pp.343–358. CRC Press, Boca Raton, FL.

Fastnaught, C.E. (2009) Barley fiber. In: *Fiber Ingredients Food Applications and Health Benefits*. (eds. S.S. Cho and P. Samuel), pp. 323–358. CRC Press, Boca Raton, FL.

IOM (Institute of Medicine), (2005) *Dietary Reference Intakes for Energy, Carbohydrate, Fiber, Fat, Fatty Acids, Cholesterol, Protein, and Amino Acids (Macronutrients)*. The National Academic Press, Washington, DC. [Online], Available: http://www.nap.edu/catalog.php?record_id=10490 [18 Nov 2012].

Jane, J. (2011) Resistant starches: RS5. *Starch Round Table*, [Online], Available: http://www.pdfio.com/k-895560.html. [18 Nov 2012].

Jones, J. (2004) Dietary fibre intake, disease prevention, and health promotion: An overview with emphasis on evidence from epidemiology. In: *Dietary Fibre Bio-active Carbohydrates for Food and Feed*. (eds. J. van der Kamp, N. Asp, J. Jones, and G. Schaafsma), pp. 143–164. Wageningen Academic Publishers, The Netherlands.

Krishnamurthy, V.M., Wei, G., Baird, B.C., Murtaugh, M., Chonchol, M.D., Raphaell, K.L., Greene, T. and Beddhu, S. (2012) High dietary fiber intake is associated with decreased inflammation and all-cause mortality in patients with chronic kidney disease. *Kidney International* 81, 300–306.

Matkovic, K. and Mehta, R. (2010) Interaction of insoluble oat fibre, dough conditioners and other dough constituents in optimizing bread formula with high whole grain and fibre content. In: *Dietary Fibre: New Frontiers for Food and Health*. (eds. J.W. van der Kamp, J. Jones, B. McCleary, and D. Topping), pp. 135–150. Wageningen Academic Publishers, The Netherlands.

McCleary, B.V., DeVries, J.W., Rader, J.I., Cohen, G., Prosky, L., Mugford, D.C., Champ, M. and Okuma, K. (2010) Determination of total dietary fiber (CODEX definition) by enzymatic-gravimetric method and liquid chromatography: collaborative study. *Journal of AOAC International*, 93, 221–233.

Mehta, R. (2005) Dietary fiber benefits. *Cereal Foods World*. 50, 66–71.

Mehta, R. (2009a) Dietary fiber I. *AIB Technical Bulletin*. 33(1), 1–7.

Mehta, R. (2009b) Dietary fiber II. *AIB Technical Bulletin*. 33(2), 1–19.

Mehta, R. (2012a) Leveraging fiber's properties to manage and improve the texture of baked goods. AACC International annual meeting, Hollywood, FL. Abstract 51–S [Online], Available: http://www.aaccnet.org/meetings/Documents/2012Abstracts/s12ma51.htm [December 1, 2012]

Mehta, R. (2012b) Gluten-free: Product development opportunities and challenges. AACC International annual meeting, Hollywood, FL. Abstract 2-S [Online], Available: http://www.aaccnet.org/meetings/Documents/2012Abstracts/s12ma02.htm [December 1, 2012]

Mehta, R., Peterson, C, Dineen, M, and Zamora, H. (2008) Leveraging fiber characteristics to functionally improve food products. In: *Consumer Driven Cereal Innovation*. (eds. P.L. Weegels, C.M. Courtin, A. Arrachid, and M. Silvestri) pp. 113–116. AACC International, St. Paul, MN.

Miller, R.A, Jeong, J., and Maningat, C.C. (2011) Effect of RS4 resistant starch on extruded ready-to-eat (RTE) breakfast cereal quality. *Cereal Chemistry* 88, 584–588.

Mintel (2012) *Fiber use as an ingredient*. http://portal.mintel.com/portal/. October 24, 2012

Nelson, A.L. (2001) *High-fiber Ingredients*. American Association of Cereal Chemists, St. Paul, MN.

Nicklas, T.A., O'Neil, C.E., Slavin, J.L., and Cho, S.S. (2012) Consumption of total fiber and types of fiber are associated with a lower prevalence of obesity and abdominal adiposity in U.S. adults. In: *Dietary Fiber and Health*. (eds. S.S. Cho and N. Almeida) pp.71–81. CRC Press, Boca Raton, FL.

Park, Y. Subar A.F., Hollenbeck, A., and Schatzkin, A. (2011) *Dietary Fiber Intake and Mortality in the NIH-AARP Diet and Health Study*. Arch Intern Med. Published online February 14, 2011. doi:10.1001/archinternmed.2011.18 http://www.skane.se/upload/Webbplatser/RCC/Dokument/Dietary%20Fiber%20Intake%20and%20Mortality.pdf

Takahashi, T. (2009) Cellulose. In: *Fiber Ingredients Food Applications and Health Benefits*. (eds. S.S. Cho and P. Samuel), pp. 263–282. CRC Press, Boca Raton, FL.

Tapola, N. and Sarkkinen, E. (2009) Oat β-glucan. In: *Fiber Ingredients Food Applications and Health Benefits*. (eds. S.S. Cho and P. Samuel), pp. 283–303. CRC Press, Boca Raton, FL.

Trumbo, P. (2012) Regulations for the food labeling of dietary fiber. In: *Dietary Fiber and Health*. (eds. S.S. Cho and N. Almeida) pp.33–41. CRC Press, Boca Raton, FL.

van der Kamp, J.W., Jones, J., McCleary, B., and Topping, D. (eds) (2010) *Dietary Fibre: New Frontiers for Food and Health*. Wageningen Academic Publishers, The Netherlands.

Addressing texture challenges in baked goods with fiber

Addendum: Textural benefits of dietary fiber – an overview and summary

Rajen Mehta, Ph.D.
Grain Millers, Inc. Eugene, Oregon, USA

Dietary fiber (DF) is added to almost every type of food, including baked goods. DF is added for nutrient content, health claim, structure/ function claim, and also increasingly for textural benefits for freshly baked and stored bakery products. Typically for low levels of fortification e.g. up to 2.5 or even 5 g/serving based on the Reference Amount Customarily Consumed (RACC) guideline set by the US FDA, single DFs can be used. However, for higher total dietary fiber (TDF) levels, and even often at lower levels, multiple DFs need to be carefully chosen and used to achieve desired textures, mouth-feel and flavors in baked goods and foods. Selection is also often a balancing act between cost and textural effects; though even if cost was not an issue, multiple DFs often need to be used for optimal sensory effects. Some DFs, notably a few soluble DFs (SDF) reach gastric distress related tolerance levels and compounded with some textural and processing limitations, cannot be used for extremely high TDF targets in baked goods (e.g. 10+ g TDF/serving); in some other foods and beverages it often is not as great a limitation. Using single source DFs for high TDF also has a deleterious effect on mixing properties, volume, and crumb structure for bread. In bagels, tortilla chips and even chocolate milk, textural defects such as lack of body, grittiness, lack of structure and textural and eating qualities were negatively affected when only one source of DF was used; however multiple DF types and sources alleviated those issues delivering excellent quality products.

When formulating, the flour, bran and other sources provide the bulk of the formulation and not only supply varying levels of TDF, but also deliver a wide range of unique textural effects; for example red wheat brans usually impart a crunchy texture (especially when stabilized) and on the other end, corn bran provides crispness or softness. Conventional and innovative commercial ingredient processes such as varying types of heat stabilization and grinding also impact absorption patterns during food processing and thus impact texture. Previous publications and reports detail that the properties of insoluble DFs (IDF) that naturally have a high level of especially lignin and hemicelluloses (e.g. oat byproducts) can be influenced by managing and lowering the levels of these DF components. These in turn dramatically impact the water holding, oil holding and other textural properties of baked goods and other foods. Using a combination of lower lignin content, opening up of the structure, and grinding to different particle sizes can provide a softer and finer mouthfeel and more effectively control syneresis in foods. On the other hand, leaving the lignin at high levels will provide a crunchy texture. Many other IDFs with varying naturally occurring compositions provide naturally varying textures more dependent on the actual raw material source which is then often controlled only by grinding to varying degrees. For low viscosity SDFs, such as inulin and fructo-oligosaccharides, the molecular weight and sometimes the type of types of bonds impact the softness, body and sometimes flowability in the food during processing; for example in drop cookies such as chocolate chip cookies, the spread and the shape are influenced by the SDF type especially in non-traditional formulations such as gluten-free cookies. Low molecular SDFs also lower water activity effectively thereby enhancing microbial shelf-life. For textural shelf-life maintainence, IDFs with longer strands typically play a greater role; high or low lignin content influences whether the product is initial crunchy or soft/chewy respectively. However, smaller particle size IDFs and with varying structures, and SDFs (both with low and high viscosity) also play a significant role. Resistant starches, because of their similarity in structure to naturally occurring starches that form the bulk of most grains, mimic to a great extent the natural texture of foods. The key challenges are balancing the TDF levels, cost per unit weight of DF, processing stability, and other characteristics of the resistant starch vs. the benefits of IDFs and SDFs. Viscous DFs add viscosity and slickness and thus can be used in bakery products for their binding characteristics as well as mouthfeel and shelf-life enhancement. To summarize, IDFs typically provide a wide range of textural impact, grains and brans bring their unique natural characteristics to the food, resistant starches bring starch-like effects – and if chemically modified, properties similar to modified starches, albeit to a more limited extent. Lastly the food effect of SDF, both viscous and low-viscosity, is influenced by their unique chemical bonds and chemistry and for low-viscosity DFs also by molecular weight.

Literature has outlined many specific examples of using various DFs for their functionality and to influence texture of foods and bakery products. Examples of such functional effects of IDF include breakage reduction, cost reduction (e.g. by formula changes, yield increases, bulk density changes, manufacturing rework control, shipping stability, product returns reduction), maintaining moisture during microwaving, shelf life, and prolonged holding prior to consumption (e.g. in food service situations). Long flexible IDF used at 4% reduced breakage of fragile baked goods by increasing the strength and flexibility of the baked goods. Long IDFs increased bowl life of cereals by 2 or more minutes; stiff high lignin fibers maintained crispiness and soft low lignin DFs retained crunch. Farinographs have demonstrated that during dough mixing, very porous IDFs have higher water absorption slowing the re-distribution of water to the flour component resulting in slower distribution of water to the flour in the dough and consequently increasing dough development time and improving dough stability; controlling the amount of porosity to varying degrees and using these at low levels in doughs can achieve a control of specific and desirable dough mixing characteristics. Properties of healthy and innovative baked goods such as whole grain, gluten-free, reduced fat for example can be effectively optimized by choosing DFs with complimentary mixing characteristics. Longer flexible IDF can hold significant quantities of water in a manner similar to natural DFs in foods with similar textures and thus can for example provide natural textures to fruit fillings.

In pasta, DF can be used to make heart healthy claims with concentrated β-glucan sources. Long flexible DF delivers *al dente* texture with resistance to over-cooking (also to replace expensive egg-whites); it also delivers high TDF and provides an opportunity to reduce bulk density thereby providing cost reduction and caloric reduction opportunities. In meats, cheeses and sauces, highly absorptive IDF used at low levels (e.g. 3%) increases yield, retains moisture and oil, and reduces purge and similar effects. Also processing effects such as machinability of substitutes and traditional products, reduction of oil/water release of pizza toppings is also improved. In beverages careful choice of combinations of selected very low levels of IDF, SDF (viscous and non-viscous) provide a balance of suspending IDF sources (including grains) and providing target mouthfeel. DFs are typically used in pet food for their crude fiber content and nutritional benefit and in weight control diets. Crude fiber content as analyzed is lower than TDF. Textural and functional effects are just starting to be leveraged. IDF has been used to reduce tartar and plaque build-up. The pet food and the human food industry should compare successes and leverage relevant benefits for their specific experiences and approaches (aggressive weight reduction and texture control respectively).

Healthy foods often do not meet the sensory expectations of consumers, sometimes such as in the case of gluten-free products, these foods have a significantly lower TDF level. Gluten free (and on occasion other healthy foods such as whole grain foods) often have varying types of textural issues such as gummy, hard, dry, dense, lack of cohesiveness/integrity, mushy, splattering on griddles, sticky dough, atypical shapes, and so on. DF can be used to address these issues which are selectively specific to a range of gluten free products such as muffins, pancakes, pizza crusts, and cookies to name a few. Carefully selected DFs alleviate the issues and make the baked goods fluffy, soft, moist, cohesive/structurally sound, crisp, no splattering, rigid, crunchy, maintenance of spread, optimizes shapes, and prevents stickiness to alleviate the respective issues by utilizing targeted specific types of IDFs. The IDFs act by strengthening while adding flexibility or rigidity as appropriate (for resiliency), opening up structure (gummy), holding water (dryness), relaxing dough (stiff/elastic dough), opening structure and adding flexibility/strength (denseness), providing structure during leavening (powdery), and hold water and add slickness (dryness).

In pizzas, IDF is used to add structure and rigidity to withstand sauces, for crunchiness, oiliness reduction from toppings, cost reduction, and to add pulpiness to the sauce. SDF controls water activity and thereby controls water migration between frozen pizza components. Viscous SDF adds slickness in low fat pizzas and combinations of all types of DF can be used to maximize TDF, reduce costs, and/or increase shelf-life. In gluten-free tortillas, carefully selected gums along with proteins provide the basic structure. In addition to traditional functionalities, DFs are used to reduce gumminess, processibility, chewiness/integrity/structure, denseness, rollability, and water activity control. DF sources have also been used in confectionery products, sweet baked goods and eggs for similar efficacies. Oat and barley fractions have been processed, fractionated, and concentrated to leverage their functionalities in various baked goods by taking advantage of the distribution and characteristics of β-glucans and other DF components. DFs have also been used to increase specific volumes of sweet baked goods and reduce shrinkage. When modified, some celluloses can be used as emulsifiers, stabilizers, dispersing agents and thickeners.

In the Tables A1 through A5 below, example formulas for DF fortification, usually to achieve a nutrient content claim of excellent source of DF (5 g/serving) or for a concurrent health claim are provided (Mehta, 2009). These examples use a variety of DFs currently commercially available. In Tables B1 through B5, examples of various functional effects of DFs are discussed (Mehta, 2009: Serrano and Valet, 2013). Table B1 details the effect of oat DF on the characteristics of white pan bread. In addition to achieving a nutrient content claim, oat DF was demonstrated to increase yield. Also a DF based dough improver was demonstrated to improve loaf volume and reduce bake loss. In Table B2 and Figure B2.1, oat DF derived hydrocolloids were demonstrated to improve the quality of pita as demonstrated by even separation of top and bottom layers, increasing strength to prevent tearing, and improving sensory qualities. When DFs are fractionated, the byproduct was effectively utilized to develop natural mold inhibitors that in contrast to chemical mold inhibitors, maintained the strength of the pita and had positive effects on sensory characteristics in addition to being natural and delaying mold growth. Table B3 demonstrates the use of oat DF to improve the textural shelf-life of oat meal cookies. Table B4 demonstrates the use of oat DF to prevent drying of sugar cookies and reducing cracking during storage. Table B5 demonstrates the use of a long flexible DF to reduce breakage in a fragile baked good by increasing the strength and flexibility of the baked good (Mehta, 2009).

Table A1 Bread – high-fiber, whole wheat (5 g TDF). (Source: SunOpta Ingredients Group.)

Bench Formulation

Ingredient	**Baker's % (Flour Basis)**
Whole wheat four	100.00
Vital wheat gluten	12.00
High fructose corn syrup	9.00
Canadian Harvest® Oat Fiber 610	7.50
Honey	5.00
Canadian Harvest® Stabilized Red Wheat Bran, Coarse	3.00
Cultured dextrose	3.00
Soy oil	3.00
Salt	2.50
Raisin juice concentrate	2.50
Sodium stearoyl lactylate	0.50
Hydrated distilled monoglycerides	0.50
Ethoxylated monoglycerides	0.50
Ascorbic acid	0.05
Dough strengthener	0.04
Yeast, compressed	5.50
Water	105.00
Total	**259.59**

Mehta (2009)

Nutritional Facts Panel

Nutrition Facts

Serving Size (50g)
Servings Per Container

Amount Per Serving
Calories 110 — Calories from Fat 10

	% Daily Value*
Total Fat 1.5g	**2%**
Saturated Fat 0g	**0%**
Trans Fat 0g	
Cholesterol 0mg	**0%**
Sodium 230mg	**10%**
Total Carbohydrate 22g	**7%**
Dietary Fiber 5g	**18%**
Sugars 3g	
Protein 5g	

Vitamin A 0% • Vitamin C 0%
Calcium 2% • Iron 6%

* Percent Daily Values are based on a 2,000 calorie diet. Your daily values may be higher or lower depending on your calorie needs:

	Calories	2,000	2,500
Total Fat	Less than	65g	80g
Sat Fat	Less than	20g	25g
Cholesterol	Less than	300mg	300mg
Sodium	Less than	2,400mg	2,400mg
Total Carbohydrate		300g	375g
Dietary Fiber		25g	30g

Calories per gram:
Fat 9 • Carbohydrates 4 • Protein 4

Table A2 White bread — fiber-enriched, heart-healthy (3 g TDF). (Source: SunOpta Ingredients Group.)

Bench Formulation

Ingredient	Baker's % (Flour Basis)
Sponge	
Spring patent four, enriched	61.20
Vital wheat gluten	10.00
Yeast, compressed	2.60
Dough conditioner, sodium stearoyl lactylate	0.50
Water	52.00
Dough	
Bread four, enriched	28.80
Canadian Harvest® Oat Fiber 300-58	4.00
Barley Balance™ beta-glucan concentrate	12.00
Sugar, granulated	8.40
Salt	2.00
Nonfat dry milk	1.60
Hydrated distilled monoglycerides GMS-90	1.00
Yeast, compressed	0.52
Calcium propionate	0.15
Dough conditioner, Dependox AXC	0.04
Ascorbic acid	0.015
Water	39.00
Total	**223.825**

Mehta (2009)

Nutritional Facts Panel

Nutrition Facts

Serving Size (50g)
Servings Per Container

Amount Per Serving	
Calories 110	Calories from Fat 5
	% **Daily Value***
Total Fat 0.5g	**1**%
Saturated Fat 0g	**0**%
Trans Fat 0g	
Cholesterol 0mg	**0**%
Sodium 210mg	**9**%
Total Carbohydrate 24g	**8**%
Dietary Fiber 3g	**10**%
Soluble Fiber 1g	
Sugars 2g	
Protein 5g	

Vitamin A 0% • Vitamin C 6%
Calcium 2% • Iron 6%

* Percent Daily Values are based on a 2,000 calorie diet. Your daily values may be higher or lower depending on your calorie needs:

	Calories	2,000	2,500
Total Fat	Less than	65g	80g
Sat Fat	Less than	20g	25g
Cholesterol	Less than	300mg	300mg
Sodium	Less than	2,400mg	2,400mg
Total Carbohydrate		300g	375g
Dietary Fiber		25g	30g

Calories per gram:
Fat 9 • Carbohydrates 4 • Protein 4

Table A3 White cake formula (5 g TDF). (Source: MGP Ingredients, Inc.)

Bench Formulation

Ingredient	Baker's % (Flour Basis)
Water	165.70
Sucrose	164.10
Cake four	100.00
Cake shortening	49.00
Cake emulsifier	6.84
Vital wheat gluten	4.56
Fibersym® RW	38.18
Dry egg whites	7.12
Dry whole eggs	11.40
Nonfat dry milk	9.23
Salt	3.99
Baking soda	2.28
SALP	2.28
Pregel™ 40	2.85
Flavor	1.71
CMC	0.28
MCP	0.28
Total	**569.80**

Mehta (2009)

Nutritional Facts Panel

Nutrition Facts

Serving Size (80g)
Servings Per Container

Amount Per Serving
Calories 230 — Calories from Fat 70

	% **Daily Value***
Total Fat 8g	**13**%
Saturated Fat 3g	**15**%
Trans Fat 0g	
Cholesterol 25mg	**9**%
Sodium 380mg	**16**%
Total Carbohydrate 40g	**13**%
Dietary Fiber 5g	**19**%
Sugars 24g	
Protein 4g	

Vitamin A 0% • Vitamin C 0%
Calcium 4% • Iron 4%

* Percent Daily Values are based on a 2,000 calorie diet. Your daily values may be higher or lower depending on your calorie needs:

	Calories	2,000	2,500
Total Fat	Less than	65g	80g
Sat Fat	Less than	20g	25g
Cholesterol	Less than	300mg	300mg
Sodium	Less than	2,400mg	2,400mg
Total Carbohydrate		300g	375g
Dietary Fiber		25g	30g

Calories per gram:
Fat 9 • Carbohydrates 4 • Protein 4

Table A4 Better-for-you brownie (5 g TDF). (Source: National Starch Food Innovation.)

Bench Formulation

Ingredient	Baker's % (Flour Basis)
Butter, unsalted	138.45
Unsweetened cocoa powder	40.86
Granulated sugar	287.21
Egg substitute	98.97
Pure Madagascar vanilla	8.24
Coffee, brewed, cooled	13.99
Hi-maize® 260	106.52
NUTRIOSE®	59.31
Baking powder	1.89
Salt	2.92
All-purpose four	100.00
Total	**858.36**

Mehta (2009)

Nutritional Facts Panel

Nutrition Facts

Serving Size (40g)
Servings Per Container

Amount Per Serving	
Calories 140	Calories from Fat 60
	% **Daily Value***
Total Fat 8g	**12**%
Saturated Fat 3.5g	**18**%
Trans Fat 0g	
Cholesterol 35mg	**12**%
Sodium 75mg	**3**%
Total Carbohydrate 25g	**8**%
Dietary Fiber 5g	**20**%
Sugars 14g	
Protein 2g	

Vitamin A 4% • Vitamin C 0%
Calcium 2% • Iron 15%

* Percent Daily Values are based on a 2,000 calorie diet. Your daily values may be higher or lower depending on your calorie needs:

	Calories	2,000	2,500
Total Fat	Less than	65g	80g
Sat Fat	Less than	20g	25g
Cholesterol	Less than	300mg	300mg
Sodium	Less than	2,400mg	2,400mg
Total Carbohydrate		300g	375g
Dietary Fiber		25g	30g

Calories per gram:
Fat 9 • Carbohydrates 4 • Protein 4

Table A5 Snack cracker (5 g TDF). (Source: Tate & Lyle.)

Bench Formulation

Ingredient	Baker's % (Flour Basis)
Pastry four	100.00
Water (85–90°F) (29–32°C)	58.03
PROMITOR™ resistant starch corn starch	36.23
X-PAND'R® food starch – modified	17.40
All-purpose trans-free shortening	12.32
PROMITOR™ soluble corn fiber 70	8.69
Granulated sugar	3.61
Monocalcium phosphate	1.15
Sodium bicarbonate	1.01
Salt	0.86
Total	**239.3**

Mehta (2009)

Nutritional Facts Panel

Nutrition Facts

Serving Size (30g)
Servings Per Container

Amount Per Serving
Calories 120 Calories from Fat 45

	% **Daily Value***
Total Fat 5g	**8%**
Saturated Fat 1g	**5%**
Trans Fat 0g	
Cholesterol 0mg	**0%**
Sodium 220mg	**9%**
Total Carbohydrate 22g	**7%**
Dietary Fiber 5g	**20%**
Sugars 1g	
Protein 1g	

Vitamin A 0% • Vitamin C 0%
Calcium 4% • Iron 2%

* Percent Daily Values are based on a 2,000 calorie diet. Your daily values may be higher or lower depending on your calorie needs:

	Calories	2,000	2,500
Total Fat	Less than	65g	80g
Sat Fat	Less than	20g	25g
Cholesterol	Less than	300mg	300mg
Sodium	Less than	2,400mg	2,400mg
Total Carbohydrate		300g	375g
Dietary Fiber		25g	30g

Calories per gram:
Fat 9 • Carbohydrates 4 • Protein 4

Table B1 Effect of oat fiber on white pan bread characteristics.

Part A. General Formula. (Source: From Grain Millers. Serrano and Valet 2013.)

Ingredient	Baker's % (Flour Basis)
Flour	100
Sugar	4
Salt	2
Active Dry Yeast	2
Water	60 to 70
Cerefava NDI 900*	0 to 0.4
Oat Fiber BCS 30*	0 to 10
Total	**168 to 188**

Serrano and Valet (2013)

Nutritional Facts Panel

Nutrition Facts

Serving Size (47g)
Servings Per Container

Amount Per Serving
Calories 100 Calories from Fat 5

	% **Daily Value***
Total Fat 0g	**0**%
Saturated Fat 0g	**0**%
Trans Fat 0g	
Cholesterol 0mg	**0**%
Sodium 190mg	**8**%
Total Carbohydrate 19g	**6**%
Dietary Fiber 3g	**12**%
Sugars 1g	
Protein 3g	

Vitamin A 0% • Vitamin C 2%
Calcium 0% • Iron 6%

* Percent Daily Values are based on a 2,000 calorie diet. Your daily values may be higher or lower depending on your calorie needs:

	Calories	2,000	2,500
Total Fat	Less than	65g	80g
Sat Fat	Less than	20g	25g
Cholesterol	Less than	300mg	300mg
Sodium	Less than	2,400mg	2,400mg
Total Carbohydrate		300g	375g
Dietary Fiber		25g	30g

Calories per gram:
Fat 9 • Carbohydrates 4 • Protein 4

Part B. Formulas testing the effect of oat fiber levels and dough conditioners on white pan bread.

	% Cerefava NDI 900 Natural Dough Improver*	% oat fiber	% water	dough wt, g.	% yield increase	loaf wt., g.	weight loss, %	loaf volume, cc
1	0	0	67	519.4	0	447	10.60%	2100
2	0.4	0	67	521.3	0	442	11.60%	2525
3	0	5	68	523.7	0.5	450.1	10.00%	1975
4	0.4	5	68	524.6	0.5	444.3	11.10%	2510
5	0	7.5	69.5	527	1.5	444.5	11.10%	1875
6	0.4	7.5	69.5	528	1.5	445.7	10.90%	2250
7	0	10	71	531.9	2	452.8	9.40%	1775
8	0.4	10	70	529.6	2	451.4	9.70%	2125

*Cerefava NDI 900 and oat fiber BCS 30 are from Grain Millers Inc.
Cerefava ingredient decleration: Natural oat fiber, calcium sulfate, natural enzymes, and ascorbic acid.
Procedure: Scale: 500 g; sheet: 9/32,5/32"; proof: 50 min., 41°, 75% RH; bake: 407°' 20 minutes; cool: 90 min.

Summary

1. Cerefava NDI 900 significantly improves bread loaf volume and reduced weight loss on baking.
2. The slight reduction in loaf volume due to fiber addition was corrected by using Cerefava dough improver.
3. The yield was increased by using oat fiber.
4. When 10% oat fiber is added, a good source of fiber is achieved.

Table B2 Effect of hydrocolloids and natural mold inhibitors on pita bread characteristics (also see Figure B2.1).

Part A. General formula. (From Grain Millers. Serrano and Valet 2013.)

Ingredient	**Baker's % (Flour Basis)**
Flour	100
Water	63–67
Instant Yeast	0.5
Salt	1
Sugar	1.5
Shortening	1
Guar, vital wheat gluten, or Oat Hydrocolloid OBF 600*	0 to 3
Natural Mold Inhibitor 1, 2, or chemical mold inhibitor*	0 to 2
Total	**167 to 176**

Serrano and Valet (2013)

Nutritional Facts Panel

Nutrition Facts

Serving Size (60g)
Servings Per Container

Amount Per Serving	
Calories 130	Calories from Fat 10
	% **Daily Value***
Total Fat 1g	**2%**
Saturated Fat 0g	**0%**
Trans Fat 0g	
Cholesterol 0mg	**0%**
Sodium 135mg	**6%**
Total Carbohydrate 26g	**9%**
Dietary Fiber 1g	**4%**
Sugars 1g	
Protein 4g	

Vitamin A 0% • Vitamin C 0%
Calcium 0% • Iron 8%

* Percent Daily Values are based on a 2,000 calorie diet. Your daily values may be higher or lower depending on your calorie needs:

		2,000	2,500
	Calories	2,000	2,500
Total Fat	Less than	65g	80g
Sat Fat	Less than	20g	25g
Cholesterol	Less than	300mg	300mg
Sodium	Less than	2,400mg	2,400mg
Total Carbohydrate		300g	375g
Dietary Fiber		25g	30g

Calories per gram:
Fat 9 • Carbohydrates 4 • Protein 4

Part B. Randomized Box-Behnken design.

		Texture Improver						Mold Inhibitor					
Test	**Random order**	**Oat hydrocolloid**		**Vital wheat gluten**		**Guar gum**		**Natural Mold Inhibitor 1**		**Natural Mold Inhibitor 2**		**Chemical Mold Inhibitor**	
A	**11**	0	1.50%	−1	0%	1	0.50%	0	1%	−1	0%	1	2%
B	**3**	−1	0%	1	3%	0	0.25%	−1	0%	1	2%	0	1%
C	**2**	1	3%	−1	0%	0	0.25%	1	2%	−1	0%	0	1%
D	**9**	0	1.50%	−1	0%	−1	0.00%	0	1%	−1	0%	−1	0%
E	**5**	−1	0%	0	1.50%	−1	0.00%	−1	0%	0	1%	−1	0%
F	**13**	0	1.50%	0	1.50%	0	0.25%	0	1%	0	1%	0	1%
G	**10**	0	1.50%	1	3%	−1	0.00%	0	1%	1	2%	−1	0%
H	**4**	1	3%	1	3%	0	0.25%	1	2%	1	2%	0	1%
I	**7**	−1	0%	0	1.50%	1	0.50%	−1	0%	0	1%	1	2%
J	**1**	−1	0%	−1	0%	0	0.25%	−1	0%	−1	0%	0	1%
K	**8**	1	3%	0	1.50%	1	0.50%	1	2%	0	1%	1	2%
L	**15**	0	1.50%	0	1.50%	0	0.25%	0	1%	0	1%	0	1%
M	**14**	0	1.50%	0	1.50%	0	0.25%	0	1%	0	1%	0	1%
N	**6**	1	3%	0	1.50%	−1	0.00%	1	2%	0	1%	−1	0%
O	**12**	0	1.50%	1	3%	1	0.50%	0	1%	1	2%	1	2%
Key:													
Level −1			0%		0%		0%		0%		0%		0%
Level 0			1.50%		1.50%		0.25%		1%		1%		1%
Level +1			3%		3.00%		0.50%		2%		2%		2%

*All ingredients are from Grain Millers, Inc.
Ingredient declerations:
Oat hydrocolloid OBF 600=oat bran, oat fiber
Natural mold inhibitor 1=cultured oat syrup solids, citric acid
Natural mold inhibitor 2=cultured oat syrup solids, citric acid, vinegar
Chemical mold inhibitor=calcium propionate and fumaric acid

Summary

1. Fresh pita breads of #5, 6, 9 and 10 that contained only natural mold inhibitors had higher pH and lower TTA values than other products due to lack of furmaric acid. All 15 treatments had water activity higher than 0.96, indicating high probability of mold growth under normal storage conditions.
2. Natural mold inhibitors extended pita bread shelf life to a maximum of 5 days under room temperature storage. NMI-2 was more effective than NMI-1 by giving pita bread one extra day shelf life. Sample #6 showed a slightly better quality and #5 had similar quality to #13. Both #9 and 10 showed a lower reliability score than #13.
3. Natural mold inhibitors worked well to prevent mold growth under refrigeration storage conditions. Based on product evaluation on days 3, 7, and 10, sample #10 gave an overall quality improvement compared to #13, while samples #5, 6 and 9 yielded similar product quality to #13.
4. Natural hydrocolloid from oat fiber showed similar functionality to VWG and guar gum in pita breads stored at room temperature and 41°F.
5. Natural mold inhibitors had the least impact on pita bread quality, while chemical mold inhibitor reduced tearing strength and made pita breads look pale over storage time at both room temperature and 41°F.
6. When both texture improver and mold inhibitors were all considered, at both room temperature and 41°F, samples #5, 6, 9, and 10 had an overall better pita bread quality than #13 because of improved tearing strength and yellowness.
7. From the observation of pita bread products (see product photos - Fig B2.1), samples #5, 6, 9 and 10 gave an even separation of top and bottom layers and natural yellow appearance.
8. From sensory testing of the products throughout storage, samples #5, 6, 9 and 10 had very natural flavor and taste, while the products containing CMI tasted bitter and sour especially after longer storage.

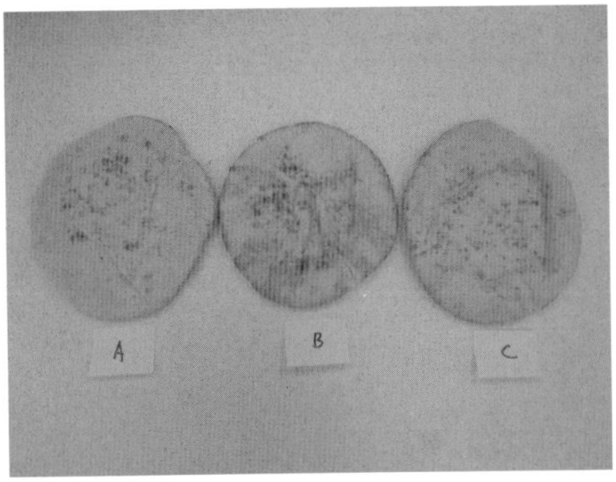

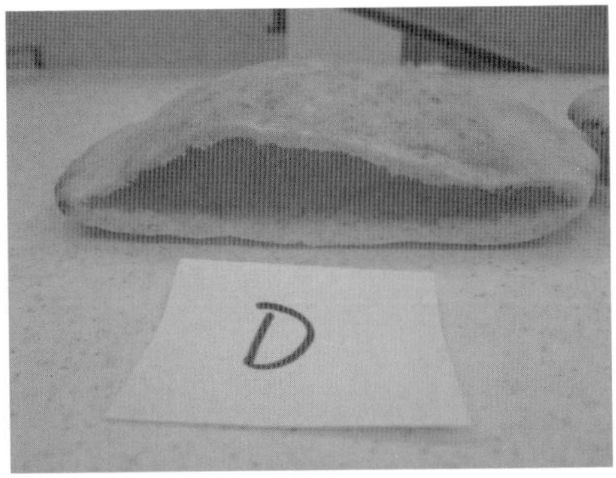

A. (11) 1.5% Hydrocolloid, .5% Guar Gum, 1% NMI 1, 2% CMI
B. (3) 3% VWG, .25% Guar Gum, 2% NMI 2, 1 % CMI
C. (2) 3% Hydrocolloid, .25% Guar Gum, 2% NMI 1, 1% CMI

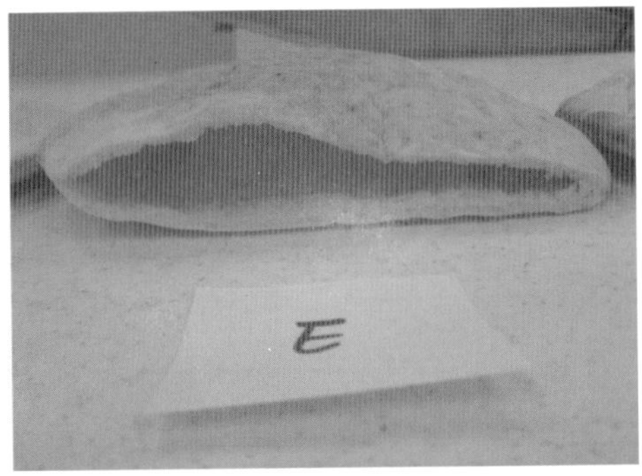

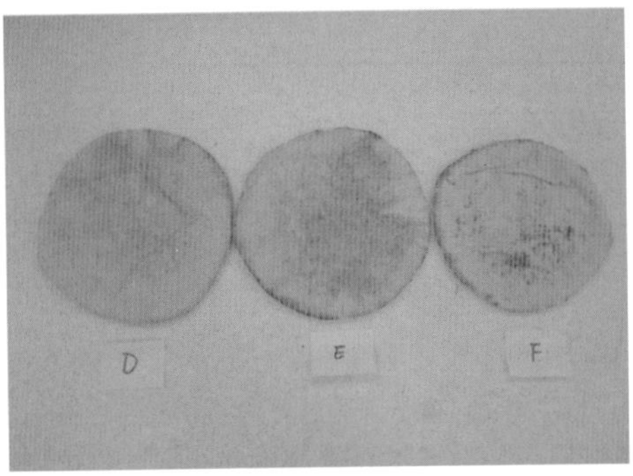

D. (9) 1.5% Hydrocolloid, 1% NMI 1
E. (5) 1.5% VWG, 1% NMI 2
F. (13)1.5% Hydrocolloid, 1.5% VWG, .25% Guar Gum, 1% NMI1, 1% NMI 2, 1% CMI

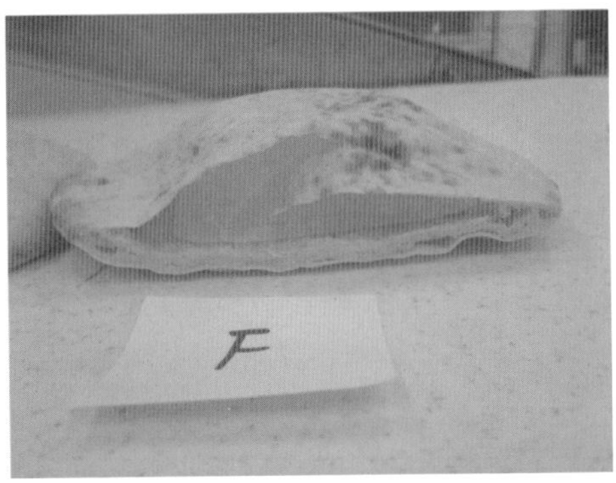

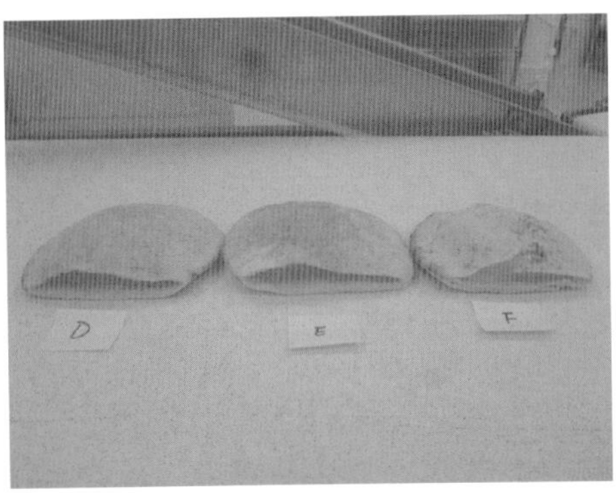

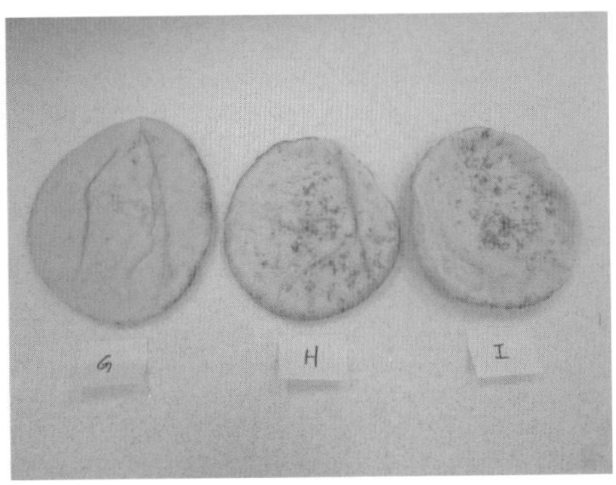

Figure B2.1 Effect of hydrocolloids and mold inhibitors on pita bread characteristics. (Source: From Grain Millers. Serrano and Valet 2013.)

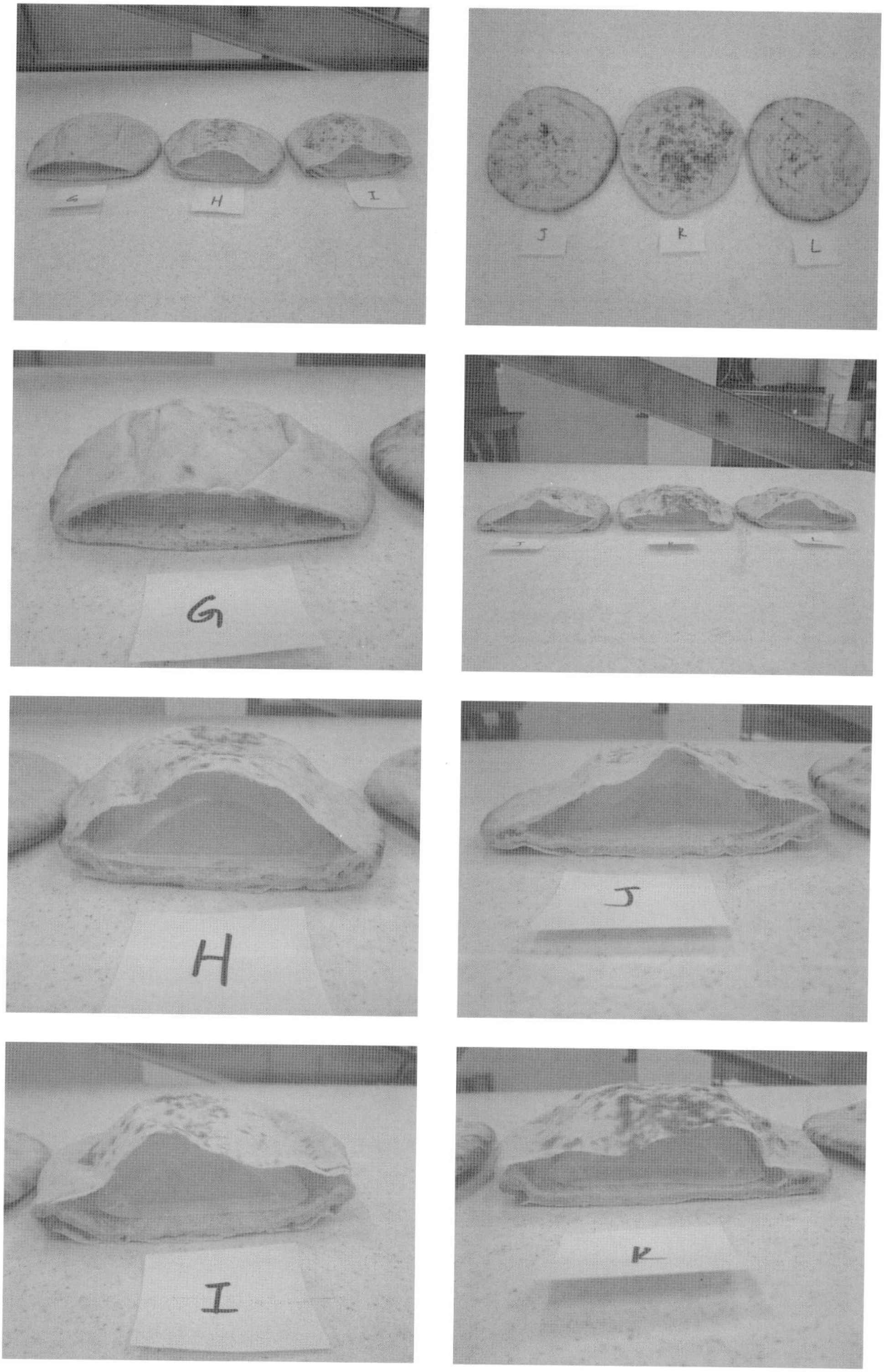

Figure B2.1 (*Continued*)

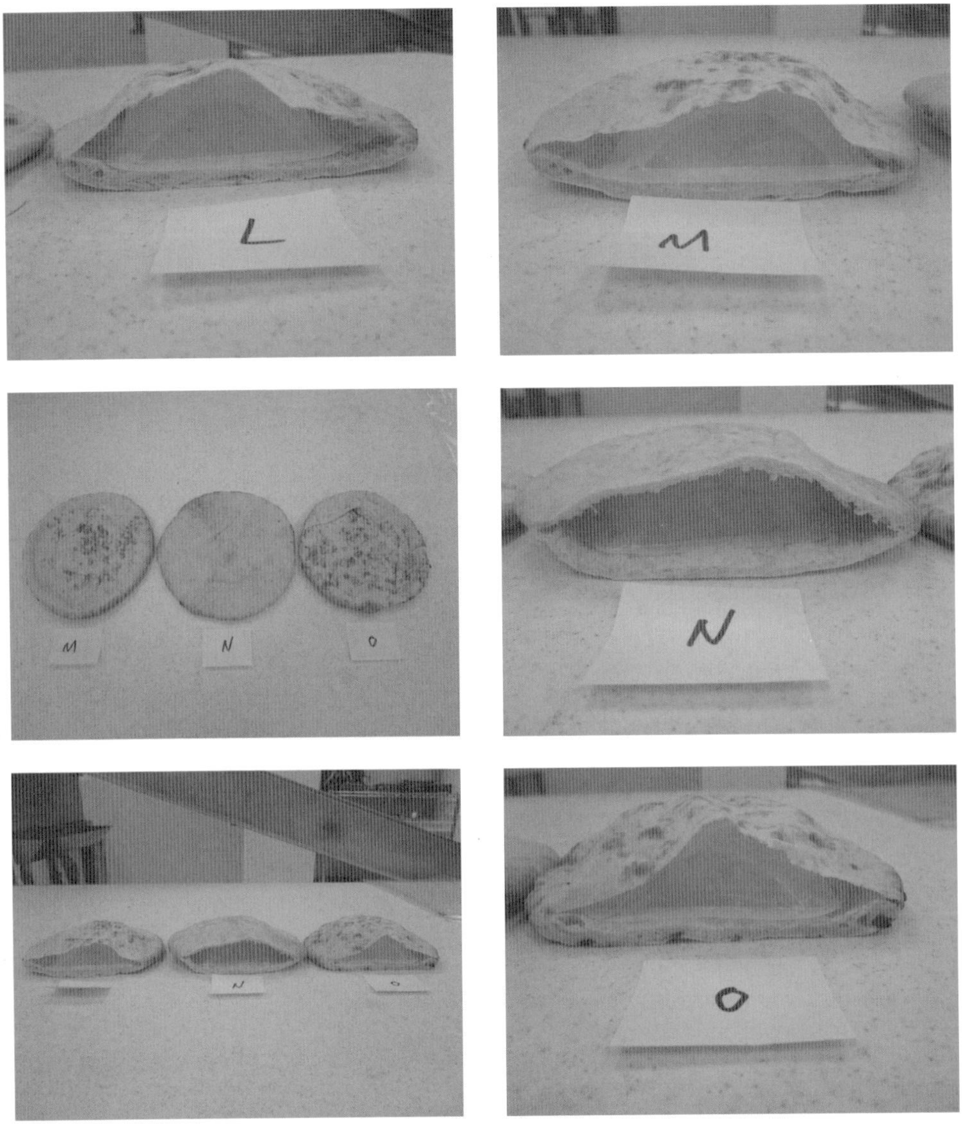

Figure B2.1 (*Continued*)

Table B3 Effect of oat fiber on oatmeal cookies. (Source: From Grain Millers. Serrano and Valet 2013.)

Ingredients	Baker's % (Flour Basis)
Flour	45
Thick Oat Flakes*	45
Oat Fiber BCS 30*	10
Salt	0.9
Baking Soda	1
Cinnamon	0.6
Nutmeg	0.4
Butter	34
Brown Sugar	60
Egg	20
Total	**216.9**

*Oat flakes and oat fiber are from Grain Millers, Inc.

Summary

1. BCS 30 oat fiber improves the textural shelf-life of oat meal cookies.
2. Thick oat flakes enhance the chewiness of the cookies.

Nutritional Facts Panel

Nutrition Facts

Serving Size (32g)
Servings Per Container

Amount Per Serving
Calories 130 Calories from Fat 145

	% **Daily Value***
Total Fat 5g	**8**%
Saturated Fat 3g	**15**%
Trans Fat 0g	
Cholesterol 20mg	**7**%
Sodium 130mg	**5**%
Total Carbohydrate 19g	**6**%
Dietary Fiber 2g	**8**%
Sugars 9g	
Protein 2g	

Vitamin A 2% • Vitamin C 0%
Calcium 0% • Iron 4%

* Percent Daily Values are based on a 2,000 calorie diet. Your daily values may be higher or lower depending on your calorie needs:

	Calories	2,000	2,500
Total Fat	Less than	65g	80g
Sat Fat	Less than	20g	25g
Cholesterol	Less than	300mg	300mg
Sodium	Less than	2,400mg	2,400mg
Total Carbohydrate		300g	375g
Dietary Fiber		25g	30g

Calories per gram:
Fat 9 • Carbohydrates 4 • Protein 4

Table B4 Effect of oat fiber on sugar cookies. (Source: From Grain Millers. Serrano and Valet 2013.)

Ingredient	Baker's % (Flour Basis)
Flour	93
Oat Fiber BCS 30*	7
Sugar	75
Baking Soda	1
Salt	1
Dry Milk	50
Vanilla Extract	3
Total	**230**

*Oat fiber BCS 30 is from Grain Millers, Inc.

Summary

BCS 30 prevents the cookies from drying out and reduces cracking during storage.

Nutritional Facts Panel

Nutrition Facts

Serving Size (28g)
Servings Per Container

Amount Per Serving
Calories 100 Calories from Fat 0

	% **Daily Value***
Total Fat 0g	**0**%
Saturated Fat 0g	**0**%
Trans Fat 0g	
Cholesterol 0mg	**0**%
Sodium 120mg	**5**%
Total Carbohydrate 21g	**7**%
Dietary Fiber 1g	**4**%
Sugars 12g	
Protein 3g	

Vitamin A 0% • Vitamin C 2%
Calcium 10% • Iron 2%

* Percent Daily Values are based on a 2,000 calorie diet. Your daily values may be higher or lower depending on your calorie needs:

	Calories	2,000	2,500
Total Fat	Less than	65g	80g
Sat Fat	Less than	20g	25g
Cholesterol	Less than	300mg	300mg
Sodium	Less than	2,400mg	2,400mg
Total Carbohydrate		300g	375g
Dietary Fiber		25g	30g

Calories per gram:
Fat 9 • Carbohydrates 4 • Protein 4

Table B5 Breakage reduction — ice cream cone with 4% oat fiber. (Source: SunOpta Ingredients Group.)

Bench Formulation

Ingredient	Control Baker's %	Model System Baker's %
Flour	100.00	100.00
Granulated sugar	54.90	54.90
Salt	0.50	0.50
Lecithin	0.30	0.30
Water	100.00	110.00
Canola oil	2.10	1.90
Canadian Harvest® Oat Fiber 780	–	4.10

Mehta (2009)

- Flat sheets of ice cream cones were prepared in the lab using a waffle cone iron. Test product contained 4% Canadian Harvest® Oat Fiber 780. A 3-point bend test with a TA.XT2 Texture Analyzer (Texture Technologies Corporation, Scarsdale, NY, www.texturetechnologies.com) was used for strength and texture analysis. The flat cones were placed in a breaker fixture and a rounded-end knife was used to apply force until the cones broke. Results were expressed as force (g) required to break the cone, reflecting strength and distance traveled (mm) by the probe prior to breaking, which reflects flexibility and fragility of the cones.
- A drop test was performed with 3 pieces of flat cones. The number of new fragments generated per flat cone is reported as broken fragments.
- Samples with Canadian Harvest® Oat Fiber 780 demonstrated a significant improvement for all attributes tested. Strength increased by 27% and flexibility increased by 34%. These impacted cone strength and fragility, which translated into a 68% reduction of broken fragments.

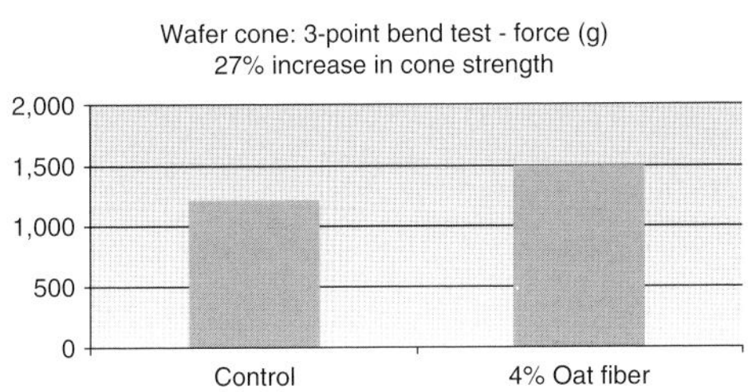

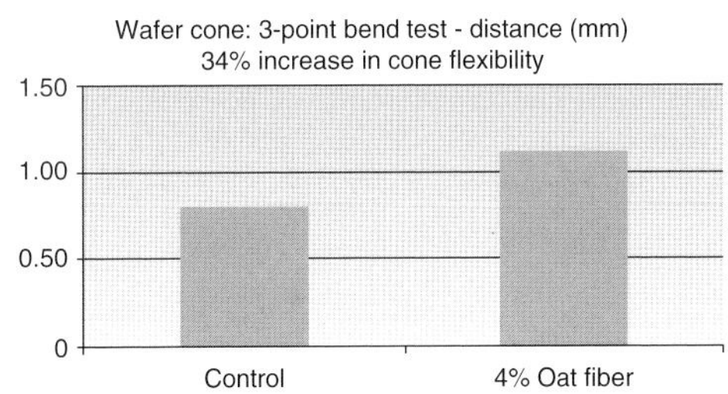

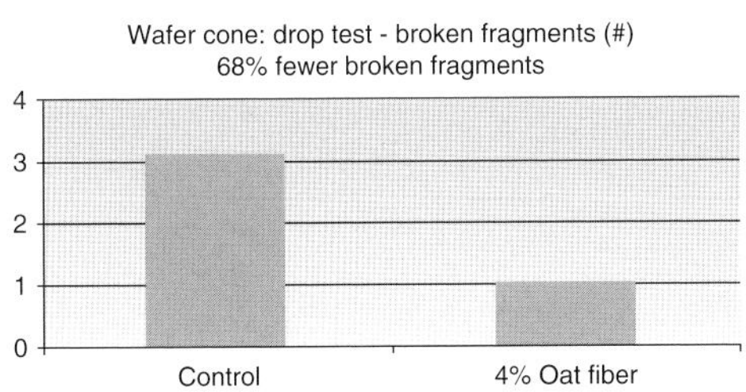

Summary

- Uses Canadian Harvest® Oat Fiber 770 or 780 (SunOpta Ingredients Group) to increase the strength of waffle ice cream cones while maintaining positive sensory attributes.
- Reduces breakage during production, shipping, and customer use.
- Increases cone strength and flexibility.
- Allows reduction in packaging costs for retail distribution.
- Maintains desired sensory attributes.
- Labeled as "oat fiber," a consumer-friendly ingredient.

The future of DF from various sources is exciting. Our current knowledge in this new area is at its infancy and is analogous to where starch was 75 years ago and proteins and fats were 30 to 40 years ago. It is a challenge to predict the direction that the science and industry is going. However, it is clear that we are going towards better characterization and understanding of the functionality of DF, making DFs more natural, enhancing the DF levels and characteristics of raw materials by a wide and varied set of processing, chemical, and physical means. The future will likely bring the functionality of various types of DFs together – often in the naturally occurring raw materials. This will start with simple blends (which is already happening) and end in multiple DFs being "created" or concentrated in one raw material such as from one grain such as oat or barley that has a high level of DF to begin with. Lastly, the functionality of other ingredients and molecules such as proteins, starches, fats and then more complex and functional ingredients such as emulsifiers, surfactants, dough conditioners, dispersing agents, thickeners and healthy components will either be grown into or delivered by processing. This again is already happening as exemplified by ingredients based on cellulose derivatives such as hydroxymethlycellulose, microcrystaline cellulose, methylcellulose, and sodium carboxymethylcellulose. All of these will become more cost effective in use and also will be simpler to use! The future for DF is exciting and bright indeed!

References

Mehta, R. (2009) Dietary fiber II. AIB Technical Bulletin. 33(2), 1–19.
Serrano, R. and Valet, C. Grain (2013) Private communication. Grain Millers, Inc., Eugene, Oregon, USA.

II Advances in texture measurements and consumer insights

This volume provides a current snapshot of the depth and breadth of efforts around the world in food texture design and optimization. It is divided into two parts. Part A provides examples of some key product development challenges that product developers and researchers around the world are dealing with. As described in Chapter 1, expertise in a variety of tools and techniques is needed to address food texture challenges. There has been significant advancement in sensory as well as analytical techniques for measuring different aspects of food texture to derive insight as well as to drive product development. Part B deals with advances in tools and techniques used for texture design and optimization.

Chapters 14 and 17 deal with techniques related to deriving consumer insight and being able to use this to focus product development efforts. Chapter 14 describes a new structured approach to uncover the underlying drivers of choice based on their preferred affinities for different food textures. Chapter 17 provides input into new understanding into mouth behavior and how that can be used to identify population groups with a predisposition towards certain types of texture. Chapter 13 deals with sensory analysis and provides a summary of advances in techniques used for the sensory characterization of food textures.

Chapters 11, 12, 15, and 16 are related to advances in instrumental techniques to advance the development of texture. Chapter 11 describes the use of electro-myography in measuring food texture. This is an important technique to perceive the direct muscular response to different textures and the ability to use this insight to optimize textures. Chapter 12 provides a good summary of studies in the role that lubrication plays in optimizing creaminess for indulgent food products. Chapter 15 provides an introduction to advanced spectroscopic techniques and how they can be used to derive insight into food texture in dairy foods. Chapter 16 describes how atomic force microscopy can be used to derive insight into designing more effective emulsions and foams.

These are just examples of some of the advances in methods and techniques that have been made in recent years but should provide a good overview of how new techniques and measurements can provide further insight for food texture design and optimization.

Food Texture Design and Optimization, First Edition. Edited by Yadunandan Lal Dar and Joseph M. Light.

11 Use of electromyography in measuring food texture

Takahiro Funami[1], Sayaka Ishihara[1] and Kaoru Kohyama[2]

[1]San-Ei Gen F.F.I., Inc., Toyonaka, Osaka, Japan
[2]National Food Research Institute, National Agriculture and Food Research Organization, Tsukuba, Ibaraki, Japan

11.1 INTRODUCTION OF ELECTROMYOGRAPHY

Food texture is perceived during mastication and can change with every chew. In the mouth, food texture dynamically changes with the breakdown of the structure, the increase in moisture from saliva and the melting at oral temperature. Such temporal changes in food texture have not been evaluated adequately by instrumental tests that can measure the physical and/or chemical properties before food consumption (Wilkinson et al., 2000). Human masticatory measurements based on oral physiology have been recently introduced for evaluation of food texture in order to integrate physical, physiological and psychophysical approaches (Wilkinson et al., 2000; Chen, 2009; Foegeding et al., 2011).

Electromyography (EMG) of human masticatory muscles is one of the most common physiological methods for the measurement of food texture (Boyar and Kilcast. 1986a; González et al., 2001; Kemsley et al., 2003; Smith, 2004; Woda et al., 2006; Foegeding et al., 2011). Mastication or chewing is rhythmically repeated by jaw movement with the jaw-opening and the jaw-closing muscles acting alternately. The masseter muscles work to close the mandible and the suprahyoid muscles that include the anterior belly of the digastric, mylohyoid and geniohyoid muscles act to open the jaw. An example of EMG signals from the right and left masseter muscles and the suprahyoid muscles is shown when a subject clenches the upper and lower teeth (Figure 11.1 left) and opens the jaw (Figure 11.1 right), respectively. Mastication is considered to be an unconscious and automatically proceeding process controlled by the central pattern generator in the brain stem (Lund, 1991; Lucas et al., 2004; Hiiemae, 2004; Yamada et al., 2005). The central command is modulated by peripheral inputs so that the masticatory motor output adapts to the textural characteristics of the food (Woda et al,. 2006; Chen, 2009; van der Bilt, 2009; Foegeding et al., 2011).

The majority of human mastication studies have been conducted on the recording of the masseter and/or the anterior temporal muscles using surface electrodes during mastication of different food samples. Since these muscles are typically used to close

Food Texture Design and Optimization, First Edition. Edited by Yadunandan Lal Dar and Joseph M. Light.

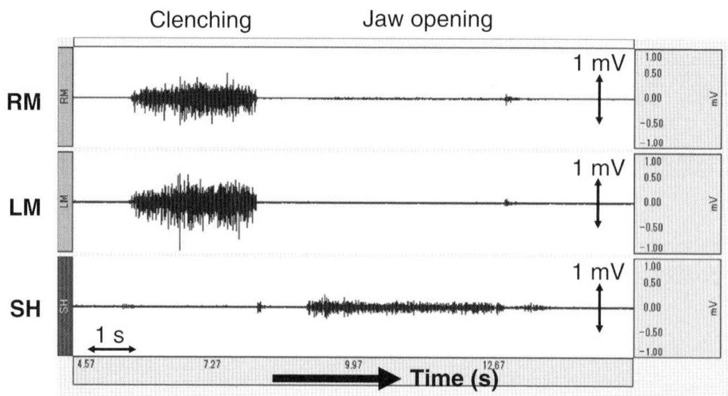

Figure 11.1 An example of electromyographic signals during clenching and jaw opening. Electromyographic signals from right masseter muscle (RM), left masseter muscle (LM), and suprahyoid muscles (SH) were recorded using a young female as a subject.

the mandible (working as the jaw-closing muscles), their EMG activities are useful in analyzing the texture of hard foods and the breakdown of their structure. The hardness of food correlates with the amplitude of the jaw-closing muscles (Edgell et al., 1987; González et al., 2001; Kemsley et al., 2003; Mioche et al., 2003; 2004; Mioche, 2004; Smith, 2004; Woda et al., 2006; Foegeding et al., 2011). Sensory assessments on crunchiness (Brown et al., 1998b), fracturability (Peng et al., 2002) and toughness and cohesiveness (Imai and Sato, 2008), which were correlated to sensory perceived hardness, relate to some variables from the masseter EMG.

Mastication of more adhesive foods such as caramels and Japanese rice cakes leads to greater activity of the suprahyoid muscles, which act in the jaw-opening phase (Sakamoto et al., 1989; Kohyama et al., 1998; 2005c; Çakir et al., 2012). Soft foodstuffs such as jellies and yogurt are eaten through compression between the tongue and the hard palate without chewing. Figure 11.2 is an example for oral processing of a thick liquid, and Figure 11.3 is that of a soft gel. The masseter activity is very weak because these food samples are not chewed using the back teeth, while the suprahyoid activity is more evident as reported by Nakazawa and Togashi (2000) and Ishihara et al. (2011). This is because the suprahyoid muscles also coordinate with tongue movements, involving the compression of the tongue against the hard palate, the protrusion of the tongue and the transportation of the bolus at the beginning or oropharyngeal phase of swallowing (Shiozawa et al., 1999a; Vaiman et al., 2004; Taniguchi et al., 2008). The suprahyoid muscles are active in swallowing water, and its activity becomes greater for thick liquids of a similar amount, which depends on the type of thickener used (Takahashi et al., 2011). The suprahyoid EMG detects the elevation of the tongue, hyoid bone and the thyroid cartilage (Dantas and Dodds, 1990; Ding et al., 2003; Huckabee et al., 2005; Jafari et al., 2003; Kohyama et al., 2007b; Reimers-Neils et al., 1994; Taniguchi et al., 2008; Vaiman et al., 2004) and is often used to monitor the swallowing behavior of food bolus. Eves and Kilcast (1987) stated that the amplitude of the suprahyoid EMG related well to the intensity of thickness assessed by subjects.

The sequence of human oral processing can be divided into three phases: 1) ingestion – transferring food between the teeth with the tongue, 2) comminution sequence – rhythmic

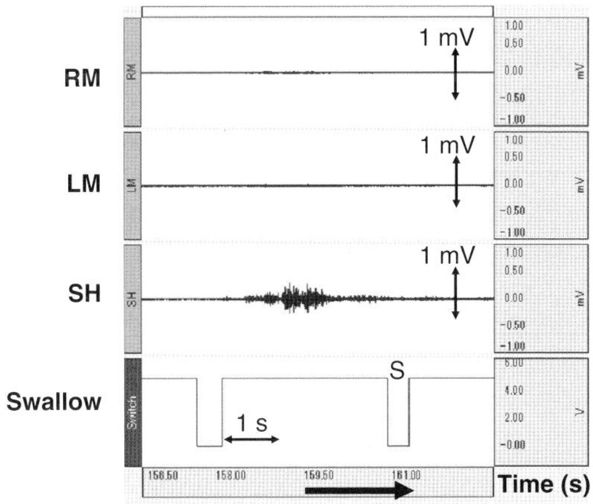

Figure 11.2 An example of electromyopgraphic signals during oral processing of liquid food. Subject and the target muscles for the EMG recording were the same as Figure 11.1. The subject was asked to push a button to indicate the initiation and the end of swallowing represented by abbreviation S in the 'Swallow' channel. As a representative of liquid food, solutions from 1 w/w% xanthan gum (VISTOP® D-3000-C, San-Ei Gen F. F. I., Inc.) were fed to the subject at the serving amount of 6 g.

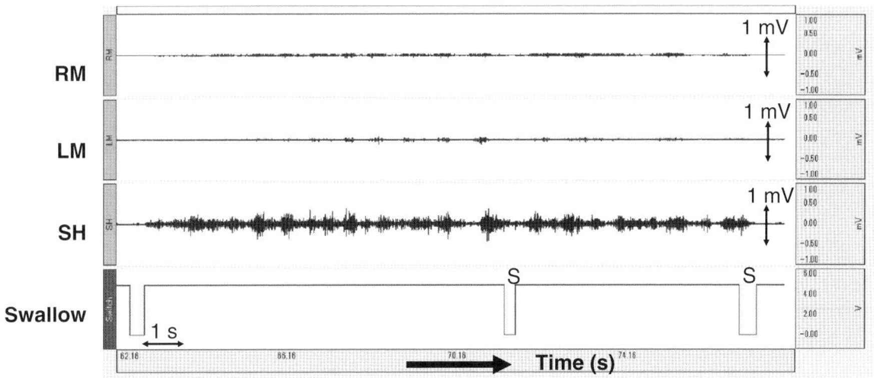

Figure 11.3 An example of electromyographic signals during eating of soft jelly.
Subject and the target muscles for the EMG recording were the same as Figure 11.1. The subject was asked to push a button to indicate the initiations of the oral processing and every swallowing. As a representative of soft jelly, gels from 0.3 w/w% de-acylated gellan gum (KELCOGEL®, San-Ei Gen F. F. I., Inc.) in the presence of 0.1% calcium lactate and 10% sugar were fed to the subject at the serving amount of 6 g. The soft jelly required the compression between the tongue and the hard palate as a main oral strategy for size reduction.

chewing in which the food is comminuted and the bolus is formed, and 3) clearance and swallowing (Heath and Prinz, 1999). Most EMG applications in solid foods have been employed for the analysis of the second phase (Boyer and Kilcast, 1986a; Sakamoto et al., 1989; Jack et al., 1993; Brown, 1994; Brown et al., 1994a, b; Duizer et al., 1994; 1996; Kohyama et al., 1998; 2002; 2003a; 2005b, c; 2007a, b; 2008; Mioche et al., 1999; 2004; Shiozawa et al., 1999a, b; 2003; Nakazawa and Togashi, 2000;

González et al., 2001; Kemsley et al., 2003; Kohyama and Mioche, 2004; Smith, 2004; Foster et al., 2006; Woda et al., 2006; Tarrega et al., 2008; Hayakawa et al., 2009; Çakir et al., 2012). The jaw-closing and the jaw-opening muscles act alternatively during chewing. Activities of the jaw-opening muscles are usually weaker than those of the jaw-closing muscles.

Although mechanical properties such as elastic modulus and viscosity as physical quantities do not depend on the mass of samples, the EMG variables depend on not only the mechanical properties but also on the amount and geometry of food samples (Nakazawa and Togashi, 2000; Foegeding et al., 2011). A larger size of sample often results in a longer period of oral processing, larger number of chewing, and higher EMG amplitude (Dantas and Dodds, 1990; Palmer et al., 1999; Miyawaki et al., 2000; Kohyama et al., 2005c; 2007a; Woda et al., 2006; Miyaoka et al., 2010; Ashida et al., 2010). Similar size-dependency is also observed in intra-oral force (Nakazawa and Togashi, 2000; Kohyama et al., 2005a) and often in texture from sensory evaluation. Therefore, a mouthful of foodstuffs with either a fixed mass or volume have been served in many mastication studies, and they could be normally swallowed at once (Kohyama et al., 2007a, b).

When humans perform completely free chewing and swallowing, more than one swallow often occurs. In the case of a hard gel, the rhythmical chewing may not be observed later than the first swallowing (Figure 11.4). During this period (T2 in Figure 11.4), the EMG activities of the jaw-closing muscle become smaller, while those of the jaw-opening muscle are greater than the corresponding data in the first sequence. The EMG activities observed in the second sequence are somewhat similar to those for a soft gel (Figure 11.3) as the nature of food inside of the mouth becomes semisolid before swallowing.

Hutchings and Lillford (1988) proposed a mouth process model for various foodstuffs, where food in the mouth changes position in the three-dimensional space with the degree of structure, the degree of lubrication and time. Solid foods are masticated

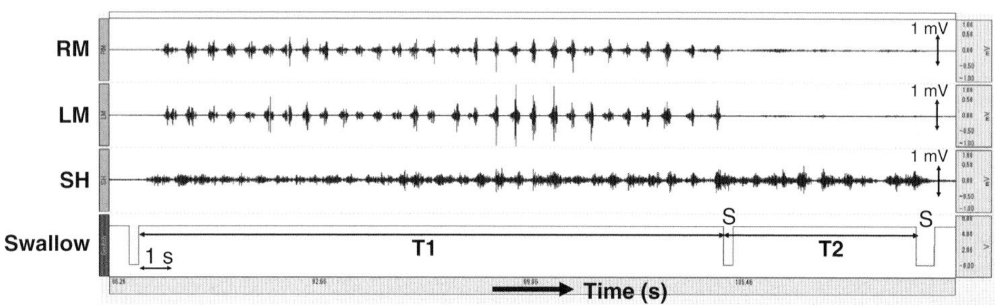

Figure 11.4 An example of electromyographic signals during eating of hard jelly.
Subject and the target muscles for the EMG recording were the same as Figure 11.1. The subject was asked to push a button to indicate the initiations of the oral processing and every swallowing. As a representative of hard jelly, gels from 2.0 w/w% de-acylated gellan gum (KELCOGEL®, San-Ei Gen F. F. I., Inc.) in the presence of 0.1% calcium lactate and 10% sugar were fed to the subject at the serving amount of 6g. The hard jelly required the chewing as a main oral strategy for size reduction. Oral processing was divided into two stages, constituting of T1 (from the start of oral processing to the first swallow) and T2 (from the end of the stage T1 until the last swallow).

to reduce the degree of structure and should be mixed with saliva to increase the degree of lubrication. They can be swallowed when the food state drops in the swallow box, where the degree of structure is below the threshold and the degree of lubrication is over the threshold. The EMG activities of the first and second sequences (T1 and T2 in Figure 11.4) during oral processing may be analyzed separately and should be related to fracture properties and bolus formation, respectively. These correspond to the degree of structure and the degree of lubrication, which are the two main factors in the oral processing of foods (Hutchings and Lillford, 1988).

11.2 ADVANTAGES OF GELS AND SOLS FROM HYDROCOLLOIDS AS MODEL FOODS FOR EMG EXPERIMENTS

Many food characteristics, including texture, flavor, color, as well as the geometry (size and shape) influence human mastication behavior. Food is difficult to reduce to a single physical dimension (Woda et al., 2006). The first step toward simplification is to focus on some mechanical or rheological properties which relate to food texture. Hardness may be the most frequently-used rheological term by researchers, and food samples of different degrees of hardness have been used in the EMG experiments. However, even if the hardness is comparable, different food samples are difficult to compare because some other textural characteristics (elasticity, plasticity, stickiness, brittleness, etc.) may obscure the effect of hardness on the masticatory response. For better control of these food properties, non alimentary models usually made with elastomers or waxes are used. Due to the impossibility to swallow, the use of these materials may modify some events in the masticatory process far from normal eating condition. This is why food hydrocolloids should be utilized for this purpose.

Hydrocolloids have been a target for the EMG studies as model foods of various textures due to the diversity in viscoelastic characters and the ability to mimic and manipulate texture with high reproducibility. It is necessary to use such a model food system with various levels of one textural attribute but similar in other textural attributes in order to identify key EMG parameters which relate to the particular textural property. Also, mechanical properties of gels and sols from hydrocolloids can be identified objectively by rheological measurements, leading to better understanding of the relationship between food texture and rheological properties. Due to the clarity of physical meaning, these outputs can be used universally despite institutions or countries, which may accelerate the development of new food products. This section deals with the EMG studies using gels and sols from hydrocolloids as test specimens.

A preliminary study was done using gelatin and carrageenan gels to understand the possibility of EMG to bridge between traditional objective and subjective texture assessments (Boyar and Kilcast, 1986b). In that study, gel samples of equivalent hardness were evaluated by a subject using the masseter muscle as the EMG target. Results show that the chewing force increases rapidly for each type of gel in 1–2 s after the onset of chewing, requiring similar oral effort during the first bite. Significant difference can be identified after the first bite in the chewing force between gelatin and carrageenan gels.

The increased force decreases for gelatin gels with prolonged time as chewing proceeds, while the increased force is maintained for carrageenan gels. This is largely due to a temperature effect and the higher rate of breakdown in the mouth for gelatin gels due to the lower melting point should be responsible for the result.

Another EMG recording by nine subjects used carrageenan gels of three different concentrations or sensory firmness as a model food (Carson et al., 2002). Results show that the EMG amplitude (relating to the biting force) the number of EMG bursts (relating to the difficulty in reducing the structure), and the chewing duration all increase with gel firmness. Firmness of the model gels correlates with three EMG variables during mastication, including total energy, peak energy, and Fourier power, and the production of these three parameters also works well as an indicator of gel firmness ($R>0.82$ at $p<0.05$).

11.3 TEXTURE DESIGN OF FOODS FOR THE IMPROVEMENT OF PALATABILITY AND THE DEVELOPMENT OF CARE FOODS FOR DYSPHAGIC PATIENTS

The chemical and physical factors influence palatability of foods. Texture, defined as mechanical properties which are perceptible by means of mechanical, tactile, visual and auditory receptors and flavor, which is mixed sensation of aroma and taste, are known as two major constituents of food palatability (Bourne, 2002). Texture is important, particularly for solid foods, because textural changes occur more drastically in solid foods than in liquid foods during oral processing (Nishinari, 2004).

Recently, the importance of texture is emphasized more in terms of the safety of food consumption in an aged society (Nishinari, 2009). In Japan, there are two working criteria using mechanical properties for nursing-care foods; when a food satisfies the criteria, manufactures can say it on the label if they follow the procedure. One is issued by the Consumer Affairs Agency of Japanese Government, intending specifically foods for swallowing difficulty, and the other is issued by the Japan Care Food Conference, which is an industrial group of food companies, intending mainly foods for mastication difficulty. The Consumer Affairs Agency uses 'hardness,' 'adhesiveness,' and 'cohesiveness,' all of which are obtained from a two-bite test performed at 10 mm/s in texture profile analysis (TPA). Three categories are identified by the range of each TPA parameter. Nishinari et al. (2013) recently warned that the 'cohesiveness' values obtained by TPA for liquid samples filled in a cup is misuse. The procedure for the test is also identified, but some practical difficulties in measurements have been pointed out (Nouchi et al., 2012). The Japan Care Food Conference uses 'hardness' and 'viscosity' obtained by separate measurements, and four categories are identified by the range of each parameter. 'Hardness' is defined as apparent stress at 67% strain through penetrating at 10 mm/s, which was previously adopted by the Ministry of Health, Labour, and Welfare (HLW) (see section 11.6.8)). Many nursing-care foods have been commercialized in the market so that the consumers can easily choose the food products by referring to the label in accordance with the degree of mastication and/or swallowing difficulty for the patients. The lower the mastication and/or swallowing performance of the patients, the smaller variation of food texture should be. The food texture is more homogeneous, and the appearance is paste-like. Although the nutritional value and taste are exactly the same as meals for healthy people, such destructured foods may lead to poor appetite and are sometimes followed by malnutrition and health decline.

The number of patients with deficient oral strategy is increasing. Dysphagia literally means difficulty with swallowing of foods or liquids. Dysphagia is commonly managed using texture-controlled foods, designed to moderate the flow speed of bolus through the pharyngeal phase (Quinchia et al., 2011). It may be the case for disphagia patients that food bolus can be misdirected to the airway and to the lung due to the delay in the closure of the epiglottis, causing aspiration and sometimes pneumonia. Aspiration-induced pneumonia is now one of the leading causes of death for the elderly in Japan and is top-ranked for males with ages over 90. Texture design of foods for dysphagia patients is thus an important challenge in the food industry. Soft gels, pastes, and thick liquids, in which food hydrocolloids are used as one of the main ingredients, have been served empirically for dysphagia patients. Soft gels which can be eaten by the compression of the tongue against the hard palate without chewing are used also for people with mastication difficulty. For soft gels, pastes, and thick liquids, the suprahyoid muscles should be the EMG target rather than the jaw-closing muscles.

11.4 TEXTURE ASSESSMENT FOR FOODS WITH HETEROGENEOUS STRUCTURE

It is often difficult to characterize textural properties of foods (not model foods) by objective mechanical or rheological measurements due to the structural heterogeneity that most foods have. If a food matrix contains other constituents dispersed within the matrix, the following points greatly impact the final results from mechanical measurements, including how large the size of plunger should be, which location the sample should be deformed (particularly when the size of plunger is smaller than the sample surface) or how to extract suitable parameters from the fluctuating and unstable data sets. Measurements of the muscle activity provide information on the dynamic breakdown of food during consumption, and from this perspective, EMG should have a great potential as an *in situ* method to obtain texture-related properties.

Heterogeneous model foods, where peanuts of different moisture contents were embedded within two types of food matrices, were used for the EMG recording to investigate the mastication behavior (Hutchings et al., 2011). In that experiment, gelatin gels, as a representative food matrix of an elastic body, were evaluated by eight subjects in comparison with chocolate as a representative food matrix of a plastic body. Results show that the food matrix causes a difference in the mastication behavior in terms of the number of chews and the chewing duration and frequency, while physical properties of embedded peanuts do not. In addition, the food matrix does not change the volume-mean diameter of peanuts before swallowing, while physical properties of embedded peanuts do. These results suggest that the food matrix influences the ratio at which particles of various sizes are selected for chewing (known as the selection function), whereas the physical properties of embedded peanuts influence the extent of breakage when the contact between the teeth and the particles occurs (known as the breakage function).

As presented here, EMG should be a powerful tool for texture assessment of heterogeneous foods because humans can manage to eat any food sample if the size, shape and amount are acceptable, giving more comprehensive results than mechanical measurements.

11.5 EXPERIMENTAL CONDITIONS OF EMG

Prior to the EMG measurements, it is necessary to get informed consent from subjects for participation. The serving size and amount, the serving order of test samples, the command of oral strategy, and so on should be based on the experimental design. Regarding the command of oral strategy, Mioche et al. (1999) reported that free style mastication required less muscle activity than side-imposed mastication, and Kohyama et al. (2010) reported that normal chewing of buckwheat noodles required less muscle work than slurping, thus, free mastication would be recorded as the most effective mastication.

The overall amplitude of the EMG signal can vary from subject to subject and also by experimental conditions like the electrode placement, and so on, although the EMG pattern is generally similar among subjects. This potential problem can be overcome by the use of a reference food serving as a calibration standard for the EMG variables. A standard sample such as chewing gum (Brown et al., 1994a, b) or a real food which resembles the test samples (Sodhi et al., 2010) may also be measured. As the standard for liquid or pasted foods, water can be used as seen in the suprahyoid EMG during swallowing (Taniguchi et al., 2008). For soft or semi-solid foods which require the tongue-palate compression for bolus formation, yogurt (either cup-set or stirred type) can be used as the standard as seen in the suprahyoid EMG during oral processing (Ishihara et al., 2011).

EMG activities are recorded using bipolar surface electrodes, in most cases from the left and right masseter muscles (LM and RM) as typical jaw-closing muscles and from the suprahyoid muscles (SH) which are active in jaw opening and tongue movement. After filtering (typically several to several hundred Hz), removal of 50 Hz noise from the power supply, and amplification to about 1,000 times the original level using EMG amplifiers, the EMG signals are saved in a computer at 1,000 Hz or faster sampling rate. The subject signals the beginning of chew, every swallow and the end of consumption by pressing a button (Figure 11.4, bottom channel). But once the subject's EMG patterns can be attributed to each eating action, those signals are not necessary to identify the actions.

Mastication time is usually defined as the period from the start of chew and the end of chew or oral processing. It can be divided into two stages: T1, where rhythmical masseter activities are shown before the first swallow; T2 from the end of stage T1 until the last swallow (Figure 11.4) (Kohyama et al., 2012). The number of chews is counted based on the number of masseter bursts on the EMG chart. In many EMG works, the masseter activities until the first or imposed swallowing have been analyzed. In the authors' works, on the contrary, subjects were asked to eat food samples as normal as

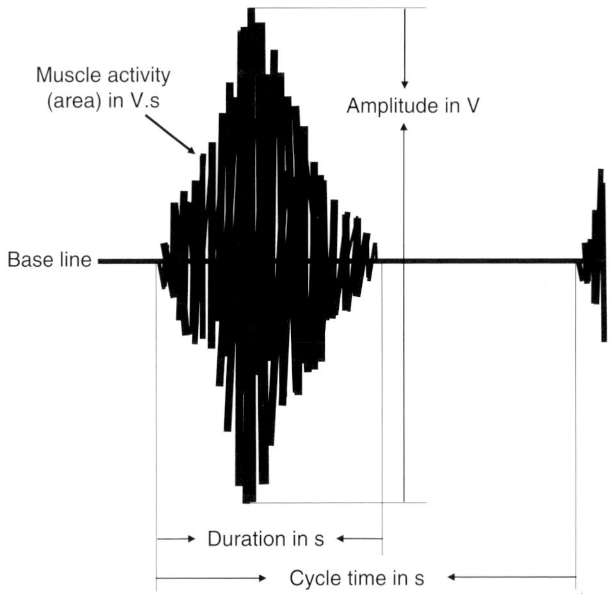

Figure 11.5 EMG parameters taken from one burst.

possible though the mouthful amount was prefixed (Kohyama et al., 1998; 2002; 2003a; 2005b, c; 2007a, b; 2008; Ishihara et al., 2011), and the following three procedures were performed for data capture and analyses: 1) calculation of amplitude as the peak to peak, duration, cycle time and muscle activity as time-integrated EMG voltages (Figure 11.5) in each burst corresponding to an action for each muscle, 2) averaging the left and right muscle activities, if both sides of the muscles were recorded, as chewing side was free, and 3) averaging those parameters for all the cycles. In some studies, the parameters at different stages of mastication, for example, the first chew, mean of the first five chewing cycles, mean of the middle five cycles and mean of the last five cycles before the first swallow, are also calculated. The EMG duration and muscle activities are summed for T1 and T2 (Kohyama et al., 2012). Mastication effort for consuming a mouthful food is evaluated as the total muscle activity by averaging both sides of the masseters until the last swallow (Kohyama et al., 2005b; 2007a).

11.6 SOME TOPICS ON TEXTURE ANALYSIS USING ELECTROMYOGRAPHY

11.6.1 Influence of age and dental status on mastication

As EMG is originally developed in physiological and medical fields, analysis of mastication behavior for different individuals is a common use of the EMG technique. Young and elder subjects were compared (Kohyama et al., 2002; 2003a; Kohyama and Mioche, 2004) using six food products (rice, beef, cheese, hard bread, apple, and peanuts) of various textures as test samples. Elder subjects require a larger number of chewing strokes and mastication time. Whatever the food type, muscle activity per chew is lower for elderly than for young subjects. One chew by elderly subjects is less effective for size reduction of food than that of young subjects.

Dental status of elder subjects can be classified according to their functional unit, which is the number of opposing post-canine (premolars and molars) teeth pairs (Hildebrandt et al., 1997). Among the elderly people of similar age (mean age 67 years), the chewing cycle, the EMG duration and the mastication time – all of which are time-related EMG parameters – increase as the functional units decrease, while the EMG amplitude decreases the same order (Kohyama et al., 2003a). Muscle activity per chew and total muscle activity, both of which are integrated EMG amplitude by time, also decrease as the functional units decrease. These tendencies are commonly observed for all foods tested and no specific food that was difficult to be eaten by the elderly group or subjects with low mastication ability is found. Subjects with weak muscle contraction may compensate for their poor chewing performance by lengthening both chewing cycle and mastication sequence. Impairment in mastication for the elderly is due to both ageing and decreasing number of functional pairs of back teeth (Kohyama et al., 2003a).

11.6.2 Relation between EMG variables and cheese characteristics

Cheese has been studied the most frequently using the EMG technique (Jack et al., 1993; González et al., 2002; Tarrega et al., 2008). Commercial cheese samples were tested by EMG, instrumental two-bite TPA, and sensory evaluation (Kohyama et al., 2012). Sensory evaluation of cheese texture was not performed by the EMG subject at the same time as the subjects could not masticate in the normal way during a difficult task such as texture evaluation. Previous works on the bite force measurement of cheese samples (Dan and Kohyama, 2007; Dan et al., 2007; 2008) reveal that the bite speed is relatively slow during sensory evaluation of texture and that the fracture force increases and fracture strain decreases with increasing bite speed due to viscoelastic character of cheese. Sensory evaluation was thus conducted separately from the EMG recording for analysis of cheese texture sensed in natural eating. Results show that sensory characteristics are essentially consistent with instrumental parameters. Cheese with higher sensory firmness is higher in hardness and Young's modulus from the instrumental test. Although degree of melting in the mouth and chewiness were evaluated after mastication started, scores of these sensory characteristics relate also to instrumental hardness.

Correlations between total muscle activity from the EMG recording and all variables from the instrumental and sensory tests were investigated. The correlation coefficient between total EMG activity of the masseter muscles and sensory score of chewiness is significantly high (Kohyama et al., 2012). The total muscle activity is also highly correlated with Young's modulus, compression load values at a fixed strain ranged less than 50% and 60–80%, and fracture strain from the instrumental test. When two samples of a similar hardness value are compared, more adhesive samples require greater mastication effort.

The highest correlation with sensory firmness is found with compression load at very large strain (80% or higher). This suggests that cheese firmness is sensed when it is compressed completely by the teeth rather than at smaller deformation. Instrumental fracture load may not influence the sensory attributes. Sensory brittleness, adhesiveness and melting in the mouth do not significantly correlate with any of the EMG and instrumental parameters. Since high chewiness and low springiness are observed in sensory and instrumental tests in hard cheese, the influence of resilience is not analyzed separately from firmness.

Principal component analysis was performed using all sets of the EMG variables, instrumental parameters, and sensory scores. Results show that the first principal component relates to many EMG variables, instrumental parameters, and sensory scores. Factor loading values of the fracture strain, cohesiveness and springiness from the instrumental test and those of springiness and melting in the mouth from the sensory test are negative. On the other hand, factor loading values of many other parameters such as a load at a defined compression strain and hardness from the instrumental test and those of firmness, brittleness, and chewiness from the sensory test are positive. The second principal component shows a large factor loading value in the fracture force and a large negative value in instrumental adhesiveness. Fracture properties have little effect on the mastication behavior and sensory evaluated texture.

Instrumental hardness of cheese correlates with many variables measured in this study, and it is difficult to choose a mechanical property that is independent of hardness but still influences the mastication behavior.

11.6.3 EMG of cooked rice prepared from different cultivars

Among various quality parameters for cooked rice, texture has the greatest effect on its palatability. Cooked rice from eight rice cultivars (*Hakuchou-mochi*, *Kogane-mochi*, *Snow Pearl*, *Milky Queen*, *Koshihikari*, *Nipponbare*, *Hoshinishiki*, and *Yumetoiro*) popularly grown in Japan, varying in amylose content (0.4–33.3%) which is a major factor determining cooked rice texture (Sodhi et al., 2010), were subjected to the EMG recording.

The number of chewing strokes is highly correlated with the mastication time, suggesting less dependency of the chewing rhythm on rice cultivars. These parameters increase as the firmness of rice increases. The similar ascending tendency with firmness is also observed for the amplitude and the muscle activities per chew on the masseter EMG. The EMG amplitude and the muscle activity are larger with increasing amylose content or firmness. The EMG amplitude of the suprahyoid muscles is usually smaller than that of the masseter muscle but is relatively large for a waxy variety due to very sticky nature (Kohyama et al., 1998).

The EMG activity of the masseter muscle generally decreases as mastication proceeds. Significant difference in the EMG parameters among different cultivars is observed mainly in the early stage of mastication but is not in the later stages (Kohyama et al., 1998; 2005c). In contrast, the suprahyoid activity does not significantly change during T1 phase (see Figure 11.4), where rhythmical chewing is performed.

Six variables: 1) mastication time, 2) duration per chew, 3) muscle activity per chew, 4) interburst time in early stage of mastication, 5) cycle time in the middle stage and 6) the EMG amplitude in the late stage were selected through cluster analysis from the masseter EMG variables during mastication. These six variables were subjected to principal component analysis. Results show that the first principal component is mainly constituted of variables 1), 2), 3) and 4) and is highly correlated with the amylose content. The second principal component is mainly represented by variables 5) and 6), which are independent of amylose content (Sodhi et al., 2010).

Instrumental texture analysis of cooked rice samples was carried out using a Tensipresser (Okadome et al., 1999), in which the two-bite TPA with 25% compression in the first bite followed by 90% compression in the second bite was performed. None of the instrumental texture parameters correlates with the second principal component, which represents attributes at the middle and late stage of mastication process. This suggests that EMG can effectively distinguish different textures in cooked rice and more importantly provides a comprehensive texture evaluation of cooked rice by representing each mastication stage (Sodhi et al., 2010).

Adding water to rice grains is a simpler method to make the texture of cooked rice softer. Rice samples made with the most popular cultivar in Japan (*Koshihikari*), cooked with various addition levels of water (1.5–4.0 times of weight), were compared using EMG. Firmer samples with lower water content exhibit longer mastication time and greater EMG amplitude and muscle activity (Kohyama et al., 2005c), the same as cooked rice prepared from different cultivars used as test specimens. These EMG observations are basically common when compared at a fixed mass, volume and solid content of the cooked rice samples although the differences are more significant in the order of mass > volume > solid content (Kohyama et al., 2005c). As cooked rice with a larger amount of water has a higher density, the mass of a fixed volume of rice increases with increasing water content of samples. The weight of a fixed solid content that is proportional to the energy increases with decreasing water content of cooked rice. As rice is a staple food, consideration on energy intake from the various preparations is important, especially in foods for patients and elderly people. Rice cooked with a larger amount of water has a greater volume and mass and requires a balance between suitable texture and nutrition for these people.

11.6.4 Effects of cutting of food

Finely cut foods are prepared the most frequently for elderly people with poor chewing performance. It is not rare that finely cut food requires more mastication effort than a block sample of one mouthful size (Kohyama et al., 2007a). This is more significant for hard or firm foods with high fracture stress but relatively low fracture strain such as raw carrot, apple and cucumbers (Kohyama et al., 2005b; 2007a). When a food sample of mouthful size is eaten, the number of chewing strokes, the mastication time, total duration of mastication and total muscle activities are significantly greater for finely cut samples than a block form of the same food product. Cutting of food does not necessarily help reduce the mastication effort although it may make mastication easier.

A visual illusion of mass affects the sensory perceived ease in mastication. Humans often overestimate the mass of finely cut food, especially fine strips, whereas they can more accurately estimate the mass of block samples (Wada et al., 2007). Finely cut carrot (7.5 g) looks equal in mass to a cube of 10.0 g when a photograph of 5 g-block sample is used as standard stimulus. This means that when a smaller mass of finely cut food is eaten, it may be recognized as more mass that should require greater mastication effort. The real magnitude of the mastication effort estimated by total muscle activity on EMG is less than what is predicted visually before being eaten; the subject also feels that it is easier to chew the food.

11.6.5 Physical properties of food bolus

When adult subjects eat foods which they have experience in eating, the mastication behavior is similar within a subject, but there are large differences in the mastication behavior among subjects. The mechanical properties of a food bolus collected from the subjects' mouth at each stage of oral processing were subjected to the two-bite TPA, which was synchronized with the EMG recording on the masseter and the suprahyoid muscles (Shiozawa et al., 1999b). Instrumental hardness, adhesiveness and cohesiveness of the bolus are significantly variable among subjects and foods. Muscle activities of five chewing strokes just before the first swallow were measured, showing that the masseter activity is greater for harder bolus, and the suprahyoid activity relates to the bolus adhesiveness.

Relationships between the mechanical properties of a food bolus and the initiation of swallowing were investigated by the two-bite TPA using boli collected from the mouth (Shiozawa et al., 2003). The boli were taken from the mouth at three different stages of mastication: the halfway stage of mastication (represented by M), immediately prior to swallowing (L), and the number of chewing strokes 20% beyond a point at which swallowing would usually occur (+20). Changes in mechanical properties during mastication (in order of M, L, and +20) were compared among food samples: rice cake, peanuts and biscuits to determine the critical condition required for swallowing. For any of the food boli collected from the mouth, instrumental hardness decreases significantly with increasing chewing number, suggesting that the decrease in hardness of a bolus is a necessary factor for the initiation of swallowing although it is not a sufficient factor by itself. For both peanuts and biscuits, instrumental adhesiveness of boli shows the maximum value at the L stage, but instrumental cohesiveness is unchanged. Swallowing of hard and brittle type foods (peanuts and biscuits) initiates when clumpy bolus is formed, while swallowing of highly adhesive food (rice cake) is triggered when adhesiveness of the bolus is sufficiently reduced to meet the swallowing threshold. These findings are consistent with the oral processing model proposed by Hutchings and Lillford (1988). Both the decrease in the degree of structure corresponding to hardness and the increase in the degree of lubrication represented by instrumental adhesiveness values over the threshold are required for the initiation of swallowing.

The combination of EMG recording during swallowing and acoustic analysis of the swallowing sound can be useful in analyzing physiological flow behavior of a bolus through the pharynx (Funami et al., 2012). It is suggested from the acoustic analysis that the duration of the swallowing sound associated with bolus flow through the pharynx can represent sensory perceived cohesiveness and ease of swallowing (Nakauma et al., 2011). The shorter the duration, the higher the intensity of sensory cohesiveness and swallowing ease are. This can be interpreted as that one cohesive bolus is more comfortable to swallow rather than scattered particles having a turbulent nature. Linkage with EMG study should be investigated to validate this interpretation.

11.6.6 Flavor and taste release: time course analysis of texture and flavor

Foods should enrich human dietary life by stimulating appetite and contribute to human health as well. Functionalities required in foods include nutrition, palatability and physiological functionality (Nishinari, 2006). Foods should never be foods if humans do not feel happiness and satisfaction during eating. From this perspective, palatability is the most important attribute of foods, which is definitely different from medicines. Food palatability is determined by some organoleptic attributes, including flavor, texture, appearance, sound and temperature; flavor and texture are two major factors determining food palatability (Nishinari, 2004). The EMG recording is used to investigate the flavor release profile from the food matrix in relation to the mastication and swallowing behaviors in a dynamic manner.

Perception of flavor intensity from gelatin gels sweetened by sucrose or aspartame and flavored with banana was assessed by sensory T-I (time-intensity) technique and synchronized with the EMG recording of the masseter and the temporalis muscles to link with the mastication behavior (Wilson and Brown, 1997). Results show that the flavor perception pattern is closely linked with the mastication pattern for each subject although large individual differences are found. The time at which the maximum intensity occurs coincides with the end of chewing sequence for most subjects. Also, the act of swallowing is often associated with a marked increase in flavor perception.

Similarly, perception of flavor intensity from sugar-based gelatin gels flavored with various concentrations of anethole was assessed by the T-I technique and synchronized with the EMG recording of the masseter and the temporalis muscles (Sprunt et al., 2002). Results show that the chewing and swallowing rates appear to be unaffected by the flavor concentration, suggesting that the rhythm of these oral strategies is primarily determined by mouth mechanical clearance rather than perceived flavor intensity.

Flavor release from food plays an important role in consumer acceptance and is influenced by changes in food structure and temperature and also the mixing with saliva during chewing. As these effects are more evident via a retronasal pathway, flavor release from solid foodstuffs during mastication has been monitored in combination with EMG (Duizer et al., 1994; 1996; Brown et al., 1996; Salles et al., 2011).

Perception of taste intensity has also been evaluated by sensory analysis in combination with the EMG recording of the masseter and the temporalis muscles. It is evidenced by a study using sugar-based gelatin gels with various concentrations of quinine, which shows that the taste cognition influences the way in which a food is broken down in the mouth (Neyraud et al., 2005). The mastication time decreases (fewer bites or less muscle effort) and the clearance time increases with increasing quinine concentration, while chewing frequency or the rate of salivation is not affected, suggesting an inherent pattern for mastication.

A similar experiment using the combination of sensory analysis and the EMG recording has been carried out with sugar-based gelatin gels containing various concentrations of quinine (Alfonso et al., 2002). Since quinine concentrations tested are lower than those in the study by Neyraud et al. (2005), chewing patterns monitored by the masseter and the temporalis EMG are not affected by quinine concentration in the gels, suggesting no feedback from taste to the motor control of mastication. Results also show that subjects who chew for longer with a greater number of chews, rate higher for perceived intensity of bitterness.

Texture and flavor release were studied when six trained panelists consumed six different gummy jelly products commercially available in Japan (Hayakawa et al., 2009). EMG from both sides of the masseter muscles run simultaneously with the T-I assessment enables clarification of the relationship between the flavor release perception and temporal texture to be analyzed. An instrumental test was also conducted to characterize the mechanical properties of gummy jellies. Results show that sourness intensity is detected to be zero at the beginning of mastication. Sourness is perceived shortly after starting to chew, reaching the maximum intensity after sufficient mastication (18–32 chewing strokes), which lasts for a relatively long time (16.8–45.9 s) after the completion of the main chewing motion. It is also evidenced that sourness is perceived mildly and late with the toughest product, strongly and early with tender and two rubbery samples, and mildly and early with the other three samples that are easy to chew. Thus, sourness and texture are related closely to each other.

11.6.7 Relationships between EMG variables and mechanical properties

Elucidation of the relationship between human eating variables measured by the EMG recording and mechanical properties of food measured by instrumental tests is sometimes difficult because change in one mechanical property usually brings about changes in other mechanical properties (Kohyama et al., 2008). For example, elastic modulus, breaking load, work for breaking, and toughness are often highly correlated with one another (Kohyama et al., 2012). Similar results have been found in the EMG variables and sensory assessments on crunchiness (Brown et al., 1998b), fracturability (Peng et al., 2002), toughness and cohesiveness (Imai and Sato, 2008), where the assessments were related to some of the masseter EMG variables but the sensory scores all correlated with sensory-perceived hardness.

Eight kinds of solid food having a wide range of physical properties (dry hard bread, elastic konjac gel, dry sausage, soft candy, raw radish, pickled radish, boiled carrot and raw carrot) were selected (Kohyama et al., 2008). Twenty-eight physical parameters were measured, and nine independent parameters (stress values at small, medium, large and extra-large compression strains, breaking stress, cohesiveness, adhesiveness, density and moisture content) were selected. The mechanical properties under larger compression strains the most importantly impact the EMG variables (Kohyama et al., 2008). In other studies, when mechanical properties measured under larger deformation do not correlate with those under smaller deformation, a parameter at larger deformation shows a higher correlation with the EMG parameters (Kohyama et al., 2007b; 2010; Hayakawa et al., 2009).

Many reports say that the EMG amplitude and the muscle activity correlate with the modulus defined under small strains, hardness measured at a relatively small deformation and fracture or breaking properties. If the EMG parameters do not relate to results of instrumental hardness under small deformation, trying correlation between the EMG results and the mechanical compression load detected under larger deformation, sometimes beyond the fracture point of sample, would be suggested. In this case, high correlation can be found between them regardless of the mode of sample fracture. The magnitude of the masseter EMG activity correlates with the compression load at such a large strain better than any mechanical parameter measured at smaller strains. Human mastication behavior is not influenced by mechanical properties of food under small deformation since the upper and lower teeth almost reach contact during mastication despite the presence or absence of sample fracture.

It has been reported that the EMG activity of the suprahyoid muscles correlates with the compression load of gel samples at very large strains (Ishihara et al., 2011). To understand the correlation between the EMG parameters during oral processing and mechanical or sensory properties of foods, two types of hydrocolloid gels having different fracture profiles were served to the suprahyoid EMG at two different concentrations for each type (Ishihara et al., 2011). Each gel sample was soft enough to be processed by the tongue-palate compression without chewing. The duration of the whole sequence is prolonged and the EMG activity of the suprahyoid muscles increases with increasing concentration for each gel type. The EMG activity of the suprahyoid muscles correlates well with the compression load of gels at extremely large strains (e.g. 90% strain) and with sensory perceived hardness. These results are similar to previous ones using model foods (e.g. gelatin gels) processed by chewing, where the increases in the duration of the whole sequence and in the EMG activity of the masseter and the temporalis with increasing Instron hardness (the compression stress at 50% strain in this case) are shown regardless of physical properties: elastic or plastic (Foster et al., 2006). It is suggested that mechanical properties at large strains should be related to physiological response during oral processing regardless of the mode of the oral strategy for size reduction in the mouth.

11.6.8 Thickeners for dysphagia patients

Three kinds of hydrocolloid sols (or pastes) from xanthan gum, guar gum and modified starch were served to the suprahyoid EMG as a representative thickener for dysphagia patients (Takahashi et al., 2011). In that study, samples of three different levels of hardness; approximately 400, 600, and 800 Pa, which were determined by the apparent stress at 67 % strain through penetrating at 10 mm/s, were evaluated using young healthy adults as subjects. Results show that in the case of xanthan gum, the harder the sample, the larger the EMG activity and the longer the EMG duration is during swallowing. This suggests that harder samples should require greater effort for transfer of the bolus from the oral cavity to the pharynx during swallowing. Samples from modified starch show a similar tendency to xanthan gum, while samples from guar gum do not. It is confirmed that the influence of hardness on the suprahyoid EMG during swallowing depends on the type of thickener.

The validity of comparison among these sol samples (i.e. thick liquids) at an equivalent hardness which is defined by the compression stress should be argued here. In Japan, food manufacturers were allowed to label 'Foods for the elderly with difficulty in masticating and swallowing' on consumer products upon acceptance by the Japanese Ministry of HLW when a food product meets mechanical and other criteria (although this institution has been replaced with a new one using different mechanical criteria and is now governed by the Consumer Affairs Agency as mentioned in section 11.3). In that regulation, hardness, which was defined as apparent stress at 67% strain through penetration at 10 mm/s, was one of the mechanical criteria. However, the regulation did not specify the range of applicable food form; all food forms, including liquids, sols and solids, should be evaluated in the same manner. This may be the reason why many researchers are still using hardness as a dominant mechanical parameter regardless of food form. The usage of one mechanical parameter in common is important from a practical point of view but may contain some problematic aspects from a scientific point of view.

Similarly, sols or pastes from a thickening agent (maybe a hydrocolloid thickener but not identified) were served to the EMG recording to monitor the signals from the anterior tongue and the suprahyoid muscles during swallowing (Inagaki et al., 2009). In that study, samples of three different concentrations were evaluated using young healthy adults as subjects. Results show that an increase in hardness or in adhesiveness increases the EMG activity for each target during swallowing, and the relationship between the food hardness and the EMG activity of each target is described by a power-law equation. The relationship between each mechanical property; hardness or adhesiveness and the duration of the target muscles activities is more complicated.

Three kinds of hydrocolloid gels from standard agar, low gel-strength agar and gelatin were served to the suprahyoid EMG as a model of food for dysphagia patients (Igarashi et al., 2002). In that study, samples of a fixed hardness; approximately 7,000 Pa which was determined by the apparent stress at 60% strain through penetrating at 1 mm/s, were evaluated by two groups of volunteers aged from 40 to 60 and from 60 to 70 as subjects. Results show that no significant difference is found in the suprahyoid EMG profile among samples (six in totals) for each group during swallowing. This suggests that the same motion should occur regardless of the nature of sample gels once the gel sample is converted into a bolus in the mouth. From a practical point of view, the result indicates the usage of agar as an alternative to gelatin having an advantage of higher melting temperature but with similar textural attributes. Texture of gelatin gels is favored due to its smoothness and rapid melting in the mouth, but is unfavorable, on the other hand, in terms of handling ease due to low melting temperature, which requires storage in a refrigerator before consumption.

11.6.9 Texture design of healthy diets

To apply the EMG measurements to a combination of foods (e.g. bread and soup, salad and dressings, dessert and cream, etc.) rather than one single food should be of another research interest in order to examine the textural characteristics and to rate the ease of eating for the combined foods. Shiozawa et al. (2007) measured eating behavior of solid food (cooked soybean curd) with or without thickening agent (1 w/w% xanthan gum). Addition of xanthan gum significantly increases the adhesiveness, prevents early swallowing of the broth and reduces total EMG activities until final swallowing.

The EMG recording is also useful for analysis of eating behavior of a whole meal. Kohyama et al. (2003b) compared four different meals (two Western and two Japanese traditional styles), consisting of a staple food (bread or rice), soup, a main dish (meat or fish) and a side dish (vegetables) with high and low mastication effort. The Japanese meal is balanced better in terms of the mastication effort, while the mastication effort of bread in Western style meals has a relatively high portion. Softly cooked foods such as gruel have low energy density due to high water content, resulting in more mastication effort to intake the required energy. This kind of study may be useful for texture design of menus particularly at public institutes like schools, hospitals, and so on.

The EMG measurements can be linked to human health. It is evidenced on the electromyogram, by detecting the number of chews per mouthful and the eating duration on food intake, that a larger number of chews per mouthful reduces food intake despite higher chewing speed and longer duration until a comfortable level of fullness is felt (Smit et al., 2011). Eating slowly is effective to reduce the energy intake of a meal with high satiety (Andrade et al., 2008). In contrast, eating until full and eating quickly are associated with being overweight (Otsuka et al., 2006; Maruyama et al., 2008) and these eating behaviors combined have a substantial impact on overweight (Maruyama et al., 2008). Thus, EMG can provide a mean for the texture design of a food/diet that controls the satiety and prevents the obesity or other health problems as well as an instruction for the improvement of dietary habit.

11.7 IN COMBINATION WITH OTHER *IN VIVO* MEASUREMENTS

In vivo measurements like EMG detect the signals from organs working during oral processing, and thus, can describe the dynamic changes in food texture. The combination of EMG and the mandibular kinematics measured at the incisor point (lower frontal tooth) is used frequently by many researchers because the EMG amplitude from the jaw-closing muscles provides information of the chewing force and the mastication movement (Brown et al., 1998a; Heath and Prinz, 1999; Kohyama and Mioche, 2004; Kohyama et al., 2008; Chen, 2009; Foegeding et al., 2011). The EMG variables are more sensitive than kinematic ones and detect differences in physical properties of foods when a fixed size ($15 \times 15 \times 10$ mm) of solid foods are measured (Kohyama et al., 2008). Kinematic parameters such as the maximum opening of the jaw and jaw-closing speed depend on food size rather than the physical properties.

Imaging techniques, including videofluorography (Boyar and Kilcast, 1986a; Pancherz et al., 1986; Dodds et al., 1990; Hiiemae et al., 1996; Logemann et al., 1998; Perlman et al., 2000; Mioche et al., 2002; Hiiemae, 2004; Clave et al., 2006; Matsuo et al., 2008; Saitoh et al., 2007; Okazawa et al., 2008), ultrasound pulse Doppler method (Nakazawa et al., 2000; Kumagai et al., 2009; Tashiro et al., 2010) and videoendoscopy (Logemann et al., 1998; Adams et al., 2008) provide visual evidence of food transformation and transportation during oral processing, both of which may be supportive to the EMG recording. B-mode and/or M-mode of ultrasonographic imaging (Shawker et al., 1983; Palmer et al., 1992; Imai et al., 1995; Peng et al., 2000; Kuhl et al., 2003; Matsumi et al., 2005; Blissett et al., 2007; Tamburrini et al., 2010) has been utilized to observe the tongue and other soft tissues of human during mastication and swallowing.

Bite force (Kohyama and Nishi, 1997; Kohyama et al., 2001a, b; 2005a) and the tongue-palate pressure measurements (Takahashi and Nakazawa, 1991; 1992; Nakazawa and Togashi, 2000; Ono et al., 2004; 2010; Hori et al., 2006) can be used in combination with EMG for texture studies of foods.

Figure 11.6 is an example of the bite force measured by a sheet sensor (Kohyama et al., 2004) and the masseter EMGs for the first five sequences. As a model food,

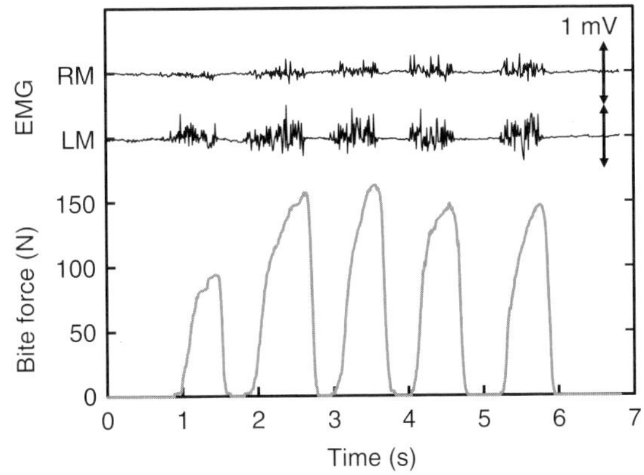

Figure 11.6 An example of simultaneous recording of the bite force and the masseter electromyography. A young female subject bit a silicone rubber on a sheet sensor five times using the left molars.

silicone rubbers (16×24×2 mm) with different hardness values were used. The EMG amplitude of the chewing side (left side LM) is much greater than that of the opposite side (RM), while the duration is slightly longer for the chewing side. The jaw-closing muscle acts before the bite force emerges because the force begins to increase when the food is compressed between the upper and lower teeth. The onset of the masseter EMG is about 20 ms earlier than that of the bite force. The maximum force appears at occlusion where the upper and lower teeth almost contact (Kohyama et al., 2001a, b; 2004; Dan and Kohyama, 2007; Dan et al., 2007; 2008), and the EMG activities decrease to the base level in about 20 ms later than that time. The bite force then reduces to zero according to the jaw opening, and the jaw-opening muscles alternate to the jaw-closing muscles at the end point of the bite force (Pancherz et al., 1986).

The EMG recording during oral processing of soft foods can be associated with the tongue-palate pressure measurements for analysis of the size reduction behavior in the mouth and in this case, both the jaw-closing and jaw-opening muscles should be the targets for the recording. For analysis of the swallowing behavior, on the other hand, the EMG recording can be linked to the tongue-palate pressure measurements and the ultrasound method and in this case, the jaw-opening muscles should be the targets for the recording.

11.8 CONCLUSION

The EMG variables recorded at each mastication stage relate to the mechanical properties of food and sensory scores of its textural characteristics. In general, the EMG activities of both sides of the masseter muscles are greater in the first stage of mastication and lesser in the late stage for solid foods. Contrary to the masseters, the suprahyoid muscles exhibit greater activity in the late stage of mastication. The masseter EMG activity is weak while soft food is eaten as the food breakdown is done by compression between the tongue and the hard palate without chewing. This is similar to the behavior of the late stage of mastication of hard food just before the last swallowing.

Harder food exhibits a longer mastication time and greater masseter EMG amplitude and muscle activity per chew. When compared at an equivalent hardness value, samples with higher adhesiveness and springiness exhibit greater EMG activities. Hardness of food correlates with many mechanical parameters, and it is difficult to choose a mechanical property, which is independent of hardness and still influences the mastication behavior. The EMG variables do not correlate with any mechanical property of foods measured under small deformation and the EMG amplitude and muscle activity at early stage of mastication relate to the breaking properties and the mastication time. Total muscle activity required until swallowing correlates well with force at a very large compression deformation, where tough food may not fracture but most of food samples do. This is true of soft food that is not chewed but compressed by the tongue and the hard palate for size reduction, where the suprahyoid activities dominate over the masseter activities.

ACKNOWLEDGMENT

This study was partly supported by the research and development projects for application in promoting new policy of Agriculture Forestry and Fisheries 22026.

References

Adams, S., Singleton, S., Juskaitis, R and Wilson, T. (2008) *In-vivo* visualisation of mouth–material interactions by video rate endoscopy. *Food Hydrocolloids* 22, 322–325.

Alfonso, M., Neyraud, E., Blanc, O., Peyron, M.A and Dransfield, E. (2002) Relationship between taste and chewing patterns of visco-elastic model foods. *J. Sens. Studies* 17, 193–206.

Andrade, A.M., Greene, G.W. and Melanson, K.J. (2008) Eating slowly led to decreases in energy intake within meals in healthy women. *J. Am. Diet. Assoc.* 108, 1186–1191.

Ashida, I., Iwamori, H., Kawakami, S.Y., Miyaoka, Y. and Murayama, A. (2010) Analysis of the pattern of suprahyoid muscle activity during pharyngeal swallowing of foods by healthy young subjects. *J. Med. Eng. Technol.* 34, 268–273.

Blissett, A., Prinz, J.F., Wulfert, F., Taylor, A.J. and Hort, J. (2007) Effect of bolus size on chewing, swallowing, oral soft tissue and tongue movement. *J. Oral Rehabil.* 34, 572–582.

Bourne, M.C. (2002) *Food Texture and Viscosity*, 2nd edn., Academic Press, San Diego.

Boyar, M.M. and Kilcast, D. (1986a) Food texture and dental science. *J. Texture Studies* 17, 221–252.

Boyar, M.M. and Kilcast, D. (1986b) Electromyography as a novel method for examining food texture. *J. Food Sci.* 51, 859–860.

Brown, W.E. (1994) Method to investigate differences in chewing behaviour in humans: I. Use of electromyography in measuring chewing. *J. Texture Studies* 25, 1–16.

Brown, W.E., Eves, D., Ellison, M. and Braxton, D. (1998a) Use of combined electromyography and kinesthesiology during mastication to chart the oral breakdown of foodstuffs: relevance to measurement of food texture. *J. Texture Studies* 29, 145–167.

Brown, W.E., Langley, K.R. and Braxton, D. (1998b) Insight into consumers' assessments of biscuit texture based on mastication analysis – hardness versus crunchiness. *J. Texture Studies* 29, 481–497.

Brown, W.E., Langley, K.R., Martin, A. and MacFie, H.J.H. (1994a) Characterisation of patterns of chewing behaviour in human subjects and their influence on texture perception. *J. Texture Studies* 25, 455–468.

Brown, W.E., Langley, K.R., Mioche, L., Marie, S., Gérault, S. and Braxton, D. (1996) Individuality of understanding and assessment of sensory attributes of foods, in particular, tenderness of meat. *Food Qual. Prefer.* 7, 205–216.

Brown, W.E., Shearn, M. and MacFie, H.J.H. (1994b) Method to investigate differences in chewing behaviour in humans: II. Use of electromyography during chewing to assess chewing behaviour. *J. Texture Studies* 25, 17–31.

Çakir, E., Koç, H., Vinyard, C.J., Essick, G., Daubert, C.R., Drake, M. and Foegeding, E. (2012) Evaluation of texture changes due to compositional differences using oral processing. *J. Texture Studies* 43, 257–267.

Carson, L., Sun, X., Setser, C. and Peng, Y. (2002) Assessing food firmness using an alectronic sensing system with a model food system. *J. Texture Studies* 33, 389–399.

Chen, J. (2009) Food oral processing – a review. *Food Hydrocolloids* 23, 1–25.

Clave, P., De Kraa, M., Arreola, V., Girvent, M., Farre, R., Palomera, E. and Serra-Prat, M. (2006) The effect of bolus viscosity on swallowing function in neurogenic dysphagia. *Aliment. Pharmacol. Ther.* 24, 1385–1394.

Dan, H., Hayawaka, F. and Kohyama, K. (2008) Modulation of biting procedures induced by the sensory evaluation of cheese hardness with different definitions. *Appetite* 158, 158–166.

Dan, H. and Kohyama, K. (2007) Interactive relationship between the mechanical properties of food and the human response during the first bite. *Archs. Oral Biol.* 52, 455–464.

Dan, H., Okamoto, M., Wada, Y., Dan, I. and Kohyama, K. (2007) First bite for hardness judgment as haptic exploratory procedure. *Physiol. Behav.* 92, 601–610.

Dantas, R.O. and Dodds, W.J. (1990) Effect of bolus volume and consistency on swallow-induced submental and infrahyoid electromyographic activity. *Braz. J. Med. Biol. Res.* 23, 37–44.

Ding, R., Longemann, J.A., Larson, C. and Rademaker, A.W. (2003) The effects of taste and consistency on swallow physiology in younger and older healthy individuals: A surface electromyographic study. *J. Speech Lang. Hear. Res.* 46, 977–989.

Dodds, W.J., Logemann, J.A. and Stewart, E.T. (1990) Radiologic assessment of abnormal oral and pharyngeal phases of swallowing. *Am. J. Roentgenol.* 154, 965–974.

Duizer, L.M., Gullett, E.A. and Findlay, C.J. (1994) The effect of masticatory patterns as measured by time-intensity and electromyography on the perception of bovine muscle tenderness. *J. Sensory Studies* 9, 33–46.

Duizer, L.M., Gullett, E.A. and Findlay, C.J. (1996) The relationship between sensory time-intensity, physiological electromyography and instrumental texture profile analysis. *Meat Sci.* 42, 215–224.

Edgell, A., Boyar, M.M. and Kilcast, D. (1987) EMG – a new concept in texture measurement. *Dairy Ind. Int.* 52(5), 27, 29.

Eves, A. and Kilcast, D. (1987) EMG – a new concept in texture measurement. Part II – Applications to dairy desserts. *Dairy Ind. Int.* 52(11), 23, 25.

Foegeding, E.A., Daubert, C.R., Drake, M.A., Essick, G., Trulsson, M., Vinyard, C.J. and van de Velde, F. (2011) A comprehensive approach to understanding textural properties of semi- and soft-solid foods. *J. Texture Studies* 42, 103–129.

Foster, K., Woda, A. and Peyron, M.A. (2006) Effect of texture of plastic and elastic model foods on the parameters of mastication. *J. Neurophysiol.* 95, 3469–3479.

Funami, T., Ishihara, S., Nakauma, M., Kohyama, K. and Nishinari, K. (2012) Texture design for products using food hydrocolloids. *Food Hydrocolloids* 26, 412–420.

González, R., Montoya, I. and Cárcel, J. (2001) The use of electromyography on food texture assessment. *Food Sci. Technol. Int.* 7, 461–471.

González, R., Sifre, S., Benedito, J. and Nogues, V. (2002) Comparison of electromyographic pattern of sensory experts and untrained subjects during chewing of Mahon cheese, *J. Dairy Res.* 69, 151–161.

Hayakawa, F., Kazami, Y., Fujimoto, S., Kikuchi, H. and Kohyama, K. (2009) Time-Intensity analysis of sourness of commercially produced gummy jellies available in Japan. *Food Sci. Technol. Res.* 15, 75–82.

Heath, M.R. and Prinz, J.F. (1999) Oral processing of foods and the sensory evaluation of texture. In: *Food Texture. Measurement and Perception.* (eds A.J. Rosenthal), pp. 18–29. Aspen, Gaithersburg.

Hiiemae, K. (2004) Mechanisms of food reduction, transport and deglutition: how the texture of food affects feeding behaviour. *J. Texture Studies* 35, 171–200.

Hiiemae, K.M., Heath, M.R., Heath, G., Murray, J., Sapper, D. and Hamblett, K. (1996) Natural bites, food consistency and feeding in man. *Archs. Oral Biol.* 41, 429–441.

Hildebrandt, G.H., Dominguez, B.L., Schork, M.A. and Loesche, W.J. (1997) Functional units, chewing, swallowing, and food avoidance among the elderly. *J. Prosthet. Dent.* 77, 588–595.

Hori, K., Ono, T. and Nokubi, T. (2006) Coordination of tongue pressure and jaw movement in mastication. *J. Dent Res.* 85, 187–191.

Huckabee, M.L., Butler, S.G., Barclay, M. and Jit, S. (2005) Submental surface electromyographic measurement and pharyngeal pressures during normal and effortful swallowing. *Arch. Phys. Med. Rehabil.* 86, 2144–2149.

Hutchings, J.B. and Lillford, P. (1988) The perception of food texture – the philosophy on the breakdown path. *J. Texture Studies* 19, 103–115.

Hutchings, S.C., Foster, K.D., Bronlund J.E., Lentlea, R.G., Jones, J.R. and Morgenstern, M.P. (2011) Mastication of heterogeneous foods: Peanuts inside two different food matrices. *Food Qual. Prefer.* 22, 332–339.

Igarashi, A., Arai, E., Watanabe, R., Miyaoka, Y., Tazawa, T., Hirano, H., Nomura, S. and Yamada, Y. (2002) Comparison of physical properties of agar, low gel strength agar and gelatin, as supplement food for people with swallowing difficulty. *J. Texture Studies* 33, 285–295.

Imai, A., Tanaka, M., Tatsuta, M. and Kawazoe, T. (1995) Ultrasonographic images of tongue movement during mastication. *J. Osaka Dent. Univ.* 29(2), 61–69.

Imai, E. and Sato, S. (2008) Electromyographic measurement for expressing food texture. *J. Home Econ. Jpn.* 59, 955–967.

Inagaki, D., Miyaoka, Y., Ashida, I. and Yamada, Y. (2009) Influence of food properties and body position on swallowing-related muscle activity amplitude. *J. Oral Rehabil.* 36, 176–183.

Ishihara, S., Nakauma, M., Funami, T., Tanaka, T., Nishinari, K. and Kohyama, K. (2011) Electromyography during oral processing in relation to the mechanical and sensory properties of soft gels. *J. Texture Studies* 42, 254–267.

Jack, F.R., Piggott, J.R. and Paterson, A. (1993) Relationships between electromyography, sensory and instrumental measures of cheddar cheese texture. *J. Food Sci.* 58, 1313–1317.

Jafari, S., Prince, R.A., Kim, D.Y. and Paydarfar, D. (2003) Sensory regulation of swallowing and airway protection: A role for the internal superior laryngeal nerve in humans. *J. Physiol.* 550, 287–304.

Kemsley, K., Defernez, M., Sprunt, J.C. and Smith, A.C. (2003) Electromyographic responses to prescribed mastication. *J. Electromyogr. Kinesiol.* 13, 197–207.

Kohyama, K., Hanyu, T., Hayakawa, F. and Sasaki, T. (2010) Electromyographic measurement of eating behaviors for buckwheat noodles. *Biosci. Biotechnol. Biochem.* 74, 56–62.

Kohyama, K., Hatakeyama, E., Sasaki, T., Dan, H., Azuma, T. and Karita, K. (2004) Effects of sample hardness on human chewing force: a model study using silicone rubber. *Archs. Oral Biol.* 49, 805–816.

Kohyama, K., Hatakeyama, E., Dan, H. and Sasaki, T. (2005a) Effects of sample thickness on bite force for raw carrots and fish gels. *J. Texture Studies* 36, 157–173.

Kohyama, K., Hayakawa, F. and Sasaki, T. (2012) Relations between mastication variables, mechanical properties and sensory scores of cheese samples with various textures. In: *Gums and Stabilisers for the Food Industry 16*. (eds P.A. Williams and G.O. Phillips), pp. 331–338, Royal Society of Chemistry, Cambridge.

Kohyama, K. and Mioche, L. (2004) Chewing behavior observed at different stages of mastication for six foods studied by electromyography and jaw kinematics in young and elderly subjects. *J. Texture Studies* 35, 395–414.

Kohyama, K., Mioche, L. and Bourdiol, P. (2003a) Influence of age and dental status on chewing behaviour studied by EMG recordings during consumption of various food samples. *Gerodontology* 20, 15–23.

Kohyama, K., Mioche, L. and Martin, J.F. (2002) Chewing patterns of various texture foods studied by electromyography in young and elderly populations. *J. Texture Studies* 33, 269–283.

Kohyama, K., Nakayama, Y., Sasaki, T. and Hatakeyama, E. (2003b) Mastication quantities for rice or bread in Japanese- and Western-style menus. *J. Jpn. Soc. Masticat. Sci. Health Promotion* (in Japanese) 12, 75–81.

Kohyama, K., Nakayama, Y., Yamaguchi, I., Yamaguchi, M., Hayakawa, F. and Sasaki, T. (2007a) Mastication efforts on block and finely cut foods studied by electromyography. *Food Qual. Prefer.* 18, 313–320.

Kohyama, K., Nakayama, Y., Watanabe, H. and Sasaki, T. (2005b) Electromyography of eating apples: Influences of cooking, cutting, and peeling. *J. Food Sci.* 70, S257–261.

Kohyama, K. and Nishi, M. (1997) Direct measurement of biting pressures for crackers using a multiple-point sheet sensor. *J. Texture Studies* 28, 605–617.

Kohyama, K., Ohtsubo, K., Toyoshima, H. and Shiozawa, K. (1998) Electromyographic study on cooked rice with different amylose contents. *J. Texture Studies* 29, 101–113.

Kohyama, K., Sakai, T. and Azuma, T. (2001a) Patterns observed in the first chew of foods with various textures. *Food Sci. Technol. Res.* 7, 290–296.
Kohyama, K., Sakai, T., Azuma, T., Mizuguchi, T. and Kimura, I. (2001b) Pressure distribution measurement in biting *surimi* gels with molars using a multiple-point sheet sensor. *Biosci. Biotechnol. Biochem.* 65, 2597–2603.
Kohyama, K., Sasaki, T. and Hayakawa, F. (2008) Characterization of food physical properties by the mastication parameters measured by electromyography of the jaw-closing muscles and mandibular kinematics in young adults. *Biosci. Biotechnol. Biochem.* 72, 1690–1695.
Kohyama, K., Sawada, H., Nonaka, M., Kobori, C., Hayakawa, F. and Sasaki, T. (2007b) Textural evaluation of rice cake by chewing and swallowing measurements on human subjects. *Biosci. Biotechnol. Biochem.* 71, 358–365.
Kohyama, K., Yamaguchi, M., Kobori, C., Nakayama, Y., Hayakawa, F. and Sasaki, T. (2005c) Mastication effort estimated by electromyography for cooked rice of differing water content. *Biosci. Biotechnol. Biochem.* 69, 1669–1676.
Kuhl, V., Eicke, B.M., Dieterich, M. and Urban, P.P. (2003) Sonographic analysis of laryngeal elevation during swallowing. *J. Neurology* 250, 333–337.
Kumagai, H., Tashiro, A., Hasegawa, A., Kohyama, K. and Kumagai, H. (2009) Relationship between flow properties of thickener solutions and their velocity through the pharynx measured by the ultrasonic pulse Doppler method. *Food Sci. Technol. Res.* 15, 203–210.
Logemann, J.A., Rademaker, A.W., Roa Pauloski, B., Ohmae, Y. and Kahrilas, P.J. (1998) Noramal swallowing physiology as viewed by videofluoroscopy and videoendoscopy. *Folia Phoniatr. Logop.* 50, 311–319.
Lucas, P.W., Prinz, J.F., Agrawal, K.R. and Bruce, I.C. (2004) Food texture and its effect on ingestion, mastication and swallowing. *J. Texture Studies* 35, 159–170.
Lund, J.P. (1991) Mastication and its control by the brain stem. *Crit. Rev. Oral. Biol. Med.* 2, 33–64.
Maruyama, K., Sato, S., Ohira, T., Maeda, K., Noda, H., Kubota, Y., Nishimura, S., Kitamura, A., Kiyama, M., Okada, T., Imano, H., Nakayama, M., Ishikawa, Y., Kurokawa, M., Sasaki, S. and Iso, H. (2008) The joint impact on being overweight of self reported behaviours of eating quickly and eating until full: cross sectional survey. *Br. Med. J.* 337, 1091–1093.
Matsumi, T., Koshino, H., Hirai, T., Yokoyama, Y. and Ikeda, Y. (2005) Evaluation of swallowing function using ultrasound diagnostic methods. *Prosthodont. Res. Pract.* 4, 1–8.
Matsuo, K., Hiiemae, K.M., Gonzalez-Fernandez, M. and Palmer, J.B. (2008) Respiration during feeding on solid food: alterations in breathing during mastication, pharyngeal bolus aggregation, and swallowing. *J. Appl. Physiol.*, 104, 674–681.
Mioche, L. (2004) Mastication and food texture perception: variation with age. *J. Texture Studies* 35, 145–158.
Mioche, L., Bourdiol, P., Martin, J.F. and Noël, Y. (1999) Variations in human masseter and temporalis muscle activity related to food texture during free and side-imposed mastication. *Archs. Oral Biol.* 44, 1005–1012.
Mioche, L., Bourdiol, P. and Monier, S. (2003) Chewing behavior and bolus formation during mastication of meat with different textures. *Archs. Oral Biol.* 48, 193–200.
Mioche, L., Bourdiol, P., Monier, S., Martin, J.F. and Cormier, D. (2004) Changes in jaw muscles activity with age: effects on food bolus properties. *Physiol. Behav.* 82, 621–627.
Mioche, L., Hiiemae, K.M. and Palmer, J.B. (2002) A postero-anterior videofluorographic study of the intraoral management of food in man. *Archs. Oral Biol.* 47, 267–280.
Miyaoka, Y., Ashida, I., Kawakami, S., Tamaki, Y., and Miyaoka, S. (2010) Activity patterns of the suprahyoid muscles during swallowing of different fluid volumes. *J. Oral Rehabil.* 37, 575–582.
Miyawaki, S., Ohkochi, N., Kawakami, T. and Sugimura, M. (2000) Changes in masticatory muscle activity according to food size in experimental human mastication. *J. Oral Rehabil.* 28, 778–784.
Nakauma, M., Ishihara, S., Funami, T. and Nishinari, K. (2011) Swallowing profiles of food polysaccharide solutions with different flow behaviors. *Food Hydrocolloids* 25, 1165–1173.
Nakazawa, F., Ohno, M., Morita, A. and Takahashi, J. (2000) Ultrasonic measurement of the velocity through the pharynx of swallowed rice boiled with differing water content. *J. Home Econ. Jpn.* (in Japanese) 51, 1067–1071.
Nakazawa, F. and Togashi, M. (2000) Evaluation of food texture by mastication and palatal pressure, jaw movement and electromyography. In: *Hydrocolloids Part 2*. (ed K. Nishinari), pp. 413–454. Elsevier, Amsterdam.
Neyraud, E., Peyron, M.A., Vieira, C. and Dransfield, E. (2005) Influence of bitter taste on mastication pattern. *J. Dent. Res.* 84, 250–254.

Nishinari, K. (2004) Rheology, food texture and mastication. *J. Texture Studies* 35, 113–124.
Nishinari, K. (2006) Polysaccharide rheology and in-mouth perception. In: *Food polysaccharides and their applications*. (eds A.M. Stephen, G.O. Phillips, and P.A. Williams), 2ndedn., pp. 541–588. CRC Press, Boca Raton, FL.
Nishinari, K. (2009) Texture and rheology in food and health. *Food Sci. Technol. Res.* 15, 99–106.
Nishinari, K., Kohyama, K., Kumagai, H., Funami, T. and Bourne, M.C. (2013) Parameters of texture profile analysis. *Food Sci. Technol. Res.* 19, 519–521.
Nouchi, Y., Ajiki, Y., Tobitsuka, K., Sasaki, T. and Kohyama, K. (2012). Bite-speed effects in two-bite texture analysis. *Nippon Shokuhin Kagaku Kogaku Kaishi* (in Japanese) 59, 96–103.
Okadome, H., Toyoshima, H. and Ohtsubo, K. (1999) Multiple measurements of physical properties of individual cooked rice grains with a single apparatus. *Cereal Chem.* 76, 855–860.
Okazawa, M., Konishi, H., Yoshida, T., Katsumata, A., Iida, Y. and Fujishita, M. (2008) Properties of rice porridge (gruel) as a meal for patients with swallowing disorder. *J. Gifu Dent. Soc.* 35, 7–12.
Ono, T., Hori, K., Masuda, Y. and Hayashi, T. (2010) Resent advances in sensing oropharyngeal swallowing function in Japan. *Sensors* 10, 176–202.
Ono, T., Hori, K. and Nokubi, T. (2004) Pattern of tongue pressure on hard palate during swallowing. *Dysphagia* 19, 259–264.
Otsuka, R., Tamakoshi, K., Yatsuya, H., Murata, C., Sekiya, A., Wada, K., Zhang, H. M., Matsushita, K., Sugiura, K., Takefuji, S., Yang, P.O., Nagasawa, N., Kondo, T., Sasaki, S. and Toyoshima, H. (2006) Eating fast leads to obesity: findings on self-administered questionnaires among middle-aged Japanese men and women. *J. Epidemiol.* 16, 117–124.
Palmer, P.M., Luschei, E.S., Jaffe, D. and McCulloch, T.M. (1999) Contributions of individulal muscles to the submental surface electromyogram during swallowing. *J. Speech Lang. Hear. Res.* 42, 1378–1391.
Palmer, P.M., Rudin, N.J., Lara, G. and Crompton, A.W. (1992) Coordination of mastication and swallowing. *Dysphagia* 7, 187–200.
Pancherz, H., Winnberg, A. and Westesson, P.L. (1986) Masticatory muscle activity and hyoid bone behavior during cyclic jaw movements in man. A synchronized electromyographic and videofluorographic study. *Am. J. Orthod.* 89, 122–131.
Peng, C.L., Jost-Brinkmann, P.G.J., Miethke, R.R. and Lin, C.T. (2000) Ultrasonographic measurement of tongue movement during swallowing. *J. Ultrasound Med.* 19, 15–20.
Peng, Y., Sun, X., Carson, L. and Setser, C. (2002) Food hardness and fracturability assessment by an electronic sensing system. *J. Texture Studies* 33, 135–148.
Perlman, A.L., Ettema, S.L. and Barkmeier, J. (2000) Respiratory and acoustic signals associated with bolus passage during swallowing. *Dysphagia* 15, 89–94.
Quinchia, L.A., Valencia, C., Partal, P., Franco, J.M., Brito-de Fuente, E. and Gallegos, C. (2011) Linear and non-linear viscoelasticity of puddings for nutritional management of dysphagia. *Food Hydrocolloids* 25, 586–593.
Reimers-Neils, L., Logemann, J. and Larson, C. (1994) Viscosity effects on EMG activity in normal swallow. *Dysphagia* 9, 101–106.
Saitoh, E., Shibata, S., Matsuo, K., Baba, M., Fujii, W. and Palmer, J.B. (2007) Chewing and food consistency: effects on bolus transport and swallow initiation. *Dysphagia* 22, 100–107.
Salles, C., Chagnon, M.C., Feron, G., Guichard, E., Laboure, H., Morzel, M., Semon, E., Tarrega, A. and Yven, C. (2011) In-mouth mechanisms leading to flavor release and perception. *Crit. Rev. Food Sci. Technol.* 51, 67–90.
Sakamoto, H., Harada, T., Matsukubo, T., Takaesu, Y. and Tazaki, M. (1989) Electromyographic measurement of textural changes of foodstuffs during chewing. *Agric. Biol. Chem.* 53, 2421–2433.
Shawker, T.H., Sonies, B., Stone, M. and Baum, B.J. (1983) Real-time ultrasound visualization of tongue movement during swallowing. *J. Clin. Ultrasound* 11, 485–490.
Shiozawa, K., Kohyama, K. and Yanagisawa, K. (1999a) Influence of ingested food texture on jaw muscle and tongue activity during mastication in humans. *Jpn. J. Oral Biol.* 41, 27–34.
Shiozawa, K., Kohyama, K. and Yanagisawa, K. (1999b) Food bolus texture and tongue activity just before swallowing in human mastication. *Jpn. J. Oral Biol.* 41, 297–302.
Shiozawa, K., Kohyama, K. and Yanagisawa, K. (2003) Relationship between physical properties of a food bolus and initiation of swallowing. *Jpn. J. Oral Biol.* 45, 59–63.
Shiozawa, K., Kohyama, K. and Yanagisawa, K. (2007) Influence of a thickening agent on the swallowing threshold. *J. Jpn. Soc. Masticat. Sci. Health Promotion* (in Japanese) 17, 27–34.
Smit, H.J., Kemsley, E.K., Tapp, H.S. and Henry, C.J.K. (2011) Does prolonged chewing reduce food intake? Fletcherism revisited. *Appetite* 57, 295–298.

Smith, A.C. (2004) Texture and mastication. In: *Texture in Food 2: Solid Foods*. (ed. D. Kilcast), pp. 53–81. Woodhead, Cambridge.

Sodhi, N.S., Suzuki, K. and Kohyama, K. (2010) Electromyographic studies on mastication of cooked rice from different cultivars. *Abstract of International Conference on Food Oral Processing*, p. 120, Leeds.

Sprunt, J.C., Raithatha, C.E. and Smith, A.C. (2002) Swallow indicator methodology as an enhancement to combined time–intensity measurement of flavour release and electromyography for monitoring mastication. *Food Qual. Prefer.* 13, 47–55.

Takahashi, J. and Nakazawa, F. (1991) Palatal pressure patterns of gelatin gels in the mouth. *J. Texture Studies* 22, 1–11.

Takahashi, J. and Nakazawa, F. (1992) Effects of dimensions of agar and gelatin gels on palatal pressure patterns. *J. Texture Studies* 23, 139–152.

Takahashi, T., Tsuge, M. and Ogoshi, H. (2011) Physical properties of semi-liquid food and swallowing behaviors through measurement of electromyography from suprahyoid musculatures during swallowing. *J. Soc. Rheol. Jpn.* 39, 93–98.

Tamburrini, S., Solazzo, A., Sagnelli, A., del Vecchio, L., Reginelli, A., Monsorro, M. and Grassi, R. (2010) Amyotrophic lateral sclerosis: sonographic evaluation of dysphagia. *Radiol. Med.* 115, 784–793.

Taniguchi, H., Tsukada, T., Ootaki, S., Yamada, Y. and Inoue, M. (2008) Correspondence between food consistency and suprahyoid muscle activity, tongue pressure, and bolus transit times during the oropharyngeal phase of swallowing. *J. Appl. Physiol.* 105, 791–799.

Tarrega, A., Yven, C., Sémon, E. and Salles, C. (2008) Aroma release and chewing activity during eating different model cheeses. *Int. Dairy J.* 18, 849–857.

Tashiro, A., Hasegawa, A., Kohyama, K., Kumagai, H. and Kumagai, H. (2010) Relationship between the rheological properties of thickener solutions and their velocity through the pharynx as measured by the ultrasonic pulse Doppler method. *Biosci. Biotechnol. Biochem.* 74, 1598–1605.

Vaiman, M., Eviatar, E. and Segal, S. (2004) Surface electromyographic studies of swallowing in normal subjects: A review of 440 adults. 1. Quantitative data: Timing measures. *Otolaryngol. Head Neck Surg.* 131, 548–555.

van der Bilt, A. (2009) Oral physiology, mastication and food perception. In: *Designing Functional Foods. Measuring and controlling food structure breakdown and nutrient absorption*. (eds. D.J. McClements and E.A. Decker), pp. 3–37. CRC Press, Boston.

Wada, Y., Tsuzuki, D., Kobayashi, N., Hayakawa, F. and Kohyama, K. (2007) Visual illusion in mass estimation of cut food. *Appetite* 49, 183–190.

Wilkinson, C., Dijksterhuis, G.B. and Minekus, M. (2000) From food structure to texture. *Trends Food Sci. Technol.* 11, 442–450.

Wilson, C.E. and Brown, W.E. (1997) Influence of food matrix structure and oral breakdown during mastication on temporal perception of flavour. *J. Texture Studies* 21, 69–86.

Woda, A., Foster, W., Mishellany, A. and Peyron, M.A. (2006) Adaptation of healthy mastication to factors pertaining to the individual or to the food. *Physiol. Behav.* 89, 28–35.

Yamada, Y., Yamamura, K. and Inoue, M. (2005) Coodination of cranial motoneurons during mastication. *Respir. Physiol. Neurobiol.* 147, 177–189.

12 Texture design for creaminess: the role of lubrication

Fred van de Velde, PhD and Els de Hoog, PhD

NIZO Food Research, Kernhemseweg, The Netherlands

12.1 INTRODUCTION

Emulsions and emulsion-based and -containing products represent an important class of food products, ranging from (semi)-solids to liquids. Products included in this class are ice creams, margarines, processed meat, sauces, desserts and liquid dairy drinks. The sensory perception of emulsions and emulsion-based foods is dominated by fat-related attributes, such as fattiness, smoothness, creaminess and velvetiness; creaminess being the most important attribute in this range. Health aspects of food products are one of the major drivers in science and industry. Health concerns related to fat intake, such as cardiovascular diseases and obesities (WHO/FAO 2002), drives the food industries to reduce the caloric density of products and to reduce the amount of saturated fatty acids (SAFAs) and trans fatty acids. Both saturated and trans fatty acids are known and used for their structure building capacity in emulsion-based foods. Therefore, the reduction of those while maintaining a good product quality is a major challenge. The optimal texture design for healthy foods requires a thorough understanding of the mechanical and physical/chemical mechanisms underlying the sensory perception of foods. For decades scientific research has been focused on building relationships between rheological measurements and the sensory perception of emulsion-based products. However, there is more to it than only the rheological behavior. This explains why correlation of rheological data with sensory data on emulsion-based foods is not always straight forward. It depends on the product group what the key parameters are for the underlying mechanisms of mouth feel descriptors. Nowadays the research in understanding mouthfeel attributes is moved more and more into the processes taking place in the mouth. Lubrication and oral coating formation are two areas that have been explored in recent years and resulted in a better understanding of mouthfeel attributes, such as creaminess. In this chapter, we describe recent findings in science that are relevant for the development of improved emulsion-based products. The focus is on lubrication and the technique to measure the lubrication behavior of foods, which is called tribology.

Food Texture Design and Optimization, First Edition. Edited by Yadunandan Lal Dar and Joseph M. Light.

12.2 CREAMINESS: A COMPLEX ATTRIBUTE

Creaminess is generally recognized as a sensory attribute that plays an important role in consumer's acceptance of food products. Creaminess perception has been demonstrated repeatedly to correlate with a positive response of consumers (Frøst and Janhøj, 2007). Often, a positive creaminess perception is connected to high fat level products. Therefore, understanding creaminess is crucial in designing successfully healthier reduced-fat products.

For decades, studies investigated the relations of creaminess with the sensory and structural characteristics of foods (Kokini, 1987; Moore et al., 1998; Kilcast et al., 2002; Akhtar et al., 2005; Jellema et al., 2005; de Wijk et al., 2006c). More recently, research was focused on the interactions between the food and humans, investigating the interaction of food ingredients with the oral cavity and the relation of oral processing with creaminess perception (van Aken et al., 2007).

Creaminess is an integrated and complex attribute, since it is correlated to a set of sensory attributes. Often mentioned in the literature are relations with thickness, smoothness, and slippery, or the absence of grittiness, astringency. These attributes explain a large part of the sensory perception of creaminess. The actual relations were found to be different between product groups.

Relationships between the physical and chemical properties of food products and their ingredients and structure have been studied for many food products (for example Fisher and Windhab, 2011; Tomashunas et al., 2012; Schuh et al., 2012). Rheological properties were correlated to creaminess, ranging from small deformation measurements (dynamic stress and frequency sweep) or large deformation properties (flow curves and steady shear rate behavior) to extensional viscosity. In addition, tribology, thin-film morphology and wetting properties were found to give relevant insight that could not be obtained by bulk rheology alone. For some product groups, the creaminess perception correlated better with thin film tribological properties rather than bulk characteristics. Several groups are working on finding relations between lubrication and sensory perception (Malone et al., 2003; de Wijk and Prinz, 2005; Dresselhuis et al., 2007 and 2008; Chojnicka-Paszun, 2009). For emulsion products in general, parameters playing a role in creaminess perception are fat content, fat globule size, thickeners, air droplet content and size, flavor concentration, and droplet stability during oral processing.

Understanding the structural changes in the food by manipulation in the mouth, increases the understanding of the sensory descriptor creaminess, and therefore will aid food product developers in optimizing the positive perception of food products.

12.2.1 In mouth mechanisms that contribute to creaminess

As discussed in the preceding section, creaminess is a complex attribute that involves different sensory aspects of the product. It is a sensory attribute that is elicited during manipulation of the food in the mouth. The manipulation of food in the mouth is often referred to as oral processing (Chen and Engelen, 2012). The physiological signals that are generated during the oral processing are converted by our brain in the perception of the texture or mouthfeel of the product. In 1988, Hutchings and Lillford described the

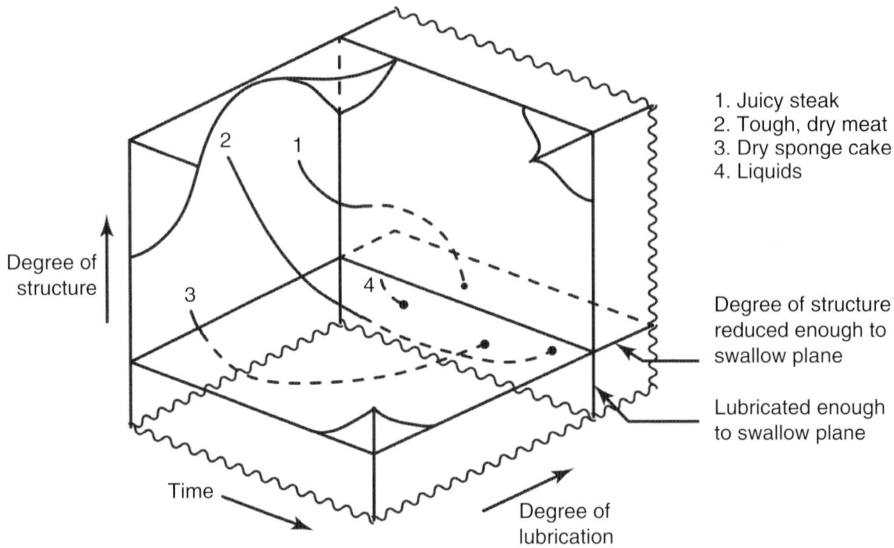

Figure 12.1 Model for in mouth processing of food. (Source: Redrawn from Hutchings and Lillford 1988.)

importance of oral processing. They described the breakdown pathway for different foods during oral processing (Figure 12.1). In the process from transforming the initial food to a bolus that is ready to be swallowed, Hutchings and Lillford indicated the degree of structure and the degree of lubrication as the two important parameters. To be able to be swallowed the bolus should fulfil the requirements of enough reduction in structure and enough lubrication to be swallowed. Tactile sensors determine the particle size reduction during the oral process, but also send information about the mechanical properties of the food product to the brains. This information is processed in the brain to the overall mouthfeel perception.

Although Hutchings and Lillford published their oral processing model already in 1988, the model received only attention in the last decade, starting with a couple of well cited publications on food physics, oral physiology and the sensory perception of texture (Lucas et al., 2002; Szczesniak, 2002). Thereafter, the model has been used to describe the influence of product and oral characteristics on swallowing (Engelen et al., 2005) and to describe the oral processing of dairy-based foods (Foegeding et al., 2010). Also the review on food oral processing in relation to the perception of gels and emulsion-filled gels started off with this oral breakdown model (Foegeding et al., 2011). Hutchings and Lillford illustrated their model with the transition curves of different food products. Figure 12.1 shows four of their curves. Curves 1 and 2 show the difference between a juicy steak and a dry, tough piece of meat. A juicy steak (1) first passes the 'lubricated enough to swallow' plane and only needs further disintegration of the structure before the bolus is ready to be swallowed. The dry, tough piece of meat (2) requires more oral processing to prepare the bolus. After the structure has been reduced, sufficient oral processing is needed to ensure the lubrication of the bolus. A dry sponge cake (3) needs only little effort to reduce the degree of structure, but requires additional oral processing to lubricate the product. Although the resident time of liquids (4) and semi-solids in the oral cavity is in general rather short, the product structure changes tremendously and consumers can distinguish very easily, for example, full fat products from low fat products. Upon oral processing, these foods are mixed with saliva, heated or cooled to body temperature, brought into contact with oral surfaces and sheared between palate and tongue. This shearing of food products or the food bolus between tongue and palate is an important oral movement to judge mouthfeel attributes, such as creaminess. In this process a frictional force is generated between palate and tongue. The liquid or semi-solid food product acts as a lubricant in the contact between tongue and palate. The importance of friction in understanding the perception of for example creaminess has been described (Malone et al., 2003; de Wijk and Prinz, 2005).

The role of oral movement (oral processing) in the perception of semi-solids has been reviewed recently by de Wijk et al. (2011). The impact of specific tongue movements on the sensory properties has been studied by training subjects to apply different tongue movements (de Wijk et al. 2003a). The movements ranged from no movement at all, to movements in which the food was smeared with the tongue over the palate in the shape of an eight and normal movements without specific instructions. By experimentally controlling the oral movements, the authors eliminated the need to actually measure the oral movements. In general, sensory perception became more intense as the complexity of the movement increased. Moreover, certain attributes, such as creaminess, benefited more from the complexity of the oral movements than other attributes, such as thickness. For all attributes the effect of the instructed oral movement was exceeded by the effect of the normal food processing, which demonstrated that the normal or free oral processing is the most efficient way to judge the sensory attributes of semi-solid foods. Analysis of spat out samples showed that the structure of the food is altered by the oral processing depending on the type of oral movements applied (de Wijk et al., 2006a). The relationship between oral processing and the perception of specific attributes is bidirectional: specific textural properties induce specific oral movements and specific oral movements are applied to judge specific sensory attributes. Ultrasonic measurements were applied to demonstrate that different movements were applied by subjects when judging different sensations (de Wijk et al., 2006b). The results of their study demonstrated that oral movements varied significantly with attribute. This verifies the importance of oral processing and oral movements in understanding and controlling specific sensory attributes in (semi)-solid foods.

12.3 PRINCIPLES OF TRIBOLOGY

Tribology is the science of friction, lubrication and wear. In the food context, tribology is interesting in relation to the experience of rubbing the tongue against the palate. This feeling can be very smooth, for example, when a high fat containing sauce is consumed, or very rough, for example, when a low pH dairy drink is taken. This feeling contributes to the overall eating experience, both to the mouthfeel during eating and to the after feel experience after swallowing a food product. As such, friction is an important property to measure, and it was found to give complementary information about the eating experience next to rheological properties.

The friction force between two surfaces in relative motion is defined as the friction coefficient times the normal force. The three parameters that influence the friction are (1) the properties of the surfaces, (2) the properties of the lubricant and (3) the properties of the adhered layers on the surfaces. The surfaces of interest here are the oral surfaces with an emphasis on the tongue and the palate. These biological tissues have their specific physiology. In relation to friction measurements it was found that the elasticity, roughness and hydrophobicity are the key relevant surface properties. In *in-vitro* measurements it is therefore crucial to choose test materials for the surfaces that have these properties in the relevant range compared to the oral surfaces. The lubricant properties, that is, the food properties, are mainly related to the viscosity, the microstructure and ingredients present in the product. The friction can therefore be influenced by changing the product composition and processing. The adhered layers are formed by the mucous layer present on the oral surfaces and the interaction of this layer with ingredients from the food product. It is known that preferential adsorption of ingredients takes place, which influences the friction measured. In particular the food coating that remains in the mouth after swallowing is important for the after feel and can be related to friction.

Friction forces are measured with tribometers. In a tribometer two surfaces are in relative motion with a food product in between the surfaces. As a function of speed of the surfaces and as a function of the load (normal force) on the surfaces, friction forces are measured with a force transducer. The friction force will change with product ingredients and processing. We found for example, that in liquid dairy emulsions an increase in fat content resulted in a decrease in friction force, due to the spreading of the fat droplets on the surfaces (Chojnicka-Paszun et al., 2012; Meyer et al., 2011). This fact can then be used to screen fat mimetics for their functionality to reduce friction in order to restore the sensory perception of low-fat products compared to the full-fat variants.

In tribology, three regimes of lubrication are distinguished, defined by the Stribeck curve (Stribeck, 1901 and 1902; Jacobson, 2003; Dresselhuis et al., 2007). In this curve, the friction coefficient is plotted against the sliding speed, see Figure 12.2. In the

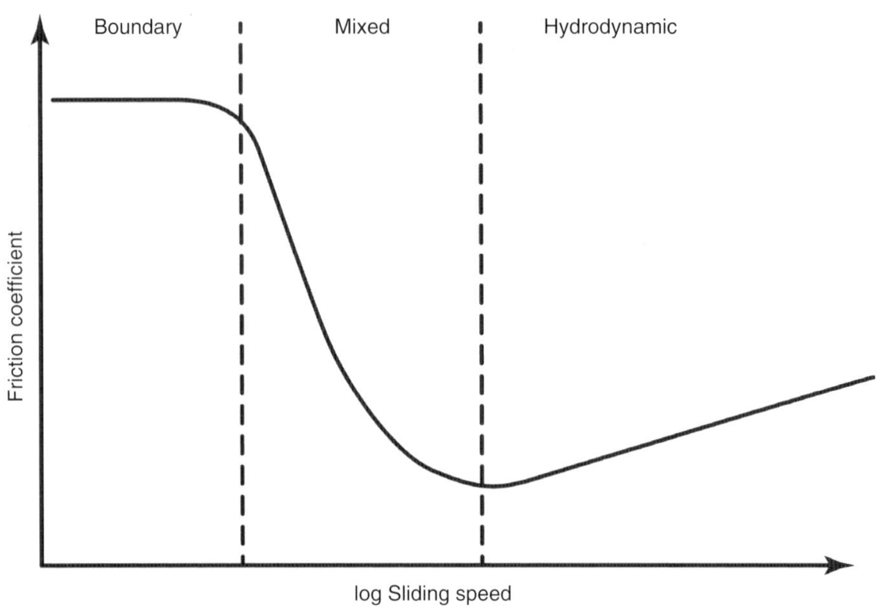

Figure 12.2 Stribeck curve with the three lubrication regimes: boundary, mixed and hydrodynamic regime.

boundary regime, the counter surfaces are in contact with each other. The properties of the surfaces and the adhered layer on these surfaces determine the friction force when the surfaces are in motion. In the hydrodynamic lubrication regime, the counter surfaces are completely separated upon sliding due to build-up of a hydrodynamic pressure as a function of speed. The ability to form a hydrodynamic film depends on the viscosity of the lubricating agents. Here, surface properties are less important. The mixed regime is in between the boundary and the hydrodynamic regime.

The regime most relevant for food manipulation in the mouth is considered to be the boundary and mixed regime. To understand how lubrication by food can influence in-mouth friction, tribometers have been used to measure friction forces between surfaces in relative motion. Tribometers were first used in material science to measure

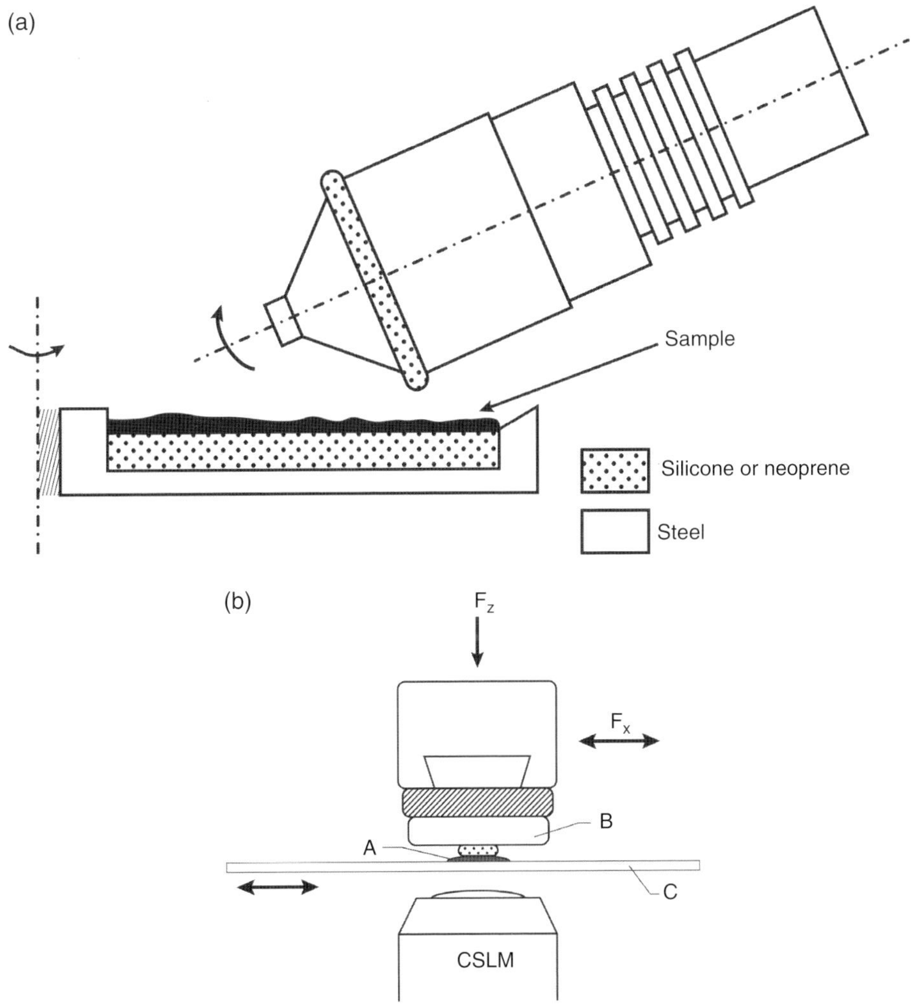

Figure 12.3 Experimental setups for the friction measurements: (a) Ball-on-disk, modified commercial equipment (MTM); (b) Pin-on-disk, home build equipment that can be mounted on top a confocal microscope (OTC).

lubrication properties in the context of avoiding wear. Often, in these cases the hydrodynamic regime is most relevant, thus high speeds and low impact of surface characteristics. To measure under mouth relevant conditions, existing tribometers were adjusted to fit with the requirements for measuring the boundary regimes. The most challenging issue was to mimic the relevant properties of oral surfaces.

Various groups used different models of tribological setups (Figure 12.3). Lee et al. (2002 and 2004) used a pin-on-disk set-up with tetrafluoroethylene and zirconia as counter surfaces. Malone et al. (2003) used a ball-on-disk set-up with a steel ball and silicon rubber disk. Bongaerts et al. (2007) used also a ball-on disk set-up but in combination with PDMS (poly(dimethylsiloxane)) surfaces. Bauer et al. used elastomers to mimic the elasticity of the surfaces. Dresselhuis et al. systematically measured the importance of the elasticity, roughness and hydrophobicity of the surfaces, they used a pig's tongue to validate their findings (de Hoog et al., 2006; Dresselhuis et al., 2007). The latest development on tribological set-ups is an *in-vivo* model (van Aken, 2013). It is based on an acoustic emission measurement ('acoustic tribology') while rubbing the tongue against the palate or against food material directly in the mouth of a subject.

12.4 THE ROLE OF MEASURING LUBRICATION IN UNDERSTANDING CREAMY PERCEPTION

In the preceding sections the role of oral processing in sensory perception and the principles of tribology have been discussed. In this section the role of lubrication in the perception of liquid emulsions is described. Perception of foods in relationship with their lubrication properties has been most intensively studied in liquid emulsions. A liquid oil-in-water (O/W) emulsion is the most general form in which fat is present in food. These emulsions may in turn be contained in protein networks, like in the case of processed meat, sausages and cheeses. These products are generally referred to as emulsion-filled gels and will be discussed separately. Liquid emulsion foods range from true liquids, such as milk, to semi-solids, such as mayonnaise. The latter being one of the first products studied by tribology (Giasson et al., 1997). The study showed a relationship between fat content (full fat, light and fat free mayonnaise) and the measured friction force. The friction force decreased with increasing fat content. Although the work of Giasson et al. (1997) showed a relationship between friction force and fat content, the authors did not connect the measurements with the sensory perception of the mayonnaises. Malone et al. (2003) were the first to relate the lubrication behavior of emulsions as measured with a tribometer to the mouthfeel attributes of the products. At the same time the lubrication and coating behavior of fat were identified as the driving factors for the creamy perception of a starch-based dairy dessert (de Wijk et al., 2003b). Thereafter, tribology and lubrication behavior of emulsions and emulsion-based food was studied intensively to obtain a better understanding of the mouthfeel of those products. Fat or emulsion droplets present in the food product contribute to the lubrication behavior of the foods and thereby an increase in the creaminess perception is observed. Several overview papers describe the role of lubrication in the perception of foods (de Vicente et al., 2006; van Aken et al., 2007; Le Révérend et al., 2010).

All publications so far concluded that the sensory properties of emulsions are not simply related to their viscosity. This is mainly due to the complexity of oral processing, whereby the food product is mixed with saliva and exposed to shear flow and rubbing between the tongue and palate. During this process of rubbing and squeezing the product between tongue and palate, the tongue and palate surfaces are in partial contact (Boundary regime of the Stribeck curve). During this process coalescence of the fat droplets may occur, which has an impact on the sensorial fat perception. In liquid emulsions, it was shown that the perception of fat-related sensory attributes (creamy and fatty) increases with increasing sensitivity of the emulsion towards coalescence. The sensitivity towards in-mouth coalescence was controlled by the amount and type of emulsifier (Dresselhuis et al., 2008). The instability of the emulsion during oral processing was increased by using a lower concentration of emulsifier or by using an amylase-sensitive modified starch as the emulsifier. The instability of the liquid emulsions was determined by microscopic imaging of spat out samples of the emulsions after oral processing. The authors showed that the increased sensitivity towards in-mouth coalescence resulted in higher sensory scores for fat-related attributes, such as creaminess, and lower scores for attributes related to oral friction, such as rough mouthfeel. The mechanism behind this phenomenon is the fact that instable emulsions adhere and spread more easily on a hydrophobic surface like the tongue surface and thereby acts as a lubricant that lowers the friction between palate and tongue. Figure 12.4 shows the friction curves of a stable and an instable emulsion. The friction determined for the instable emulsion is the lowest showing a better lubrication behavior. The impact of partial coalescence on the sensory properties of liquid emulsions was also studied for emulsions prepared with a series of different fats and oils that differ in their solid fat content (Benjamins et al., 2009). By selecting fats with different melting curves, the authors developed a set of

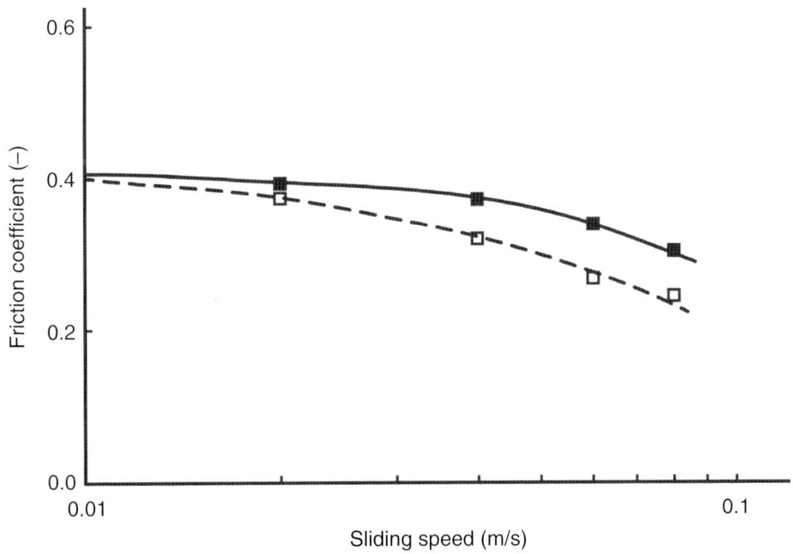

Figure 12.4 Friction coefficient measured at different speed for a stable emulsion (closed symbols and solid line) and an instable emulsion (open symbols and dashed line).

emulsions with varying sensitivity to partial coalescence. When fat crystals formed inside the droplets are able to enter or penetrate the adsorbed layer at the surface of another droplet emulsion droplets can partially coalesce. The authors used sensory panels to demonstrate that an increase in partial coalescence in the mouth resulted in an increase in most fat-related sensory attribute scorings, such as creaminess, fattiness and after feel coating.

Emulsion-filled gels represent a large group of food products, including for example dairy desserts, cheese and processed meat. In contrast to liquid emulsions, emulsion-filled gels need to break down before their lubrication properties can be investigated. The sensory properties of emulsion-filled gels are controlled by three variables: the properties of the gelled matrix, the concentration (volume fraction) oil or fat, and the interactions between the emulsion droplets and the gelled matrix. The latter variable is of crucial importance as it affects how the presence of emulsion droplets contributes to the overall properties of the product. Emulsion droplets can either be bound to the matrix or unbound. Bound emulsion droplets interact with the matrix (also called active fillers) and generally reinforce the matrix. Unbound droplets do not interact with the matrix (also called inactive fillers) and generally weaken the matrix. The interaction of the emulsion droplets with the matrix is controlled by their surface properties, such as their charge. Proteins are often applied in food products as they have excellent emulsifying properties. Moreover, the pH of the solution determines the net charge of a protein. Thus the interaction between a protein-stabilized emulsion droplet and the food matrix can be varied by changing the pH of the product. On the other hand, non-charged emulsifiers give the opportunity to create emulsion-filled gels with unbound droplets. In general products with unbound emulsion droplets are perceived as creamy. In model systems the sensory scores for creaminess were related to the liberation of emulsion droplets during oral processing. Unbound droplets were liberated from all gel matrices studied, whereas for gels that melt during oral processing also bound emulsion droplets were liberated (Sala et al., 2008). The authors hypothesized that for emulsion-filled gels the perception of mouthfeel attributes related to the presence of oil droplets could be mediated by the release of the droplets from the gel matrix upon oral processing. However, the release of the oil droplets is not the only mechanism explaining the perception of creaminess and fattiness in emulsion-filled gels. Therefore, the authors conducted and reported an additional study on the lubrication properties of sheared emulsion-filled gels (Chojnicka et al., 2009). Shearing the emulsion-filled gels was applied as an experimental procedure representing the breakdown of food during oral processing. To mimic the oral processing further, the sheared gels were mixed with saliva. To a large extent the lubrication behavior of the emulsion-filled gels was determined by the matrix and the breakdown behavior of the gel matrix. The lubrication properties of the matrix were remarkably altered by the inclusion of oil droplets in the gel matrix. And again the interactions between oil droplets and gel matrix controlled the impact of the emulsion droplets. For bound droplets the lubrication behavior of the sheared emulsion-filled gels related to the 'apparent viscosity' of the sheared gel mass. For unbound droplets the increase in lubrication with increasing oil content is likely to be the result of the individual contribution of gel matrix and released oil droplets. Despite of the two mechanisms affecting the lubrication behavior of sheared emulsion-filled gels, the overall effect of the oil droplets on lubrication does not differ significantly between gels with bound droplets and gels with unbound droplets.

Unfortunately studies on the relationship between lubrication of foods and their sensory properties are limited to liquid emulsions and emulsion-filled gels as described above. Some additional studies on model systems consisting of polysaccharides (Malone et al., 2003; Zinoviadou et al., 2008) and protein aggregate dispersions (Chojnicka et al., 2008) has been published. Surface friction has been studied by Chen et al. (2004) for whey protein isolate gels. In our view, tribology is an important method to measure and quantify the mouthfeel of various foods, including both liquids and semi-solids.

12.5 ENGINEERING TEXTURE FOR CREAMINESS PERCEPTION

Learnings on the oral processing, the impact on the structural changes and the effect on the sensory perception are used to design and optimize food textures. For food developers, ingredient formulation and processing conditions are the tools or the parameters that can be tuned, given the conditions in the oral cavity. The challenge is to engineer the products in such a way to have the largest impact on creaminess.

Pyett and Watsuk (2009) showed that the principle of destabilizing emulsion droplets in the mouth to increase creaminess perception is a valid mechanism in a salad dressing. In this product, an octenyl succinic anhydride (OSA)-modified starch was used as the emulsifier. During oral processing, the OSA-starch breaks down due to the amylase activity in saliva. The result is the destabilization of the emulsion droplets, and coalescence into bigger droplets, and possibly also coalescence on the oral surface stake place (Dresselhuis et al., 2008). To destabilize the emulsion droplets in the mouth while keeping the stability during storage on the shelf was also achieved in model emulsion systems by Dresselhuis et al. (2008). The destabilization effect was achieved by using OSA-starch or having an incomplete coverage of the emulsion droplets with milk proteins. This research showed that the mechanism indeed has a positive influence on creaminess perception. But in dairy products destabilization is more difficult to achieve, since dairy products contain high concentrations of proteins which can act as emulsifiers and increase the stability of the emulsion droplets.

Lowering the friction force was often found to have a positive effect on the creaminess perception. This implies that measuring relevant friction forces can be an aid in product development for fast screening of potential ingredients. The effect of adding inulin to skimmed milk on the tribology and sensory profiles was investigated by Meyer et al. (2011). Inulin is a carbohydrate and is defined as a soluble fiber. Meyer et al. demonstrated that the addition of a relatively low amount of inulin to skimmed milk can lower the friction force measured to values in the range of those of full-fat milk. However, correlation analysis of the tribology and the sensory data did not reveal a significant relation. This implies that the creaminess perception for these specific products is not fully restored by only restoring the lubrication properties. In addition, parameters, such as the aroma release mechanism, should be optimized further.

Chojnicka-Paszun et al. (2012) investigated the relevance of tribology in food textures and perception as a function of the fat content. It appeared that below 1% of fat in milk samples cannot be distinguished. Above 1% fat content strong correlation is found between a number of attributes and the friction coefficient.

Often, optimization of creaminess is connected to reduced-fat products. The multiple functions of fat in food products need to be restored. Bulk viscosity is one of them, and especially in products where thickness is relatively more important, using thickeners is successful in optimizing creaminess. However, in products where lubrication is more important, like in thinner, watery products, friction issues need to be repaired. In addition, the flavor balance is disturbed when the ratio between water and oil is changing due to the partition coefficients of the aromas. The approach to find solutions for restoring creaminess should therefore be a multidisciplinary approach.

12.6 CONCLUSIONS AND OUTLOOK

Studies describing the relationship between lubrication of foods and their sensory properties are limited to liquid emulsions and emulsion-filled gels. Tribology clearly is complementary to rheology in understanding the physical/chemical origins of food perception. In our view, tribology is an important method to measure and quantify the mouthfeel of various foods, including both liquids and semi-solids. In addition to tribology, the approach to finding solutions for restoring creaminess in low fat products should therefore be a multidisciplinary approach, including aspects such as aroma release.

References

Akhtar, M., Stenzel, J., Murray, B.S. and Dickinson, E. (2005) Factors affecting the perception of creaminess of oil-in-water emulsions. *Food Hydrocolloids* 19, 521–526.

Benjamins, J., Vingerhoeds, M.H., Zoet, F.D., de Hoog, E.H.A. and van Aken, G.A. (2009) Partial coalescence as a tool to control sensory perception of emulsions. *Food Hydrocolloids* 23, 102–115.

Bongaerts, J.H.H., Fourtouni, K., Stokes, J.R. and Jin, Z.-M. (2007) Soft-tribology: lubrication in a compliant pdms-pdms contact. *Tribology International* 40, 1531–1542.

Chen, J. and Engelen, L. (eds) (2012) *Food oral processing – Fundamentals of eating and sensory perception*. Wiley-Blackwell, Oxford.

Chen, J.S., Moschakis, T. and Nelson, P. (2004) Application of surface friction measurements for surface characterization of heat-set whey protein gels. *Journal of Texture Studies* 35, 493–510.

Chojnicka, A., de Jong, S., de Kruif, C.G. and Visschers, R.W. (2008) Lubrication properties of protein aggregate dispersions in a soft contact. *Journal of Agricultural and Food Chemistry* 56, 1274–1282.

Chojnicka, A., Sala, G., de Kruif, C.G. and van de Velde, F. (2009) The interactions between oil droplets and gel matrix affect the lubrication properties of sheared emulsion-filled gels. *Food Hydrocolloids* 23, 1038–1046.

Chojnicka-Paszun, A. (2009) Tribology of food dispersions and their sensory attributes (thesis). Utrecht University, the Netherlands.

Chojnicka-Paszun, A., de Jongh, H.H.J. and de Kruif, C.G. (2012) Sensory perception and lubrication properties of milk: Influence of fat content. *International Dairy Journal* 26, 15–22.

Fischer, P. and Windhab, E.J. (2011) Rheology of food materials. *Current Opinion in Colloid & Interface Science* 16, 36–40.

de Vicente, J., Stokes, J.R. and Spikes, H.A. (2006) Soft lubrication of model hydrocolloids. *Food Hydrocolloids* 20, 483–491.

de Hoog, E.H.A., Prinz, J.F., Huntjens, L., Dresselhuis, D.M. and van Aken, G.A. (2006) Lubrication of oral surfaces by food emulsions: the importance of surface characteristics. *Journal of Food Science* 71, E337–E341.

de Wijk, R.A., Prinz, J.F. and Engelen, L. (2003a) The role of intra-oral manipulation in the perception of sensory attributes. *Appetite* 40, 1–7.

de Wijk, R.A., van Gemert, L.J., Terpstra, M.E.J. and Wilkinson, C.L. (2003b) Texture of semi-solids; sensory and instrumental measurements on vanilla custard desserts. *Food Quality and Preference* 14, 305–317.

de Wijk, R.A. and Prinz, J.F. (2005) The role of friction in perceived oral texture. *Food Quality and Preference* 16, 121–129.

de Wijk, R.A., Prinz, J.F. and Janssen, M.A. (2006a) Explaining perceived oral texture of starch-based custard desserts from standard and novel instrumental tests. *Food Hydrocolloids* 20, 24–34.

de Wijk, R.A., Wulfert, F. and Prinz, J.F. (2006b) Oral processing assessed by M-Mode ultrasound imaging varies with food attribute. *Physiology & Behavior* 89, 15–21.

deWijk, R.A., Terpstra, M.E.J., Janssen, A.M. and Prinz, J.F. (2006C) Perceived creaminess of semi-solid foods. *Trends in Food Science & Technology* 17, 412–422.

de Wijk, R.A., Janssen, A.M. and Prinz, J.F. (2011) Oral movements and the perception of semi-solid foods. *Physiology & Behavior* 104, 423–428.

Dresselhuis, D.M., Klok, H.J., Cohen Stuart, M.A., de Vries, R.J., van Aken, G.A. and de Hoog, E.H.A. (2007) Tribology of o/w emulsions under mouth-like conditions: determinants of friction. *Food Biophysics* 2, 158–171.

Dresselhuis, D.M., de Hoog, E.H.A., Cohen Stuart, M.A., Vingerhoeds, M.H., and van Aken, G.A. (2008) The occurrence of in-mouth coalescence of emulsion droplets in relation to perception of fat. *Food Hydrocolloids* 22, 1170–1183.

Engelen, L., Fontijn-Tekamp, A. and van der Bilt, A. (2005) The influence of product and oral characteristics on swallowing. *Archives of Oral Biology* 50, 739–746.

Foegeding, E.A., Çakır, E. and Koç, H. (2010) Using dairy ingredients to alter texture of foods: Implications based on oral processing considerations. *International Dairy Journal* 20, 562–570.

Foegeding, E.A., Daubert, C.R., Drake, M.A., Essick, G., Trulsson, M., Vinyard, C. and van de Velde, F. (2011) A comprehensive approach to understanding textural properties of semi- and soft-solid foods. *Journal of Texture Studies* 42, 103–129.

Frøst, M.B. and Janhøj, T. (2007) Review: Understanding creaminess. *International Dairy Journal* 17, 1298–1311.

Giasson, S., Israelachvili, J. and Yoshizawa, H. (1997) Thin film morphology and tribology study of mayonnaise. *Journal of Food Science* 62, 640–652.

Hutchings, J.B. and Lillford, P.J. (1988) The perception of food texture – the philosophy of the breakdown path. *Journal of Texture Studies* 19, 103–115.

Jacobson, B. (2003) The Stribeck memorial lecture. *Tribology International* 36, 781–789.

Jellema, R.H., Janssen, A.M., Terpstra, M.E.J., de Wijk, R.A. and Smilde, A.K. (2005) Relating the sensory sensation 'creamy mouthfeel' in custards to rheological measurements. *Journal of Chemometrics* 19, 191–200.

Kilcast, D. and Clegg, S. (2002) Sensory perception of creaminess and its relationship with food structure. *Food Quality and Preference* 13, 609–623.

Kokini, J.L. (1987) The physical basis of liquid food texture and texture-taste interactions. *Journal of Food Engineering* 6, 51–81.

Lee, S., Heuberger, M., Rousset, P. and Spencer, N.D. (2002) Chocolate at a sliding interface. *Journal of Food Science* 67, 2712–2717.

Lee, S., Heuberger, M., Rousset, P. and Spencer, N.D. (2004) A tribological model for chocolate in the mouth: General implications for slurry-lubricated hard/soft sliding counterfaces. *Tribology Letters* 16, 239–249.

Le Révérend, B.J.D., Norton, I.T., Cox, P.W. and Spyropoulos, F. (2010) Colloidal aspects of eating. *Current Opinion in Colloid & Interface Science* 15, 84–89.

Lucas, P.W., Prinz, J.F., Agrawal, K.R. and Bruce, I.C. (2002) Food physics and oral physiology. *Food Quality and Preference* 13, 203–213.

Malone, M.E., Appelqvist, I.A.M. and Norton, I.T. (2003) Oral behaviour of food hydrocolloids and emulsions. Part 1. Lubrication and deposition considerations. *Food Hydrocolloids* 17, 763–773.

Meyer, D., Vermulst, J., Tromp, R.H. and de Hoog, E.H.A. (2011) The effect of inulin on tribology and sensory profiles of skimmed milk. *Journal of Texture Studies* 42, 387–393.

Moore, P.B., Langley, K., Wilde, P.J., Fillery-Travis, A. and Mela, D.J. (1998) Effect of emulsifier type on sensory properties of oil-in-water emulsions. *Journal of the Science of Food and Agriculture* 76, 469–476.

Pyett, S.P. and Watsuk, K. (2009) Weinig vet en toch romig. *VMT* 2009, issue 7, 22–24.

Sala, G., de Wijk, R.A., van de Velde, F. and van Aken, G.A. (2008) Matrix properties affect the sensory perception of emulsion-filled gels. *Food Hydrocolloids* 22, 353–363.

Schuh, V., Allard, K., Herrmann, K., Gibis, M., Kohlus, R. and Weiss, J. (2012) Impact of carboxymethyl cellulose (CMC) and microcrystalline cellulose (MCC) on functional characteristics of emulsified sausages. *Meat Science* 93, 240–247.

Stribeck, R. (1901) Kugellager für beliebige Belastungen. *Zeitschrift des Vereines deutscher Ingenieure* 45, 73–79 (part I) & 118–125 (part II).

Stribeck, R. (1902) Die wesentlischen Eigenschaften der Gleit- und Rollenlager. *Zeitschrift des Vereines deutscher Ingenieure* 46, 1341–1348 (part I); 1432–1438 (part II) & 1463–1470 (part III).

Szczesniak, A.S. (2002) Texture is a sensory property. *Food Quality and Preference* 13, 215–225.

Tomaschunas, M., Hinrichs, J., Köhn, E. and Busch-Stockfisch, M. (2012) Effects of casein-to-whey protein ratio, fat and protein content on sensory properties of stirred yoghurt. *International Dairy Journal* 26, 31–35.

van Aken, G.A., Vingerhoeds, M.H. and de Hoog, E.H.A. (2007) Food colloids under oral conditions. *Current Opinion in Colloid & Interface Science* 12, 251–262.

van Aken, G.A. (2013) Acoustic emission measurement of rubbing and tapping contacts of skin and tongue surfaces in relation to tactile perception. *Food hydrocolloids* 31, 325–331.

WHO/FAO (2002) Joint WHO/FAO Expert consultation on diet, nutrition and the prevention of chronic diseases 2002, Geneva Switzerland, WHO technical report series 916.

Zinoviadou, K.G., Janssen, A.M. and De Jongh, H.H.J. (2008) Tribological properties of neutral polysaccharide solutions under simulated oral conditions. *Journal of Food Science* 73, E88–E94.

13 Descriptive analysis of food texture: advances in the sensory characterization of food textures

Gail Vance Civille and Joanne Seltsam

Sensory Spectrum, Inc., New Providence, New Jersey, USA

13.1 INTRODUCTION

In the past decade, the characterization of food texture has evolved. The development of the Texture Profile Method at General Foods in the early 1960s integrated the texture evaluation of all foods by both instruments and sensory descriptive panels (Szczesniak, 1963; Szczesniak, Brandt and Friedman, 1963; Szczesniak, Loew and Skinner, 1975). The further development of the Spectrum Descriptive Analysis Method enabled expansion of the Texture Profile Method to include more attributes with related references and an expanded rating scale. The need to better understand consumer liking and description of food textures coupled with shortened product development timelines has pushed the sensory field to more accurately capture textural sensory experience. By doing so, sensory and product development teams can work quickly to uncover key attributes that contribute to consumer liking and create products that tap into these needs with higher accuracy and efficiency.

While texture is an obvious part of the sensory experience, in recent years its deeper importance has only begun to be recognized. Flavor is most often believed to be the focus of consumer liking and purchase decisions. However, it may well be the textural experience of certain foods that consumers unconsciously crave. Food companies that understand these unspoken needs will have a competitive edge in creating texturally "addictive" new products for their consumers.

13.2 COMPONENTS OF TEXTURE DESCRIPTIVE ANALYSIS

Descriptive analysis approaches textural description in a structured and linear manner. Because there are stages to how a food is consumed (e.g. first bite/ chew, chewdown, swallow, etc.) this approach is appropriate. The different stages that are included in texture evaluations are shown in Table 13.1.

During each stage of consumption, many textural attributes are commonly evaluated (see Table 13.2).

Food Texture Design and Optimization, First Edition. Edited by Yadunandan Lal Dar and Joseph M. Light.

Table 13.1 Stages of texture in descriptive analysis.

Solid Foods	Semi-Solid Foods	Beverages
• Surface • First Bite/ First Chew • Chewdown • Residual	• Surface • First Compression • Manipulation • Residual	• In Mouth–Immediate • Manipulation • Residual

Table 13.2 Common texture attributes.

Solid Foods	Semi-Solid Foods	Beverages
• Surface ◦ Oiliness ◦ Wetness ◦ Roughness • First Compression ◦ Springiness • First Bite/Chew ◦ Hardness ◦ Cohesiveness ◦ Denseness ◦ Fracturability/Crispness/ Crunchiness ◦ Uniformity of Bite/Chew • Chewdown ◦ Juiciness ◦ Moisture Absorption ◦ Dissolvability ◦ Cohesiveness of Mass ◦ Moistness of Mass ◦ Roughness of Mass ◦ Adhesiveness to Palate ◦ Fibrous Between Teeth ◦ Persistence of Crisp/Crunch • Residual ◦ Mouthcoat ◦ Loose Particles ◦ Toothpack ◦ Toothstick	• Surface ◦ Slipperiness • First Compression ◦ Firmness ◦ Semi-Solid Cohesiveness ◦ Semi-Solid Denseness • Manipulation ◦ Mixes with Saliva ◦ Rate of Breakdown ◦ Roughness of Mass ◦ Cohesiveness of Mass ◦ Adhesiveness • Residual ◦ Mouthcoat	• First Sip ◦ Viscosity ◦ Shape ◦ Heaviness ◦ Carbonation • Manipulation ◦ Mixes with Saliva ◦ Slipperiness ◦ Cohesiveness ◦ Rate of Breakdown ◦ Amount of Particulates • Residual ◦ Mouthdrying ◦ Mouthcoat

This list captures common attributes but can never be considered complete. As long as product developers create unique sensory experiences for consumers, texture will continually evolve. Viewing texture descriptive analysis as dynamic will ensure that sensory scientists continue to explore and develop new attributes and techniques to accurately describe those experiences.

13.2.1 Creating a new attribute and scale: the truth is in the sample

When a new attribute is discovered during tasting, it is helpful to carefully define that attribute and even more advantageous to create a scale demonstrating that attribute at different intensities. The following outlines the aspects most important to adequately defining the new attribute.

Attribute Name – The attribute must be discrete. In other words, it must describe only one characteristic of the sample's texture. An attribute that describes two or more characteristics is considered an 'integrated' term and makes for difficult and inaccurate interpretation. The term should be as descriptive of the attribute as possible. It should not include consumer or hedonic terms. Many consumer based texture terms are integrated terms. That is they are a combination of different discrete attributes. Terms, such as 'creamy' 'rich' and 'tender/tough' are not simple discrete attributes. Creamy includes the texture terms smooth (not gritty or grainy or particulate), thick, and fatty/oily. Creamy may also include flavor terms, such as milky, dairy, buttery and/or vanilla. The use of integrated terms in descriptive analysis will ultimately lead to confusion in guiding product development effort and when trying to interpret consumer response.

Definition – The attribute definition will explain the attribute to the rest of the descriptive and non-descriptive world. It can be somewhat technical in language but should be understandable by people outside of the sensory field. It should provide a clear and concise definition, easily understood by the descriptive panel as well as those who will use the data generated by the panel. All efforts should be concentrated on reducing the risk of misinterpretation.

Technique – Possibly the most difficult of the three, the technique will dictate how panelists will evaluate the product for the attribute. Some considerations that should be included when developing the technique are:

- **When** is the attribute evaluated (e.g. first chew)?
- **How much** sample should be used for the attribute's evaluation?
- **At exactly what point** in the first chew/chewdown/residual is the attribute evaluated?
- What **additional techniques** are used to capture the attribute's intensity (e.g. move tongue around the outside of the mass)?
- Does the technique adequately capture **how consumers typically consume** the product? When possible, this should be taken into consideration, especially if the data will be used to interpret consumer response.

More information about new methods in developing and refining texture evaluation techniques is provided in Section 13.3.1 (New Views on Technique).

Scale – All scales, regardless of the attribute, range from 'none' to 'extremely high' for the attribute. This should be kept in mind when creating a new attribute. The low end of the scale will always be anchored by zero, not 'low.' The absence of a characteristic is an important part of the description of a product. For example, if the Crispness scale's lowest point was 'low' or '1.0,' how would panelists indicate when a potato chip has gone stale to the point of becoming limp? Surely understanding that the potato chips no longer have Crispness is vital to understanding the product's quality.

References – Many texture attributes have scales developed to aid in the teaching of the attribute. In other words, specific products are referenced within the scale to demonstrate a specific intensity. While this is necessary for training new descriptive panelists, and certainly helpful in clearly defining your scale, it is by no means necessary when a new term is created. Understanding the low and high ends of the scale for any attribute should allow panelists to accurately rate the intensity of that attribute. The less training a panel has, the more references will be necessary to ensure that the attribute and scale are being used correctly. References should be products or materials that are consistent, easy to obtain, easy to prepare, and easy to keep for up to one week.

13.2.2 Creating a new attribute and scale: a practical example

When considering a new food product for texture descriptive analysis, understanding how the product is typically consumed will lead to what techniques and attributes will best document its textural properties. It is absolutely crucial to have the product(s) available when developing a lexicon. A rough draft lexicon may be created based on prior experience with the product. However, the only way to accurately capture all aspects of a product's texture is to build the lexicon while experiencing the product itself. If multiple products are to be evaluated, all products should be included in the lexicon development. The panel needs to review 6–12 products in the broad category to be studied. The products need to be prepared according to normal usage directions and the panelists need to evaluate all of the products for each textural stage – surface, first bite or chew, chewdown, and residual. Each panelist records his or her terms for each stage and the panel then discusses the appropriate procedure for testing, the resulting attributes, and the clear definition for each attribute.

While many texture attributes are already defined for food products, they are not always appropriate for a given sample. Additionally, with advances in food product texture, it is sometimes necessary to develop new attributes to adequately capture unique characteristics. The provided list of textural attributes (Table 13.2) is by no means exhaustive and should be used as a guide for developing an appropriate lexicon for each product's unique textural characteristics. If a unique texture is discovered, choosing the correct attribute name, definition, and technique are all equally important.

Upon reviewing samples of a new type of semi-solid product, it was discovered that a key differentiating textural characteristic was the amount of gritty particulates perceived within the matrix. The size of the particles was not different amongst samples.

Step 1: Agree on Attribute Name, Definition, and Technique
Because the panel did not have a term to describe this difference, it was necessary to create a new attribute, definition, and technique. The following was agreed upon:

Attribute Name: Amount of gritty particles in the matrix
Definition: The amount of gritty particles (small abrasive/sharp-edged particles) perceived in the matrix during manipulation.
Technique: Place approximately 1/2 teaspoon on tongue and manipulate between tongue and palate several times; use tongue to feel number of gritty particles between tongue and palate.

Step 2: Develop a Scale
To best train the panel to accurately rate this attribute, the panel leader decided to develop references for this new attribute. In other words, this attribute was fully developed into a new scale.

Because all scales range from 0.0 (none) to 15.0+ (very high), it was necessary to find or create samples that demonstrated different levels on the scale. As there were not enough commercial products that demonstrated this attribute, the reference samples had to be created.

The base matrix decided upon was store bought vanilla pudding that was stabilized enough not to separate upon standing. Corn meal was used to create the 'gritty particles.' This was a good choice as it did not change in textural character after being incorporated into the pudding. Other grains were found to hydrate in the matrix and change in character from 'gritty' to 'grainy' and often became larger in size the longer they stayed in the matrix.

For initial development, one cup of pudding was used as the base. To each separate cup of pudding, a different amount of corn meal was added (e.g. 1 teaspoon, 1 tablespoon, ¼ cup, etc.). Once the corn meal was fully mixed into the matrix, the panel was instructed to evaluate each of the samples and come to consensus on the rating for the new attribute. Because the panel understood that the scale would range from 0.0 (none) to 15.0 (very high), they were able to use basic ratios to rate each sample for the attribute. In other words, a sample with no gritty particles would be rated a 0.0. A sample where no matrix could be perceived amongst the gritty particles would be rated a 15.0. A sample where 50% of the sample was gritty particles and 50% was matrix would rate a 7.5, and so on. Using the initial sample ratings as guidelines, additional samples were created and evaluated, and several references were agreed upon by the panel.

Step 3: Develop a Protocol

It is necessary to capture the information developed in a written protocol so that the scale references can be accurately duplicated in the future. The protocol will serve as a guide for whoever prepares the exercise and will ensure that the references will be exactly as originally established. Include in this protocol an overview of the samples to be prepared with their scale numbers, detailed instructions for sample preparation and storage, a complete shopping list and serving instructions.

Place the attribute name (e.g. Amount of Gritty Particles in the Matrix) in the header of a word document. On the first page list the reference samples to be shown with their corresponding rating on the 15 point scale. Indicate the type of container needed for serving. (Table 13.3)

Texture Exercise
Amount of Gritty Particles in the Matrix

Next, include in the protocol the exact preparation method for the samples. Indicate how far in advance they can be prepared and the proper storage of the samples (Table 13.4).

Table 13.3 Protocol overview (See Appendix 1).

Reference Samples	Scale Value Code (Label on cup)	Container
Brand X pre-made refrigerated vanilla pudding with 1 tsp. Brand X corn meal	2.0	approx. 3 oz. sample cup with lid
Brand X pre-made refrigerated vanilla pudding with 1 Tbsp. Brand X corn meal	5.0	approx. 3 oz. sample cup with lid

Table 13.4 Preparation (See Appendix 2).

CODE 2.0	**Brand X pre-made refrigerated vanilla pudding with 1 tsp. Brand X corn meal** For 1 cup of pudding add 1 teaspoon corn meal and mix thoroughly. Provide 2 tablespoons of sample per participant in coded, lidded sample cups. This may be prepared 24 hours in advance and refrigerated.
CODE 5.0	**Brand X pre-made refrigerated vanilla pudding with 1 Tbsp. Brand X corn meal** For 1 cup of pudding add 1 tablespoon corn meal and mix thoroughly. Provide 2 tablespoons of sample per participant in coded, lidded sample cups. This may be prepared 24 hours in advance and refrigerated.

Table 13.5 Shopping list.

Brand X Refrigerated vanilla pudding
Brand X Corn Meal
General tasting supplies:
3 oz. sample cups with lids (approx. size)
Opaque cups (Styrofoam or coated paper) with lids (spit cups)
Paper napkins
Plastic spoons
No/low flavored water
Water cups

Serving Instructions (See Appendix 3)

Bring samples to room temperature for serving. Provide one sample per participant. Include at each station, a spit cup/lid, rinse water/cup, crackers, napkins, spoons (Table 13.5).

File the word document so that it can be easily retrieved for future use.

Keep in mind that protocols need to be reviewed periodically to determine that they continue to be accurate. A change in the products used may necessitate updating the numbers or identifying product substitutions.

13.3 NEW THINKING ON SPECTRUM TEXTURE DESCRIPTIVE ANALYSIS

13.3.1 New views on technique

Older texture techniques are very specific in the size of the sample used for evaluation and the number of chews completed before assigning an intensity (for certain attributes). Unfortunately, controlling these variables may lead to wide ranges in panelist response due to differences in physiology (e.g. mouth size, salivation rate, body temperature, etc.). Understanding panelist differences and incorporating this 'noise' into the technique will lead to more consistent ratings and ultimately a more accurate texture profile. The attributes these differences affect most are the chewdown (solid food) or manipulation (semi-solid food) attributes.

For example, in the past, panelists were trained to take a ½ "bite of the sample and chew it for 12–15 chews before evaluating for Cohesiveness of Mass. However, for a panelist with a small mouth, that ½" piece was very large. For a panelist with a larger mouth, that same piece was quite small. It is not surprising that 12–15 chews resulted in very different textural experiences for those two panelists and therefore very different ratings for that attribute. Similarly, if a panelist had a high salivation rate, they may not have any sample left at 12 chews while others may still have an almost fully intact bolus.

Newer thinking allows more freedom for the panelists to appropriately alter the technique to their specific physiological differences.

13.3.1.1 Bite/chew/spoonful size

Instead of indicating the size of the bite or spoonful, panelists are now instructed to take an 'average bite' of a sample. Panel leaders will need to observe panelists initially to manage this as one panelist's 'average' may be different from another's. With practice, this method will become natural for panelists and may even relieve frustration incurred by older methods.

While the size can be adjusted for panelists, the 'type' of bite should be controlled (or not) depending on the project objective. Often the ends of products (e.g. bread crust, candy bar ends, chicken finger edges, etc.) are avoided as they have very different textural properties than the inside of the products. Sampling from the middle is still a common practice in textural evaluations. When practical, the ends of these types of products can be evaluated separately to capture the total textural experience. For more on managing variable products, see Section 13.3.2 (Managing Variable and Complex Products).

13.3.1.2 Chewdown

After taking a bite, panelists are instructed to chew until 'phase change.' The phase change for a product is the point at which the sample has come to bolus and then begins to dissolve or breakdown past the point of bolus. For many consumers, this is the point at which they would typically swallow and for panelists the point at which they would expectorate. However, swallow/expectoration timing is also quite variable amongst consumers and panelists and can only be used as a guideline when training.

Many products have obvious phase changes. Bakery products (e.g. cookies, cakes) are good examples of products that have obvious phase changes. During chewdown, a bakery product typically starts as a solid; mixes with saliva as it is mechanically broken down (i.e. chewed); and then 'comes together' into a mass. Upon further chewing, however, it begins to break down or dissolve into the saliva making the mass much less cohesive and more spread out in the mouth. That point of change is the product's phase change. Not all products have obvious phase changes (e.g. carrots). For products that do not begin to dissolve, a discussion should be held with the panel to better understand what each panelist is experiencing (e.g. the bolus is getting less rough or the pieces in the bolus are getting smaller). Often this conversation is critical in helping each panelist understand the technique appropriate for their unique physiology. It will also allow the panelists to understand if they are chewing too long – a common occurrence in descriptive paneling.

Chewdown attributes that depend highly on phase change timing for accurate ratings include: Cohesiveness of Mass, Roughness of Mass, Moistness of Mass, and Moisture Absorption.

(Note: When rating Moisture Absorption, the phase change is described as the point at which the sample no longer absorbs saliva. Once this point is reached, the sample begins to comingle with the saliva and two separate phases are experienced in the mouth.)

Once the panelists have adjusted their technique, the panel will have greater success in reaching agreement on an attribute and ultimately a sample's textural profile. These discussions will also give the panel leader a better understanding of his/her panelists and will allow for a more accurate interpretation of the data.

13.3.2 Managing variable and complex products

More and more panel leaders are challenged to accurately capture the texture of variable and complex products. Product variability can range from natural variability (e.g. nuts) to intended variability (e.g. trail mix). Product complexity can range from a single grain flake cereal in milk to multi-component sandwich builds. The approach to capturing the sensory textural data will depend on the objective of the test and how the data will ultimately be used.

If a product has natural variability, texture attributes are often reported as an average of several bites. In other words, each panelist is instructed to rate three different bites of the product for an attribute (e.g. Hardness) and then report their score for that sample as an average of those three ratings. If the range of variability is large and ranges across different parts of the scale, the range can simply be reported as a range (e.g. Hardness: 4.0–7.0). However, a more accurate method of rating the texture, and one that allows for statistical analysis, is to identify which attributes have large ranges and rate them each for low, high, and mode. For example, when rating almonds, there is often a large range for Hardness. Therefore, each panelist records the rating for their lowest and highest experience during the evaluation for each of those attributes as well as the rating for what they experienced most often (i.e. mode). For both consensus and individual rating panels, the data for each attribute should be reported as the 'extreme' for low and high (not the average) and the average of the mode. Table 13.6 shows this type of data and how it is reported.

Table 13.6 Reporting large range of variability.

Panelist	Sample	Hardness: Low	Hardness: High	Hardness: Mode
12	246	11.0	13.5	12.0
14	246	10.0	14.0	11.0
17	246	9.5	13.0	10.5
20	246	10.5	13.5	12.5
24	246	11.3	14.5	12.0
25	246	10.5	12.0	11.3
31	246	11.5	14.0	11.8
22	246	11.0	13.8	12.0
	Min	9.5	12.0	10.5
	Max	11.5	14.5	12.5
	Average	10.7	13.5	11.6
	Score Reported	**9.5**	**14.5**	**11.6**

For products with multiple variable components (e.g. trail mix), it makes more sense to capture the texture of handfuls of product as this is how these types of products are typically eaten by consumers. It is not recommended that panelists sort through the product to ensure they have each component in each handful as this is not representative of consumer behavior. Instead, the 'low/high/mode' technique can be used but with the full handful instead of an individual component. Expecting large ranges and accepting that the product is variable will help the final interpretation of the data. If using this data to interpret consumer response, it may be necessary to look at larger trends across attributes when drawing conclusions about a product's texture (e.g. Product A has the smallest range in Hardness, Product B has the greatest range in Crispness, etc.).

When descriptive data is being collected with the intention of linking it statistically with consumer data, it is important that the descriptive approach simulates the consumer usage as closely as possible. For example, if the product of interest is cereal, the sensory scientists must develop a protocol for the consumer test as well as for the descriptive analysis. Will consumers taste the cereal samples dry only? In milk only? Both ways? Whatever is decided, the descriptive protocol must mimic the consumer protocol as closely as possible. If the texture of the cereal over time in milk is important, a timed consumer evaluation may be necessary (e.g. consumers evaluate cereal immediately after pouring milk over cereal; after 1 minute; after 3 minutes, etc.). The texture descriptive protocol will follow the consumer protocol. If there is not a set protocol already developed, the sensory scientists can use descriptive analysis to identify critical points in the timing to recommend for the consumer test. Again, it is important to look at the whole study to best determine what protocol(s) to use.

13.3.3 Consensus vs. individual ratings

There is an ongoing debate over whether reporting consensus ratings are 'as good as' reporting the mean of individual ratings. This question is particularly important when rating textural attributes. The simple answer is yes, consensus ratings are equally accurate when collected by a panel leader trained in consensus balloting. It can be argued that consensus ratings collected in such a manner are even more accurate as consensus panels are constantly discussing their individual experiences to ensure that the final score recorded for each attribute accurately describes all of those experiences. Too often individual rating panels forgo the orientation time needed to truly understand the product differences that will be seen across all samples. Because of this, when confusion occurs during evaluations, panelists are forced to interpret their experience the best they can, as the ballot is already set and the sessions are typically not open for discussions. It is then the panel leader's responsibility to make sure that the final data makes sense and to revisit any samples where confusion occurred. Unfortunately, panelists do not always report these incidences and data are often compromised as panelists force fit their experiences onto the existing static ballot. Even a thorough orientation, including all products to be in the study, can miss some textural nuance and cause similar inaccuracies in the data.

Conversely, consensus balloting with an inexperienced panel leader can also lead to inaccurate data. The panel leader must be skilled in hearing all the panelists input and finding a way to interpret those experiences into an agreed upon answer. Simply reporting the mean is not sufficient. If half the panel rates the sample as a '4.0' and the other half as an '11.0', reporting the final score as a '7.0' does not represent any of the panelists' responses. Clearly there is a bigger issue at hand if this range is observed because the panel is not seeing the sample the same way. The panel leader's job is now to ask questions to uncover where the inconsistency lies. Is everyone using the same technique? Does everyone understand the attribute definition? Is everyone sampling from the same part of the product? The conversation must continue until the answer is clear and everyone agrees on the final rating. A trained consensus panel leader can do this with ease. The final result is an accurate textural profile. If the consensus panel session is replicated, it is possible to analyze the data statistically by using the replicates as the variables in the data.

13.4 SPECIAL ISSUES IN TEXTURE TERMINOLOGY

13.4.1 Understanding crispy and crunchy

13.4.1.1 Crispy and crunchy-descriptive analysis perspective

The terms crispy and crunchy stem from the original term Fracturability that measures the response of foods to applied forces. Some foods absorb the forces that are applied and bend and deform. These foods are cohesive. Those products that break under the force are fracturable and include products that are crumbly, crispy, crunchy or brittle. The force is NOT the force to break the product but the force ***with which*** the product breaks (rather than deform). To add to the confusion about fracturability, many sensory scientists are unclear about the differences between crispy and crunchy textures (Vickers, 1984 and 1985). The simplest explanation for the crispy/crunchy dilemma is that crispy products (both wet and dry) have many small breaks and make a high pitched noise when breaking (Examples are baked extruded cheesy corn puffs, lettuce, and buttery crackers). Crunchy products break with fewer larger breaks and are often denser than their crispy counterparts (Examples are fried extruded corn snacks, carrots, peanuts).

While this more simplistic view of the crispy/crunchy is helpful in describing the difference between the two to a naïve audience, crispy and crunchy are truly complex attributes to measure. Some previous training techniques taught panelists to focus completely on the pitch. Unfortunately, not all panelists can differentiate between pitch and therefore find this approach inaccurate and frustrating. Additionally, many products produce multiple pitches during the first bite/chew making this approach even more unreliable. Focusing on the force ***with which*** the product breaks is the only way to assure accuracy of these attributes. As mentioned before, discussion amongst panelists led by a highly trained panel leader can help address any confusion of the attributes or aid in accurately describing any product variability, which is quite common in crispy/crunchy products.

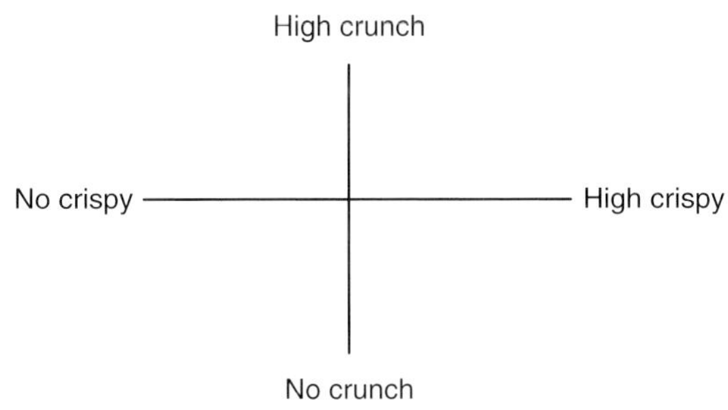

Figure 13.1 Qualitative panel activity grid.

13.4.1.2 Crispy & crunchy-consumer perspective

To better understand how the concepts of crispy and crunchy are interpreted by consumers, an exercise was conducted with a qualitative community narrative panel[1] utilizing the references on the Fracturability Scale as well as other 'crispy' & 'crunchy' food products.

Session Outline:

- The qualitative panel was asked to define crispy and crunchy, outlining the differences between them in their own words based on their own previous experiences.
- Products were shown (Fracturability Scale references and other products) and panelists evaluated whether each was considered crispy, crunchy or both.
- Each panelist placed the product within a grid similar to the one in Figure 13.1 where they felt each fit.
- Once each product was evaluated and plotted, the panel discussed reasons for placement.
- Common characteristics between samples that were within the same quadrants were discussed.

Key Findings:

- Denseness, fracturability, noise level/pitch and how *labor intensive* the product was determined crispiness or crunchiness for consumers.
- Persistence of crunch (i.e. the length of time during chewdown that the product remained crispy/crunchy) was a factor for consumers in selecting whether a product was crispy or crunchy – the longer the experience, the more it was considered crunchy as opposed to crispy.

[1]The Spectrum Community Narrative Panel (SCAN) is an effective solution to better understand how to bridge the gap between the company, and/or descriptive data and the consumer. The panel is comprised of creative, articulate consumers who have built a community within their group. This process allows for honest communication, enhanced creativity, and increased description of the target product category and/or concept.

Table 13.7

Crispy	Crunchy
Light	Low and harsh sound
Airy	Deeper
Seems thinner	More substantial
Less pressure to chew	Denser
Sounds like 'snap, crackle, pop'	Harder
Snap and shatter	Have to work at it
Higher pitch	Sticks around
Fractures like a potato chip	More molar packing
Explodes	Louder
Goes into many pieces	Have to use back teeth
Hard to control	May tend to eat less of it
More shards	Time consuming
Larger pieces	Involves more chewing
Disappear more quickly than something crunchy	Complex
Flows, you put it in your mouth and it's gone	Satisfying
Mindless	Lasts longer
	Crushes into more solid pieces rather than shards
	More control over the pieces

Table 13.7 shows the terms the panel assigned to 'crispy' and 'crunchy' after having experienced the range of products and placing them on the grid.

It has been hypothesized that the consumer experience of 'crispy' versus 'crunchy' may be highly correlated to how long/how many chews the product remains crispy/crunchy. One panelist described the difference in this way:

> *'I think the initial bite has more to do with crispy; I think the crunchy stays crunchy longer. The crispy for me is all in the first or second bite.'*

This may be related to the comment on 'labor intensiveness' seen in the key findings as products that stay crunchy longer require more time to completely chew. While this is most likely not the only factor that influences consumer views on these attributes, it may help direct product developers when creating new products in these categories. From a descriptive perspective, a strategic panel leader can relate these consumer observations and needs. The panel's ratings for persistence of crispy or crunchy may be as important as the initial crispy or crunchy ratings. The ability to quickly and inexpensively 'understand' the consumers' observations and stated needs provides real strategic value to product development.

13.4.2 The issues with annoying textures

Product developers are often challenged by consumer test results that are unfavorable but unclear as to the reasons for consumer low satisfaction or liking. Often the source of consumer dislike is annoying but subtle textures. During the development of many high fiber cereals, consumers were aware of small fibers that collected along the gum line after swallowing the cereal. Often the researchers failed to ask about afterfeel or cleared palate. Once product development saw some of the descriptive data they became aware of the issue of residual fibers and introduced the food equivalent of vacuum cleaners in the form of raisins and other dried fruit.

A key design element of good texture in a product is the unified breakdown so that all the elements disappear at the same time. The more lubricating or moistening elements (fruit, caramel) carry the drier components (nuts or grains).

Perhaps the huge success of the grape tomato stems from the fact that it can be closed into the mouth and will not burst like a cherry tomato. Control of the food is essential to consumer satisfaction and delight.

13.4.3 Texture as a signal of wholesomeness and freshness

Although the flavor of products can change over time to indicate that the product is past its shelf life – sour milk and rancid oils and fats – texture is often the first sign that the product is aging. The texture parameters that are most indicative of changes in product freshness are Cohesiveness and Hardness.

Products that are crispy or crunchy when fresh (they rupture and break when bitten or chewed) begin to soften (less hard) and bend and deform (less fracturable and more cohesive) as they age. Consumers describe this change as 'getting soggy'. Snacks (crackers, cookies, chips) and fruits and vegetables (lettuce, carrots, peppers) are firm and rigid when fresh (see Figure 13.2 for Freshness). Upon aging these crispy, crunchy (fracturable) foods move to a softer and more deforming section on the Freshness map (see Figure 13.3). This freshness model is not the same for all foods. Those foods that start cohesive and soft when fresh, such as a slice of white bread, a stick of chewing gum, or a marshmallow, age by becoming more rigid/less deforming and harder. (See Figures 13.2 and 13.3 to see the changes).

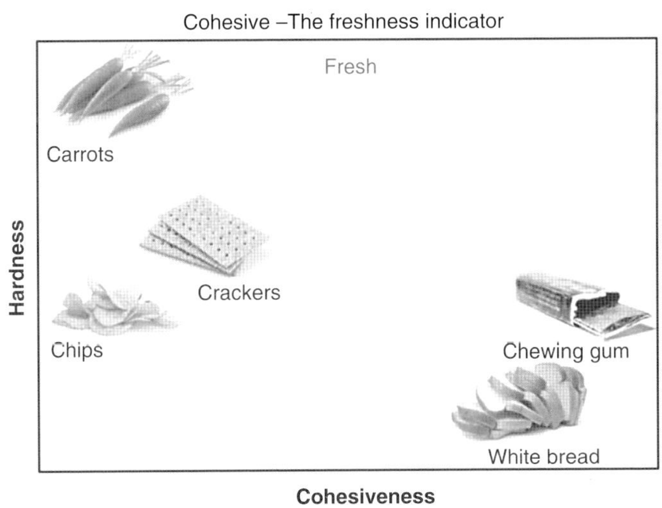

Figure 13.2 Hardness and cohesiveness indicate freshness in several foods. For instance carrots are hard and crisp (not cohesive) when fresh whereas white bread is soft (not hard) and cohesive when fresh.

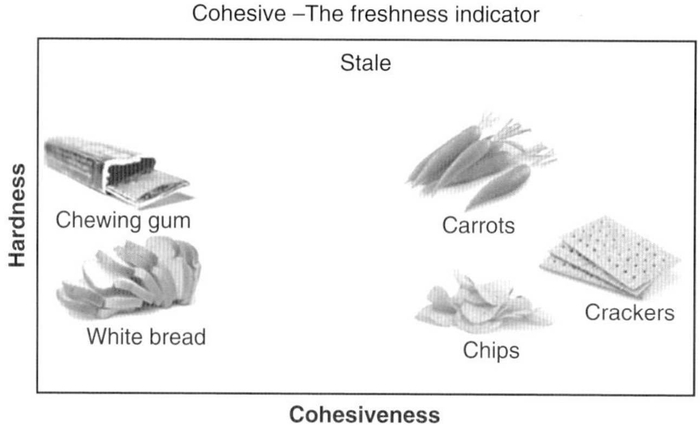

Figure 13.3 Changes in hardness and cohesiveness can indicate staling. Carrots become less hard and more cohesive (less crisp) when stale. White bread becomes harder and more fracturable (less cohesive) with staling.

13.5 APPLICATION AND USE OF TEXTURE DESCRIPTIVE ANALYSIS

Understanding a product's descriptive textural profile is comparable to taking a snapshot of that product at any given time. The quality of the description method (how anchored the attributes and scales are to references) and the quality of the panel (how calibrated and validated are the panelists) will provide technical data that is both valid and reliable documentation – as any scientist expects from an instrument. It allows for comparisons for reasons such as tracking changes during shelf-life, understanding production variability, and understanding process or ingredient changes. While it is not always possible or practical to have the 'idea' or 'gold standard' product available, that descriptive 'picture' tells sensory analysts exactly what textural properties that product had at that point in time. A trained descriptive sensory scientist can bring that experience back to mind by simply looking at the profile. From there, they can accurately describe how the product in front of them differs from that original sample.

13.5.1 Understanding products similarities and differences

A primary reason for documenting products' textural characteristics is to highlight product similarities and differences. Whether it is to support new product development efforts, to investigate improvement opportunities, to substantiate cost reduction and ingredient substitution initiatives, to set specifications for product quality and maintenance or to understand product sensory stability over time or under stress conditions, using trained descriptive panels allows food manufacturers to better understand their own portfolio of products as well as that of their competitors.

Consider as an example the strawberry yogurt category and the challenges it faced to keep in line with the consumer's ever-growing interest for a healthier lifestyle. Initially, yogurt manufacturers expanded their lines to include non-fat and lite yogurt as alternatives to the existing low-fat products that were more traditionally used as dessert or snacks. The category then increased to include probiotic cultures promoted for their positive impact on digestive health. Those new products were originally designed to provide a different experience to consumers, and positioned as breakfast alternatives for a healthier start of the day. They soon started to be used not only at breakfast but also throughout the day for snacking. More recently, trends showed the arrival on the market of new players: the Greek-style yogurts, also known as strained yogurt. With double the protein and often times half the carbohydrates of regular yogurt, they are marketed as a complete, convenient, satisfying and satiating choice for breakfast or afternoon snack.

A descriptive panel was used to gather insights about the category and describe product differences from a textural standpoint across the category. Once the data was collected, it was analyzed using a multivariate statistical technique called factor analysis. In a nutshell, the technique allows researchers to summarize the information by bundling attributes that are correlated with one another into sensory dimensions. So rather than looking at product similarities and differences along 12, 15 or more individual attributes, sensory scientists can summarize key differences and similarities among the samples tested by looking only at three, sometimes four key sensory dimensions. Results of the analysis are presented as two-dimensional maps (Figures 13.4 and 13.5). The analysis returned four key dimensions of variability. Samples that are close to one another along a dimension are similar in the characteristics defined by the dimension, samples that are far apart on the dimension have opposite characteristics.

A quick look at the maps reveals the following insights about the category:

- Greek yogurts offer a very different textural experience than their traditional counterparts: less slippery on the surface, they are typically firmer than even the thick and creamy yogurts and tend to leave the mouth more coated.
- In comparison to traditional yogurt made with Lactobacillus cultures, dairy specialties made with probiotic cultures tend to be slightly more cohesive at first compression with lower ability to mix with saliva, resulting in a 'slimier' perception than both the Greek-style and regular yogurt offerings.
- From a competitive standpoint, brands clearly differentiate themselves with regards to amount, size and firmness of fruit inclusions. Fruit inclusions become part of the brand's sensory identity while matrix properties are more aligned with product type and positioning.
- Within the traditional yogurt category, for some brands, low-fat and non-fat product offerings deliver a very consistent texture, with little sensory trade-offs for the added health benefit. While for other brands, non-fat offerings are more mouthdrying, and in contrast to their low-fat counterpart, are associated with a chalky mouthfeel rather than a dairy film.

In as little as a few hours of panel time, one can generate very powerful insights about sensory cues and their alignment with product positioning, consumer benefit delivery, brand strategies and opportunities for improvement, new product development and/or product differentiation.

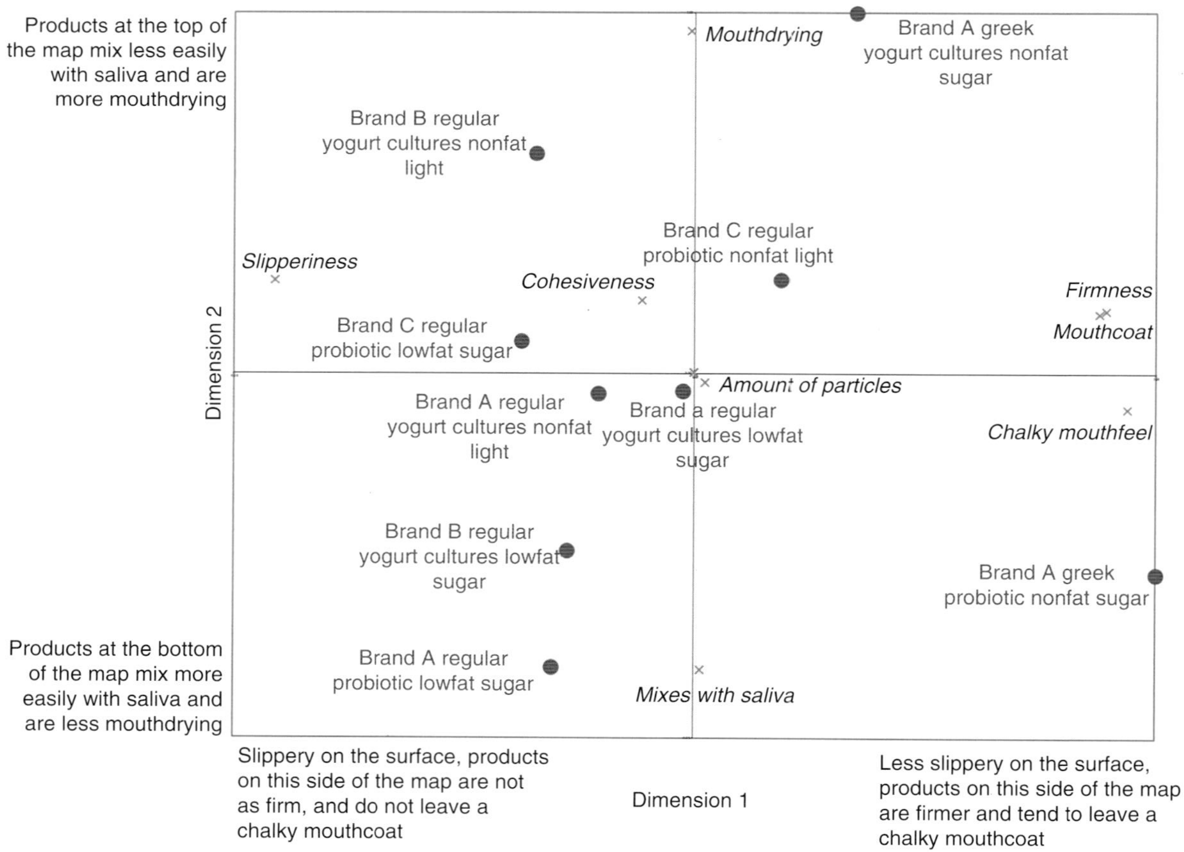

Figure 13.4 Perceptual Map of the first and second sensory dimensions: Strawberry Yogurt.

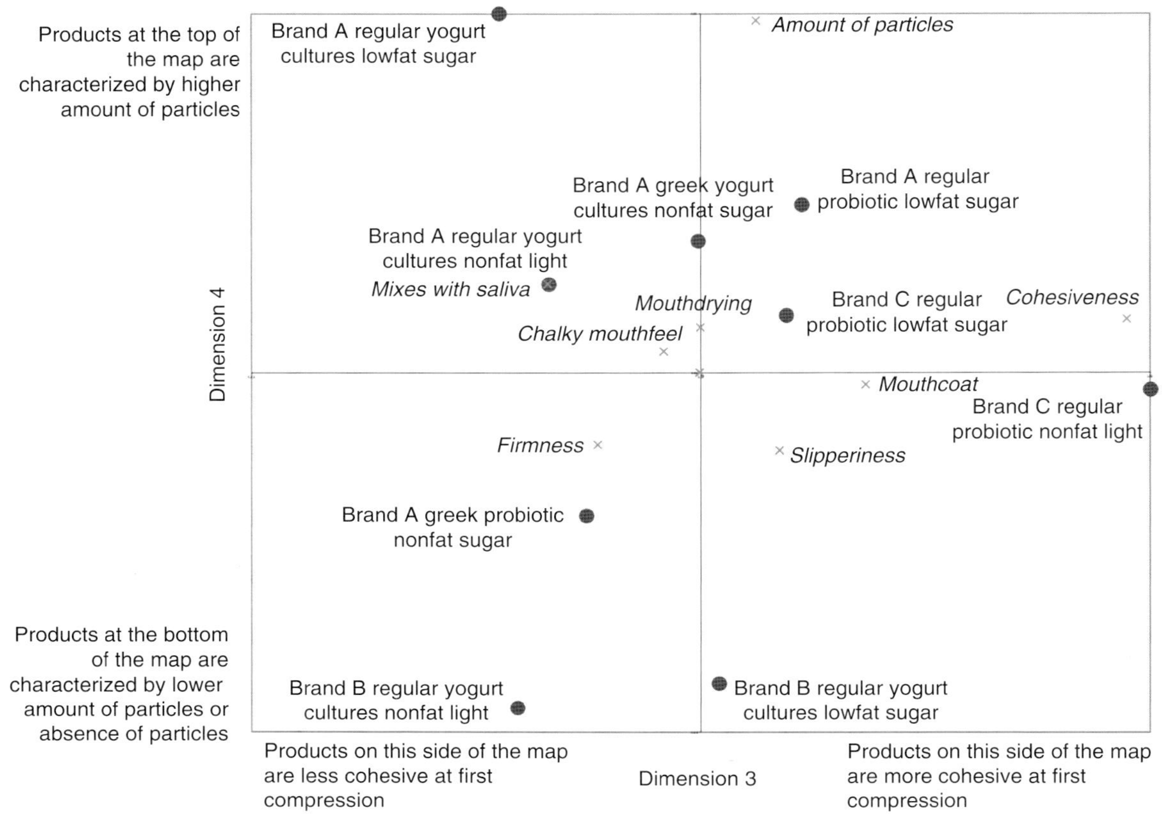

Figure 13.5 Perceptual Map of the third and fourth sensory dimensions: Strawberry Yogurt.

13.5.2 Correlation of sensory and instrumental texture measurements

Texture can only be measured by humans; texture is a sensory property. Any attempt to measure texture instrumentally needs to keep the sensory aspects of product texture in mind. When developing an instrumental method to simulate human texture measurement, the scientist needs to consider the full range of products that are included in the product set under the study. When looking at snacks it is not sufficient to include crispy snacks alone, when chewy snacks are considered part of the snack array. If only one set of products (crispy snacks) are considered, the correlations are easier but may not reflect all the texture and rheological variables necessary to create an instrumental test that predicts human texture measures. In addition to considering the scope of the product category to include in the correlation work, the scientist must also pay strict attention to the way that the food is destroyed in the mouth or hand – the force application used in the texture instrument should simulate: A) the type of force application (compression – tongue to palate; shear – incisors; or shear/compression – molars; B) the direction of force application – which face of the product should receive the deforming force; C) the rate of force application – most food is bitten or chewed at a specific rate. Changes in the rate can alter a deforming behavior (slow shear) into a shattering behavior (fast compression). If the data is to be correlated with human consumption data, only when the rheologist has consumed several samples of each of 10–20 varieties or brands of a product can the scientist begin to develop an instrumental approach to manipulate the product.

13.5.3 Linking consumer and product understanding

One very powerful reason to accurately describe a product's texture is to use it to decipher consumer liking and perception of benefits. While it is commonly accepted that flavor plays a major role in consumer acceptance, the role of textural properties is often underappreciated. A cursory look at the texture of successful products in today's North American marketplace highlights that while successful products tend to vary widely in their texture properties at first bite and chew, one of their commonalities is that they typically blend fairly quickly into a rather smooth or lumpy and somewhat moist mass, and disappear quickly, leaving little to no toothpack and chalky, gritty or grainy mouthfeel. This is, of course, a broad generalization and exceptions certainly exist.

Furthermore, it is interesting to notice that across numerous food categories, consumer segments exist that typically put more emphasis on texture characteristics than they do on flavor. Texture becomes their primary liking driver: with little tolerance for deviation from the expected ideal texture, texture becomes the 'make or break' decision for these texture sensitive consumers, while a broader range of flavor characteristics might be acceptable to them.

As described earlier in this chapter, texture descriptive analysis is used to objectively, precisely and accurately document the discrete sensory properties of foodstuffs. However, when consumers experience foods and beverages, the characteristics are often integrated into overall impressions. The consumer perception of juiciness in meat, for example, can be delivered through the combination of optimal levels of initial firmness, moisture release at first bite, moisture absorption, moistness of mass, and awareness of fibers during chewdown, and so on. Similarly, consumer perception of spongy texture in cakes is related to optimal levels of initial springiness, hardness, airiness (i.e. low denseness), and cohesiveness at first bite, among other textural characteristics. Combining product knowledge with consumer feedback allows the researcher not only to identify which sensory characteristics are key to the perception of category-specific benefits (juiciness in meat, creaminess in ice-cream, decadent texture in brownies, etc.) and ultimately liking, but also to determine the relative importance of those characteristics and their optimal level to fully satisfy the consumer.

To further complicate the matter, any two consumers might not necessarily agree on the importance and/or optimal levels of each characteristic needed to optimize their acceptance of the product. An obvious example is in the cookie category, where consumers can be broadly separated into three distinct groups: those who prefer crunchy cookies, those who prefer chewy cookies, and those who enjoy both chewy and crunchy cookies. Consider the crunchy cookie lovers. Within that segment of the population, some might prefer a harder, denser, more fracturable initial bite with some persistence of crisp, while others might be more favorable to a slightly softer, less dense, less fracturable texture somewhat reminiscent of 'sables' or 'shortbread' cookies.

To decipher consumer perceptions and liking and provide sound and precise directions for product development and improvement, consumer information can be interpreted in light of objective descriptive properties. Numerous techniques may be used to establish directions for texture optimization based on qualitative consumer feedback, quantitative consumer results or a combination of both.

The discussion provided earlier (Understanding Crispy & Crunchy Section 13.4.1) is a great example of how data from qualitative discussion can be linked to data from descriptive evaluation to fully understand what consumers mean by crunchy or crispy. Similar techniques may be used to obtain qualitative feedback on products and, by interpreting consumer feedback in light of the descriptive results, provide direction for product improvement. This is often done in early development in the context of iterative co-creation or rapid prototype development with consumers.

When quantitative consumer techniques are used and a sufficient number of products are evaluated (typically more than eight), elaborate statistical techniques, such as clustering, multivariate reduction methods (Principal Component Analysis, Factor Analysis) and regression (Principal Component Regression, Partial Least Square Regression) can often be used to:

- Identify segments in the population with different preference patterns for texture properties, if they exist.
- Model their preference against sensory features.
- Identify target profiles and opportunities for product development.

Such methods are described in detail in *Sensory Evaluation Techniques* (4th Edition) (Meilgaard, Carr and Civille, 2006) and *Multivariate and Probabilistic Analyses of Sensory Science Problems* (Meullenet, Xiaong and Findlay, 2008) among others.

13.6 APPENDICES

Appendix 13.1

PROTOCOL FORMAT			
NUMBER OF PARTICIPANTS			
DATE			
REFERENCE SAMPLES **PRODUCT**	**SCALE VALUE CODE**	**CONTAINER**	**# TO MAKE**
		3 oz. (approx.) plastic sample cup	1/person

Appendix 13.2

ADVANCE PREPARATION **Please read and make a schedule for completing the work to be done prior to presenting this exercise. Include obtaining supplies and coding containers**
REFERENCE SAMPLES
CODE
CODE
CODE
CODE

Appendix 13.3

DAY OF EVALUATION **Consult the instructor for the time the exercise should be ready. Fill in the time in the area provided.**
Return all samples to room temperature before serving
TIME
REFERENCE SAMPLES
CODE
CODE
CODE
CODE

Any samples not prepared in advance should be prepared within two hours of evaluation.

All refrigerated samples should be removed from the refrigerator 1–2 hours before evaluation.

Please present all prepared samples listed on the Protocol Format at the same time, in the order listed, at the appropriate times.

You may prepare individual trays for each participant containing one of each evaluation sample.

OR

Alternately, you may group like samples together on large cafeteria trays and pass the trays around the evaluation room for each participant to take one of each sample.

At the evaluation area, have the following materials available for each participant, including the instructor:

- Styrofoam or coated paper cup with lid (spit cup)
- Plastic or paper water rinse cup
- Spring water serving pitchers (at least 1 per 4 people)
- Napkins
- Plastic forks/spoons if necessary

References

Meilgaard, M.C., Carr, B. T., Civille, G. V. (2006) *Sensory Evaluation Techniques*, Fourth Edition. Boca Raton: CRC Press.

Meullenet, J.F., Xiong, R., Findlay, C. (2008) *Multivariate and Probabilistic Analyses of Sensory Science Problems*, Ames, Iowa: Wiley-Blackwell.

A.S. Szczesniak, A.S. (1963) 'Classification of textural characteristics,' *Journal of Food Science*, 28, 4, 397.

Szczesniak, A.S., Brandt, M.A. and Friedman, H.H. (1963) 'Development of standard rating scales for mechanical parameters of texture and correlation between the objective and the sensory methods of texture evaluation,' *Journal of Food Science*, 28, 4, 397–403.

Szczesniak, A.S., Loew, B.S. and Skinner, E.Z. (1975) 'Consumer texture profile technique,' *Journal of Food Science*, 40, 1243.

Vickers, Z.M. (1984) 'Crispness and crunchiness – a difference in pitch?' *Journal of Texture Studies*, 15, 157–163. doi: 10.1111/j.1745-4603.1984.tb00375.x

Vickers, Z.M. (1985), 'The relationships of pitch, loudness and eating technique to judgments of the crispness and crunchiness of food sounds' *Journal of Texture Studies*, 16, 85–95. doi: 10.1111/j.1745-4603.1985.tb00681.x

14 Mind genomics® and texture: the experimental science of everyday life

Howard R. Moskowitz

Moskowitz Jacobs, Inc., White Plains, New York, USA

14.1 INTRODUCTION

When we think about texture, how do we react? Do we think about foods, about experiences, or about the meal? How can we understand the mind of the consumer with respect to the texture of foods?

The foregoing question is a broad one, neither a trick question nor one that can be readily answered. Those of us who are scientists have been schooled to think about problems in a structured manner, in a one-at-time fashion. We are taught from our earliest days in school that real scientists 'isolate' the problem to study it, to understand it, and then finally to discover a bit more about how the world works.

The sheer number of food experiences we have over a lifetime makes the study of texture both easy (we know intuitively what it is) and difficult (because there's so much going on). In the everyday world, we confront texture as many different sensory inputs, creating a veritable kaleidoscope of sensory impressions that our brain kneads and shapes into basic sensory impressions. These basic sensory impressions are the building blocks. They underlay our deeper, compound and complex percepts of appearance, aroma, taste, texture, as well as internal sensations.

14.2 ENTER SCIENCE

For the most part, the one-at-a-time method works quite well to drive scientific understanding. When it comes to sensory perceptions, specifically those involved with texture, we begin by identifying the test stimuli, that 'stuff of reality' which embodies what we want to understand, the variable that is called by the custom-crafted word 'texturogen.' Then we make our measurements. Those interested in the history of science are invited to read E.G. Boring's *Sensation and Perception in the History of Experimental Psychology* (1929). This 80+year-old book by Harvard's grand old man of psychology recounts the early efforts by psychologists to understand the basic sensory percepts underlying the experience of texture.

Food Texture Design and Optimization, First Edition. Edited by Yadunandan Lal Dar and Joseph M. Light.

Our efforts may begin with real foods, complex stimuli given to us by nature. We may stretch, deform, and otherwise 'torture' the stimulus, measuring the resultant deformations, attempting always to correlate those fundamental measures to consumer perception. Or, our efforts may begin with so-called 'model systems,' simplistic stimuli that we ourselves develop to isolate one variable and study it. We may create a set of stimuli that vary along one key factor, for example, a set of rubber foams varying in the force needed to make an indentation, and instruct respondents to rate these different stimuli on some scale, for example, perceived hardness or perceived springiness. In either case, working with nature's products or with our own model systems, we end up searching for a relation between the physical stimulus and the rating. Our goal is to uncover the existence of relations by correlations between variables, or even better, use curve fitting, regression, to create a predictive equation telling us how changes in one physical domain correspond to changes in the perceived texture of the test stimulus.

14.3 ENTER THE CONSUMER AND THE QUANDARY OF MULTI-ATTRIBUTE PERCEPTION

In this chapter, we are going to go back into the mind, to understand how we think about texture, and how some of us may be wired to respond to texture strongly, whereas others may be wired to respond to taste, or to appearance, or aroma. What we will learn is that people do respond to messages about texture, but that there are some people who seem to pay more attention to it than others. We will discover how to identify these people. In the end, we will learn about the *mind of the consumer* when it comes to sensory perception. Our tool will be language, the study of how we respond to messages about the characteristics of products, and the role of texture within this study of messages.

14.4 A NOTE ABOUT THE RESEARCH STRATEGY WE EMPLOY IN THIS CHAPTER

This chapter is unorthodox, presenting new ideas developed about how the mind works. The approach, known collectively as Mind Genomics®, deals with the experimental science of everyday life, the way we respond to the world around us when we think about our ordinary experiences (Moskowitz, German and Saguy, 2005). Mind Genomics® lets us understand what's important to people, and how people may differ fundamentally from each other in the way they respond to ideas.

Texture is one small part of these person-to-person differences. The tools of Mind Genomics® focuses on those differences, expanding them to show us a whole world of sensory perception that we did not even know existed. Of course, marketers have used segmentation, dividing people in differences, often by their values and life-styles, so-called psychographic segmentation (Wells, 1975). Typically a topic as narrow as texture and other sensory impressions would not occupy center stage in these traditional methods. However, Mind Genomics® allows the focus to sharpen, moving from the general to the particular, moving from food in general, to sensory percepts in particular.

14.5 THE TOOLS OF MIND GENOMICS®

Mind Genomics® IdeaMap®.net is a proprietary form of conjoint analysis utilizing Rule Developing Experimentation (RDE), a systematized solution-oriented business process of experimentation that designs, tests, and modifies alternative ideas, packages, products, or services in a disciplined way using experimental design, so that the developer and marketer discover what appeals to the customer, even if the customer can't articulate the need, much less the solution.

RDE was developed by Moskowitz Jacobs, Inc. in cooperation with Professor Jerry (Yoram) Wind (Wharton School of Business at the University of Pennsylvania). The term was initially coined by Dr. Howard R. Moskowitz and Dr. Alex Gofman in a series of articles and conference papers. The paradigm for systematic design and developing/using the rules in various applications was formalized in their book *Selling Blue Elephants: How to Make Great Products That People Want Before They Even Know They Want Them* (Wharton School Publishing, 2007).

Some aspects of RDE had previously been used during the past 25 years for the design and optimization of physical products, product development concepts, marketing communications, as well as packaging. Beginning in 1982, RDE was variously described in different books and articles by a number of different terms (e.g., product optimization), depending on the product area and the topics. RDE has been effectively utilized over the years by many Fortune 500 companies around the world.

Initially, RDE was used for product optimization followed by message optimization for advertising, promotions, and so on. New applications of RDE include new product development (NPD), combinatorial innovation, multivariate landing page optimization, political elections, stock markets, crisis communications management, package design, magazine cover, flyer optimization, social science, public policies, and so on – virtually any field that involves people, decisions and choices, emotional perceptions of messages and other stimuli. RDE is the foundation for the new science of Mind Genomics® modeled on the emerging science of genomics and the technology of informatics.

The Three Roots of RDE:

1. Experimental psychology – RDE is founded on the realization that perception and behavior are linked in a two-way exchange. RDE uses a variation of conjoint analysis as its statistical basis.
2. Driving power of business – a need to create new products and services that people like, doing so fast and inexpensively.
3. The world-view of social science – RDE is related to a field called adaptive experimentation (AE), or adaptive management. The most publicized cases of AE are very lengthy, large-scale, even monumental projects in ecology, theoretical science, or the sociology/environmental area. However, AE doesn't generate rules.

The Seven Basic RDE Steps:

Step 1. Identify groups or classes of features that constitute the target product (offering, etc.) For example, in the case of a credit card offer, the variables could be APRs (annual percentage rates), Rewards Options, and so on. Every such variable (also called a 'silo' or a 'bucket' of ideas) comprises several APRs, Rewards Options, and so on.

Step 2. Mix and match the elements according to an experimental design to create a set of prototypes. The second step is usually done automatically by a tool that creates a unique individual design plan for each respondent, allowing for individual models of utilities for each respondent.

Step 3. Show the prototypes to consumers and collect their responses on a rating question (e.g., 'How likely would you be to buy this product?').

Step 4. Analyze results using a regression module. One of the key differentiating points of RDE is individual models of utilities for each respondent. This allows patterns discovered in the data, across elements and respondents (Step 6) to generate rules for more targeted optimization as well as uncover all meaningful two-way synergisms and suppressions between the elements. RDE uses dummy variable regression to estimate the absolute values of the utilities, which in turn allows for databasing the results. The database makes for meaningful comparison of the element utilities in one study, and across studies, which then track the consumers' minds across studies and time.

Step 5. Optimize. Uncover the optimal product, find the best combination that has the highest sum of utilities.

Step 6. Identify naturally occurring attitudinal segments of the population that show similar patterns of the utilities.

Step 7. Apply the generated rules to create new products, services, offerings, and so on.

Key Advantages of RDE:

- The structure of RDE imposes the discipline of thinking resulting in a higher success rate than other methods. Many users report that just preparing for RDE (seven steps above) helps them to better understand the problem and find possible solutions.
- Individual models for each respondent allows for more robust and rich data to generate rules. Some of the advantages of individual models include:
  - Pattern-based discoveries (Step 6, segmentation) to generate rules for more targeted optimization (also called attitudinal segmentation or latent segments).
  - Simulate market share and choice analysis.
  - Discover any and all meaningful pair-wise interactions between the elements.
  - Analyze how one element 'drives' responses to another element using scenarios (e.g., in case of a category containing brands names, the researcher can compare the utilities of other elements when each of the brands are kept constant).
- RDE uses a variation of dummy variable regression to estimate the absolute values of the utilities, which in turn allows databasing the result for tracking the consumers' minds and easy comparison of the utilities across the categories.

14.6 PART 1 – JUST WHAT IS IMPORTANT ABOUT FOODS – AND WHERE DOES TEXTURE FIT IN?

We begin our exploration of texture by looking at the relative importance of texture versus other characteristics of foods. In 2001, we began a series of studies to understand how people respond to different aspects of food. The first part of the study dealt with messages. The second part of the study involved self-profiling. In one of the questions in the self-profiling questionnaire, we instructed the respondents to select the top three factors that drove their interest in the product. The respondent was presented with a variety of factors and asked to select three. Their choices told us a lot about the importance of texture.

The first study, Crave IT!™, dealt with the factors that drive people to crave different foods (Moskowitz, Porretta and Silcher, 2005; Moskowitz, German and Saguy, 2005; see additional references). The actual study investigated 30 different foods. Each food was examined in its own study, which was comprised of two parts, the experiment (systematic design), and then the self-profiling classification. In the first part of the study, respondents evaluated 60 different vignettes or combinations of elements. In the second part, the respondent answered a series of self-profiling questions, telling us who the respondent was, and his/her attitudes towards food issues.

Question #7 in the second part of the study, the self-profiling classification, instructed respondents to identify the three product attributes that influenced 'craving.' The question was stated as follows:

Which 3 product attributes MOST influence your craving for <Product Name>? [check three]

M1. Appearance	*M2. Aroma*	*M3. Texture*
M4. Taste	*M5. Memories*	*M6. Associations*
M7. Brand	*M8. Advertising*	*M9. Packaging*
M10. Portion size	*M11. Social situation*	*M12. Mood*

The frequency with which the respondents selected *texture* from the set of 12 told us the importance of texture as a perceived driver of 'craving'. Table 14.1 shows us the frequency of selection for the four sensory factors, appearance, aroma, taste/flavor, and texture, respectively.

1. Across the four sensory attributes, not unexpectedly, taste comes first, followed by aroma. Appearance and texture come in third and four places, respectively. (The table does not show data about the remaining eight, non-sensory factors.)
2. Texture is not always in third place in the world of self-described sensory drivers of craving. For some products, texture moves up in importance, from third and fourth place to second place (e.g., for cheesecake, and salad).
3. For the beverages studied in the Crave IT!™ Study, 'texture,' that is, mouthfeel, is virtually irrelevant, at least at the conscious level.

The second set of studies, also dealing with the selection of texture as a driver of food choice, comes from another 30-product study, Healthy You!™ This mega-study dealt with products that are good for you.

Table 14.1 Frequency (proportion of times) that each of four sensory factors was chosen as being among the top three factors out of 12 as drivers of 'craving' for a product. (Source: Data from the 2001 Crave IT!™ mega-study, courtesy of IT! Ventures, LLC. Reproduced with permission.)

Crave IT!™ 2001	Taste	Aroma	Appearance	Texture
Mean choice across products	0.65	0.42	0.19	0.15
Cheesecake	0.66	0.15	0.41	0.36
Mashed Potato	0.69	0.30	0.13	0.33
Gravy	0.63	0.55	0.17	0.31
Bread	0.64	0.70	0.20	0.30
Chocolate Chip Cookie	0.67	0.55	0.21	0.24
Cheese	0.71	0.24	0.18	0.24
Ice Cream	0.72	0.05	0.19	0.22
Fresh Fruit	0.65	0.34	0.50	0.21
Pretzels	0.63	0.38	0.24	0.20
Tortilla Chips	0.68	0.21	0.19	0.18
Snack Mix	0.59	0.18	0.23	0.18
French Fries	0.64	0.49	0.15	0.17
Potato Chips	0.64	0.23	0.10	0.17
Bacon	0.62	0.76	0.24	0.16
Chocolate Candy	0.67	0.22	0.07	0.15
Nuts	0.65	0.37	0.13	0.15
Donuts	0.65	0.60	0.30	0.14
Meatloaf	0.65	0.58	0.18	0.13
Shellfish	0.64	0.36	0.19	0.12
Steak	0.62	0.48	0.23	0.12
Salad	0.60	0.09	0.43	0.11
Pizza	0.63	0.45	0.22	0.10
Cinnamon Rolls	0.58	0.71	0.32	0.10
Chicken	0.63	0.57	0.20	0.08
Tacos	0.67	0.40	0.14	0.08
BBQ Ribs	0.64	0.62	0.21	0.08
Popcorn	0.64	0.78	0.06	0.05
Hot Dogs	0.55	0.45	0.13	0.04
Hamburgers	0.63	0.46	0.18	0.04
Coffee	0.71	0.80	0.01	0.04
Cola	0.71	0.05	0.02	0.03
Iced Tea	0.71	0.21	0.02	0.02

We again go to the self-profiling portion of the study. The questionnaire followed the same general format as presented above for Crave IT!™ The key question instructed the respondents to select the three factors out of 16 that were important to the respondent as a driver of 'purchase.' The question and the different answers were phrased as follows:

Which attributes MOST influence your purchase of < Food Name > ? [please check 3]

M1. Appearance *M2. Aroma* *M3. Amount of fat*
M4. Calories *M5. Vitamin and Mineral content* *M6. Texture*
M7. Taste *M8. Memories/Associations* *M9. Brand*
M10. Advertising *M11. Packaging* *M12. Portion Size*
M13. Other Members Of The Family *M14. Price*
M15. Like To Explore New Foods *M16. It's Healthful*

Table 14.2 Frequency with which each sensory factor is chosen as being among the top three factors out of 16 as drivers of 'purchase' for a product. (Source: Data from the 2001 Crave IT!™ mega-study, courtesy of IT! Ventures, LLC. Reproduced with permission.)

Healthy You™	Taste	Texture	Appearance	Aroma
Average	0.73	0.22	0.19	0.13
Chocolate	0.17	0.90	0.17	0.09
Bread	0.68	0.46	0.27	0.17
Salsa	0.82	0.40	0.33	0.18
Peanut Butter	0.80	0.36	0.16	0.28
Cheese	0.87	0.33	0.27	0.16
Juice – Citrus	0.79	0.33	0.12	0.06
Yogurt	0.79	0.30	0.13	0.07
Pasta Sauce	0.77	0.26	0.21	0.20
Veggie Burgers	0.68	0.25	0.20	0.12
Pasta	0.56	0.24	0.25	0.07
Frozen Fish	0.69	0.23	0.33	0.10
Salad Dressing	0.82	0.23	0.20	0.11
Canned Fruit	0.70	0.21	0.32	0.05
Milk Shakes	0.73	0.20	0.17	0.10
Crackers	0.85	0.20	0.16	0.06
Pretzels	0.72	0.18	0.21	0.07
Soup	0.82	0.18	0.21	0.19
Energy Bars	0.70	0.17	0.14	0.06
Canned Beans	0.74	0.17	0.17	0.15
Margarine	0.75	0.17	0.14	0.04
Cold Cereal	0.77	0.16	0.17	0.03
Flavored Rice	0.77	0.16	0.21	0.16
Nuts	0.76	0.14	0.21	0.14
Soft Chews	0.52	0.11	0.12	0.07
Frozen Meals	0.76	0.09	0.34	0.07
Juice – Non Citrus	0.80	0.07	0.12	0.07
Milk	0.59	0.07	0.10	0.08
Coffee	0.87	0.07	0.09	0.61
Tea	0.81	0.05	0.10	0.40
Water	0.70	0.02	0.15	0.04

As shown in Table 14.2 in this second study, the role of texture increases as a driver of purchase (rather than craving). Instead of being the lowest of the sensory factors, chosen on average 15% of the time, texture is chosen 22% of the time as a driver of product purchase, more frequently on average than appearance and aroma. The improved performance of texture may be due to the nature of the study (focus on purchase, rather than on craving. In both studies, taste remained factor #1, as a driver of craving (Crave IT!™), and as a driver of purchase intent (Healthy You!™).

14.7 PART 2: BEYOND SIMPLE STATISTICS TO THE GRANULARITY OF EXPERIENCE

Tables 14.1 and 14.2 show us that texture is important, but not as important as taste, at least when respondents are instructed to select the factors that drive craving or purchase. Yet, simple tables like Tables 14.1 and 14.2 fail to give us a sense of the experience of texture. To get a better sense of the experience, we turn now to descriptions about texture in different foods and how they perform as drivers in interest in a concept when the descriptions are embedded in the concept together with other elements. Or, in simple terms, we now move to texture as an element in the description of a product experience.

The task of understanding texture through Mind Genomics® uses the statistical method of experimental design, particularly that of conjoint analysis. We focus on the consumer response to compound concepts about a food experience, with the concept comprising different 'messages' or elements. Some of those elements describe the 'texture' of the food, in language that a person can understand, that paints a '*mental word picture of the experience*' (Box, Hunter and Hunter, 1978; Moskowitz, 1994; Moskowitz, Gofman, Itty, Katz, Manchaiah and Ma, 2001; Wittink and Cattin, 1989).

The conjoint measurement research tool traces its history to work on basic measurement theory done a half century ago (Luce and Tukey, 1974). Applied researchers in the world of mathematical psychology quickly recognized conjoint measurement as a way to understand decision processes, the implicit 'weighting rules' by which a person decides which of several alternatives to choose. Conjoint measurement soon ended up becoming one of the favorite tools of the researcher, migrating from the field of experimental psychology to marketing, and then on to a host of different specialty areas where 'decision rules' were the topic of study (see Johnson, 1974; Green and Wind, 1973 for early applications, and Wittink and Cattin, 1989 for an exposition of how conjoint measurement was being using just 15 years later).

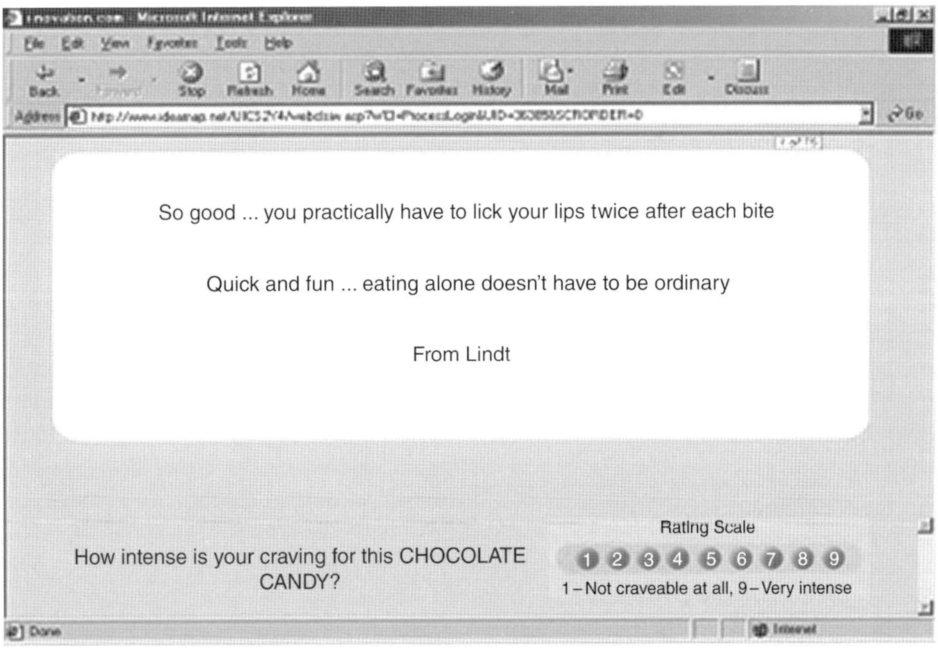

Figure 14.1 Example of a test concept for chocolate, from the CraveIT!™ mega study on chocolate. Each respondent evaluated 60 different vignettes, created by experimental design. The experimental design was permuted for each respondent, allowing the evaluation of many different combinations across the respondents. (Source: Courtesy of IT! Ventures, LLC. Reproduced with permission.)

The fundamental stimulus for Part 2 is a concept, a vignette, a combination of elements, shown in Figure 14.1. The elements come from descriptions of the product, from emotions, from brands, and so forth.

We see examples of the elements in Table 14.3. The specific topic food was chocolate, but the structure holds for the other products as well in the Crave IT!™ mega-study.

Table 14.3 The 36 elements for chocolate from the 2001 Crave IT!!™ study. Texture-relevant phrases are shown in italicized font. (Source: Courtesy of IT! Ventures, LLC. Reproduced with permission.)

	Silo A – product description (including statements about texture)
A1	A *smooth, dense* piece of chocolate
A2	Smooth appearance with a light chocolate flavor and a creamy texture
A3	*Crispy wafers* coated in thin layers of milk chocolate
A4	Real chocolate made with ingredients like chocolate, cocoa butter, vanilla, and sugar
A5	White chocolate with *crunchy cookie pieces* throughout
A6	*Heavy dense chunk of chocolate* with complex flavors, velvet appearance ... enticing aroma
A7	*Dense chocolate* with swirls of dark chocolate and chocolate sprinkles on the surface
A8	Clusters of chocolate and nuts, with caramel and marshmallow throughout
A9	Golden milk nougat with whole almond pieces on top, caramel drizzled over them and enrobed with semi-sweet chocolate
	Silo B – usage situation, emotions
B1	When it's cold outside, chocolate is cozy and inviting
B2	With a hot cup of coffee, tea, hot cocoa ... or carbonated beverage
B3	Bite size pieces; ready for a fast taste ... with a chocolate truffle filling
B4	Premium quality ... that great classic taste, like it used to be
B5	You can just savor it when you think about it during school and work
B6	100% natural ... and new choices every month to keep you tantalized
B7	With fruit fillings in any flavor you want
B8	You can imagine the taste as you walk in the door
B9	So good ... you practically have to lick your lips twice after each bite
	Silo C – emotions and rewards
C1	Quick and fun ... eating alone doesn't have to be ordinary
C2	When you think about it, you have to have it ... and after you have it, you can't stop eating it
C3	Fills that empty spot in you ... just when you want it
C4	When you're sad, it makes you glad
C5	Now you can escape the routine ... a way to celebrate special occasions
C6	A joy for your senses ... seeing, smelling, tasting
C7	An outrageous experience ... shared with family and friends
C8	Pure ecstasy
C9	It feeds THE HUNGER
	Silo D – brands and reassurances
D1	From Hershey's®
D2	From M&M/Mars®
D3	From Nestlé®
D4	From Cadbury®
D5	From Lindt®
D6	From Godiva®
D7	Made fresh ... especially for you
D8	Simply the best chocolate in the whole wide world
D9	With the safety, care and cleanliness that makes you trust it & love it all the more

14.8 ANALYZING THE DATA TO UNDERSTAND THE CONSUMER MIND

The objective of Mind Genomics® is to understand how consumers react to phrases about their ordinary, everyday lives. In this case, we were studying chocolate and other foods. To the degree that the elements comprise messages dealing with texture, the analysis showed how texture, among other messages, 'drives' consumer responses.

The analysis followed a specific series of steps, listed below. The steps, comprising an algorithm, began with the response to compound messages, and ended up deconstructing the compound message into the contributions of the components.

1. Each respondent evaluated 60 different concepts, with every element appearing five times in the array of 60, and absent 55 times from the 60 concepts. The elements were present in an experimental design, so that the 36 elements appeared as statistically independent ideas, that is, 'free agents,' that could be mixed and matched. This statistical independence of the elements was important for subsequent modeling by OLS (ordinary least-squares) regression analysis.
2. The respondent read each test concept and rated the concept on an appropriate rating scale, here 'craveability,' (how intense is your craving?) Typically, the rating scale comprised 9-points, although other ratings scales would have done as well.
3. The ratings from the 9-point scale were transformed to create binary responses. Ratings of 1–6 were transformed to 0 to represent 'not craved,' and ratings of 7–9 were transformed to 100 to represent 'craved.' This transformation comes from the origin of these conjoint studies in the world of consumer research and before that in the world of sociology. In both fields interest focuses on membership in a class (e.g., 'craved versus not craved'), and not on the degree of feeling (degree of craving).
4. For each respondent, OLS regression related the presence/absence of each element to the binary, 0/100. OLS regression analytics can be found on most commercially available statistical packages (e.g., Systat, 2007).
5. The equation was: Binary Rating $=k_0+k_1$(Element A1)$+k_2$(Element A2)..k_{36}(Element D9).
6. The coefficient, or impact value of the equation, k_i, showed the conditional probability of the vignette being rated 'craveable' when the element 'i' was incorporated in the test concept.
7. We compared the values of the impact values across elements and studies, as we have in other mega-studies.

14.9 UNDERSTANDING TEXTURE IN THE CONTEXT OF THESE CONJOINT 'MEGA-STUDIES'

Now that we have explicated the approach to discover the contribution of the various elements to the binary rating of craving (1–6 $\rightarrow$ 0 or do not crave; 7–9 $\rightarrow$ 100 or do crave), we move to looking at the contribution of the various elements, that is, the impact values. We concentrate only on the elements from the first silo, A1–A9, the elements that describe the characteristics of the products. We have removed all the elements that lack a phrase that can be construed as textural. We have approximately half of the elements left.

Table 14.4 shows us different foods (first column), the phrases from Silo A describing the product, (second column), the impact value of the description from the total panel (third column), and impact value for the description from those respondents who said that 'texture' is one of the three key factors. The elements were sorted in descending order of impact value. Only the very strong performing elements are shown, with these elements defined as generating an impact value of +15 or higher for either the total panel (first data column) or for those respondents who said that texture was one of the three major factors driving craveability (second data column).

Table 14.4 Impact value of elements from the 2001 Crave IT!™ mega study. The table shows the high performers (impact value > 15), comparing the total panel to respondents who say that texture is critical as a factor driving cravability. The words describing texture are in *italics*. (Source: Courtesy of IT! Ventures, LLC. Reproduced with permission.)

		Total Sample	Texture critical
Steak	Juicy filet mignon cooked medium rare to melt in your mouth	20	19
Chicken	*Plump*, juicy chicken breast, marinated in a special sauce and cooked over an open-fire for a smoky grilled taste	20	26
Tacos	Homemade *soft* taco shells wrapped around warm simmered meat and topped with *chunks* of tomato and *shreds* of lettuce and cheese	20	7
Cheese cake	Cheesecake with swirls of raspberry, *chunks* of white chocolate, baked in a *crunchy* crust and garnished with pecans	19	13
Ice cream	Sundae with scoops of ice cream, *thick* sauce, chopped nuts, real *whipped* cream, and a bright red cherry	19	16
Cheese cake	*Dense* cheesecake swirled together with ribbons of chocolate chips in a chocolate crust ... served with a raspberry sauce drizzled across your plate	18	17
Cinnamon rolls	Big 3 inch spiraled rounds of *dense chewy* pastry like a donut with sweet cinnamon inside, covered with sweet icing	18	29
Cheese cake	Cheesecake so rich it melts slowly to release delicate, intense flavor	17	18
Hamburger	Lots of *crispy* bacon and cheese on a juicy grilled hamburger on a lightly toasted bun	17	26
Cheese cake	Cheesecake with a *smooth* appearance, light flavor, *fluffy creamy* texture and a graham cracker crust	16	17
Cinnamon rolls	Bite size spiraled rounds of *flaky* pastry with sweet cinnamon inside, dripping with sweet white icing	16	19
Pretzel	Big *soft* pretzels with small shiny *nuggets* of salt	15	18
Chicken	Chicken coated in your favorite batter and spices, then deep fried for a *crunchy* good taste	14	20
Cinnamon rolls	Huge, *thick*, 4 inch spiraled rounds of light *flaky* pastry with sweet cinnamon inside, covered in a cream cheese frosting	13	20
Chocolate	Heavy *dense chunk* of chocolate with complex flavors, *velvet* appearance ... enticing aroma	12	16
Tortilla chips	Tortilla chips cooked with seasoning ... every bite gives you a unique Mexican flavor and *crunch*	11	15
French fries	Long, skinny fries lightly salted and *crispy* on the outside	11	15

Table 14.4 shows that:

1. There are three elements with exceptionally high impact values, of coefficients in the regression model, 20 or higher. The impact value of 20 means that when the particular element was incorporated into the vignette, an additional 20% of the respondents would be expected to change their rating from not interested (rating of 1–6) to interested (rating 7–9).
2. The strong performing elements came from a variety of different foods, not just one food.
3. The strong performing elements incorporated texture as part of the description, rather than relying on texture alone to carry the entire element. In fact, for these strong performing elements, texture seemed to be an integral part of the description, complementing the ingredient, rather than seizing and occupying the reader's attention. A good example is *Juicy filet mignon cooked medium rare to melt in your mouth.*
4. Recall from Part 1 that the second part of the interview comprised a self-profiling classification. The second data column came from those respondents who checked 'texture' as one of the three factors which drive craveability. These respondents defined themselves as texture-sensitive, with sensitivity defined as assigning texture a degree of importance. The other factors were taste, appearance, aroma, but also packaging, price, and so forth. For most elements, but not for all, those who said that texture was important for the particular ended up with higher impact values for texture-relevant elements than did the total panel, even though it would be impossible to engineer this result at a conscious level.
5. Our data from these early mega-studies suggested that the sensory descriptions of a product drive craveability because elements with some mention of texture were found in the very strong performing group (impact>15). Those respondents who said that texture is important often appreciated the foods more than the total panel; they ended up showing higher impact values than the total panel showed, for many of the strong-performing elements.

Our first two data tables told us how people thought about products but they don't tell us about how people responded to the experience. Psychophysics can tell us about the intensity of the sensation, but don't give us a sense of the qualitative nuances. Descriptive analysis by experts can tell us about the attributes of the experience, but they are often limited to dry 'checklists,' on which experts are trained, and which are used to create a fingerprint of the product. We get a sense of the experience by looking at the results from conjoint measurement, especially the Crave IT!!™ mega-study. Table 14.3 suggests that while texture was often a key driver of acceptance, typically it took a back seat to flavor.

14.10 PART 3: MIND GENOMICS® – BEYOND THE GENERAL TO THE SPECIFIC EXPERIENCE

The term 'Mind Genomics®' is a metaphor, a phrase that conveys the notion that our external variation in responses to the world of what we perceive come from a limited set of basic mind-sets. The task of the Mind Genomics® science is to identify those mind-sets for each particular domain being studied. The essence of Mind Genomics® is to study everyday life, to identify what's important, and then to identify mind-set segments.

Mind Genomics® expands the insights provided by conjoint analysis. The objective of Mind Genomics® is to understand everyday experience through science, to understand how perception organizes our lives, how our values drive responses to stimuli, and how people differ from each other.

Part 3 of this chapter applies the principles of Mind Genomics® to thinking about texture, and other sensory experiences a person has with a product. Mind Genomics® works with descriptions of experience, looking at how the typical consumer responds to these descriptions, and of course how the components of the descriptions drive ratings. Thus in the next two studies we present here, Mind Genomics® will tell us how people responded to descriptions of texture, first in a not-well-defined 'healthy snack' (Part 3) and then in a very well defined healthy yogurt (Part 4). Through experiments, we discover the nature of the sensory experience, in terms of the dollar value of the experience (sensory economics of texture), the emotions underlying the experience, and finally the existence and nature of mind-sets, and specifically, groups of individuals who are specifically oriented towards texture.

14.11 HOW TO LEARN ABOUT THE RICHNESS OF THE EXPERIENCE – THE NOTION OF SILOS AND ELEMENTS

Part 2 of this chapter introduced the notion of silos and elements in an experimental design, presenting the results from a large-scale study regarding what descriptions make a person say that a food is craveable. We now move into a specific investigation of the different sensory aspects of food, and specifically the role of texture. We built test concepts comprising elements specifically talking about texture, as well as elements talking about other sensory experiences, such as appearance, taste, aroma, and even swallowing. Our rationale was that we would discover much more when we explored the mind's reaction to texture in a designed experiment where texture was one of several key variables. That is, we moved from attitudes towards food (Part 1), to attitudes about food description with texture as an occasional component (Part 2), to experiments dealing with texture and other sensory descriptions (Parts 3 and 4).

14.12 CREATING THE 'SYNTHETIC' PRODUCT IDEAS TO STUDY EVERYDAY LIFE

Rather than studying full concepts about a food, including brand, emotion, and product description in the way we did it in Section 14.2, we created systematically varied concepts about a food, incorporating in those concepts a description of the way the product appeared, smelled, tasted, felt, and even the way one felt when chewing and swallowing, respectively. These elements were not simply bare-boned phrases of the type one might get in a scientific or marketing briefing. Rather, the elements were phrased in the way people might describe their experiences, phrasing that isn't in literature perhaps, but phrasing that might come from the pen of an advertising agency whose job it was to describe the food, communicate the experience, and entice the prospective buyer.

We began with the first experiment in this section, whose 36 elements are shown in Table 14.5. The six categories or silos were set up to provide a verbal walk-through of a

Table 14.5 The 36 elements for the healthful snack, arranged into six categories or silos, each comprising six elements. (Source: Courtesy of IT! Ventures, LLC. Reproduced with permission.)

	Category A: See it
A1	Imagine every drop of honey blended with brown sugar syrup... a mouthwatering experience unlike any other!
A2	All-natural goodness you can see with a sweet yogurt coating
A3	Ingredients you can see...know exactly what you're eating!
A4	Wholesome goodness you can see... seeing is believing
A5	You can see every ingredient...that's what makes it so good
A6	100% organic ingredients you can see ... gets your appetite going
	Category B: Hear it
B1	As you chew, it makes a distinct sound to signal that your energy is going again
B2	This is no ordinary sound... experience the sound of the pure texture that nature has to offer
B3	Munchy and crispy...just thinking about it makes you want to run out to get it
B4	Hear every bit of the crunch in every bite! Absolutely delicious...naturally nutritious
B5	Soft & slippery... like a stream of running water
B6	Indescribable sensation... crackling and snapping...like fireworks exploding in your mouth
	Category C: Smell it
C1	Enticingly nutty aroma... hazelnuts, almonds, pistachios all crushed and blended together...
C2	The aroma of roses and jasmine...will have you coming back for more
C3	A delicious sweet and fruity mango and papaya aroma...faint enough to remind you of summer...strong enough to have you asking for more
C4	Smells like the old fashioned butter cookies Grandma used to make!
C5	Invigorating, delectable aroma...as if you just walked into a bakery
C6	Has an inviting peanutty aroma that leaves you feeling satisfied all over....
	Category D: Chew it and taste it
D1	Mmm...sticky yet light...and just a tiny bit salty
D2	Soft & chewy with a savory aftertaste
D3	Chew it to your heart's content...absolute pleasure!
D4	So chewy...with a faint buttery aftertaste
D5	Mouthwatering chewiness you'll relish right from your first bite...simply heaven
D6	Extra chewy.....for extra enjoyment
	Category E: Swallow it
E1	So light and smooth...close your eyes and enjoy!!
E2	Glides down like honey...unforgettably satisfying
E3	Tingles your taste buds as you swallow... and for a moment you'll think you're out of this world!
E4	Swallow slowly ... appreciate the moistness and richness as you take another bite
E5	Like silk inside your mouth
E6	If you've had trouble with your little ones swallowing other wholesome foods, don't worry... your trouble days are over
	Category F: Feel something... fullness, satisfaction
F1	Not filling... just enough to keep you going
F2	Feels like you are one with nature and all is well with the world
F3	Satisfy your hunger a little different this time
F4	Leaves you with a feeling of nostalgia... just like when you were a kid
F5	Doesn't just give you a quick boost of energy...makes you feel like you can do just about anything
F6	Relaxes every muscle in your body... worth taking a break for

possible sensory experience. The product was a healthful snack; the precise nature of the snack was deliberately left vague. The specifics of the healthful snack beyond the actual descriptions were to be conjured up in the mind of the respondent, who was reading the test concepts and rating them.

14.13 EXPERIMENTAL DESIGN – SYSTEMATIZING THE STRUCTURE OF THE TEST CONCEPT

We are taught that in order to understand how a phenomenon works, indeed how a specific part of the world functions, we are best off identifying that factor that we want to study, and then creating conditions by which we can isolate that factor and focus on it. Many researchers approach the problem of understanding a set of stimuli by the strategy of isolating the variables, suppressing the noise, and then studying the variables, a three-step approach that comes from a philosophy (and strategy) of science popular during the past several hundred years.

The reality of consumer research is quite different. The world of food excites several senses at once. Understanding reactions to one sense alone, especially messages about the perceptions occasioned by that sense, misses the rewarding complexity of a real world experience. For example, just knowing that a product looks 'dark and dappled' tells us only part of what we need to know to make a judgment of the product. We'd like to know what it smells like, or perhaps even more important, what it tastes like. We find increased richness and verisimilitude, reality, when the description comprises statements about the experience. Indeed, most good literature describing food doesn't stop with one sensory aspect, but looks at the kaleidescope of different experiences occasioned by a single product.

Our experiment works by experimental design, the systematic variation of messages, to create a more realistic description. We work only with the elements in Table 14.5, mixing/matching them into combinations with the property that a single test concept comprises at most one element from each category or silo, and at most four elements. This stratagem of presenting the respondent incomplete yet reasonably detailed combinations of descriptions makes it possible later on to deconstruct the response to the test concepts into the contribution of each element.

Experimental designs provide a 'recipe book' for the test combinations. The experimental design provides the numerical representation that will be needed for the subsequent statistical analyses. Table 14.6 shows part of a statistical design for one respondent.

Table 14.6 An example of the data that would be used both to represent the concept structure, as well as code the responses. The table shows part of the experimental design for four concepts, as well as the ratings, and the transformed values of the ratings in a form that could be readily used by statistical analysis software. (Source: Courtesy of IT! Ventures, LLC. Reproduced with permission.)

	Conc1	Conc2	Conc3	Conc4
A1 (from Category A)	0	0	0	0
A2 (from Category A)	0	1	0	0
A3 (from Category A)	1	0	0	1
F4 (from Category F)	0	1	0	0
F5 (from Category F)	0	0	0	0
F6 (from Category F)	0	0	0	0
Question #1: Rate1 (1–9)	2	3	2	7
Question #2: Select a price from the set of prices	1	2	2	4
Transformed Question #1 (Binary 1–6 →0; 7–9 →100)	0	0	0	100
Transformed Question #2: Price$	6	8	8	12

The table is necessarily incomplete; to show the coding for all 36 elements requires six times the space. However, we can get a good idea of what can be done from looking at the table.

1. The first set of rows, A1 – F6, correspond to the elements. We see only a partial set of rows, six to be specific, three from Category A/Silo A, three from Category F/Silo F, respectively. In actuality, there were 36 rows, one row for each of the 36 elements. The experimental design called for six silos, A-F, each with six elements.
2. Each respondent evaluated a set of 48 concepts from a unique design, that unique design being an isomorph of the basic, underlying design. The structure of the basic design was the same for all respondents. *However, the elements were permuted.* This stratagem made the results more robust by presenting respondents with an exceptionally wide range of different concepts, rather than repeating the same 48 concepts, respondent after respondent (Gofman, 2006; Moskowitz and Gofman, 2004).
3. The columns Conc1 – Conc4 show the structure of the first set of concepts, concepts 1–4. These concepts constituted four of the 48 concepts that a respondent evaluated.
4. In a row, the number '0' represents the fact that the element was absent from the concept. The number '1' represented the fact that the element was present in the concept. If we were to have presented all 36 elements, that is, shown the 36 rows, each column would have comprised 3–4 '1 s' and the remainder 0 s. In fact, the average of each column, that is, each concept, would have been exactly 3.75, as dictated by the experimental design.
5. The row labeled as Question #1 was the 1–9 scale, for likelihood the respondent would buy the product.
6. The row labeled as Question #2 was the price selected from a list of prices. There were seven prices from which to select. A respondent was instructed to read the concept, rated the overall feeling (Question #1), and then selected a price (Question #2).
7. The row labeled as Transformed Question #1 is the binary recode of the first rating question, Q1. When Q1 is low, between 1 and 6, the value is defined as 0. When Q1 is high, between 7 and 9, the value is defined as 100.
8. The row labeled Transformed Question #2 shows the dollar value corresponding to the actual selected number in Question #2. Since each number to be selected in Question #2 corresponds to a dollar value, we merely needed to substitute the dollar value.

14.14 THE ACTUAL INTERVIEW—THE ORIENTATION PAGE AND AN EXAMPLE OF A TEST CONCEPT

The essence of the experiment is the orientation and the test concept. Figure 14.2 shows the orientation page. Respondents have an easier time doing the interview when they have a sense of what is expected of them. Figure 14.2 tells the respondent what the experiment is about, but tries not to 'lead' the respondent in terms of the expected or appropriate answer. The best orientation pages are those that instruct, telling the

We are interested in your opinion about a totally NEW HEALTH snack for Today's hectic life.

On the following screens you will be presented with concepts (vignettes) and will be asked to rate each one on 2 questions:

1) Based on what you see and read here, how much do you want to eat this snack?
1=Not at all .... 9=Very much
and
2) *What do you think is a fair price to pay for a package of 48 at a club store?*
1=$6, 2=$8, 3=$10, 4=$12, 5=$14,
6=$16, 7=$18, 8=$20, 9=$22

Please make sure to use the entire 1-9 scale.
At the end. we will ask just a few questions to help us understand your needs.

Figure 14.2 The orientation page for the healthy snack study.

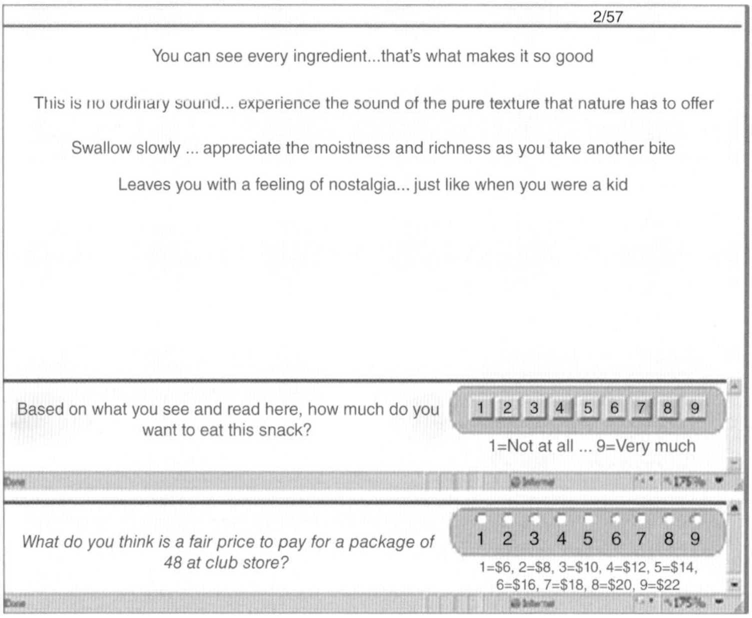

Figure 14.3 A test concept and the two rating scales.

respondent what s/he is supposed to do, instructing the respondent regarding how to read the test concept (i.e., as an entirety, not as a set of disparate, disconnected elements), and finally how to rate the concept and the number of minutes that the survey will take. It is important to put the respondent into the appropriate framework of mind, yet not to lead the respondent.

An example of the actual concept and the two rating scales appears in Figure 14.3. The respondent read one concept and rated the concept on the two attributes. Other studies of the same type intermixing concepts and rating scales, that is, with a concept read and rated on one of the two scales, ended up being an onerous task. It is easiest for the respondent to read one concept and rate the concept on two scales.

14.15 BUILDING THE CONCEPT MODEL AT THE INDIVIDUAL RESPONDENT LEVEL

The key to Mind Genomics® is to understand how the individual elements 'drive' the ratings, that is, the possible arithmetic mechanism underlying the ratings. The test concepts are merely vehicles by which one can test the elements in a more naturalistic way. By embedding these elements in concepts, one creates a system that cannot be 'gamed.'

In these concepts, the elements compete with each other to drive the response, rather than appearing in isolation. Combinations of elements prevent the respondent from changing her/his strategy in a conscious or unconscious way, when the element changes from one dealing with taste to one dealing with texture. Such conscious or unconscious shifts in the criteria of judgment occur all the time; simply think of evaluating brand name, then price, then sensory description of a product. The criteria must change, as the respondent copes with demand to put a 'comparable' number on each test stimulus, no matter the type of element it is.

At the level of the individual respondent, the experimental design prescribed what elements went together in each test concept. We knew from the design that each element appeared independently, and appeared five times in the set of 48. Such independence enabled us to relate the presence/absence of the elements to the rating, whether the rating was acceptance rated on the 1–9 scale (question #1), or the rating a dollar value selected from the set of different dollar values (question #2).

A simple linear equation expressed the (possible) relation between rating and the presence/absence of the element in a concept. There could be many other expressions, that is, equations, depending upon our theory of how elements combine to generate a rating or a dollar value, but the simplest in theory and expression is that the elements each contributed a fixed amount to the overall rating.

We express the relation as: Rating $= k_0 + k_1$(Element A1) $+ k_2$(Element A2) ... k_{36} (Element F6). The full set of data comprised 37 numbers, the additive constant, along with the individual coefficient or impact value of each of the 36 elements.

To create the models at the level of the individual respondent required the following four straightforward, simple steps, made possible because of the provision that each respondent's unique set of 48 test concepts was constructed by an experimental design.

Step 1 – *Preparation*: Prepare the data for analysis by regression modeling, following the coding shown in Table 14.6.

Step 2 – *Persuasion Model*: For each person, one at a time, create the so-called Persuasion Model. This Persuasion Model uses the 9-point rating as the dependent variable, and the presence/absence of the 36 elements as the independent variables. OLS (ordinary least-squares) regression returns with an estimate of the additive constant and the 36 coefficients, one coefficient per element. Often the Persuasion Model is used for segmentation. For some equations (e.g., question #2; price), OLS might estimate the 36 coefficients, but force the equation through the origin by forcing the additive constant, k_0, to be 0. Typically such 0-values for the additive constant are the rule when we deal with price, with the selection of an emotion from a list of emotions, or when we deal with graphical elements. In all three cases, no elements mean no price to be paid, no emotion to be selected, or no visual package to be inspected.

Step 3 – *Interest Model*: Another model, one that we use more frequently, is the Interest Model. Do the same analysis as described in Step 2, except that this time the dependent variable is the binary transformed value, which, as Table 14.6 shows, takes on the value 0 when the rating was 1–6, and takes on the value of 100 when the rating was 7–9. It will be these binary values and the Interest Model, which furnish us with the numbers we will discuss in the results.

Step 4 – *Dollar Model*: Use OLS regression to relate the presence/absence of the 36 elements to the dollar value. Do not, however, estimate the additive constant. Rather, use the option of forcing the regression model 'through 0.' The rationale is that respondents would not be willing to pay for something they know nothing about, that is, would not pay for something that is not described, other than being a dessert.

14.16 THE RESULTS

The simple experiment with an undefined healthful dessert produced a wealth of information, primarily because each of the 36 elements by themselves was 'cognitively rich,' conveying information, arousing some level of interest. We learned from the response to each element, as well as learning from the total panel. Table 14.7 shows the full set of 36 elements, with the elements in descending order of dollar value, from the total panel.

1. When it comes to simple attitude, basic acceptance, the additive constant was 27, meaning that for this unspecified healthful dessert, 27% of the respondents were prepared to rate the concept 7–9, highly acceptable, even without any elements.
2. The elements did not perform well, with exception of two elements which painted very clear word pictures:
   a. *Enticingly nutty aroma... hazelnuts, almonds, pistachios all crushed and blended together...* One could almost sense the product (impact = 11)
   b. *Smells like the old fashioned butter cookies Grandma used to make!* (impact = 7)
3. Just because an element was liked does not mean that it could demand a high price. Respondents distinguished between what they liked and the dollar value. Dollar value was driven by the contents of the element, the specific product inside.
4. We didn't find high interest in the elements that talked about texture. These elements are shown in bold.
5. As a first finding, therefore, texture by itself did not appear to drive acceptance nor did texture command high dollars, at least for this first analysis of a generalized healthy snack product. The importance of texture may be in segmentation, in different mind-sets, smaller subsets of respondents who do care about texture, but whose individual results get 'washed out' when simply part of a large data set.

Table 14.7 Deconstruction of the data for the otherwise unspecified 'healthy snack,' showing the contributions of the elements to interest, and to the dollar value of the snack. Texture relevant elements are shown in bold type. The results were developed by OLS (ordinary least-squares) regression. (Source: Courtesy of IT! Ventures, LLC. Reproduced with permission.)

	Total panel, ranked by decreasing dollar value of the elements	Interest	Dollar
	Additive constant (base level, without any elements)	27	NA
C1	Enticingly nutty aroma...hazelnuts, almonds, pistachios all **crushed** and blended together...	11	$ 3.43
F5	Doesn't just give you a quick boost of energy...makes you feel like you can do just about anything	2	$ 3.08
A2	All-natural goodness you can see with a sweet yogurt coating	1	$ 3.08
A3	Ingredients you can see...know exactly what you're eating!	1	$ 3.08
A1	Imagine every drop of honey blended with brown sugar syrup...a mouthwatering experience unlike any other!	4	$ 3.07
C3	A delicious sweet and fruity mango and papaya aroma...faint enough to remind you of summer...strong enough to have you asking for more	1	$ 3.07
A6	100% organic ingredients you can see...gets your appetite going	1	$ 3.05
C4	Smells like the old fashioned butter cookies Grandma used to make!	7	$ 3.04
C5	Invigorating, delectable aroma...as if you just walked into a bakery	3	$ 3.03
D4	So **chewy**...with a faint buttery aftertaste	1	$ 2.98
F2	Feels like you are one with nature and all is well with the world	0	$ 2.98
A5	You can see every ingredient...that's what makes it so good	–1	$ 2.98
F3	Satisfy your hunger a little different this time	4	$ 2.94
F6	Relaxes every muscle in your body...worth taking a break for	–1	$ 2.94
F4	Leaves you with a feeling of nostalgia...just like when you were a kid	2	$ 2.92
B4	Hear every bit of the **crunch** in every bite! Absolutely delicious...naturally nutritious	1	$ 2.91
D6	**Extra chewy**...for extra enjoyment	0	$ 2.91
D5	Mouthwatering **chewiness** you'll relish right from your first bite...simply heaven	1	$ 2.89
D3	Chew it to your heart's content...absolute pleasure!	1	$ 2.88
B3	**Munchy** and **crispy**...just thinking about it makes you want to run out to get it	–2	$ 2.87
F1	Not filling... just enough to keep you going	–1	$ 2.86
B6	Indescribable sensation... **crackling** and **snapping**...like fireworks exploding in your mouth	–3	$ 2.86
C6	Has an inviting peanutty aroma that leaves you feeling satisfied all over....	–1	$ 2.85
A4	Wholesome goodness you can see... seeing is believing	–1	$ 2.84
B2	This is no ordinary sound... experience the sound of the pure texture that nature has to offer	–3	$ 2.83
B1	As you chew, it makes a distinct sound to signal that your energy is going again	–3	$ 2.79
E1	So **light** and **smooth**...close your eyes and enjoy!!	2	$ 2.72
E3	Tingles your taste buds as you swallow... and for a moment you'll think you're out of this world!	2	$ 2.71
E4	Swallow slowly ... appreciate the **moistness** and **richness** as you take another bite	1	$ 2.71
D2	**Soft & chewy** with a savory aftertaste	0	$ 2.70
E6	If you've had trouble with your little ones swallowing other wholesome foods, don't worry... your trouble days are over	0	$ 2.61
D1	Mmm...**sticky** yet light...and just a tiny bit salty	–1	$ 2.58
E2	Glides down like honey...unforgettably satisfying	3	$ 2.56
E5	Like **silk** inside your mouth	2	$ 2.52
B5	**Soft & slippery**... like a stream of running water	–5	$ 2.50
C2	The aroma of roses and jasmine...will have you coming back for more	–5	$ 2.50

14.17 MIND-SETS: IS THERE A GROUP WILLING TO PAY A LOT MORE FOR TEXTURE?

Although there was no evidence from the total panel that respondents either liked texture a great deal, or were willing to pay more, mind-set segmentation often uncovers groups of respondents showing radically different preference patterns. These respondents might have 'looked like' all the others in the study; they responded strongly to a common set of elements, with these elements 'painting a picture' of a specific way of thinking about the product or service being studied by RDE.

Conjoint analysis is well suited to discovering these segments because it deconstructs the respondent's ratings to the part-worth contributions. Applying the statistical armory of clustering reveals groups of respondents whose patterns of utility values (coefficients) suggest a common and meaningful theme (see Green and Krieger, 1991; Moskowitz, 1996). We applied the same segmentation 'thinking' to our data on the healthy snack, basing our clustering on the dollar value of each item that a respondent would pay. The RDE study gave us rich data with which to work; each respondent generated 36 such dollar values from the OLS regression applied to the data of the respondent.

1. Recall that from Table 14.5, we were working with six different categories or silos, each silo dealing with elements that were presumed to excite 'seeing,' 'hearing,' 'smelling,' 'chewing,' 'swallowing,' and 'inside sensation,' respectively. The elements were expressed in the way a person might be expected to see in an advertisement.
2. We created a dollar model for each respondent showing the dollar value that would be assigned by that respondent to the 36 elements. The ingoing experimental design of the concepts at the level of the individual respondent, coupled with OLS regression, created this individual-level 'dollar model.'
3. We clustered the respondents by the pattern of their 36 dollar values, one dollar value for each element.
4. At the start of the segmentation, we didn't know what we would find. Segmentation through cluster analysis is a mathematical procedure. It is up to the researcher to name the clusters, that is, our mind-set segments. We did so by finding a limited number of segments (parsimony), and were able to 'tell a story' about each segment so that the story was simple and compelling (interpretability). We had a third criterion; we were looking for a segment that was driven by texture phrases.
5. We did three clusterings, using the clustering program (Systat, 2007). We first divided our 240 respondents into two complementary groups, then into three complementary groups, and finally into four complementary groups. This created a so-called 'two-segment' solution, then a three-segment solution, and finally a four-segment solution. Every respondent had to fall into one of the two groups when we created a two-segment solution, or fell into one of the three groups when we created a three-segment solution, and so forth.
6. We looked for a segment that was willing to pay substantially more for the chewing/hearing elements. When we found this segment, we had evidence of a group of respondents who are texture-sensitive.

The foregoing strategy for discovering a 'texture-minded' segment ended up being successful, but only because it divided people into more homogeneous groups, whose 'driving elements' emerged clearly. The results were compelling with the four-segment solution, which showed a segment willing to pay a lot of money for elements that described the chewing experience.

14.18 HOW THE CLUSTERING REVEALS THE SEGMENTS, AND HOW TO MAKE THE DECISION (TABLE 14.8).

Clustering to generate mind-set segments is a hallmark of Mind Genomics®. As such, it is worth looking at in a bit more numerical detail. Table 14.8 shows the sequence of steps:

Table 14.8 The search for the 'texture segment,' i.e., the segment that values the experience of chewing as described by the elements. The dollar values are the average for the segment, of the six elements in the category or silo. The discriminating capability is the F ratio from discriminant function analysis. A 'texture segment' emerged almost immediately in the two segment solution, and became sharper in the three and four segment solutions. (Source: Courtesy of IT! Ventures, LLC. Reproduced with permission.)

Sensory 'silo'	**Average dollar value for the six elements in the sensory silo – by the segment**				**Discriminating Capability**
Base	240				
A: See	**$3.02**				
B: Hear	$2.79				
C: Smell	$2.99				
D: Chew	$2.82				
E: Swallow	$2.81				
F: Internal	$2.95				
	Seg1	**Seg2**			**F ratio**
Base	126	114			
Sensory silo					
A: See	**$3.72**	$2.33			58.62
B: Hear	**$3.24**	$2.39			18.32
F: Internal	$2.65	**$3.36**			13.74
D: Chew	$2.42	**$3.09**			13.45
E: Swallow	$2.61	**$3.05**			5.09
C:Smell	**$3.09**	$2.77			2.59
	Seg1	**Seg2**	**Seg3**		**F ratio**
Base	89	77	74		
B: Hear	$2.72	$1.98	**$3.86**		34.74
C: Smell	**$3.60**	$2.82	$2.26		18.30
E: Swallow	$2.17	**$3.29**	**$3.09**		14.81
A: See	**$3.66**	$2.54	$2.88		12.29
F: Internal	$2.97	**$3.53**	$2.45		10.35
D: Chew	$2.54	**$3.19**	$2.51		5.71
	Seg1$	**Seg2$**	**Seg3$**	**Seg4$**	**F ratio**
Base	76	53	60	51	
A: See	**$3.98**	$2.76	**$3.13**	$1.94	23.03
B: Hear	$2.76	$1.75	**$3.88**	$2.86	21.28
E: Swallow	$2.22	**$3.71**	**$3.26**	$2.26	17.18
C: Smell	**$3.60**	$2.99	$2.27	$2.68	10.30
F: Internal	**$3.04**	**$3.62**	$2.35	$2.99	7.07
D: Chew	$2.55	$2.91	$2.30	**$3.36**	6.07

1. *Operational Definition*: We 'operationally defined' the texture-minded segment as a group of respondents who were willing to pay $3.00 or more, on average, for the elements dealing with chewing and swallowing. Not all the elements had to generate a dollar value of $3.00 or more, but the average for the texture elements had to be higher than $3.00. This operational definition would tell us if/when we had discovered the texture segment.
2. *The four sections*: Table 14.8 divides into four results sections, the first for the total panel, the second through the fourth for the 2–4 segment solutions, respectively.
3. *The six silos*: The first column shows the six sensory silos. These silos are sorted by the F ratio (described by step 4, below).
4. *Segmenting power*: The second column shows the F ratio, the statistic which shows the ability of the particular category or silo to drive the segmentation.
5. *Dollar value*: The remaining columns show the average dollar value for each category for each segment.
6. *Criterion*: We looked for texture-relevant elements in the categories which described the sensory impression that we get when we chew and swallow. The categories showing the greatest dollar value appear in bold type.

As we began the process of segmenting respondents, Segment 2, the 'texture segment' emerged almost immediately for this particular product. As the clustering algorithm moved from creating two segments, to three segments, to four segments, the segmentation became more focused. However, Segment #2 remained the most likely 'texture segment.'

14.19 PART 4 – MIND GENOMICS® – TEXTURE AND EMOTIONS

Let us go one step further with Mind Genomics®, moving beyond liking and economics, to emotions and economics. At the time of this writing, mid-2011, the notion of emotion in the product experience has become increasingly important. Beginning in the world of consumer research with the idea that an 'engaging experience' of a positive nature is to be desired, sensory researchers have recognized that perhaps this idea of 'engaging' can be translated into a positive 'emotional experience' while eating.

In this fourth section we use Mind Genomics® to explore the relation between statements about sensory experience generated by a yogurt, dollar value (as we did for the healthful snack) and the selection of feelings/emotions (new). We were guided by early work on yogurt, using the same approach, but done ten years earlier, when Mind Genomics® was less developed (Luckow, Moskowitz, Beckley and Hirsch, 2005).

The approach is straightforward, following the test structure in Part 3:

1. *Select the food and the elements*. Here we studied responses to statements about a specific product, healthy yogurt, a dairy product that excited a plethora of sensory impressions, textural impressions being very important. (We need only witness the current wild popularity of heavy texture Greek yogurt to recognize how important texture is.)
2. *Create the test concepts according to an experimental design.*

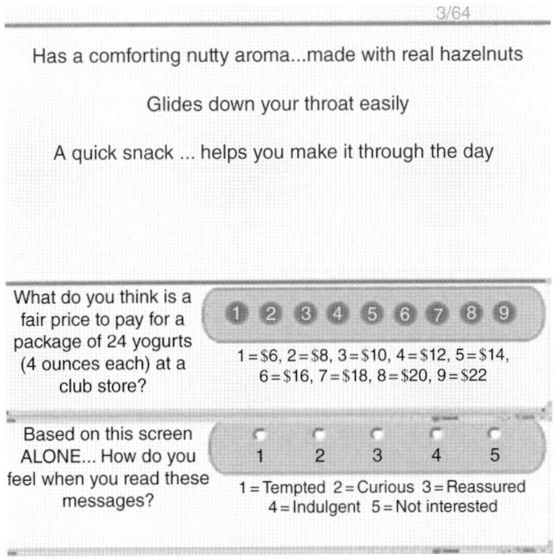

Figure 14.4 Screen shot of a concept about a healthy yogurt. The respondent first selects the dollar value for 24 - 4 oz. containers at a club store, and then selects the feeling/emotion experienced when reading the concept.

3. *Instruct respondents to rate these concepts on the response attributes of interest.* Here the first rating attribute was price (see Figure 14.4 for one concept with the price questions), the second rating question was the selection of a feeling/emotion (see the bottom of Figure 14.4 for the feeling/emotion question).
4. *Select five different feelings/emotions, from a list of several dozen.* With each respondent ending up rating 48 concepts, it was critical to keep the number of feelings/emotions within a reasonable limit.

14.19.1 The dollar value of texture in yogurt – total and among the four ethnic groups (Table 14.9)

Do texture elements end up being 'prized' by respondents, or were they just 'there' as factoids?

1. In these RDE studies, the respondent selects a particular response which best describes a vignette. The response can be a rating of acceptance on a scale, in which case the respondent would be instructed to select the appropriate level of acceptance on a scale of graded levels. The response can be select a price, for example, a dollar value, that would be most appropriate for the vignette being read. In this latter case, we provide a set of graded dollar values, all within a reasonable range that would be expected for the product. The respondent reads the vignette, and selects the appropriate dollar value, which dollar value then becomes the dependent variable in a regression model. Therefore, we related the presence/absence of the 36 elements to the total dollar value selected. This analysis used OLS (ordinary least-squares) regression, following the approach we used before, in Part 3. The 'dollar model' had no additive constant. Table 14.9 shows the dollar values for the total panel and for the four mind-set segments.

Table 14.9 Dollar value for elements. Texture elements are shown in bold type. Elements commanding $3.30 or higher are shown in bold type. The table shows the performance of elements for total panel, and for the four ethnic groups, recruited specifically to produce a study with 'readable' base sizes of the four ethnic groups. (Source: Courtesy of IT! Ventures, LLC. Reproduced with permission.)

		Total Sample	White	Black	Hispanic	Asian
	Base size	200	57	45	46	52
	Average dollar value across 36 elements	$3.11	$3.18	$2.96	$3.04	$3.22
A1	Imagine... fruit in the bottom, cream on top... just mix it all together and enjoy!	**$3.55**	$3.10	**$3.64**	**$3.30**	**$4.17**
A6	Looks thicker than your average yogurt	**$3.39**	$3.08	**$3.37**	$3.22	**$3.89**
A3	Once you open it, you will see that you are eating something healthy	**$3.38**	$3.18	**$3.50**	$3.22	**$3.63**
A2	All-natural goodness you can see... A creamy yogurt delight!	**$3.35**	$3.08	**$3.65**	$2.81	**$3.85**
F5	Satisfies your cravings for sweets	**$3.35**	**$3.41**	**$3.39**	$2.93	**$3.61**
C2	A delicious berry aroma... as if you have a basket full of freshly-picked berries right in front of you	**$3.32**	**$3.48**	$3.05	**$3.37**	**$3.35**
E4	Goes down like a regular yogurt	**$3.31**	**$3.73**	$2.76	**$3.54**	$3.14
F6	A quick snack ... helps you make it through the day	**$3.30**	$3.05	**$3.37**	**$3.33**	**$3.50**
F3	Satisfy your hunger in a healthy way this time	$3.28	$2.99	**$3.33**	$2.94	$3.85
D4	Healthy yogurt that tastes like a real treat	$3.27	**$3.76**	$3.17	$3.16	$2.90
F4	Pleases your stomach and mind... worth taking a break for	$3.24	**$3.32**	$3.09	$3.23	$3.27
A4	Wholesome goodness you can see... seeing is believing	$3.19	$3.28	$2.99	$2.98	**$3.47**
F1	Very light and not filling... won't sit in your stomach	$3.18	$3.15	$2.82	$3.02	**$3.66**
B2	Feels like you're eating soft ice-cream!	$3.15	**$3.35**	$2.87	$2.72	**$3.55**
D2	Savory flavor... perfect for any time of day	$3.14	**$3.34**	$2.84	$2.66	$2.66
B3	An absolute mouth-coating pleasure of yogurt will have you wanting more	$3.13	$3.23	$3.08	$3.13	$3.07
B6	Looks delicious	$3.13	**$3.37**	$2.32	$3.13	$2.80
C3	So smooth and fresh... feels like silk in your mouth	$3.13	$3.08	$2.66	$3.23	$2.54
E5	Fruit chunks pleasantly tickle your tongue	$3.13	$3.15	$3.04	$3.00	$3.14
E1	Take your time swallowing... appreciate the comforting creamy texture	$3.09	$2.93	$2.59	$3.02	$2.77
C5	Pleasant and comforting... the perfect bedtime snack	$3.08	$2.84	$2.75	$3.24	$2.84
E3	Has a pleasant aroma of vanilla beans and fruit... like walking into an old-fashioned dairy shop	$3.05	$3.06	$2.67	$3.05	$3.14

(continued)

Table 14.9 *(continued)*

		Total Sample	White	Black	Hispanic	Asian
B1	Close your eyes, smell it – our new flavor will carry you away to vanilla land	$3.01	$3.13	$2.89	$2.92	**$3.53**
D1	Has a very subtle yogurt smell	$2.99	$3.04	$2.62	$3.15	$3.03
F2	Glides down like mousse or pudding	$2.99	$3.23	$3.22	$2.76	$3.08
D5	Pleasantly tangy.... enjoy it as a dressing for your favorite salads	$2.97	$3.30	$2.67	$2.78	$2.75
E6	Leaves you with a nice aftertaste...	$2.97	$3.29	$2.75	$2.98	$2.69
B4	Easy to swallow... even little kids won't have trouble	$2.94	**$3.37**	$2.75	$3.26	$3.07
C6	Has a pleasantly tingling flavor...just what the doctor ordered	$2.93	$2.97	$2.97	$3.06	$2.95
E2	Velvety & moist... a refreshing feeling	$2.93	$2.96	$2.25	$2.64	$2.54
D3	A decadent yogurt with an invigorating aroma	$2.91	$3.14	$3.28	$3.11	$3.05
D6	Has a real 'dairy' taste...brings back sweet memories from your childhood days	$2.90	**$3.33**	$2.77	$2.79	$2.92
B5	Not bland... taste each unique delicious flavor in the yogurt	$2.89	$2.82	$2.29	$2.97	**$3.57**
C4	Has a comforting nutty aroma...made with real hazelnuts	$2.88	$2.98	$3.01	$2.58	**$3.43**
C1	Glides down your throat easily	$2.83	$2.78	$3.12	$3.23	$3.12
A5	Cooling sensation while you are swallowing	$2.62	$3.22	$3.13	$2.83	**$3.30**

2. The most valued elements, defined operationally as those elements commanding dollar values of $3.30 or higher, are also shown in bold. The dollar value of $3.30 was chosen as a reasonable cut-off value for illustration purposes only. One could do a statistical analysis, using the principles of inferential statistics in order to derive cut-off points associated with probabilities (e.g., $p<0.05$), but in these particular studies we look for patterns, rather than for strict statistical significance. The value $3.30, or values thereabouts was therefore chosen as a heuristic, to help us highlight the underlying pattern.
3. *The texture-relevant elements are presented in bold.* Texture elements appeared among the elements commanding high dollar values, but texture elements also appeared among the elements commanding low dollar values:
   a. Thickness and creaminess were prized, commanding high dollar values among the total sample.
   b. But not all texture-relevant elements were so prized. A cooling feeling, a product that glided down (easy swallowing) generated low value, perhaps because the cooling feeling was irrelevant to yogurt.
   c. Blacks and Asians valued texture more than White and Hispanic respondents, respectively.

14.19.2 Linking feelings/emotions to texture elements – it's about 'curious'

We don't typically think about linking emotions to some products, especially yogurt. Yet, the experimental design helped us create models that showed these linkages, just as we showed those between elements and dollar value. The goal here was to identify the 'strong' linkages between elements and feelings/emotions, especially texture elements. Here is how we uncovered the linkages:

1. For each test concept, we created five new variables, one variable for each of the five feelings/emotions that were set up as the selections in Question #2. Thus for the yogurt concepts studied here, the five feelings/emotions and their selection code were: 1 = tempted, 2 = curious, 3 = reassured, 4 = indulgent, 5 = interested.
2. Looking at each concept, one at a time, we inserted the value '100' for the feeling/emotion that was selected, and inserted the value '0' for the remaining four feelings/emotions, those that were not selected. Any row of data, that is, any concept, did not have exactly one '100' and four '0's for these five variables. For computation purposes, we added a very small random number to each assignment of '0' or '100.' The random number wouldn't make any material difference to the results, but ensured that the statistical analysis by OLS regression ran smoothly.
3. Keep in mind that our objective was to link together elements, especially elements dealing with texture, with these five feelings. Just as in the previous section we linked together dollars and elements, here we linked feelings and elements.
4. We used the workhorse technique OLS (ordinary least-squares) regression to relate the presence/absence of the 36 elements about yogurt to each feeling/emotion. We ran a simple model, without going through the origin, just as we did for dollars and elements. The results of the linkage appear in Table 14.10. If there were merely random linkages between feelings/emotions and elements, we expected that the coefficient for the element for any feeling/emotion would be around 5.33. The calculation is:
   (100% allocation)/(5 feelings/emotions x average of 3.75 elements per concept) = 5.33.
5. We looked at strong linkages, defined as a linkage of 10 or higher, substantially higher than the expected 5.33 value from OLS regression for random linkages. Looking at the texture elements in Table 14.10, we see that the main feeling/emotion, not only for texture, but for other sensory inputs, was 'curious.' Or, for some statements, the feeling was 'reassured.' And in one case, just not interested, as in the case of a tingling sensation.

14.19.3 Is there a texture mind-set for yogurt, and if so, then how do we discover it?

We followed the same clustering approach for yogurt that we followed for the healthful snack, creating a r model for each respondent showing the dollar value of each element.

In the previous section on a healthful snack, we used price as a method by which to segment respondents, recognizing that the act of dividing the population of respondents

Table 14.10 Linkage between elements and feelings/emotions for yogurt. (Source: Courtesy of IT! Ventures, LLC. Reproduced with permission.)

Total Sample	**Curious**	**Tempted**	**Not interested**	**Reassured**	**Indulgent**
Average linkage across 36 elements (random = 5.33)	8	5	5	5	4
Once you open it, you will see that you are eating something healthy	14	3	4	5	3
Has a very subtle yogurt smell	12	-4	8	5	4
A decadent yogurt with an invigorating aroma	12	3	3	2	5
Velvety & moist… a refreshing feeling	12	6	1	4	2
Goes down like a regular yogurt	10	2	6	8	1
Has a comforting nutty aroma…made with real hazelnuts	10	2	6	1	7
An absolute mouth-coating pleasure of yogurt will have you wanting more	10	3	4	7	3
A quick snack … helps you make it through the day	10	6	2	8	1
Easy to swallow… even little kids won't have trouble	10	3	7	6	0
Close your eyes, smell it – our new flavor will carry you away to vanilla land	10	2	4	8	1
Looks thicker than your average yogurt	9	1	5	8	6
All-natural goodness you can see… A creamy yogurt delight!	9	5	3	7	6
Cooling sensation while you are swallowing	9	5	5	6	-1
So smooth and fresh… feels like silk in your mouth	9	5	4	5	3
Pleases your stomach and mind…worth taking a break for	9	7	1	5	5
Imagine… fruit in the bottom, cream on top… just mix it all together and enjoy!	9	3	3	4	9
Wholesome goodness you can see… seeing is believing	9	5	4	4	5
Healthy yogurt that tastes like a real treat	8	11	3	2	3
A delicious berry aroma… as if you have a basket full of freshly-picked berries right in front of you	8	5	0	8	7
Looks delicious	8	5	4	7	3
Has a pleasant aroma of vanilla beans and fruit… like walking into an old-fashioned dairy shop	8	6	4	4	4
Savory flavor… perfect for any time of day	7	8	4	7	1
Feels like you're eating soft ice-cream!	7	8	2	5	5
Glides down like mousse or pudding	7	6	4	5	3
Pleasant and comforting… the perfect bedtime snack	6	6	2	8	5
Satisfy your hunger in a healthy way this time	6	6	5	8	3
Fruit chunks pleasantly tickle your tongue	6	4	5	7	4
Take your time swallowing… appreciate the comforting creamy texture	6	2	6	7	5
Satisfies your cravings for sweets	6	6	6	6	5
Glides down your throat easily	6	6	7	5	1
Has a real 'dairy' taste…brings back sweet memories from your childhood days	6	9	4	4	2
Leaves you with a nice aftertaste…	6	5	9	4	3
Pleasantly tangy…. enjoy it as a dressing for your favorite salads	5	5	10	2	4
Not bland… taste each unique delicious flavor in the yogurt	5	9	8	0	5
Has a pleasantly tingling flavor…just what the doctor ordered	4	2	10	5	5
Very light and not filling… won't sit in your stomach	4	3	8	9	3

into like-minded groups does not necessarily uncover groups that are homogeneous with respect to single sensory inputs. The segmentation method, cluster analysis, does not interpret the clusters; these groups of respondents are defined mathematically by the similarity of their response patterns.

We don't need to use dollar value; we could cluster the respondents on the patterns of emotions. Or, if we were to use liking as a rating, then we could cluster the respondents on the basis of the coefficients of the liking model (liking versus the presence/absence of elements). In this study, we clustered on the basis of dollar value.

As in the previous example with an otherwise unspecified healthful snack, the clustering program created two groups, then created three groups, and finally created four, five and six groups or segments, respectively. We looked for a segment solution that would suggest to us a group of respondents willing to pay more for texture sensations; this would be our texture mind-set. From the point of view of looking at the data, we went into the segmentation exercise looking for the answer, a 'texture segment', and did not simply accept a plausible pattern in the segmentation.

A segment that we would call 'texture' did not emerge until we divided the 205 respondents into six segments. The respondents in Segment 6 responded to texture. However, the texture mind-set was not very dramatic. The 'internal gravity' of the texture segment was so low that we had to create five other segments first. Only after peeling away five other mind-sets, the five segments, did we end up with the sixth segment, the one we were looking for.

We are not interested in the first five segments that emerged from the RDE study on yogurt. We were interested only in the 41 respondents assigned to Segment 6, our 'texture' segment. We classified the remaining 164 respondents as belonging to the non-texture segment. We recognized that the remaining 164 respondents constituted several different mind-sets, but for our purposes they were simply in the non-texture mind-set.

Table 14.11 shows the dollar value of all elements. As we did before, we show the texture and swallowing elements in bold type. This strategem allows us to look down a column and quickly identify relevant elements for texture and mouthfeel. To the right are the dollar values. We bold all dollar values exceeding \$3.30. A texture segment would manifest itself by showing high dollar values for a relatively large number of the texture elements. Keep in mind that the dollar value emerged from the reaction to the vignettes, so there was really no possibility that a respondent could 'game' the system.

When we defined the product as yogurt, we didn't end up with a simple texture segment, that is, with a segment that responded to texture elements, and that was all. That is, we didn't end up with one segment where all the elements having to do with texture rose to the top, generating the highest dollar value. Texture was only a part of the distinguishing characteristics of this segment. The segmentation depended upon the full integrated experience, of seeing, smelling the product, tasting, and swallowing, along with the features. Texture was not the main player, but it played a stronger role than it did in the other segments.

When it came to a specific food like yogurt, the segmentation strategy informed us that texture was simply information, the respondent, what the product will be like, but not the center of attention. It was the experience itself, the WiiFM (what's in it for ME), and not the sensory attribute that drove segmentation. Our defined texture segment (Segment 6/ was the closest group willing to pay extra for the texture benefits, or perhaps willing to pay more for the texture and taste benefits.

Table 14.11 Uncovering the 'texture' segment (Segment 6, 41 respondents) and the complementary non-texture segment (segments 1–5, 164 respondents). The elements are sorted by the dollar value of the elements according to Segment 6 (Texture Segment). (Source: Courtesy of IT! Ventures, LLC. Reproduced with permission.)

	Seg6 Texture	**Remainder Non-Texture**
Base size	41	164
Average across all 36 elements	**$3.34**	$3.05
Goes down like a regular yogurt	**$4.10**	$3.11
Has a comforting nutty aroma...made with real hazelnuts	**$3.75**	$2.65
Imagine... fruit in the bottom, cream on top... just mix it all together and enjoy!	**$3.69**	**$3.51**
Looks thicker than your average yogurt	**$3.66**	**$3.32**
All-natural goodness you can see... A creamy yogurt delight!	**$3.53**	**$3.30**
Take your time swallowing... appreciate the comforting creamy texture	**$3.47**	$2.99
Feels like you're eating soft ice-cream!	**$3.46**	$3.07
So smooth and fresh... feels like silk in your mouth	**$3.44**	$3.05
Once you open it, you will see that you are eating something healthy	**$3.39**	**$3.38**
Has a pleasantly tingling flavor...just what the doctor ordered	**$3.32**	$2.83
An absolute mouth-coating pleasure of yogurt will have you wanting more	$3.22	$3.10
Easy to swallow... even little kids won't have trouble	$3.22	$2.87
Close your eyes, smell it – our new flavor will carry you away to vanilla land	$3.19	$2.96
Fruit chunks pleasantly tickle your tongue	$3.17	$3.12
Glides down your throat easily	$3.14	$2.75
Has a pleasant aroma of vanilla beans and fruit... like walking into an old-fashioned dairy shop	$3.12	$3.03
Velvety & moist... a refreshing feeling	$3.08	$2.89
Healthy yogurt that tastes like a real treat	$3.03	**$3.33**
Cooling sensation while you are swallowing	$3.03	$2.51
Glides down like mousse or pudding	$3.01	$2.98
Satisfy your hunger in a healthy way this time	$3.01	**$3.35**
Wholesome goodness you can see... seeing is believing	$2.97	$3.25
Has a very subtle yogurt smell	$2.96	$2.99
Not bland... taste each unique delicious flavor in the yogurt	$2.91	$2.89
Has a real 'dairy' taste...brings back sweet memories from your childhood days	$2.88	$2.90
A decadent yogurt with an invigorating aroma	$2.87	$2.92
A delicious berry aroma... as if you have a basket full of freshly-picked berries right in front of you	$2.83	**$3.45**
Very light and not filling... won't sit in your stomach	$2.78	$3.28
Pleasant and comforting... the perfect bedtime snack	$2.77	$3.16
Pleasantly tangy.... enjoy it as a dressing for your favorite salads	$2.70	$3.04
A quick snack ... helps you make it through the day	$2.68	**$3.47**
Looks delicious	$2.63	$3.26
Savory flavor... perfect for any time of day	$2.63	$3.28
Pleases your stomach and mind...worth taking a break for	$2.62	**$3.40**
Leaves you with a nice aftertaste...	$2.34	$3.14
Satisfies your cravings for sweets	$2.22	**$3.64**

Table 14.12 How Segment 6, the Texture Segment defined itself by selecting what was important. Numbers in the body of the table are percents. (Source: Courtesy of IT! Ventures, LLC. Reproduced with permission.)

	Total	Seg6 Texture	Remainder Non-Texture
A: Select the ONE ATTRIBUTE that is Extremely Important to you when it comes to buying HEALTHY food for you or your family?			
Taste	50	46	50
Healthy Ingredients	26	20	28
Price	14	20	13
Amount of Calories	7	10	6
Amount of Protein	4	5	4
B: Choose ALL the ATTRIBUTES that are EXTREMELY IMPORTANT to you when it comes to CHOOSING a HEALTHY yogurt to purchase and eat.			
Taste	80	78	80
Price	67	66	67
Healthy Ingredients	58	56	59
Texture	57	56	57
Amount of calories	53	49	54
Amount of fat	42	37	43
Aroma	32	24	34
Appearance	32	29	33
Brand	29	17	32
Preserved for longer freshness	18	29	15
Packaging/Storage	16	12	17
Shape of packaging (i.e., cup, container, tube, etc.)	14	7	15
From the list below, please select the ONE ATTRIBUTE that is Extremely Important to you when it comes to buying HEALTHY food for you or your family?			
Taste	50	46	50
Healthy Ingredients	26	20	28
Price	14	20	13
Amount of Calories	7	10	6
Amount of Protein	4	5	4

14.19.4 What do these complementary segments tell us about themselves?

When we instructed these respondents to tell us about themselves, especially with respect to attitudes about what's important, we discovered, however, that these 41 respondents in our Texture Segment, Segment 6 didn't really differ from the remaining 164 respondents, in terms of most attitudes. The key difference was that our Texture Segment was selected by parsing the respondents into complementary groups until a group emerged that we could call the Texture Segment.

Table 14.12 suggests that respondents in Segment 6, our Texture Segment could be characterized as follows:

1. No different at all from the complementary segment when it came to the one attribute that was extremely important, in general, when buying healthy food (Part A, top of table).
2. No different from the complementary segment in terms of thinking about *texture* when it came to selecting yogurt. About half of Segment 6 and about half of the

respondents in the complementary segment felt that texture was important, but neither group felt that texture was as important as taste, price, or healthful ingredients (Part B).
3. Believed that preservation was more important than did the complementary segment (Part B).
4. Brand was less important than for the complementary segment (Part B).

14.19.5 Finding these segments in the population of consumers

Mind Genomics® gives us a way to identify what's important, at least conceptually, for a total group of people, for an individual, and even for newly discovered groups of people who share the same viewpoint about a particular, often limited part of reality. For example, in our study of yogurt, in Part 4 shows that there may be a segment that is more responsive to texture percepts, at least in terms of what it is willing to pay. The question now is finding these people in the population. An analysis of the demographics and even the self-explicated beliefs of this texture segment suggest that it may be hard to find people belonging to the Texture Segment, as this group and its complementary groups, end up looking alike.

Then, how do we find these mind-set segments in the population? More than likely, the traditional methods for finding people in segments won't work. Segments don't align by age, gender, ethnicity, or even product purchased, and attitude confessed. From time to time, segments may align with those easy-to-measure variables, but more than likely the segments exist, well hidden in the body of the general population.

Instead of looking for external cues that predict membership in a mind-set segment, one can adopt a more direct way, intervention, the way today's medicine is practiced for diagnosis. In medicine, one can attempt to predict that a person has a disease by asking lots of questions, by looking at family history and, in general, by epidemiological methods. Or, one can create a short intervention test, whose data will suggest that a person either does not have the disease or has the disease.

We adopt this thinking for Mind Genomics®. The operative idea is that we know from our respondents which individuals belong to the Texture Segment, and which do not. We also know, for each respondent, the model relating the presence/absence of each of the 36 elements and the dollar value. That is, we know the dollar value of each element. Statistical methods such as DFA (discriminant function analysis) can provide us with a simple set of mathematical expressions, classification functions. These classification functions are used to assign a prospect to one of the two segments, Texture or Non-texture.

We began with the actual classification tool as a respondent might see it, and then uncovered the analysis beneath. Figure 14.5 shows a simple screen, the 'typing tool' The screen comprised four questions, each to be answered by the respondent selecting one of three answers. The four questions came from the 36 elements. The three answers came from the range of prices tested in the RDE portion of the study. The actual test was quite simple, taking about 30 seconds.

What do you think is a fair price to pay for a package of 24 (4 ounces each) decadent yogurts at a club store that...

1 = Less than $2.85
2 = Between $2.85–$4.15
3 = More than $4.15

Have a delicious berry aroma... as if you have a basket full of freshly-picked berries night in front of you?	2
Has a comforting nutty aroma... made with real hazelnuts?	3
Has a savory flavor... perfect for any time of day?	3
Goes down life a regular yogurt?	1

Figure 14.5 The 'typing tool', which when combined with the classification function allows one to assign a now person to one of the two mind-set segments, the Texture Segment, or the other segment.

14.19.6 Developing and using the typing tool for yogurt – what's underneath the 'hood' (Table 14.13)

The essence of the typing tool is the well known statistical procedure, DFA, discriminant function analysis. DFA operates as follows on the data emerging out of Mind Genomics® and the tool RDE, rule developing experimentation.

1. Each of the 205 respondents generated a dollar model, showing the dollar value of every element. OLS, ordinary least-squares regression, estimated this dollar value for each element for each respodent. With 205 respondents, and with 36 elements per respondent whose dollar value is estimated, we ended up with 7,380 dollar values ($36 \times 205 = 7{,}380$).
2. We identifed three ranges across this 7,380; the lowest quartile (<$2.85) where ¼ of the dollar values are lower, the middle two quartiles, and the highest quartile (>$4.15), below which ¾ of the dollar values are lower.
3. For each respondent we replaced the 36 dollar values by the number 1 (when the dollar value for the element<$2.85), or by the number 2 (when the dollar value for the element lies between $2.85 and $4.15), or by the number 3 (when the dollar value for the element lies above $4.15).
4. For each respondent we also knew the segment to which the respondent belonged, whether this was the Texture Segment (Segment 6), or the Non-Texture Segment (Segments 1–5, now coalesced).
5. DFA created two so-called classification functions (A in Table 14.13). The classification functions comprised an additive constant and a set of weights to be applied to the ratings assigned in the typing tool.
6. Each person who participated in the typing tool rated four different questions (see Figure 14.5), using a 3-point scale, anchored with specific dollar amounts.

Table 14.13 A worked table DFA (discriminant function analysis). The table shows the four elements, the classification function, the response patterns from five hypothetical individuals, the values of both classification functions for each person, segment assignment (bold type), and what to say to the segment. (Source: Courtesy of IT! Ventures, LLC. Reproduced with permission.)

	A. Classification Functions		**B. How five hypothetical people might have assigned ratings using the typing tool**				
	Segment 1 Non-Texture	**Segment 2 Texture**	**Per1**	**Per2**	**Per3**	**Per4**	**Per5**
Additive constant	–5.855	–5.963					
A delicious berry aroma… as if you have a basket full of freshly-picked berries right in front of you	1.556	0.572	1	3	1	2	1
Has a comforting nutty aroma… made with real hazelnuts	1.283	2.083	2	2	2	3	1
Savory flavor… perfect for any time of day	1.562	0.849	3	2	3	1	1
Goes down like a regular yogurt	1.289	2.072	2	3	1	1	2
C. Value of the classification function for each segment, and segment assignment based on the classification function showing the higher positive value		Seg1	5.5	8.4	4.2	4.0	1.1
		Seg2	5.5	7.8	3.4	4.3	1.7
D. Strongest elements – Non-Texture Segment							
Satisfies your cravings for sweets						$3.64	
Imagine… fruit in the bottom, cream on top… just mix it all together and enjoy!						$3.51	
A quick snack … helps you make it through the day						$3.46	
A delicious berry aroma… as if you have a basket full of freshly-picked berries right in front of you						$3.45	
E. Strongest elements – Texture Segment							
Goes down like a regular yogurt						$4.10	
Has a comforting nutty aroma…made with real hazelnuts						$3.75	
Imagine… fruit in the bottom, cream on top… just mix it all together and enjoy!						$3.69	
Looks thicker than your average yogurt						$3.66	

Section B shows how five hypothetical people might have assigned ratings using the typing tool.

7. Section C shows the value of the classification function for each segment and segment assignment based on the classification function showing the higher positive value. For Person 1 the two classification functions ended up with equal numbers, so the person belonged to both segments. For Persons 2–5 the classification function generated a clear assignment. The classification function is 76% accurate, based upon its performance when predicting segment membership of the 205 respondents.
8. Sections D and E showed what to say to each mind-set segment, that is, the elements that generated the highest dollar value.

14.20 SUMMING UP

Texture studies began in earnest about 50 years ago, with the development of instruments to measure texture and efforts to discover the language of the subjective texture experience. Over the decades, researchers became increasingly aware of the importance of texture as one of the key sensory inputs to the appreciation of food and drink. However, the inner experience of texture was not particularly of interest. Mind Genomics®, the science of everyday life, allows us to peer more deeply into the inner experience of texture.

This chapter began with the standard type of research, to determine the importance of texture for the appreciation of food, either in terms of making food craveable, or driving a person to purchase the food. However, simply asking a person whether texture is important or not, with whatever scale, does not go deeply into the actual inner experience.

The chapter then introduced the notion of experimental design, not of physical features, but rather the experimental design of ideas. It is here that Mind Genomics® comes into play. When we allow these ideas to encompass the sensory experiences we have, with the ideas being descriptions of the type people would want to read about, we move towards understanding the inner experience.

Our first foray with Mind Genomics® showed that when we describe foods by phrases, and test these phrases in combination, a fair number of texture-related phrases emerge as important. But it's not texture per se in those phrases, but rather texture as an addition to the description of the product, texture not as a main actor but as an amplifier of the experience.

Our second foray into Mind Genomics® was done to determine whether we could uncover a mind-set segment focusing on texture. We uncovered this mind-set segment by instructing respondents to rate test concepts both on interest, and in terms of the amount of money the respondent would pay for the product (a healthful dessert) in a club store. OLS (ordinary least-squares) regression revealed that each element of the dessert product could be associated with a dollar value. Further analysis of the pattern of dollar values revealed quickly that one segment would pay more money for those sensory experiences associated with texture and swallowing. This segment emerged quickly from the data.

In our second foray in Mind Genomics® we continued the approach, working this time with yogurt, a popular dairy product undergoing renewed interest with the increased interest in Greek-style yogurt. The respondents rated the different test concepts on two attributes, first selecting a price (as was done in the previous study), and then selecting a feeling/emotion. This second study revealed that the predominant feeling/emotion attached to texture-relevant elements was curiosity, suggesting that texture elements add information, but probably don't increase one's drive to eat the product. Texture is informational, rather than strongly motivating. The second study also revealed a texture mind-set, but a mind-set that emerged only after six segments were identified by clustering.

Mind Genomics® is only at its beginning. The studies reported here can shed more profound light on texture. Fortunately, the technology to acquire the data is already here. It remains only a matter of 'doing the experiment,' a task that is now within the financial and intellectual reach of just about any researcher interested in doing so. Rich intellectual rewards await.

References/for further reading

Beckley, J, H., Gillette, M. and C. Marketo. (2002) Changes in Consumer Mindsets Since 9/11 — The Crave It!™ Foundational Study (Sponsored By Mccormick & Company). Food & Beverage Innovation (2002) — Trend–Focused Strategies For Product Success. IIRUSA, Chicago, Nov. 2002.

Beckley, J. and Moskowitz, H.R. (2002) Databasing The Consumer Mind: The Crave It!!, Drink It!, Buy It! & Healthy You!™ Databases. Institute of Food Technologists Annual Meeting, Anaheim, Ca, July 2002.

Boring, E.G. (1929) *Sensation and Measurement in the History of Experimental Psychology*. New York: Appleton Century Crofts.

Box, G.E.P., Hunter, J. and Hunter, S. (1978). *Statistics for Experimenters*. New York: John Wiley.

Gofman, A. (2006). Emergent scenarios, synergies, and suppressions uncovered within conjoint analysis. *Journal of Sensory Studies*, 21, 373–414.

Green, P.E. and Krieger, A.M. (1991) Segmenting markets with conjoint analysis. *Journal of Marketing*, 55, 20–31.

Green, P.E. and Wind, J. (1973) *Multiattribute Decisions in Marketing: A Measurement Approach*. Hinsdale, IL: Dryden Press.

Johnson, R.M. (1974) Trade-off analysis of consumer values. *Journal of Marketing Research*, 11, 121–127.

Luce, R.D. and Tukey, J.W. (1964) Simultaneous conjoint measurement: A new type of fundamental measurement. *Journal of Mathematical Psychology*, 1, 1–27.

Moskowitz, H.R. (1994) Concepts: Their systematic design, development and optimization. Chapter 1. In: *Food Concepts and Products: Just In Time Development*, (pp. 1–69). Westport, CT: Food and Nutrition Press.

Moskowitz, H.R. (1996) Segmenting consumers worldwide: An application of multiple media conjoint methods. Istanbul, Turkey: *Proceedings of the 49th ESOMAR Congress*, (pp. 535–552).

Moskowitz, H.R. and Beckley, J.H. (2005). *The Handbook of Food Science, Technology, and Engineering, Volume 2. Chapter 59 – Large Scale Concept–Response Databases for Food and Drink Using Conjoint Analysis, Segmentation, and Databasing*. Edited by Y.H. Hui. Boca Raton, Fl: CRC Press, 2005.

Moskowitz, H.R., German, J.B. and Saguy, I.S. (2005) Unveiling health attitudes and creating good-for-you foods: The genomics metaphor, consumer innovative web based technologies. *CRC Critical Reviews in Food Science and Nutrition*, 45, 165–191.

Moskowitz, H.R. and Gofman, A. (2004) 'System and method for performing conjoint analysis'. Provisional patent application, 60/538,787, filed January 23, 2004.

Moskowitz, H.R., Gofman, A., Itty, B., Katz, R., Manchaiah, M. and Ma, Z. (2001) Rapid, inexpensive, actionable concept generation and optimization – the use and promise of self-authoring conjoint analysis for the foodservice industry. *Food Service Technology*, 1, 149–168.

Moskowitz, H.R., Poretta, S. and M. Silcher (2005) *Concept Research in Food Product Design and Development*, Chapter 20 – 'Creating An Integrated Database From Concept Research: The It!® Studies', pp. 409–430. (Credit to J. H. Beckley & H. Ashman); and Chapter 24 – 'Emotions in Concept', pp. 497–511 (Credit to H. Ashman). Ames, IA: Blackwell Publishing Professional, Ames, Iowa.

SYSTAT (2007) SYSTAT for Windows, Version 11. Chicago, ILL: SYSTAT Software Inc.

www.mindgenomics.org

Wells, W.D. (1975) Psychographics, A critical review. *Journal of Marketing Research*,12, 196–213.

Wittink, D.R. and Cattin, P. (1989) Commercial use of conjoint analysis: An update. *Journal of Marketing*, 53, 91–96.

15 The use of advanced spectroscopic techniques to understand texture in dairy foods

Marcela Alexander and Milena Corredig

Department of Food Science, University of Guelph, Ontario, Canada

15.1 INTRODUCTION

During the processing of food, the components assemble in structures at various length scales. Molecules interact and form structures of only a few nanometers in size, which then assemble at the microscale level, forming what is macroscopically visible. The dynamics occurring during assembly of the components to form food matrices are complex, and often difficult to follow and describe at the molecular level. However, a better understanding of how to fine-tune the formation of structure is fundamental to be able to control processes, as well as create more stable foods. The characteristics of the final structure also affect how we perceive texture and other sensory attributes of the food. Finally, evidence is becoming available that the way food molecules assemble, disassemble and re-assemble during transit in the digestive tract is also critical to the bioefficacy of the foods we ingest.

In recent years, the consumer demand for foods with benefits beyond simple nutrition and convenience has challenged product developers to re-evaluate traditional manufacturing practices and formulations. The next generation of food products designed to deliver, for example additional health benefits, will contain ingredients that may affect the way molecules assemble into structures. Bioactive molecules added to traditional formulations might affect the interactions between ingredients and breakdown during processing causing a decrease in colloidal stability. Nutritionally important concentrations of micro or macro nutrients may result in thermodynamically incompatible formulations, or in changes of the interfacial composition when competing with the functional emulsifiers present in the formulations. These are just a few of the many structure-affecting processes that may evolve from changes in formulations or manufacturing practices to create healthier food matrices. It is therefore increasingly important to develop analytical tools allowing the observation of structure formation at the early stages of the reactions. A better understanding of the initial stages of these reactions would lead to the optimization of processes and formulations with the final outcome of obtaining optimal product quality.

Food Texture Design and Optimization, First Edition. Edited by Yadunandan Lal Dar and Joseph M. Light.

The dynamics of change of colloidal systems during formation of structure are challenging to investigate. Due to the complexity of foodstuffs, ingredients can be in metastable equilibrium and in order to get a better understanding of the changes occurring to the supramolecular structures during formation of texture, it is imperative to keep the systems as close as possible to the conditions of the final product (pH, temperature, ionic strength, volume fractions), as changes in any of these conditions will undoubtedly affect the structure of the food being investigated. Bulk techniques such as rheology can precisely identify the point when the structure has formed, but are limited in their ability to describe the processes occurring during the early stages of structure formation. Techniques such as diffusing wave spectroscopy and ultrasonic spectroscopy have been employed to probe into the aggregation dynamics of dairy systems because of their non-invasive nature and their ability to measure changes occurring *in situ*. These techniques have shown great promise in the characterization of the structure of food matrices as complementary tools to conventional techniques such as rheology or various types of microscopy.

This chapter will attempt to explain and describe the work which has been published to date in the area of structure formation and manipulation at the nano to micron-level, as investigated by diffusing wave spectroscopy (DWS) and ultrasound spectroscopy (US). Most of the work available in the literature using these techniques in foods has dealt with dairy emulsions and milk matrices. DWS and US are ideally suited for application in food systems for three main reasons: the length and time-scale of resolution by light and ultrasound are in the same order of magnitude of the particles and aggregates to be tested, and they are both non-invasive and non-disruptive techniques; hence, no extensive sample preparation is required such as freezing, drying, or mechanical deformation and finally, they can be used in or near the concentrations in which the foods are consumed, that is, no dilution is necessary.

The following pages will present how DWS and US have been successfully used to investigate the formation of structure of colloidal systems subjected to different destabilizing mechanisms. It is not the intention of this work to be an exhaustive treaty of the dynamics and processes of structure-formation, but rather to offer a glimpse into the ways in which structure development can be successfully observed with these two techniques. After a basic introduction of the techniques, the chapter will bring examples of the application of DWS and US in the study of structure formation of dairy systems. Specifically, the chapter will describe how DWS and US have been used to follow the development of structure in various types of milk gels and oil-in-water emulsion stabilized by milk proteins. It will become clear that the application of these techniques in food systems is just at its infancy, and some of the challenges ahead will be highlighted.

15.2 SPECTROSCOPIC METHODS

15.2.1 Diffusing Wave Spectroscopy

A very suitable and long-established methodology for investigating the colloidal aspects of food systems has been light scattering, both in the traditional photon correlation mode, called dynamic light scattering (DLS), but sometimes also used in its integrated, static form (ILS). One of the clear advantages of using light scattering has always been the non-invasive nature of the measurements, as the passage of the laser beam through the sample does not generally affect it. The technique is relatively user friendly and easy to interpret, and the speed at which detectors and correlators now function enables the user to get immediate results such that measurements may be made in a matter of only a few seconds, if necessary. Numerous important advances have been made in the field of food science by the utilization of light scattering (Dalgleish and Hallet, 1995; Alexander and Dalgleish, 2006). However, one of the critical shortcomings of traditional light scattering is the need to work in what is called the 'single scattering regime', that is, photons must be scattered only once before they are detected. This, of course, entails high dilution of the samples (between 500 and 10,000 times or more), thereby severely limiting the applicability of the technique, especially when trying to investigate phenomena like gelation, phase separation or depletion flocculation, for example, which are highly dependent on volume fraction (or concentration).

In 1987, a model was proposed that extended the limits of light scattering to turbid media (Maret and Wolf, 1987). This technique, called Diffusing Wave Spectroscopy, was also based on the detection of spatial correlations of scatterers as a function of time, much like DLS. However, while DLS requires single scattering events, DWS treats the scattering of photons and their paths through the sample as a diffusive phenomenon. In order to attain this 'random-walk' condition DWS requires a relatively high turbidity of the sample which makes the technique ideal for the study of 'real-life" food systems and their structure-forming, destabilizing requirements such as gelation, sol–gel transitions and glassy states.

The technique of DWS, just like DLS, follows the intensity fluctuations of a speckle of scattered light with time. As the particles move, the interference pattern produced by the scattered light will change over very short time scales and this will depend on the dynamics of the scattering particles. In a free-diffusing medium, small, rapidly diffusing particles will have fast fluctuations, whereas larger particles will generate slow fluctuations. The time-scale of these intensity changes is inversely proportional to the particle diffusion, thereby providing dynamic information about the scattering colloids. As the systems become more concentrated or destabilized by physical or chemical forces, the interactions between neighboring particles becomes important and begin to affect the nature of the fluctuations of the scattered light. For example, a small particle, which under stable conditions, would have had higher-frequency fluctuations would appear to have slower movement once trapped in a developing gel. There is another very important parameter which can be obtained from DWS and this is the photon transport mean free path or l^*. This parameter relates to the optical turbidity of the samples and is solely dependent on the static properties of the system. In general, it is an indication of the spatial distribution (or correlations) between the colloidal particles in a system and again, it can show changes in a destabilized system, as the average spatial distribution of the scatterers will change in a sample undergoing gelation, for example.

DWS can be utilized in two different modes, backscattering or transmission, depending on whether the scattered light is collected on the same side as the laser source (backscattering), or if the scattered light is collected after the photons traversed the whole sample thickness (forward scattering, or transmission). Figure 15.1 shows a schematic representation of the different geometries. There is, of course, a formal similarity between the two methods, but there are also some differences in the interpretation of the results. The backscattering

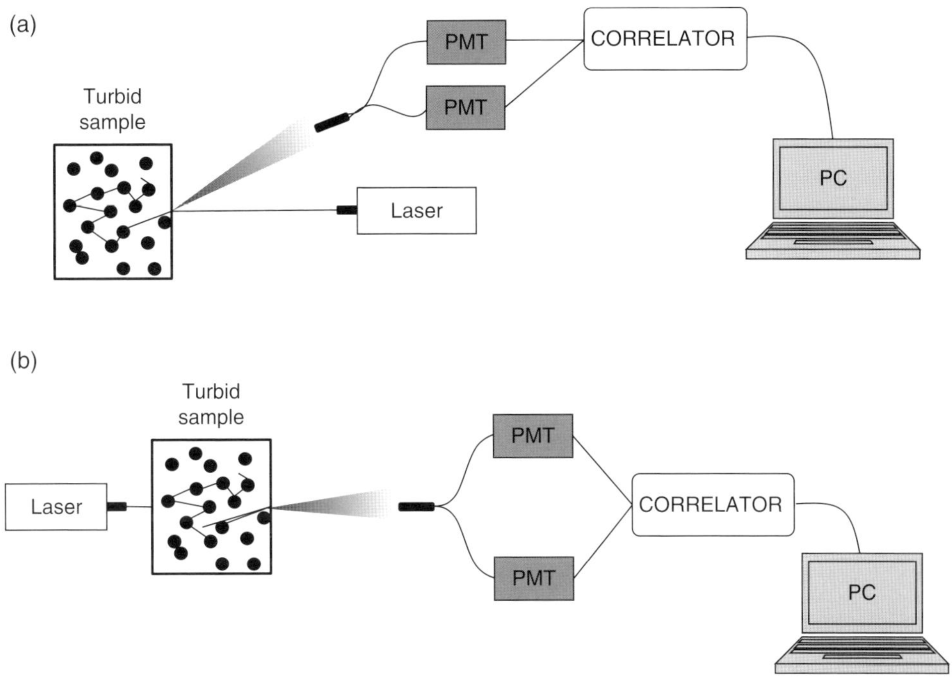

Figure 15.1 A schematic representation of a typical diffusing wave spectrometer set-up in backscattering (a) or forward scattering (b). 'PMT' stands for Photo Multiplier Tube (detector).

mode is logistically the simplest geometry of the two. Light source and detection systems are placed on the same side of the sample; therefore it has the advantage that only one side of the sample needs open access. This can be particularly useful in industrial settings, for example for process monitoring on-line (De Kruif et al., 2004). The disadvantage of this setup lies in the difficulty to fully interpret the results, because collecting data in backscattering mode implies probing many length scales at once: photons that have gone deep into the sample are detected together with those which got backscattered after only a few or even just one scattering event. Because the vast majority of the results presented in this chapter deal with the forward scattering technique, we will only expand on this latter theory. A complete description and explanation of backscattering DWS can be found elsewhere (Weitz and Pine, 1993; De Kruif et al., 2004).

In a light scattering experiment, the intensity of the light reaching the detector at any instant depends on the interference pattern created by the scattered light from all of the particles in the scattering volume. A full derivation and exhaustive treatment of light scattering theory can be found elsewhere (Berne and Pecora, 1976). As the particles move, the interference pattern will fluctuate, and so the detected intensity will change over very short time-scales. These fluctuations in the intensity of the scattered light can be expressed by their temporal autocorrelation function which is given by:

$$g_{(2)}(t) \equiv \left[\frac{1}{\beta} \left(\frac{\langle I(t) I(0) \rangle}{\langle I \rangle^2} - 1 \right) \right] \tag{15.1}$$

where β is a constant determined primarily by the collection optics and $I(t)$ is the intensity at a given instant in time (Berne and Pecora, 1976).

When the thickness of the sample L is much larger than l^* (where l^* is the transport mean free path), in general, when $\frac{L}{l^*} > 10$ and the time $t \ll \tau_o$ (the characteristic correlation time of the sample), the correlation function can be described by (Weitz and Pine, 1993):

$$g_{(1)}(t) \approx \frac{\left(\frac{L}{l^*}+\frac{4}{3}\right)\sqrt{\frac{6t}{\tau}}}{\left(1+\frac{8t}{3\tau}\right)\sinh\left[\frac{L}{l^*}\sqrt{\frac{6t}{\tau}}\right]+\frac{4}{3}\sqrt{\frac{6t}{\tau}}\cosh\left[\frac{L}{l^*}\sqrt{\frac{6t}{\tau}}\right]} \tag{15.2}$$

This correlation function is nearly exponential in time, with a characteristic decay time of

$$\tau = \tau_o\left(\frac{l^*}{L}\right)^2 \tag{15.3}$$

As before, it is assumed that completely diffusive particle motion occurs. Furthermore, since the number of scattering events is large, there is no appreciable angular dependence of the light emerging from the sample and measurements at a precisely-defined angle are not necessary.

For a monodisperse sample of spherical, uncharged particles, τ is related to the diffusion coefficient, D (related to the flow of particles per unit time), of the particles by

$$\tau = \frac{1}{Dk_o^2} \tag{15.4}$$

where k_o is the scattering vector defined by $k_o = 4\pi n/\lambda$ and n is the index of refraction of the medium and λ is the wavelength of the laser light (Weitz and Pine, 1993). In the case of free-diffusing particles, the diffusion coefficient can be transformed into the apparent hydrodynamic radius of the particles using the Stokes-Einstein equation (Berne and Pecora, 1976).

In the case of gelling media when the system ceases to have a unique characteristic decay time (Equation 15.4), the interpretation of the correlation function in terms of diffusion coefficients and particle sizes is no longer valid. The correlation function can be related to the average mean square displacement (MSD) of the particles. This MSD is related to the average two-dimensional excursions that a particle makes per unit time and it will be directly affected in a destabilization process. For example, a system containing colloidal particles which undergo aggregation and eventual gelation will have mean square displacements which will evolve as a function of time. The MSD values will depend linearly with correlation time for free-diffusing motion, but will display an asymptotic behavior, reaching a maximum when undergoing the formation of a network and eventual arrested motion. Figure 15.2 illustrates the evolution of the MSD values at

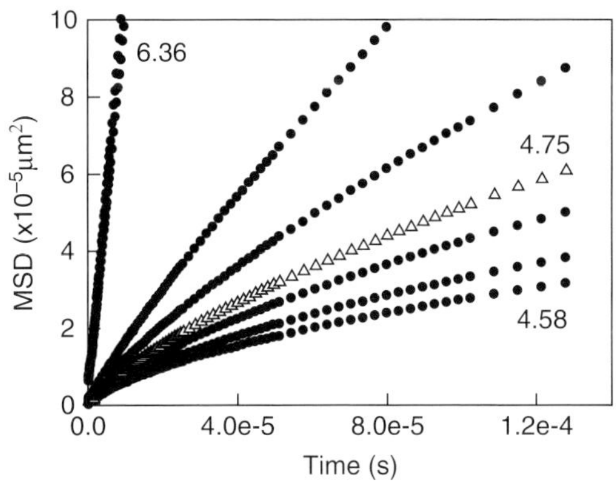

Figure 15.2 Evolution of the mean square displacement of skim milk after addition of glucono-delta-lactone. Values in the graph correspond to the pH values measured during the acidification process. At the regular pH of milk (~6.7), the displacement of the micelles is unimpeded in time whereas at a low pH when the micelles form a network, their displacement is restricted to under 3 x 10^{-5} µm^2.

different pH during the gelation of casein micelles in skim milk induced by the formation of acid by addition of glucono-delta-lactone. While at the beginning of the reaction the MSD values are linear with correlation time, later there is a clear deviation from linearity, because of the arrested motion of the casein micelles forming a gel network.

The turbidity of a multiple scattering system is related to a quantity called the photon transport mean free path or l*. This is the distance over which the direction of the scattered photon has been completely randomized and it is related to the total scattered intensity of the system. The l* is a function of the form factor, $F(q)$, related to the size, shape and contrast of the scatterers, described in the Mie and Rayleigh-Gans-Debye (Bernie and Pecora, 1976; van de Hulst, 1957) in reference to dilute suspensions of scattering particles. In addition, the l* is also dependent on the structure factor, $S(q)$, which describes the positional correlations between the scatterers. The general expression is given by (Weitz and Pine, 1993):

$$l^* \propto \left(\int F(q) \cdot S(q) \cdot q^3 dq\right)^{-1} \tag{15.5}$$

where q is the wave scattering vector related to the incoming and scattered waves. In concentrated and optically dense suspensions the spatial positions of the particles and their correlation given by $S(q)$ may become significant, especially if the particles are charged or can otherwise affect one another. These correlations in position affect the angular distribution of the scattered light and hence the turbidity and l*.

In certain cases this dependence can yield information on the development of the structure of a system. For example, in samples where the volume fraction and refractive index contrast between the particles and the continuous phase can be assumed constant throughout a process, as could be milk gelation for example, a change in the value of l* will be directly related to a change in the interactions between particles (Alexander et al., 2006). Therefore, changes in this parameter can be taken as indications of structural changes within the suspension.

15.2.2 Ultrasonic spectroscopy

Another important technique which is underutilized in the studies of structure formation in food matrices is ultrasonic spectroscopy. Like DWS, US and electroacoustic methods can analyze concentrated and optically opaque systems *in situ*. US experiments can be carried out in a range of frequencies, from less than 1 MHz to about 100 MHz. The detailed description of an ultrasonic spectrometer can be found elsewhere (Dukhin and Goetz, 2000). Figure 15.3 shows a schematic diagram of an ultrasound set-up.

In the case of food colloids, the particles are generally of sub-micron size, and much smaller than the wavelength of the ultrasound waves. During ultrasonic spectroscopy, a planar wave propagates through the sample, and due to the physico-chemical properties of the dispersion and its density and compressibility inhomogeneities, the intensity and phase of the sound wave will change.

The velocity of the sound is dependent on the chemical composition of the system and temperature, and in suspensions or emulsions can be predicted with the Urick equation for homogeneous materials (Urick, 1947),

$$v = (m * \rho)^{-1/2} \tag{15.6}$$

where v is the velocity of the sound wave through the material, m is the compressibility of the material and ρ is the density of the material. The equation is valid for diluted systems, where the particles are very small compared to the wavelength of the sound and multiple scattering can be ignored. For suspensions like food colloids, which are weak scatterers, this equation has been shown to be valid for volume fractions as high as 40% (McClements and Povey, 1987). In dispersions where viscous and thermal transport processes occur at the boundary layer, the Urick equation [15.6] would estimate a lower ultrasonic velocity than that obtained experimentally; however, in general for food colloids, the density gradient between the phases is very small, and the inertial term can be ignored (Mc Clements and Povey, 1987). Ultrasonic velocity is quite sensitive to changes in chemical composition and temperature. For example, an increase in the salt concentration of about 1 M would increase the sound speed by 100 m/s, and each °C would also cause a raise in sound speed of about 2.4 m/s (Duckin et al., 2005).

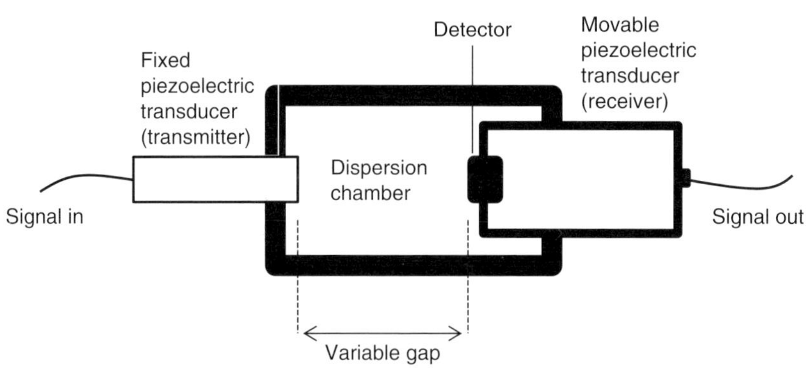

Figure 15.3 A schematic representation of the ultrasound set-up used in this work. The incoming signal produces ultrasound excitation of the sample over a wide range of frequencies.

During the propagation of the ultrasonic wave, there is a reduction in the energy carried by it. This effect is measured as attenuation. There are four major mechanisms that contribute in an additive fashion to ultrasonic attenuation, and the extent of the contribution depends on the material properties of the sample. These can be summarized by the equation below:

$$A_{total} = A_{intrinsic} + A_{scattering} + A_{inertial} + A_{thermal} \tag{15.7}$$

The intrinsic attenuation of a system is a sum of the various intrinsic attenuations of the components (Allegra and Hawley, 1972). Scattering losses can also substantially contribute to the ultrasonic wave attenuation; when the particles redirect the energy of the ultrasonic wave away from the propagation path. This is the case for larger particles, and the scattering component of ultrasonic attenuation becomes increasingly larger with increasing frequency (Allegra and Hawley, 1972; Dukhin and Goetz, 2002). Scattering losses are not important in food colloids, as their sizes are quite small compared to the wavelength of the sound wave. The energy of the ultrasonic wave can also be lost because of viscoinertial or thermal losses. Viscoinertial losses derive from density differences between the colloidal particles and the solution, while thermal losses are caused by the generation of a temperature gradient at the boundary layer between the colloidal particle and the solution during the compression and decompression cycles (Dukhin and Goetz, 2002; Dukhin et al., 2005).

In dairy systems such as milk and emulsions, thermal losses play a major role in attenuating the ultrasonic wave, indeed, there is very little density difference between the colloids and the aqueous solution (hence there is a minimal viscoinertial term) and the size of the colloidal particles is very small compared to the wave (hence the scattering losses are low).

The relationship between ultrasonic attenuation and frequency of the wave varies greatly with the particle size of the dispersion, as illustrated in Figure 15.4. This relationship can be employed to derive the size distribution of the particles, when their heat

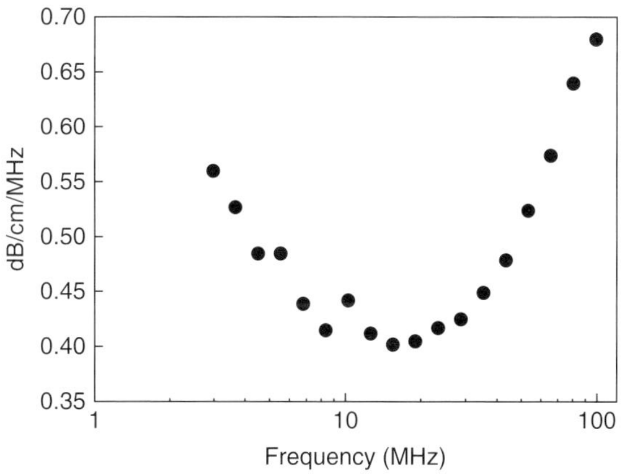

Figure 15.4 An example of the frequency dependence of the ultrasonic attenuation for a 10% corn oil in water emulsion stabilized with whey protein isolate, with a 1 μm average diameter (as measured by light scattering).

conductance, thermal expansion and heat capacity are known (Coupland and McClements, 2001). The changes occurring to the colloidal system during aggregation can then be followed measuring ultrasonic attenuation (see for example, Chanamai et al., 1998).

As ultrasonic measurements can be applied to optically opaque systems, they have shown potential for monitoring on-line processes. In addition to particle size determination, ultrasonic spectroscopy can be applied to creaming and crystalization studies (McClements and Povey, 1992). Indeed, ultrasound spectroscopy has been shown to be very effective at investigating lipid phase transitions in emulsion droplets (McClements et al., 1993).

The amount of solid fat content is an important factor that determines the structure and texture of fat-containing matrices. The ultrasonic velocity of the wave propagating through a sample is directly proportional to the solid fat content, and the results can correlate well with low resolution NMR (Martini et al., 2005a; 2005b). With ultrasonic spectroscopy the fat crystalization can be followed in real time, placing transducers in the pipe walls during manufacturing (Martini et al., 2005a). It has also been demonstrated that ultrasonic attenuation increases as crystals form and grow, suggesting that variations in the attenuation may be related not only to solids fat content, but also to the changes in microstructure of the samples as crystalization proceeds, especially at low solids fat content (Martini et al., 2005b).

By combining ultrasonic spectroscopy with electroacoustics, it is also possible to monitor changes occurring to the structure of food colloids. When a sample is subjected to an ultrasonic pulse, the vibrations generated by wave-induced compressions cause the motion of the particles relative to the others present in the surrounding bulk phase. This motion perturbs the electrical double layers of the spatially-correlated particles, and creates a current (colloidal vibration current (CVI)). This current can be measured, and from its signal the electrophoretic mobility of the particles can be derived. By knowing the density difference between the colloidal particles and the solution, it is possible to measure their electrokinetic properties. Details of the models used in the calculations can be found elsewhere (Dukhin and van de Ven, 1994; Dukhin and Goetz, 2000; 2002). This method has been shown to measure with high accuracy the ζ-potential of polymer particles to volume fractions as high as 50% (Dukhin and Goetz, 2002) and the ζ-potential of casein micelles in milk up to 40% volume fraction (Gülseren et al., 2010). The electrophoretic mobility of the particles can be a very sensitive measure of the changes occurring during the destabilization of colloidal systems, and with electroacoustics, it can be measured *in situ*, without dilution (Gülseren and Corredig, 2011).

15.3 FOLLOWING STRUCTURE FORMATION WITH DWS AND US

The employment of DWS and US to investigate structure formation in food systems is still very much in its infancy, although these techniques have been employed for some time to characterize the properties of food colloids (Alexander and Corredig, 2007; Corredig and Alexander, 2008; Nicolas et al., 2003).

In the case of colloidal systems such as milk and food emulsions, there are a number of ways to achieve structuring. In general, structures can be obtained by the destabilization of thermodynamically (meta)-stable systems by letting the system progress to a more favorable, less energetic state. The use of the prefix 'meta' is necessary here, as nearly none of the food systems available are in a true state of equilibrium. Creaming of emulsions or separation of thermodynamically incompatible protein and polysaccharides are examples of such states, as often the tendency of these colloidal systems to develop into more entropic conditions is somewhat kinetically limited. This can be achieved by simple interference with the naturally-destabilizing forces by slowing down their effects, for example, by increasing the viscosity of the background medium or by gelling it.

15.3.1 Milk gelation

Gelation of milk is usually caused by the destabilization of protein particles. The reactions involve the casein micelles in milk, colloidal particles of about 200 nm of diameter (Dalgleish, 2011; De Kruif et al., 2012). By changing the casein micelles' properties, either by changes to the environment (i.e. pH) or by the use of specific enzymes (i.e. rennet), these protein particles will aggregate upon collision. The pathways leading to the formation of the aggregates will ultimately affect the type of structures formed. Figure 15.5 is a schematic of these two processes. As previously mentioned, very few techniques can be employed to follow the dynamics of destabilization preceding gelation. US and DWS have shown potential as tools to obtain a better understanding of these meta-stable stages, to be able to exploit this fragile equilibrium in food systems and manipulate desirable final bulk properties in food products.

The colloidal properties of skim milk and the reactions leading to gelation have been extensively studied in the past using traditional light scattering methods. Under diluted conditions, it has been possible to investigate the molecular details leading to the instability of the casein micelles during rennet or acid induced coagulation (Dalgleish, 1980; Horne, 1986). Bulk rheology measurements have also been employed to describe the

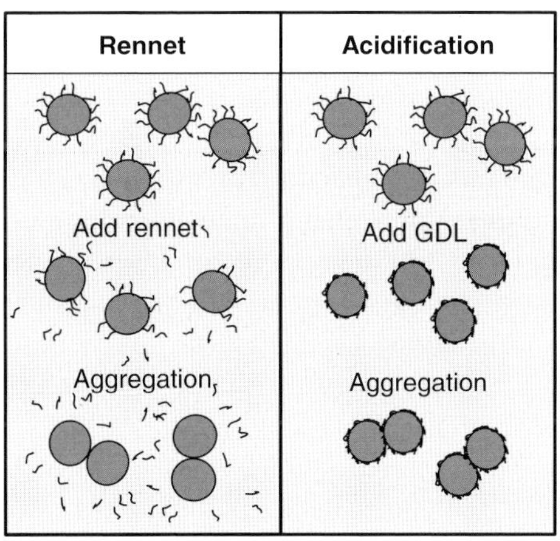

Figure 15.5 Schematic representation of the possible structuring of casein micelles in skim milk after enzymatic reaction by rennet or after acidification of the system by addition of Glucono-δ-lactone (GDL). In the first case, the hairy layer is cleaved by enzyme action and released into solution whereas in the second case the decrease in pH of the milk causes the hairy layer to collapse onto itself. In both cases, the decrease of steric stabilization causes the micelles to aggregate.

process of milk coagulation (Horne, 1999; Lucey, 2002; van Hooydonk and Walstra, 1987). However, the need to derive information on the gelation behavior *in situ* of milk systems at their natural concentration have been satisfied only recently, after the employment of DWS and US techniques (see for example, Alexander and Dalgleish, 2004; Corredig et al., 2004a; Hemar et al., 2004).

The first report of the use of DWS to study undiluted milk was by Horne and Davidson (1990). A backscattering set up was employed to investigate rennet-induced gelation of the casein micelles. Further investigations showed that the equipment could accurately measure the coagulation time and also qualitatively correlate the intensity of the scattered light to the elasticity of the milk systems. Furthermore, it was concluded that DWS could provide qualitative information on the different structures of gels (either acid or enzymatically-driven) (Dalgleish and Horne, 1991). After these initial reports, an important body of work ensued, each one expanding our understanding of the dynamic and structural behavior of milk gelling systems.

The first reports on gelation studies of milk with ultrasonic spectroscopy showed very little changes of ultrasonic velocity and attenuation with acidification of milk (Benguigui et al., 1994). Ultrasonic shear waves were used to characterize the milk gels (Kudryashov et al., 2000; 2001; Smyth et al., 2001). Recent work has demonstrated that during the early stages of milk aggregation induced by rennet, there is an increase in the attenuation of sound and, once the gel is formed, the value of attenuation no longer changes (Corredig et al., 2004a). In acid-induced gels, it is also possible to follow the dissolution of the calcium phosphate from the casein micelles by observing changes in the ultrasonic velocity (Dalgleish et al., 2005). In addition, the attenuation of sound shows a small increase since the early stages of acidification (very close to the native pH of milk), and increases more rapidly after pH 5.8, reaching a plateau at the gelation pH (Dalgleish et al., 2004; Azim et al., 2010a). This increase in attenuation has been attributed, at least in part, to the changes in composition of the serum phase (Dalgleish et al., 2005), but a gelation point can be clearly detected when observing the structural changes of the milk gels using frequencies between 10–100 MHz (Azim et al., 2010a).

Using DWS it is possible to follow with much precision the formation of the structure in the preliminary stages of the aggregation. The development of the l^* value is very different depending on the mode of destabilization of the casein micelles. The signature of the l^* parameter is quite unique depending on the pathway of destabilization, and it can also show large differences during acid-induced gelation of heated and unheated milk (Alexander and Dalgleish, 2005). In particular it is possible to measure the interactions between heat-induced soluble complexes of whey proteins and casein micelles during acid coagulation (Alexander and Dalgleish, 2005). Interestingly, although the changes in the value of $1/l^*$ seem to parallel values of attenuation for acid milk gels measured by US (Dalgleish et al., 2004; Azim et al., 2010b), experiments with heated and unheated milk show very similar attenuation patterns in US, although the initial states of the proteins in heated and unheated milk are very different (Anema and Li, 2003; Lucey, 2002). Again, this would confirm that the changes in attenuation may be related mostly to changes in the composition of the serum phase during the stages preceding gelation.

The addition of different whey protein fractions (native or denatured) on the acid gelation of milk has been studied by a combination of DWS, rheology and microstructure

(Vasbinder et al., 2004). The kinetics of aggregation were followed using DWS and demonstrated, for the first time, that although the gelation point shifts to a higher gelation pH when denatured whey proteins are present, whey proteins may be present as separate soluble aggregates or attached to the casein micelles, and the gelation behavior, at the beginning of structure formation, seems to be identical. Clear differences can be observed using rheology. Milk with heat-induced whey protein aggregates present in solution shows the highest gel modulus, while milk containing aggregates mostly adsorbed on the surface of the casein micelles shows the lowest gel modulus.

In summary, these initial reports show that there is potential for the use of DWS and, in some instances, US for investigating the underlying mechanisms of aggregation and structuring in turbid food systems and soft materials. The early successes obtained by the application of DWS and US to milk systems encouraged the extension of the use of these techniques into further more complex, mixed food systems.

15.3.2 Whey protein aggregation

Ultrasonic spectroscopy has been employed to follow the heat-induced denaturation and aggregation of whey proteins. Protein aggregates show a very different dependence of the attenuation as a function of frequency compared to native whey proteins (Bryant and McClements, 1999). Both velocity and attenuation of sound can clearly identify the critical temperature of denaturation of the proteins (Corredig et al., 2004b; Wang et al., 2006). Unlike other spectroscopic techniques often employed to observe protein structural changes, ultrasonic spectroscopy can be employed in highly concentrated and opaque solutions. By analyzing whey protein isolates heated at increasing temperature, it was demonstrated that in a concentrated solution of whey proteins structural changes occur at temperatures below 60 °C (Corredig et al., 2004b).

The gelation at room temperature of a concentrated β–lactoglobulin solution (12%) in the presence of fatty acid salts (3.6% sodium caprate) has also been investigated using ultrasonic spectroscopy. Gelation occurs after several hours of incubation, and ultrasonic attenuation increases with time for a β-lactoglobulin solution containing sodium caprate, while the velocity changes suggest an increase in the compressibility of the gel (Yuno-Ohta and Corredig, 2007). It was also demonstrated by observing the changes in ultrasonic velocity and attenuation that β-casein can be used as a texture modifier of these β-lactoglobulin gels induced by the addition of fatty acid salts (Yuno-Ohta and Corredig, 2011).

15.3.3 Structure formation in systems containing polysaccharides

DWS has been coupled to microscopic techniques to observe the formation of protein polysaccharide coacervates. By a combination of DWS and confocal microscopy it was possible to determine the difference in the aggregates formed between β-lactoglobulin and acacia gum at pH 4.2 (Schmitt et al., 2001). DWS results were used to determine the stability of the systems towards sedimentation. By analyzing the scattered intensity of the backscattered light and shape of the correlation function, three distinct regimes were determined for this mixed system. When equal amounts of protein and polysaccharide are used, the correlation function obtained are well-defined, showing a decrease in the mobility of the particles over time, characteristic of the coarsening of coacervates. When the amount of protein is large relative to the polysaccharide, the mixtures exhibit a rapid decrease of the intensity with time, with an ill-defined correlation function. This is indicative of rapid sedimentation, presumably through coalescence of particles. Lastly, at a protein-to-polysaccharide ratio of 2:1, the correlation functions shift towards larger correlation times and the aggregates undergo coalescence and sedimentation. With DWS it is then possible to fine-tune the conditions, leading to the creation of different structures in coacervated systems.

The presence of polysaccharide molecules in a colloidal system may result in structure formation due to the interactions (associative or dissociative) between the polymer and the protein particles, namely, the casein micelles in milk systems (Doublier et al., 2000; Turgeon et al., 2007). DWS, and to some extent US, have been employed to follow the formation of structure in the presence of polysaccharides. Their addition often leads to competing destabilization forces within the food matrix. By enhancing one driving force over another, different structuring arrangements can be obtained, and with these novel *in situ* spectroscopic techniques it is possible to follow the changes, ultimately allowing for the fine tuning of structures and macrostructure formation.

When adding a negatively-charged polysaccharide such as pectin to milk above a critical concentration, phase separation can occur. DWS and ultrasound have been employed to closely observe the formation of structure when two competing destabilization mechanisms are present in skim milk: the aggregation of casein micelles induced by renneting together with the segregation of casein micelles induced by the addition of high methoxyl pectin (Acero Lopez et al., 2009). DWS and US can be used to closely observe the mixtures during renneting of milk containing different concentrations of high methoxyl pectin at different stages throughout the sol–gel transition. In this case also, similarly as for the coacervate solutions, it was found that the level of high methoxyl pectin added to the milk defines three very distinct regions when the mixtures are observed with these techniques. At very low concentrations of pectin (<0.1%), there is no visible phase separation in the absence of rennet. At medium concentrations, and although the values of $1/l^*$ and attenuation are lower than in the original skim milk suggesting microscopic phase separation, the changes occurring to these parameters during rennet-induced gelation are very similar to those shown in control skim milk. However, at 0.12% pectin concentration in skim milk, although no visible phase separation is observed in the bulk solutions under quiescent conditions, there is a very different time-dependent behavior of the $1/l^*$ and attenuation parameters during renneting, as measured by DWS and US, respectively. The instrumental analysis clearly indicates depletion effects at the nano-scale, with implications on the microstructure of the gels formed with renneting (Acero-Lopez et al., 2009). This published work clearly illustrates how these techniques can improve the ability to follow structure formation and aid in the manipulation of the microstructure in mixed systems. In addition, some information can be derived on the types of gel formed.

When investigating the rennet gels formed in the presence of high methoxyl pectin (Acero-Lopez et al., 2009), DWS and ultrasound measurements indicated that although the initial values of the turbidity parameter and the attenuation may be similar between the milk mixtures, the final values after gelation change dramatically depending on the

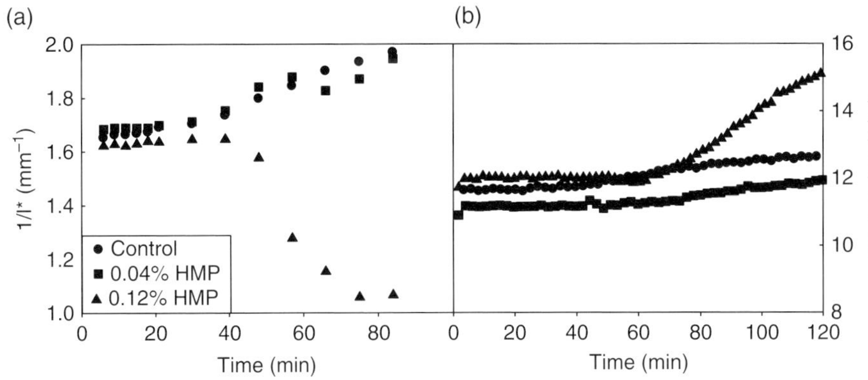

Figure 15.6 Example of results obtained for 1/I* (a) and ultrasonic attenuation (b) as a function of time for the rennet-induced gelation of milk containing two different amounts of high methoxyl pectin (HMP). At low HMP concentration (■) the general distribution of the micelles in skim milk as not affected by the presence of the polysaccharide and do not differ from that of the control (•). At high concentrations of HMP (▲), phase separation induces the structuring of the micelles as evident by the changes seen in both turbidity and attenuation measurements.

amount of polysaccharide present, as seen in Figure 15.6. Hence, networks with structural and dynamic characteristics different from control milk can be created by the interplay between two destabilizing mechanisms, phase separation by pectin addition and gelation by addition of rennet.

The same experimental approach described above for high methoxyl pectin and milk was extended to a more complex system, containing an additional polysaccharide, κ-carrageenan. To this mixture, gelation was again induced by rennet and the changes observed by DWS (Acero-Lopez et al., 2010). It is well known that κ-carrageenan is able to interact with milk casein micelles and improve stability by inhibiting visible phase separation (see for example, Spagnuolo et al., 2005). Although visible phase separation was inhibited at all levels of high methoxyl pectin in the presence of κ-carrageenan, the measurements with DWS demonstrated that phase separation still occurs at the nanometer scale. Furthermore, the presence of this polysaccharide does not affect the mobility of the casein micelles at short length-scales (in the order of a few nanometres). This work clearly demonstrated that the data derived from DWS experiments are very complementary to those observed by rheology, showing that different gels can form with subtle differences in the structure at the microscopic level. By using a multi-technique approach it may then be possible to closely follow the competing destabilizing and stabilizing effects occurring in mixed colloidal matrices, and open the possibilities of development of new structures and textures by carefully controlling the aggregation of casein micelles.

The same principles were also demonstrated using another polysaccharide, flaxseed gum. This polysaccharide was added to skim milk and using DWS and US, its effect on the formation of structure was measured, with or without κ-carrageenan (Chappellaz et al., 2010). In agreement with the results shown for pectin (Acero-Lopez et al., 2009), above a certain concentration of flaxseed gum, visible phase separation occurs, but both US and DWS parameters indicated that micro-depletion effects are present at much lower concentrations than those causing visible separation. Again, as already mentioned above, the segregative effects at low polymer concentrations still affect the structure of the gels. The MSD and the turbidity parameter, 1/l*, measured by DWS, as well as the attenuation values measured by US demonstrated that structure in milk forms via clustering of the polysaccharide-depleted domains. By combining DWS results with confocal microscopy it was possible to show that, although milk mixtures containing flaxseed gum alone or both flaxseed gum and κ-carrageenan show a similar structural organization, with clustering of the micelles due to the segregative effect of flaxseed gum, the presence of κ-carrageenan 'freezes' the structure in place and arrests the displacement of the clusters. In this case, the system was described as a loose network formed with the helices of κ-carrageenan and the casein micelles present at the periphery of the polysaccharide-depleted microdomains. This network is responsible for preventing macroscopic destabilization of the milk mixture.

The details of the acid induced aggregation of milk in the presence of modified waxy maize starch have also been described by a combination of DWS, US and rheological measurements (Azim et al., 2010a; 2010b). Interestingly, under the conditions used in these experiments, neither the starch granules nor the bacterial cells could be observed by US or DWS. The presence of starch affects the kinetics of aggregation of the caseins, resulting in an increase in the stiffness of the gel, as well as in an earlier pH of aggregation. The results from both DWS and US experiments suggest that although the rheological properties of the acid gels are different, the mechanisms involved in the aggregation of the casein micelles remain unchanged in the presence of starch.

15.3.4 Recombined milk

The presence of fat globules in protein gels strongly affects the structure of the final matrix. The effect of different types of oil droplets recombined with skim milk has been investigated using DWS. At present, no work is available on ultrasonic spectroscopy studies on the gelation of recombined milk with oil droplets. This is a particularly interesting system, when studied with DWS, as the oil droplets, providing a larger fraction of the scattering signal compared to the casein micelles (Gaygadzhiev et al., 2008) can be employed as probes to follow during structure formation of the milk gels.

A system comprising of skim milk recombined with soy oil droplets stabilized either with sodium caseinate or whey proteins has been used as a model for DWS observation of structure-formation during renneting (Gaygadzhiev et al., 2009; 2011). All recombined milk samples were dialyzed against milk containing $CaCl_2$, to ensure that the ionic conditions in the serum phase were kept constant when comparing their renneting behavior. These studies clearly demonstrate the potential of DWS as a technique complementary to rheology in characterizing the structure formation of reconstituted milk gels. The behavior of the milk samples during the gelation with rennet showed very different responses depending on the type of protein used. The aggregation behavior of milk with NaCas used as a stabilizer showed that both the values of G' (rheology) and apparent radius (DWS) remained unchanged even 3 hours after the addition of rennet. The presence of NaCas-stabilized fat globules largely hindered the ability of the casein micelles to form a rennet gel. The samples containing WPI-stabilized fat globules, on the other hand, showed aggregation behavior quite similar to the coagulation profile of control skim milk. It seems that the WPI-stabilized fat globules did not interfere with the formation of the casein network and that gel network formation was completely dominated by the aggregation of the casein micelles. These findings suggested that the WPI-stabilized fat globules fill the spaces within the casein gel strands but otherwise remain uninvolved in their backbone of the network. Their dynamics of movement, probed by measuring the diffusion coefficient and the MSDs, clearly demonstrated that the droplets remained free to move within the pores created by the rennet-induced protein network. The same behavior was observed by increasing the concentration of the casein micelles. There was a steeper increase of the apparent radii with time after rennet addition due to an enhanced arrest in the motion of the micelles due to crowding. Furthermore, the gel was significantly stiffer (with a higher elastic modulus) and this can be attributed to the presence of a higher number of gel-forming moieties resulting in more bonds per unit space.

Similarly, the effect of concentration of oil droplets in structuring of rennet gels was also evaluated in milk systems (Gaygadzhiev et al., 2009). The turbidity parameter ($1/l^*$) measured by DWS suggested differences in the structure of the gels with different amounts of emulsion droplets. The change from the initial turbidity parameter to that measured after gelation decreases with increasing volume fraction of oil droplets in the gel. While at the low volume fractions the droplets have freedom of movement within the pores of the gel, at the high fat volume fractions the droplets are immobile within the pores. By increasing the volume fraction, there is less space available for the emulsion droplets within the casein matrix and, as gelation progresses; there is a progressive squeeze of the fat globules within the casein gel pores. This effect is reflected by an increase in the gel modulus of these gels, in spite of the non-interacting nature of the oil droplets.

Similarly to what was described above for recombined milk, studies have been carried out also on rennet and acid gelation of homogenized milk (Ion Titapiccolo et al., 2010). Gelation studies using DWS can also show the subtle differences in the mobility of the oil droplets in reconstituted milk containing emulsions with similar interfacial composition but different colloidal state. DWS analysis of rennet induced gelation of reconstituted skim milk containing non-interacting whey protein stabilized emulsion droplets clearly demonstrates that the mobility of the oil droplets can differ depending on the colloidal state of the emulsion droplet (stable or flocculated in milk) (Corredig et al., 2011). Before reconstitution, the oil droplets were prepared by homogenizing whey proteins dispersed in buffer or in milk serum (obtained by ultrafiltration of skim milk). This difference in methodology permitted the manipulation of the colloidal state of the emulsion system from stable to flocculated. In spite of the same amount of oil droplets being added to the reconstituted milk, vastly different structures were created. While rheological measurements clearly show that, in spite of the identical interfacial composition, the oil droplets cause different types of gels, using DWS it is possible to describe the details of the interactions. MSD measurements demonstrate that the small, stable, buffer-prepared droplets reside within the pockets formed by the casein micelles' network and are free to move within the gel pockets. On the other hand, when the same droplets are flocculated, their size approximates that of the mesh size, and the rate of gel-formation and the stiffness of the gel increases dramatically, because as the network develops, the flocculated droplets become trapped inside the formed cages. Again this can be clearly demonstrated by MSD measurements carried out by DWS. Figure 15.7 shows results obtained from these experiments.

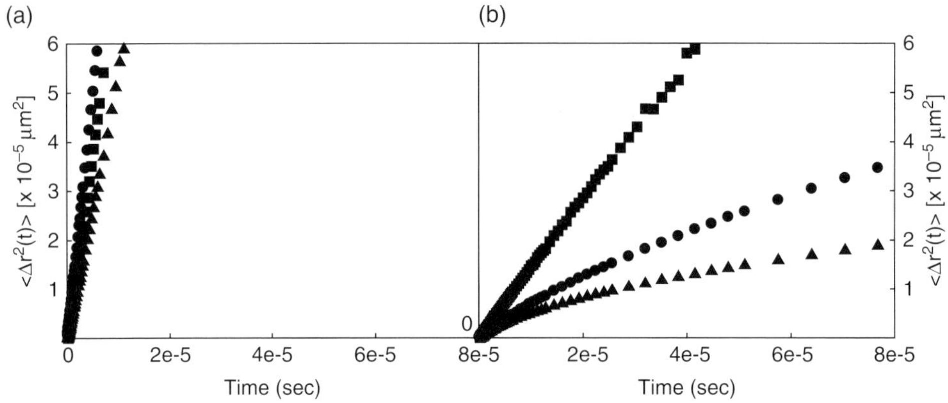

Figure 15.7 Mean square displacement (MSD) as a function of correlation time for control skim milk (•) and recombined milks containing whey protein isolate-stabilized oil droplets prepared in low ionic strength buffer (■) resulting in highly stable particles and in permeate (▲) resulting in aggregated particles. Panel (a) shows the system immediately after, and panel (b) 180 min, after addition of rennet. The stable particles remained mobile throughout the renneting process but the presence of aggregated droplets showed a dramatic decrease in mobility. This was caused by the progressive constraint experienced by the fat globules within the pores of the casein network.

These examples once again demonstrate that DWS allows for a better understanding of the micro-structural details occurring during the formation of the gel structures, and for a better explanation of the rheological properties of the gels.

15.3.5 Structure formation in dairy emulsions

DWS and US have been also employed to gain in-depth understanding of the different mechanisms driving the destabilization in emulsion systems, mainly of the water-in-oil type. Early work investigated the kinetics of destabilization by Ostwald ripening of α_{s1}-casein-stabilized emulsions using backscattering DWS (Hemar and Horne, 1999). This work unequivocally demonstrated that DWS is an effective tool in determining the colloidal behavior of emulsions *in situ* – very critical in systems where destabilization mechanisms need to be studied without dilution or any other disturbance.

Similar efforts on studying oil in water emulsions have also been published in the area of ultrasonic spectroscopy. This technique also demonstrated potential for its employment in the characterization of the colloidal properties of oil droplets dispersions, and in the study of the pathways leading to destabilization (Coupland and Mc Clements, 1997; Chanamai et al., 1998).

The destabilization of emulsions by depletion flocculation can be followed using DWS. In a combination of backscattering and transmission DWS the stability of sodium caseinate emulsions containing excess protein has been studied (Hemar et al., 2003). The photon transfer mean free path parameter, l*, was employed to follow the state of correlation of the particles as destabilization by flocculation progressed. Recently, the same system, a sodium caseinate emulsion destabilized by excess protein in solution, was also studied using electroacoustics (Gülseren and Corredig, 2011). Destabilization of the emulsion shows a reduction in ultrasonic attenuation and a decrease in the electrophoretic mobility of the oil droplets. Electroacoustics is particularly sensitive to monitoring structural rearrangements similar to those induced by polymer depletion flocculation (Gülseren and Corredig, 2011).

Structure formation of emulsion systems in the presence of polysaccharide was also investigated, for example, in whey protein stabilized emulsions in the presence of exopolysaccaride produced from lactic acid bacteria (Ten Grotenhuis et al., 2000). By varying the concentration of the polysaccharide the authors determined that the amount of EPS present modulates the interactions occurring between droplets from a stable system at low concentrations, to phase separating and finally, at high enough EPS amounts, the formation of a gel. By using a multi-technique approach they could detect differences in the signal between phase separating and gelling, especially by calculating the MSD values of the particles. As previously mentioned, the nature of the displacement of the particles is very different depending on whether they are an integral part of a gel or whether they are still single entities in a gelling system. Similar principles were also demonstrated using a whey protein stabilized emulsion system containing dextran and excess protein in solution (Blijdenstein et al., 2003). Again, by use of DWS, these experiments could be performed at industrially relevant concentrations and with no experimentally-induced stress, emphasizing that structure formation can be very well followed using this technique, allowing for better control of formulation and processes.

Using a combination of DWS, rheology and confocal microscopy it is also possible to observe emulsion systems where more than one destabilization mechanism is occurring. This multitechnique approach aids in pointing out the differences between the depletion flocculation of the whey protein stabilized emulsions caused by dextran addition and the bridging flocculation caused by the addition of carboxylmethylcellulose (Blijdenstein et al., 2004). DWS measurements can help distinguish between depletion and bridging type flocculation in emulsions. This was clearly illustrated with a system containing whey protein stabilized emulsions with flaxseed gum (Khalloufi et al., 2009). This work showed that, at pH 7.0, flaxseed gum does not interact with the oil droplets and, at sufficiently high concentrations it causes phase separation by depletion flocculation mechanisms. At this pH and at low polysaccharide concentration, the overall spatial distribution of the droplets does not change compared to the control emulsion. At intermediate concentrations, rapid phase separation is observed, while at concentrations $>0.2\%$ there is a substantial increase in the background viscosity, thereby slowing down the mobility of the domains containing the emulsion droplets and retarding phase separation (Khalloufi et al., 2009). By using DWS it was possible to show that at low polysaccharide concentrations, the system remains free diffusing, and this is a clear indication of a non-aggregating system. On the other hand, at pH 3.5, the negatively charged polysaccharide fraction from flaxseed interacts with the protein adsorbed at the interface leading to bridging flocculation. Values of the mean square displacement measured by DWS demonstrate that in this system, the mobility of the emulsion droplets is arrested. The study of this system using DWS allowed for a clear understanding of the droplets dynamics in flocculated systems, and it was possible to distinguish between depletion and bridging mechanisms of destabilization.

DWS measurements have been shown to bring complementary information to rheology when studying sodium caseinate stabilized emulsions during acidification in the presence of excess of sodium caseinate (Ruis et al., 2007). By following the changes in the turbidity parameter as well as the correlation times, it is possible to prove that different gel structures can form depending on the concentration of protein and the pH, ranging from a droplet network (made up of emulsion droplets) to a protein network in which the droplets are embedded. As previously mentioned for recombined milk studies, rheological measurements are not able to detect these fine changes in the structure of the emulsions, while DWS can clearly demonstrate these differences.

US and DWS have been extensively used to investigate the dynamics of structuring of destabilized emulsion systems (Bonnet et al., 2005; Gancz et al., 2005; Khalloufi et al., 2009; Liu et al., 2008). One of the initial reports of a combined study by US and DWS involved the investigation of the effects of addition of high methoxyl pectin on WPI-stabilized emulsion at pH 7.0 and pH 3.5 (Gancz et al., 2005; 2006). At neutral pH, no interactions were observed between the protein-covered droplets and the pectin molecules, since repulsion between both negatively-charged molecules prevented close approach. At acidic pH, however, the charge of the protein is reversed and bridging flocculation occurs via electrostatic interactions. Making use of the parameters obtained by both US and DWS, it is possible to describe in great detail the interactions occurring at the molecular level causing differences in the structuring behavior of the polysaccharide at different pH values.

Destabilization of sodium caseinate emulsions during acidification was also studied with a combination of US and DWS spectroscopy (see Figure 15.8). At a pH close to the isoelectric point of the protein, the emulsion droplets lose their colloidal stability and form aggregated structures. The details of the sol gel transitions in these systems can be followed using DWS, US or electroacoustics (Liu et al., 2008; Gülseren and Corredig, 2011). The addition of a negatively charged polysaccharide such as high methoxyl pectin or soy soluble polysaccharide to the acidifying sodium caseinate emulsion can cause a shift of the sol gel transition to a lower pH, or even inhibit it if sufficient polysaccharide molecules adsorb to the interfacial layer of the emulsion. DWS and US are very

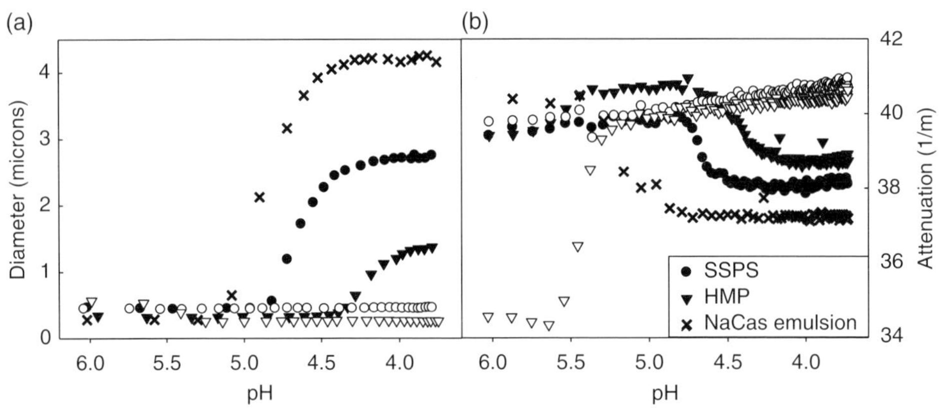

Figure 15.8 Changes in diameter measured by DWS (a) and ultrasonic attenuation measured by ultrasonic spectroscopy (b) during acidification of a NaCas-stabilized emulsion (**×**) with added polysaccharide, (•) SSPS, or (▾) HMP. The presence of small amounts of polysaccharide, 0.05% (filled symbols) delays the onset of aggregation as seen by an increase in diameter or a decrease in attenuation. When large amounts, 0.2%, of either SSPS or HMP are added (empty symbols), the sol–gel transition is inhibited.

effective tools to fine tune the amount of polysaccharide needed to control the sol gel transition (Bonnet et al., 2005; Liu et al., 2008). Both the turbidity parameter, 1/l*, and ultrasonic attenuation parameters can be used successfully to study the rearrangements in structure occurring to the oil droplets at the relevant concentrations (Liu et al., 2008).

15.4 CONCLUSIONS

The previous sections have described, in a very general way, the different works published in literature that make use of DWS and US to investigate the development of microstructure in different, usually destabilized systems in the field of food science. In general, as is often the case, these techniques were applied as one amongst a group of many, to try to interpret and reveal the mechanisms governing the interactions between components and better explain the resulting microstructure. A multi-technique approach is always a better, more thorough way of examining a problem, as different techniques probe different attributes of a sample and by assembling the results obtained by each one of them, a more complete picture of the true nature of the changes occurring to the complex matrix can be obtained. There is, however, an important point to discuss, particularly given the title and objective of this book. It is well established that food structure has a direct and crucial influence in food acceptability and is a key aspect to consider in the design of new foods. It is very important to be able to understand which factors play a role in determining the pathways leading to the formation of structure in colloidal systems, so that we can control and manipulate the final outcome. The two techniques described in this chapter are only able to probe and describe behaviors and interactions at the stages preceding macroscopical structure formation. Challenges still remain in developing the framework necessary to be able to fully understand and model the parameters that are measured by DWS and US for complex food matrices. These non-invasive, non-disruptive techniques have already resulted in great advances in our fundamental understanding of the destabilization behavior of colloidal food systems.

It becomes very clear by examining the topics covered by the present book, that the design, optimization and delivery of food texture are by no means trivial. The application of advanced or novel spectroscopic techniques such as diffusing wave spectroscopy and ultrasound are simply two more instruments in the suite of tools available to cover the wide field of variables that make up the relation between structure and texture. They are perhaps the more 'basic' determinants of structure, as these two techniques can only assess organization up to a few microns in size. It is hoped that they will continue to be used in the future to observe the behavior of the building blocks during their assembling and disassembling in food structures.

References

Acero-Lopez, A., Corredig, M. Alexander, M. (2009) Diffusing wave and ultrasonic spectroscopy of rennet-induced gelation of milk in the presence of high-methoxyl pectin. *Food Biophysics* 4, 249–259.

Acero-Lopez, A., Alexander, M. and Corredig, M. (2010) Diffusing wave spectroscopy and rheological studies of rennet-induced gelation of skim milk in the presence of pectin and kappa-carrageenan. *International Dairy Journal* 20, 328–335.

Alexander, M., and Dalgleish, D.G. (2004) Application of transmission diffusing wave spectroscopy to the study of gelation of milk by acidification and rennet. *Colloids Surfaces B* 38, 83–89.

Alexander, M. and Dalgleish, D.G. (2005). Interactions between denatured milk serum proteins and casein micelles studied by diffusing wave spectroscopy. *Langmuir* 21, 11380–11386.

Alexander, M. and Dalgleish, D.G. (2006) Dynamic light scattering techniques and their applications in food science. *Food Biophysics* 1, 2–13.

Alexander, M. and Corredig, M. (2007) Spectroscopic methods to determine in situ changes in dairy systems – ultrasonic and light scattering. *Lait* 87, 435–442.

Alexander, M., Corredig, M. and Dalgleish, D.G. (2006) Diffusing wave spectroscopy of gelling food systems – the importance of the photon transport mean free path (l*) parameter. *Food Hydrocolloids* 20, 325–331.

Allegra, J.R. and Hawley, S.A. (1972) Attenuation of sound in suspensions and emulsions. *Journal of Acoustic Society of America* 51, 1545–1564.

Anema, S. Li, Y. (2003) Effect of pH on the association of denatured whey proteins with casein micelles in heated reconstituted skim milk. *Journal of Agricultural and Food Chemistry* 51, 1640–1646.

Azim, Z., Corredig, M., Koxholt, M. and Alexander, M. (2010a) Sol gel transitions during acid gelation of milk containing modified waxy maize starch. Differences between chemical and bacterial acidification measured using rheological and spectroscopic techniques. *International Dairy Journal* 20, 785–791.

Azim, Z., Alexander, M., Koxholt, M. and Corredig, M. (2010b) Influence of cross-linked waxy maize starch on the aggregation behavior of casein micelles during acid-induced gelation. *Food Biophysics* 5, 227–237.

Benguigui, L., Emery, L., Durand, D. and Busnel, J.P. (1994) Ultrasonic study of milk clotting. *Lait* 74, 197–206.

Berne, B.J. and Pecora, R. (eds.) (1976) *Dynamic Light Scattering*, John Wiley, New York.

Blijdenstein, T.B.J., van Vliet, T., van der Linden, E. and van Aken, G.A. (2003) Suppression of depletion flocculation in oil in water emulsions: a kinetic effect of beta-lactoglobulin. *Food Hydrocolloids* 17, 661–669.

Blijdenstein, T.B.J., van der Linden, E., van Vliet, T. and van Aken, G. (2004) Scaling behavior of delayed demixing, rheology, and microstructure of emulsions flocculated by depletion and bridging. *Langmuir* 20, 11321–11328.

Bonnet, C., Corredig, M. and Alexander, M. 2005. Stabilization of caseinate-covered oil droplets during acidification with high methoxyl pectin. *Journal of Agricultural and Food Chemistry* 53, 8600–8606.

Bryant, C.M. and McClements, D.J. (1999) Ultrasonic spectrometry study of the influence of temperature on whey protein aggregation. *Food Hydrocolloids* 13, 439–444.

Chanamai, R., Herrmann, N., and McClements, D.J. (1998) Ultrasonic spectroscopy study of flocculation and shear-induced floc disruption in oil-in-water emulsions, *Journal of Colloids and Interface Science* 204, 268–276.

Chappellaz, A., Alexander, M. and Corredig, M. (2010) Phase separation behavior of caseins in milk containing flaxseed gum and k-carrageenan: A light-scattering and ultrasonic spectroscopy study. *Food Biophysics* 5, 138–147.

Corredig, M. and Alexander, M. (2008) Food emulsions studied by DWS: recent advances. *Trends in Food Science and Technology* 19, 67–75.

Corredig, M., Alexander, M. and Dalgleish, D.G. (2004a). The application of ultrasonic spectroscopy to the study of the gelation of milk components. *Food Research International* 37, 557–565.

Corredig, M., Verespej, E., and Dalgleish, D.G. (2004b) Heat induced changes in the ultrasonic properties of whey proteins. *Journal of Agricultural and Food Chemistry* 52, 4465–4471.

Corredig, M., Ion Titapiccolo, G., Gaygadzhiev, Z. and Alexander, M. (2011) Rennet-induced aggregation of milk containing homogenized fat globules. Effect of interacting and non-interacting fat globules observed using diffusing wave spectroscopy. *International Dairy Journal* 21, 679–684.

Coupland, J.N. and McClements, D.J. (1997) Physical properties of liquid edible oils. *Journal of the American Oil Chemists' Society* 74, 1559–1564.

Coupland, J.N. and McClements, D.J. (2001) Droplet size determination in food emulsions: comparison of ultrasonic and light scattering methods. *Journal of Food Engineering* 50, 117–120.

Dalgleish, D.G. (1980) A mechanism for the chymosin-induced flocculation of casein micelles. *Biophysical Chemistry* 11, 147–155.

Dalgleish, D.G. (2011) On the structural models of bovine casein micelles – review and possible improvements. *Soft Matter* 7, 2265–2272.

Dalgleish, D.G. and Horne, D.S. (1991) Studies of gelation of acidified and renneted milks using diffusing wave spectroscopy. *Milchwissenschaft* 46, 417–419.

Dalgleish, D.G. and Hallet, F.R. (1995) Dynamic light scattering applications to food systems. *Food Research International* 28, 181–193.

Dalgleish, D.G., Alexander, M., and Corredig, M. (2004) Studies of the acid gelation of milk using ultrasonic spectroscopy and diffusing wave spectroscopy. *Food Hydrocolloids* 18, 747.

Dalgleish, D.G., Verespej, E., Alexander, M., and Corredig, M. (2005) The ultrasonic properties of skim milk related to the release of calcium from casein micelles during acidification. *International Dairy Journal* 15, 1105–1112.

de Kruif, G.C., Huppertz, T., Urban, V.S. and Petukhov, A.V. (2012) Casein micelles and their internal structure. *Advances in Colloid and Interface Science* 171–172, 36–52.

de Kruif, C.G., Grotenhuis, E.T., Weitz, D.A., Nishimura, G. and Yodh, A.G. (2004) Arrangement and method to apply diffusing wave spectroscopy to measure the properties of multi-phase systems, as well as the changes therein. US Patent Number 6,831,741 B1 assigned Dec 14, 2004.

Doublier, J.L., Garnier, C., Renard, D. and Sanchez, C. (2000) Protein polysaccharide interactions. *Current Opinion in Colloid and Interface Science* 5, 202–214.

Dukhin, A.S. and Goetz, P. (2000) Particle size distribution and zeta potential using acoustic and electroacoustic spectroscopy. US Patent number 6,109,098 issued August 29, 2000.

Dukhin, A.S. and Goetz, P.J. (2002) *Ultrasound for characterizing colloids. Particle sizing, zeta potential, rheology*. Elsevier, Amsterdam.

Dukhin, A.S. and van de Ven, T.G.M. (1994) Electrokinetic characterization of polydisperse colloidal particles. *Journal of Colloid and Interface Science* 165, 9–18.

Dukhin, A.S., Goetz, P.J. and Travers (2005) Use of ultrasound for characterizing dairy products. *Journal of Dairy Science* 88, 1320–1334.

Gancz, K., Alexander, M. and Corredig, M. (2005) Interactions of high methoxyl pectin with whey proteins at oil/water interfaces at acid pH. *Journal of Agricultural and Food Chemistry* 53, 2236–2241.

Gancz, K., Alexander, M. and Corredig, M. (2006) In situ study of flocculation of whey protein-stabilized emulsions caused by addition of high methoxyl pectin. *Food Hydrocolloids* 20, 293–298.

Gaygadzhiev, Z., Corredig, M. and Alexander, M. (2008) Diffusing wave spectroscopy study of the colloidal interactions occurring between casein micelles, emulsion droplets and their mixtures: a comparison to hard sphere behaviour. *Langmuir* 24, 3794–3800.

Gaygadzhiev, Z., Corredig, M. and Alexander, M. (2009) The impact of the concentration of casein micelles and whey protein-stabilized fat globules on the rennet-induced gelation of milk. *Colloids and Surfaces B: biointerfaces* 68, 154–162.

Gaydadzhiev, Z., Corredig, M., and Alexander, M. (2011) Rennet coagulation properties of milk in the presence of oil droplets stabilised by a combination of sodium caseinate and whey protein isolate. *Dairy Science and Technology* 91, 719–737.

Gülseren, I. and Corredig, M. (2011) Changes in the colloidal properties of oil in water emulsions stabilized with sodium caseinate observed by electroacoustic spectroscopy *Food Biophysics* 6, 534–542.

Gülseren, I., Alexander, M. and Corredig, M. (2010) Probing the colloidal properties of skim milk using acoustic and electroacoustic spectroscopy. Effect of concentration, heating and acidification. *Journal of Colloid and Interface Science* 351, 493–500.

Hemar, Y. and Horne, D.S. (1999) A diffusing wave spectroscopy study of the kinetics of Ostwald ripening in protein-stabilised oil/water emulsions. *Colloids and Surfaces B: biointerfaces* 12, 239–246.

Hemar, Y., Singh, H. and Horne, D.S. (2004) Determination of early stages of rennet-induced aggregation of casein micelles by diffusing wave spectroscopy and rheological measurements *Current Applied Physics* 4, 362–367.

Hemar, Y., Pinder, D.N., Hunter, R.J., Singh, H., Hébraud, P. and Horne, D. (2003) Monitoring of flocculation and creaming of sodium-caseinate-stabilized emulsions using diffusing-wave spectroscopy. *Journal of Colloid and Interface Science* 264, 502–508.

Horne, D.S. (1986) Steric stabilization and casein micelle stability. *Journal of Colloid and Interface Science* 111, 250–259.

Horne, D.S. (1999) Formation and structure of acidified milk gels. *International Dairy Journal* 9, 261–268.

Horne, D.S. and Davidson, C.M. 1990 The use of dynamic light scattering in monitoring rennet curd formation. *Milchwissenshchaft* 45, 712–715.

Khalloufi, S., Corredig, M. and Alexander, M. (2009) Interactions between flaxseed gums and WPI-stabilized emulsion droplets assessed in situ using diffusing wave spectroscopy. *Colloids and Surfaces B: biointerfaces.* 68, 145–153.

Kudryashov, E., Smyth, C., Duffy, G. and Buckin, V. (2000) Ultrasonic high resolution longitudinal and shear wave measurements in food colloids: moinitoring gelation processes and detection of pathogens. *Progress in Colloid and Polymer Science* 115, 287–294.

Kudryashov, E., Hunt, N.T., Aikainen, E.O. and Buckin, V. (2001) Monitoring of acidified milk gel formation by ultrasonic shear wave measurements. High frequency viscoelastic moduli in milk and acidified milk gel. *Journal of Dairy Science*, 84, 375–388.

Ion Titapiccolo, G., Alexander, M. and Corredig, M. (2010) Rennet-induced aggregation of homogenized milk: impact of the presence of fat globules on the structure of casein gels. *Dairy Science and Technology*, 90, 623–639.

Liu, J., Verespej, E., Corredig, M. and Alexander, M. (2008) Investigation of interactions between two different polysaccharides with sodium caseinate-stabilized emulsions using complementary spectroscopic techniques: Diffusing wave and ultrasonic spectroscopy. *Food Hydrocolloids* 22, 47–55.

Lucey, J.A. (2002) Formation and physical properties of milk protein gels. *Journal of Dairy Science* 85, 281–294.

Maret, G. and Wolf, P.E. (1987) Multiple light scattering from disordered media – the effect of Brownian motion of scatterers. *Zeitschrift fur physik B- Condensed Matter* 65, 409–413.

Martini, S., Herrera, M.L. and Marangoni, A. (2005a) New Technologies to determine solid fat content on line. *Journal of the American Oil Chemists' Society* 82, 313–317.

Martini, S., Bertoli, C., Herrera, M.L., Neeson, I. and Marangoni, A. (2005b) Attenuation of ultrasonic waves: influence of microstructure and solid fat content. *Journal of the American Oil Chemists' Society* 82, 319–328.

McClements, D.J. and Povey, M.J.W. (1987) Ultrasonic velocity as a probe of emulsions and suspensions. *Advances in Colloid and Interface Science* 27, 285–316.

McClements, D.J. and Povey, M,J.W. (1992) Ultrasonic analysis of edible fats and oils. *Ultrasonics* 30, 383–388.

McClements, D.J., Povey, M.J.W. and Dickinson, E. (1993) Absorption and velocity dispersion due to crystallization and melting of emulsion droplets. *Ultrasonics* 31, 433–437.

Nicolas, Y., Paques, M. van den Ende, D., Dhont, J.K.G., van Polanen, R.C., Knaebel, A., Munch, J.P., Bljdenstein, T.B.J. and van Aken, G.A. (2003) Microrheology: new methods to approach the functional properties of food. *Food Hydrocolloids* 17, 907–913.

Ruis, H.G.M., van Gruijthuijsen, K., Venema, P. and van der Linden, E. (2007) Transitions in structure in oil-in-water emulsions as studied by diffusing wave spectroscopy. *Langmuir* 23, 1007–1013.

Schmitt, C., Sanchez, C. , Lamprecht, A., Renard, D., Lehr, C., de Kruif, G.G. and Hardy, J. (2001) Study of beta-lactoglobulin/acacia gum complex coacervation by diffusing-wave spectroscopy and confocal scanning laser microscopy. *Colloids and Surfaces B, Biointerfaces* 20, 267–280.

Smyth, C., Kudryashov, E.D. and Buckin, V. (2001) High-frequency shear and volume visocelastic moduli of casein particle gel. *Colloids and Surfaces A* 41, 19–47.

Spagnuolo, P.A., Dalgleish, D.G., Goff, H.D., and Morris, E.R. (2005). Kappa-carrageenan interactions in systems containing casein micelles and polysaccharide stabilizers. *Food Hydrocolloids* 19, 371–377.

Ten Grotenhuis, E., Paques, M., and van Aken, G.A. (2000) The application of diffusing-wave spectroscopy to monitor the phase behavior of emulsion-polysaccharide systems. *Journal of Colloid and Interface Science* 227, 495–504.

Turgeon, S.L., Schmitt, C. and Sanchez, C. (2007) Protein polysaccharide complexes and coacervates. *Current Opinion in Colloid and Interface Science* 12, 166–178.

Urick, R.J. (1947) A sound velocity method determining the compressibility of finely divided substances. *Journal of Applied Physics* 18, 983–987.

Vasbinder, A.J., de Velde, F.V. and de Kruif, C.G. (2004) Gelation of casein-whey protein mixtures. *Journal of Dairy Science* 87, 1167–1176.

Van de Hulst, H.C. (1957) *Light Scattering by Small Particles* Wiley, New York.

Van Hooydonk, A.C.M. and Walstra, P. (1987) Interpretation of the kinetics of the renneting reaction in milk. *Netherlands Milk and Dairy Journal* 41, 19–47.

Wang, Q., Tolkach, A. and Kulozik, U. (2006) Quantitative assessment of thermal denaturation of bovine a-lactalbumin via low-intensity ultrasound, HPLC, and DSC. *Journal of Agricultural and Food Chemistry* 54, 6501–6506.

Weitz, D.A. and Pine, D.J. (1993) Diffusing wave spectroscopy. In: *Dynamic light scattering: the method and some applications*. (ed. W. Brown), 49, 652–720. Oxford University Press, Oxford, UK.

Yuno-Ohta, N. and Corredig, M. (2007) Characterization of β-lactoglobulin A gelation in the presence of sodium caprate by ultrasound spectroscopy and electron microscopy. *Biomacromolecules* 8, 2542–2548.

Yuno-Ohta, N. and Corredig, M. (2011) β-casein aids in the formation of a sodium caprate-induced β-lactoglobulin B gel. *Colloids and Surfaces B: biointerfaces* 54, 442–445.

16 Atomic force microscopy for determining surface interactions of relevance for food foams and emulsions

Marta Krasowska, Clive A. Prestidge and David A. Beattie

Ian Wark Research Institute, University of South Australia, Mawson Lakes, Australia

16.1 INTRODUCTION

Foams and emulsions are central to many foods, including dairy products, spreads, dressings, and desserts. Both foams and emulsions are mixtures of one phase dispersed within another continuous phase. Food foams are dispersions of a gas in a continuous liquid phase; food emulsions are most commonly dispersions of oil within a continuous aqueous phase. From the standpoint of physical chemistry, foams and emulsions are unlikely systems due to the inherent instability of so much oil/water or air/water surface area. Without the presence of significant amounts of surfactants, polymers, or proteins, which are instrumental in lowering the surface tension of the oil/water or gas/water interfaces, foams and emulsions rapidly collapse through the aggregation of individual bubbles/drops, and coalescence due to the rupture of the intervening liquid films.

Aggregation and coalescence of emulsions and foams are critical processes in food systems. First (and in most cases foremost), destruction of the emulsion/foam system is detrimental to the long term storage of the food product, as destruction of the dispersed structure affects and enhances microbial activity resulting in early product spoilage. In addition, the product sensory characteristics are also significantly affected, including both visual and textural attributes. It is therefore no surprise that a large amount of all emulsion and foam research is dedicated toward controlling foam and emulsion stability. Second, although textural characteristics of food foams and emulsions are destroyed due to aggregation and coalescence during storage, the aggregation and coalescence of these systems in the mouth is an essential part of the food sensory experience (Vingerhoeds et al., 2005).

Given the importance of the processes of drop/bubble aggregation and coalescence to food stability and texture for foam/emulsion products, it is essential that food scientists and physical chemists understand the parameters of the systems that control these processes. At the core of these two instability processes are the attractive or repulsive forces between droplets or bubbles; whether droplets/bubbles are attracted to one another (or not) dictates their aggregation and coalescence (McClements, 2004). As well as understanding the forces present between oil droplets and between air bubbles, it is equally

Food Texture Design and Optimization, First Edition. Edited by Yadunandan Lal Dar and Joseph M. Light.

important to understand how one modifies these forces. Serendipitously, most of the droplet/bubble formation aids (surfactants, polymers, or proteins) – termed foaming agents or emulsifiers for foams and emulsifiers, respectively – also have a controlling influence on the inter-droplet and inter-bubble forces.

In spite of this core aspect of emulsion/foam stability and texture, food scientists and product formulators rarely use techniques that can actually measure these surface forces. This chapter focuses on a description of a technique that allows one to determine directly the strength of attraction or repulsion between oil droplets or air bubbles: colloid probe atomic force microscopy (AFM) (Ducker et al., 1992 and Ducker et al., 1994). Before a detailed description of the methodology of the technique, we will outline the basic underlying theory of surface forces of relevance to food foams and emulsions. The last component of the chapter will be a summary of recent insights gained into the behavior of emulsion/foam systems as determined using colloid probe AFM.

16.2 SURFACE FORCES

16.2.1 Surface forces in liquid environment – DLVO components

The DLVO theory (named after Derjaguin, Landau, Verwey and Overbeek) was developed to describe the force between charged surfaces interacting through a liquid medium (Derjaguin and Landau, 1941; Vervey and Overbeek, 1948) The theory combines the effects of the van der Waals and the electrostatic interactions in order to explain the stability (or its lack) of dispersions of colloidal particles, the total force, F_{TOT}, between particles in electrolyte solutions is simply given by the sum of van der Waals attraction, F_{vdW}, and electrical double-layer repulsion, F_{el}:

$$F_{TOT} = F_{vdW} + F_{el} \tag{16.1}$$

The nomenclature **van der Waals forces** includes three forces of different origin. The dominant contribution to van der Waals interactions comes from dispersion (London) force, which originates from non-zero instantaneous dipole moments of all atoms and molecules. The second contribution comes from the Keesom force due to the attraction between rotating permanent dipoles. The third contribution, the Debye force, accounts for the interaction between permanent dipoles and corresponding induced dipoles. Of these three, Keesom and Debye forces are only found among molecules which have permanent dipole moments. The dispersion force, however, is universal and since all materials are polarizable this component of van der Waals forces is the most important.

The Van der Waals forces in liquids are important from a distance of a few Angstroms to tenths of nanometers and decay as a function of separation cubed. Other important features of Van der Waals are: (i) the forces are anisotropic and (ii) Van der Waals forces are non-additive. The forces between two molecules are affected by other molecules nearby, since they behave like a medium. Van der Waals forces are geometry dependent. The most relevant geometries and respective formulas in studies involving bubbles and droplets are listed in Table 16.1.

Table 16.1 Van der Waals interaction forces between macroscopic bodies of different geometries.

Geometry of bodies, $l<<R$		Force (per unit area)
Two flat surfaces (per unit area)	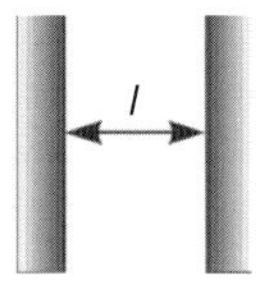	$F_{vdW} = -\frac{A}{6\pi l^3}$
Sphere of radius R near flat surface	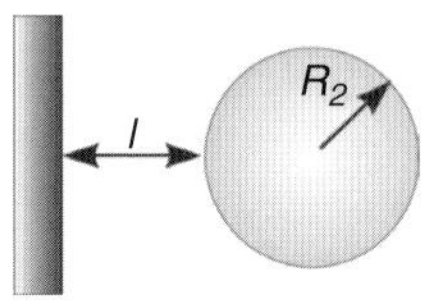	$F_{vdW} = -\frac{AR}{6l^2}$
Two spheres of radii R_l and R_2	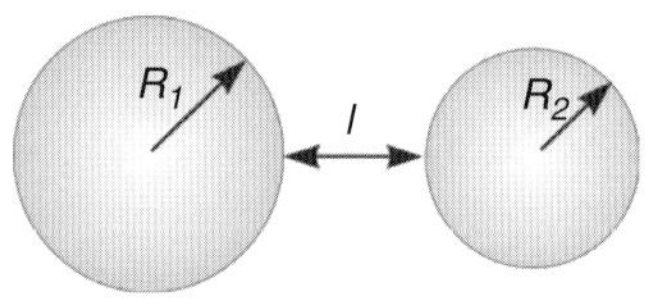	$F_{vdW} = -\frac{A}{6l^2}\frac{R_1R_2}{R_1+R_2}$

One common component in all the equations is the Hamaker constant, A. The Hamaker 'constant' in reality is not a constant but a function of separation, *(l)*, and reflects the strength of the van der Waals interaction for two bodies 1 and 2 interacting in medium 3. Because the Hamaker constant is separation dependent, its value is only accurate at near-zero separations. At greater distances, this value will be smaller in magnitude. Lifshitz theory may be used to calculate the true Hamaker parameter at any separation (Dzyaloshinskii et al., 1961).

For two identical materials (1) and (2), where permittivities $\varepsilon_1 = \varepsilon_2$ and refractive indices $n_1 = n_2$, across any medium (3) of a permittivity ε_3 and refractive index n_3, A is always positive resulting in the attractive van der Waals forces. This is the case for food foams and emulsions. If ε_3 and n_3 are intermediate to ε_1, n_1 and ε_2, n_2, then A is negative, and as a consequence, van der Waals forces are repulsive. For the situation, in which ε_3 and n_3 equal the permittivity and index of refraction of either of the two bodies, A vanishes. In such scenario van der Waals forces equal zero. Since van der Waals forces are ubiquitous, they have been applied successfully to understand interactions within a vast range of colloidal systems, including emulsions and foams.

When immersed in a polar liquid such as water, a surface charge on the surface of a solid, other immiscible fluid, or gas may be induced by the dispersing fluid. This may occur by either ionization or dissociation of the surface species, or by adsorption of ions from solution. To maintain electrical neutrality, ions of the opposite charge (counterions) accumulate near such surface forming so-called Stern layer, while others form a diffuse layer which stretches out for some distance (of order of nanometres) from the particle, droplet or bubble surface. These two layers together comprise the electrical double layer.

When two charged surfaces are brought into close proximity, the overlap of diffuse layers will give rise to an electrostatic attraction or repulsion, depending on the sign and magnitude of the surface potentials. The surface potential, ψ_0, is a potential precisely situated at the particle/droplet/air bubble – liquid interface and cannot be measured directly. A good approximation of $\psi_{0,}$ in the case of indifferent electrolytes, is the zeta potential, ζ. Zeta potential is determined at the slipping plane, that is, a plane that separates mobile fluid from fluid that remains attached to the surface. The slipping plane is usually located 3–5 Å from the surface. The surface potential can be derived from the zeta potential via:

$$\psi_0 = \zeta\left(1+\frac{z}{R}\right)e^{kz} \tag{16.2}$$

where: R is the radius of the surface (of solid particle/droplet/air bubble), z is the distance from the particle/droplet/air bubble – liquid interface and κ^{-1} is so-called Debye length:

$$\kappa^{-1} = \sqrt{\frac{\varepsilon\varepsilon_0 \kappa_B T}{2N_A e^2 I}} \tag{16.3}$$

dependent on ionic strength of the aqueous solution, I. ε and ε_0 are dielectric constants of the aqueous solution and vacuum permittivity, respectively, k_B is the Boltzman constant, T is the temperature (in K), N_A is the Avogadro number and e is the elementary charge.

For gas bubbles and oil droplets, the surface potential most probably arises from adsorption of ions from the aqueous solution. Specifically, the negative charge at a bubble or hydrocarbon droplet surface is believed to be due to specific adsorption of OH^- ions (Beattie and Djerdjev, 2004).

The **electrostatic double electric layer force** can be calculated from the continuum treatment given by Gouy (1910) and Chapman (1913) and Debye and Hückel (1923). The approach ignores the molecular nature of the aqueous phase and treats the aqueous phase as a continuous medium of a certain dielectric constant. In continuum theory the potential distribution is determined from the Poisson–Boltzmann equation with either constant charge (upon approach the surface charges remain constant) or constant potential (during approach the surface potentials remain constant) boundary conditions. The force in a spherical particle – flat geometry, assuming constant potentials of the flat surface ψ_s and the particle ψ_p, is given by: (Hogg et al., 1966)

$$F_{el}^{(cp)} = 2\pi R\varepsilon\varepsilon_0\kappa\left[2\psi_s\psi_p e^{-l\kappa} - \left(\psi_s^2+\psi_p^2\right)e^{-2l\kappa}\right] \tag{16.4}$$

and, for constant charge conditions the electrostatic double layer force is given by:

$$F_{el}^{(cc)} = \frac{2\pi R}{k\varepsilon\varepsilon_0}\left[2\sigma_s\sigma_p e^{-l\kappa} + \left(\sigma_s^2+\sigma_p^2\right)e^{-2l\kappa}\right] \tag{16.5}$$

where σ_s and σ_p are the surface charges of the flat surface and particle, respectively.

In both cases it is assumed that: (i) the surface potentials are low, that is, below $|50|$ mV, (ii) radius of curvature has to be large ($R \gg \kappa$) and (iii) $l \geq \kappa$. The choice of the boundary condition depends on the materials used.

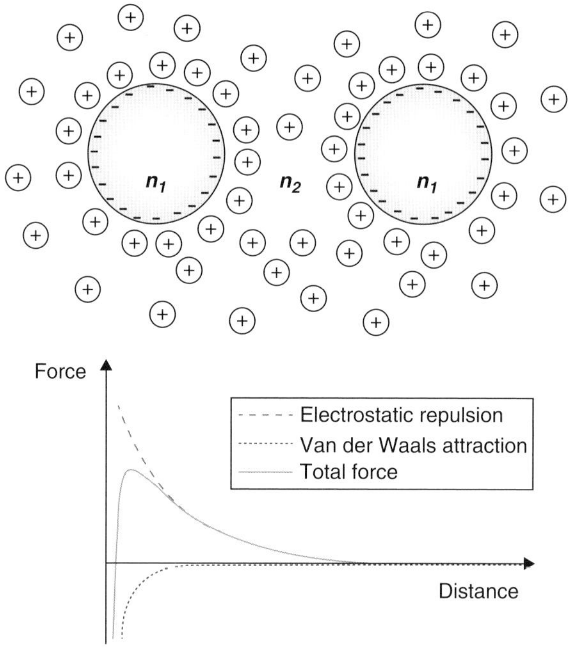

Figure 16.1 Schematic representation of the interaction forces between two oil droplets of refractive index n_1, interacting through aqueous solution of refractive index n_2. Only counter ions were marked for the simplicity. A typical force vs distance curve showing the contribution of van der Waals attraction (dotted line), electrostatic repulsion (dashed line) to the total force of interactions (solid line).

Figure 16.1 presents a schematic of forces acting between two oil droplets across aqueous solution. The electrostatic interaction is always repulsive between surfaces of the same material, and the van der Waals is always attractive in such a case. Since the exponential law is dominant at intermediate separations and the power law is always dominant at small separations, the resulting force between the two oil droplets is repulsive at greater separations, but turns attractive as the droplets get closer.

16.3 NON-DLVO FORCES

The continuum theories of van der Waals forces and electrical double layer fail when applied to describe the mutual interaction of two surfaces approaching at distances smaller than few nanometers. It is because: (i) they are not valid at small separations, and (ii) other forces arise and are dominant over such small distances.

Solvation forces arise when liquid molecules are ordered and form nearly discrete layers between two surfaces. In most of the cases, these forces exhibit an oscillatory behavior with a period that, with a good approximation, equals the molecular diameter d_m and with a range of few molecular diameters. **Hydration forces** occur when two hydrated surfaces are brought into contact in water. The water molecules near the interface are more ordered than those in the bulk. The hydration forces are repulsive and relatively short-ranged (between 1 and 3 nm) with characteristic decay lengths between 0.2–1.5 nm. The hydration forces become stronger and longer ranged with increasing salt concentration. This is why they are easily distinguished from electrostatic forces. **Steric forces** arise when the particle's, droplet's or bubble's surfaces are completely covered with a layer of polymer. The polymer adsorbs at the interface and extends out into solution. In a 'good' solvent, as the separation distance increases, and eventually becomes less than two lengths of the adsorbed polymer, the polymer layers begin to overlap and a repulsive entropic force arises. The polymers most effectively creating this steric barrier are those of low to moderate molecular weight.

In a 'poor' solvent the non-adsorbing polymer collapses and forms a globular structure with a radius of gyration, R_g. Such polymer coils are usually excluded from the region near the surface, and thus such a zone is called the depletion zone. When the two surfaces approach each other and the depletion zones overlap the total volume accessible to the polymer coils increases resulting in a strong osmotic-based attraction between the surfaces.

When polymers chains adsorb loosely at the surface, a polymer chain may bridge two interacting surfaces. This so-called **bringing force** is attractive resulting in bridging flocculation. These three different mechanisms, involving polymers and being responsible for emulsion, foam or suspension stability (or its lack), are schematically presented in Figure 16.2.

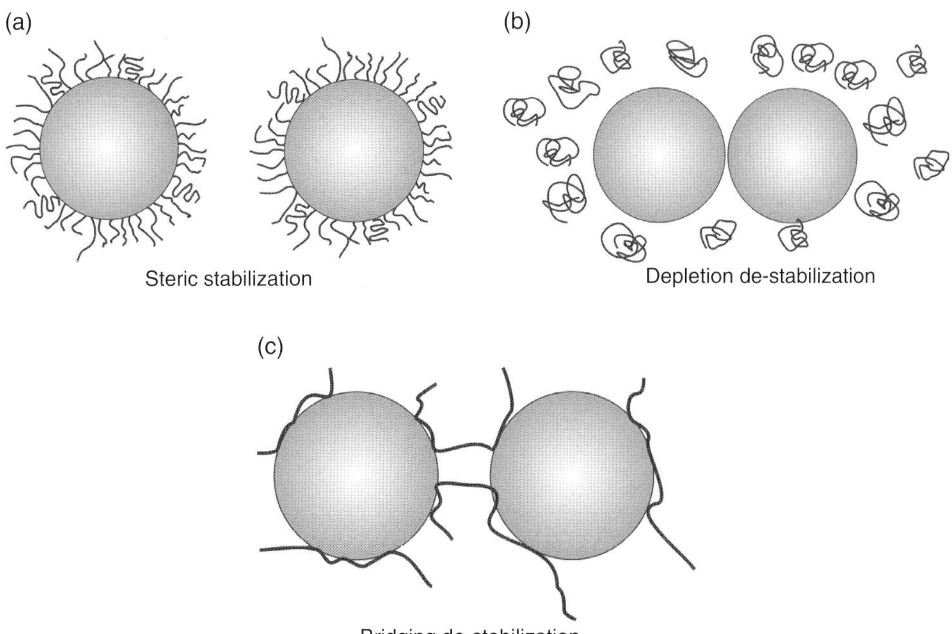

Figure 16.2 Schematic representation of three mechanisms of polymer-mediated droplet-droplet interactions: (a) the prevention of droplet aggregation through steric stabilization; (b) encouragement of droplet aggregation due to depletion forces; and (c) the encouragement of droplet aggregation due to droplet bridging.

In summary, if we consider two surfaces immersed in water, the overall force between them is a direct result of the intermolecular forces between those surfaces. A number of forces can act between surfaces in a liquid: (Israelachvili, 1992)

$$F_{tot}(l) = F_{el}(l) + F_{vdw}(l) + F_H(l) + F_{st}(l) + F_{solv}(l) + F_{hydr}(l) \qquad (16.6)$$

where F_{tot} is the total force and depends on separation distance l. F_{el} is the electrostatic force which is dominant at low salt concentration and large separations between interacting surfaces; F_{vdw} are the van der Waals forces which occur between all atoms, molecules and surfaces and is dominant at very short separations; F_H is the hydrodynamic drag force which depends on the viscosity of the intervening liquid as well as the velocity of approach and increases with the size of the approaching surfaces; F_{st} is the steric force that occurs when, for example, polymers are present on the surfaces, F_{solv} is the hydration force originating from the adsorption of solvent molecules onto the surfaces and F_{hydr} is the hydrophobic force.

16.4 ATOMIC FORCE MICROSCOPY

As two macroscopic surfaces approach each other the liquid between them drains and eventually a thin film between the interacting bodies is formed. The stability of such a thin film is governed by the total force acting between the surfaces. A variety of techniques have been used to measure forces between two solids, both in air and in liquid environment.

The direct measurement of surface forces was pioneered by Derjaguin and co-workers (as described in Larson et al., 1997). They used a convex glass lens and a flat plate to measure the van der Waals interactions in vacuum. However, the early measurements were not very precise and were limited to separations greater than about 100 nm (Larson et al., 1993). More accurate measurements of interaction forces were obtained by Tabor and co-workers (Tabor et al., 2011; Israelachvili and Tabor, 1972), using crossed mica cylinders in air. This technique was further improved by Israelachvili and Adams (1972) and Ishida and Higashitani, 2006, giving the beginning of the now well-known surface force apparatus (SFA). Although the SFA revolutionized our understanding of surface forces it is limited to materials that are smooth over ~ 1 cm^2, which restricted SFA measurements to muscovite mica or sheets of silica and sapphire (Nguyen et al., 2004). Additionally, the surface separation is measured using interferometry imposing one additional condition – the surface needs to be optically transparent.

16.4.1 Normal force measurements – colloid probe Atomic Force Microscope (AFM)

The invention of the colloid probe AFM technique (Ivanov and Kraichevsy, 1988; Ivanov, 1980; Butt, 1991) dramatically increased the applicability of the AFM for quantitative measurements of interaction forces as the technique allows for a direct measurement of the interaction between single micron-sized particles of any material. In this approach a micron-size particle (so-called colloid probe) is mounted at the end of an AFM cantilever (see Figure 16.3). Such a probe is brought into contact with a flat

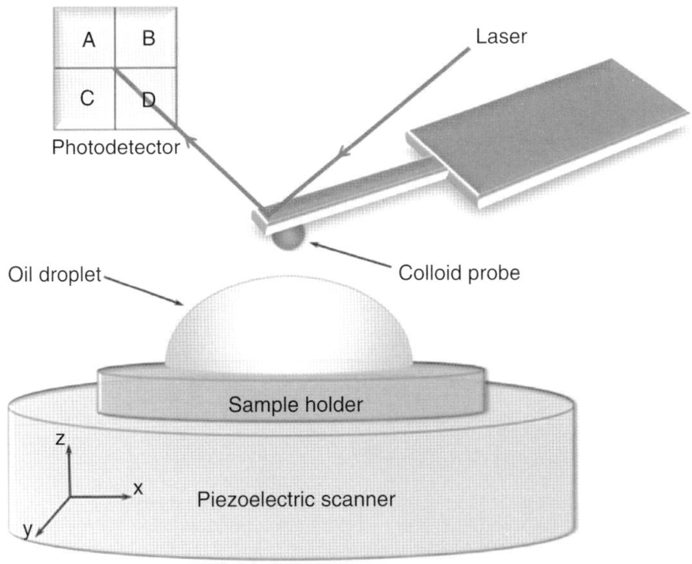

Figure 16.3 Schematic diagram representing the major components of an atomic force microscope when configured for colloid probe force measurements.

surface or a second particle. The travel distance of the surface is controlled by a piezo-actuator as it expands and contracts in a very precise manner in response to an applied potential. The information about force acting between two surfaces is obtained from the cantilever deflection. The deflection is measured by the laser reflected off the back of the AFM cantilever and directed onto a photodiode detector. The reflected laser light is then converted into a voltage signal. Cantilever deflection measured in volts as a function of the scanner displacement is referred to as raw data.

Initially, when the surface and the colloid probe are far away, there is no interaction. As the separation between these two decreases, the cantilever deflects in response to the forces it experiences. In the case of repulsive interactions, the cantilever deflects away from the approaching surface. In contrast, for attractive interactions the cantilever bends towards the surface. At small separations, when the attractive van der Waals force exceeds the cantilever's spring constant, the colloid probe 'jumps-into-contact' with the surface and, from now on, two surfaces move together. This intimate contact is called constant compliance region, and its slope is used to convert the cantilever deflection data measured in volts into cantilever deflection measured in nm. Such data can be easily converted into a normal force, F, using Hooke's Law:

$$F = -k_s \cdot \Delta x \tag{16.7}$$

where k_s is the cantilever's normal spring constant.

A very simplistic force diagram is presented in Figure 16.4. It shows one of the possible characteristics of colloid probe interacting with a solid surface in liquid environment: (i) zero force at large colloid probe-solid separations, (ii) 'jump-into-contact' due

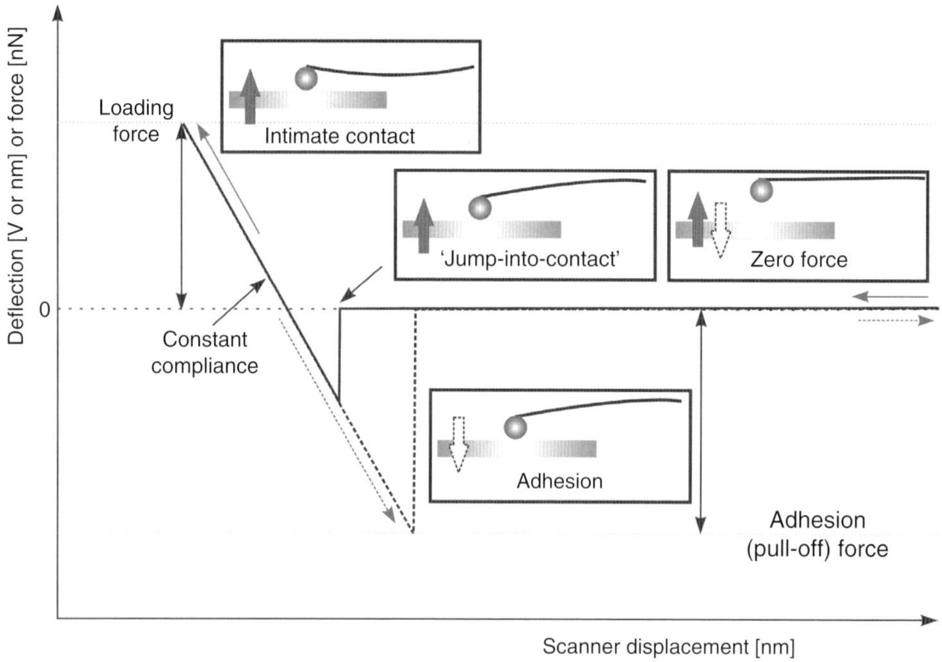

Figure 16.4 Deflection/Force *vs* scanner displacement diagram representing approach (solid line) and retraction (dotted line) of a colloid probe from a solid surface in liquid. Inserts represent the cantilever position relative to the solid substrate.

to attractive forces (e.g. van der Waals) upon approach, (iii) linear compliance region, and (iv) large pull-off force hysteresis related to adhesion (upon the colloid probe and solid surface separation).

16.4.2 Normal spring constant measurements

A crucial consideration for AFM cantilevers is the spring constant, k_s. This parameter defines how flexible the cantilever is. If the spring constant is low, the cantilever may be too flexible which results in large cantilever deflection even if the low force is applied. Such cantilevers will also jump-into-contact too fast under the action of the attractive forces, not providing the information about force profile at small separations. In contrast, a high spring constant of the cantilever adds extra stiffness which will result in the cantilever being not sensitive enough.

Quantitative force measurements require an accurate value of the cantilever spring constant. In early AFM work, nominal values supplied by the manufacturers were used. However, taking into account the manufacturing process the nominal spring constant should be only treated as a rough guide and never used in precise AFM measurements. There are several methods to calculate or measure the spring constant of the cantilevers. In principle, the spring constant can be calculated from the cantilever's geometry and the knowledge of material properties (Neumeister and Ducker, 1994). However, the material properties, such as for example, the Young modulus, have to be measured by other techniques and even though the width and length of the cantilever beams can be determined quite accurately from scanning electron microscopy (SEM) images, the thickness of the cantilevers is seldom uniform, and thus is difficult to assess.

The Sader method (Sader et al., 1995) is based on a finite element analysis of the static deflection of V-shaped cantilevers and presents exact numerical results for the spring constant for a variety of cantilevers. Once again, the thickness and modulus need to be known. In addition, the method has been shown to miscalculate spring constants for certain materials (Levy and Maaloum, 2002). Cleveland suggested a method of measuring the resonant frequency of the cantilever before and after addition of a tungsten microsphere (Cleveland et al., 1993), the accuracy of this method is limited to the accuracy of mass of the sphere measurement. Direct measurement of the spring constant is the optimal solution. By measuring the response of the cantilever to thermal noise and assuming the cantilever behaves as a damped harmonic oscillator the determination of the spring constant becomes fast, easy and reliable (Hutter and Bechhoefer, 1993).

16.4.3 Derjaguin approximation

The Derjaguin approximation relates the force F between two bodies at a given surface separation distance, D, to the interaction free energy per unit area, $W(D)$, between two planar surfaces. This makes this approximation a very useful tool, since it is usually easier to derive the interaction energy for two planar surfaces rather than for curved surfaces and has a form:

$$F(D) = 2\pi R_{eff} W(D) \tag{16.8}$$

where R_{eff} is the effective curvature radius and is given by: $\frac{1}{R_{eff}} = \frac{1}{R_1} + \frac{1}{R_2}$, where R_1 and R_2 are the curvature radii of both bodies at the point of closest approach.

The Derjaguin approximation is applicable as long as the range of the interaction and the separation distance is small compared to the radii of curvature. Rentsch et al. (2006) tested the validity of this approximation in AFM force measurements by measuring interaction forces between colloidal particles of different sizes, but of identical composition. They proved the Derjaguin approximation was robust and held at all distances, including those comparable with the characteristic length scales of the surface roughness or the surface charge heterogeneities.

16.4.4 Normal force measurements between solids

Numerous AFM studies are available on the interaction between two solid surfaces (Ivanov, 1980; Larson et al., 1993; Toikka and Hayes, 1997; Farrokhpay et al., 2006). These studies have covered electrostatic (Rutland and Senden, 1993) hydration (Rutland and Senden, 1993), van der Waals (2001), steric (Rutland and Senden, 1993), and so-called hydrophobic forces (Carambassis et al., 1998) as well as steric or depletion forces caused by the presence of both adsorbing and non-adsorbing polymer molecules (Zauscher and Klingenberg, 2000; Milling and Biggs, 1995). In these cases interpreting the data is generally straightforward since both force and separation data can be derived from the raw data. Recently, Ducker et al. (2004, 2005) improved the colloid probe AFM technique by measuring the scattering of visible light from an evanescent field by colloid probe. This was the first time the separation distance between two rigid surfaces was measured independently.

16.5 AFM MEASUREMENTS FOR BUBBLES AND DROPLETS: CURRENT STATE OF THE ART

The case when either one or two of the surfaces are soft, that is, bubbles or droplets, and undergo deformation has received much less attention than when both surfaces are solids (Fielden et al., 1996; Hartley et al., 1999; Nguyen et al., 2003; Vakarelski et al., 2003; Englert et al., 2009). An understanding of the interaction between deformable surfaces is critical for food systems, where foams and emulsions are key components, as their stability relies on the careful manipulation of surface forces (Butt, 1991). The interplay between the intervening film thickness and degree of the soft surface deformation is absolutely crucial in these systems. Hence, a critical issue for force measurements involving at least one deformable body is the accurate determination of the separation between the interacting surfaces.

A modified SFA technique has afforded very precise separation information between deformable surfaces (Connor and Horn, 2001; Connor and Horn, 2003; Connor and Horn, 2003; Aston and Berg, 2002). The separation can be measured at any time of the experiment by using an optical fringe-counting method (fringes of equal chromatic order, FECO), however the SFA does not provide direct force information. In addition, despite the fact that these experiments have provided important information regarding the role of deformation in interactions between deformable surfaces, the surfaces explored had large radii of curvature (>1 mm). In particular, they showed that deformable surfaces are more sensitive to surface forces, than for solid-solid interactions, and highlighted the interplay between deformability and equilibrium forces and their importance in these interactions.

Studies of drops and bubbles with radii of curvature an order of magnitude smaller require other techniques, such as AFM, to probe surface interactions. This size regime is particularly important since many of the food products involve deformable surfaces of micrometre dimensions. In these cases the Laplace pressure in the bubble or drop will be high and subsequent deformation low. Nonetheless deformation will occur particularly when the two surfaces are in very close proximity (<100 nm). Reliable experimental data on the absolute real-time separation at the nano-scale in AFM surface force measurements involving deformable surfaces currently do not exist.

There have been attempts to interpret experimental results involving deformable surfaces using various theories. Early efforts approximated the deformable interfaces as elastic springs (Ducker, 1992). However, this approach did not take into account that geometry changes due to deformation of the bubble or droplet affect the interaction energy. Gillies et al. (2001) suggested obtaining 'zero' separation by fitting force curves in the presence of interactions weak enough to exclude significant deformation of the drop or bubble. A fit of long range electric double layer forces requires knowledge of surface potentials and these are not always known precisely. This is especially problematic for the air bubble interfaces. Since the assumption of negligible deformation is usually not valid, Attard and Miklavcic (2001, 2002, 2003) used a simplified analysis for a colloid probe approaching a large droplet or bubble, where the interactions are dominated by electrical double layer interactions. Chan et al. (2001) developed a more advanced model that enabled extraction of the interaction energy from the AFM force data. The model was restricted to the repulsive case and required knowledge of the surface potentials. Dagastine and White (2002) extended Chan's analysis to the attractive case using a similar approach which also requires knowledge of the surface potential at both interfaces. This model was further developed, by introducing time as a variable, and accounting for fluid flow by lubrication theory to describe the dynamic interactions between droplets and/or bubbles (Krasowska et al 2011, and Manor et al 2008). Yet again the models require knowledge of surface potentials and this is not always possible, and indeed, the force measurements are often the most appropriate method for obtaining the surface potential values (Larson et al., 1993; Johnson et al., 1995; Larson et al., 1997). Hence, it becomes a circular argument if the separation is not measured independently. One of the first approaches to measure separation between hydrocarbon droplets was taken by Tabor et al (2011). They incorporated confocal fluorescence microscopy to directly image the position of oil droplets in AFM experiments. However, this approach provided only the initial microscale (several μm) separation of the droplets before interaction, while the real-time nano-scale separation during interaction was again extracted using a theoretical model (Krasowska et al., 2011).

Despite the fact that separation cannot be measured many systems involving polymeric or hydrocarbon droplets and/or air bubbles have been explored. Gillies et al. (2002) have studied interaction forces between a spherical colloid probe and a cross-linked poly(dimethylsiloxane), PDMS, droplet of Young's moduli in the range 10^6 N/m^2. They observed that as the colloid probe approached the PDMS droplet, as a result of droplet deformation, the repulsive force increased much less rapidly than expected for the electrical double layer interaction of rigid particles. During surface separation, a hysteresis was recorded and attributed to the viscoelastic response of the PDMS droplet.

In the very first attempts with bubbles ~ 500 μm air bubble was immobilized at the solid surface and the colloid probe was pressed against the bubble. In such an arrangement, the forces of interactions between air bubble and colloid probe interacting across a liquid film between them are measured as the colloid probe approaches the air bubble. During the approach the thin liquid film starts to drain until it reaches the thickness where the DLVO (van der Waals and electrostatic) and non-DLVO (for example hydrophobic, steric, etc.) interactions are operative. The stability of such thin liquid film depends on whether the total force is repulsive or attractive. For instance only repulsive forces occur between a hydrophilic, negatively charged colloid probe and air bubble (Fielden et al., 1996; Assemi et al., 2008; Englert et al., 2009; Krasowska et al., 2011) where the electrostatic, van der Waals and hydrodynamic components of the interaction are all repulsive. The wetting film ruptures when the total force between the interacting surfaces is attractive. Film rupture is indicated by a 'jump-into-contact' recorded at short to medium separations on the approach part of the AFM force curve. Such 'jump-into-contact' indicates presence of the attractive force and occurs when, for example, a hydrophobic particle is pressed against an air bubble (Ishida and Higashitani,

2006; Nguyen et al., 2004). Depending on particle hydrophobicity, a partial or even complete engulfment of the particle occurs. Since the three phase contact line between the particle and the air bubble is then established all interactions from this point on are dominated by the capillary force.

Dagastine et al. (2005) measured the interaction force between a colloid probe and a butyl- or octyl-acetate droplet in aqueous electrolyte solutions without addition of any stabilizers. They found the force behavior was dependent on the type of electrolyte: for some electrolytes (sodium nitrate) the force measured upon approach was monotonically repulsive at all concentrations, while for other electrolytes (calcium nitrate or sodium perchlorate) the force upon approach was repulsive, but at smaller separations it become attractive and 'jump-into-contact' was observed.

Similar experiments have been conducted for a colloid probe interacting with hydrocarbon droplet. It has been proved that in general the initial curvature of the hydrocarbon droplet interface has a relatively small effect on the force profile in such experiments (Aston and Berg, 2002). Since the solid colloid probe has a much smaller radius than the radius of the droplet it will have much greater impact on the deformations, for example, for a smaller colloid probe, the hydrocarbon droplet interface deforms to a greater extent in a more localized region and this translates into: (i) greater cantilever deflection (and thus force) and (ii) the equilibrium film thickness will be reached faster than for a larger colloid probe.

The measurement of surface forces involving bubbles and droplets is complex due to interfacial fluidity and deformability. These two features significantly affect the hydrodynamic interactions and, as a result, kinetic stability of foam and emulsions against coalescence (Ivanov, 1980; Ivanov and Kralchevsky, 1988).

Further experiments in systems involving bubbles and droplets have indeed confirmed that the forces measured between a particle and droplets were highly dependent on the interaction velocity (Aston and Berg, 2002). Measuring interactions between bubble(s) and/or droplet(s) at higher velocities has allowed the hydrodynamic forces acting during their approach to be explored (Aston and Berg, 2002; Krasowska et al., 2011; Manor et al., 2008; Lockie et al., 2011). These have included an air bubble colloid probe (~ 50–100 μm air bubble attached to the end of AFM cantilever) interacting with the solid (Vakarelski et al., 2008; Manor et al., 2008), the interactions between two hydrocarbon droplets in the absence (Lockie et al., 2011) and in the presence of surfactant, proteins or polymers (Gromer and Gunning, 2010). The most recent advancement of AFM force measurements includes the interactions between an air bubble and hydrocarbon droplet (Tabor et al., 2011).

Understanding the hydrodynamic interactions is crucial in understanding and controlling bubble and droplet coalescence. Vakarelski et al. (2010) used a high ionic strength in order to screen the electrostatic repulsion between two air bubbles and to explore different ways in which coalescence may occur. As the bubbles were pushed against each other, the thin liquid film between thinned and, at certain points, reached the thickness at which van der Waals attraction took over. At this moment coalescence occurred. The coalescence can also occur upon retraction. This is caused by the hydrodynamic suction that arises from the resistance of the liquid to enter the thin gap formed between two air bubbles. The liquid re-entering can only happen at a finite rate, leading to an attractive force between the bubbles if they are pulled apart faster than the liquid can refill the gap. Such a mechanism can be extended to the system containing droplets. This phenomenon

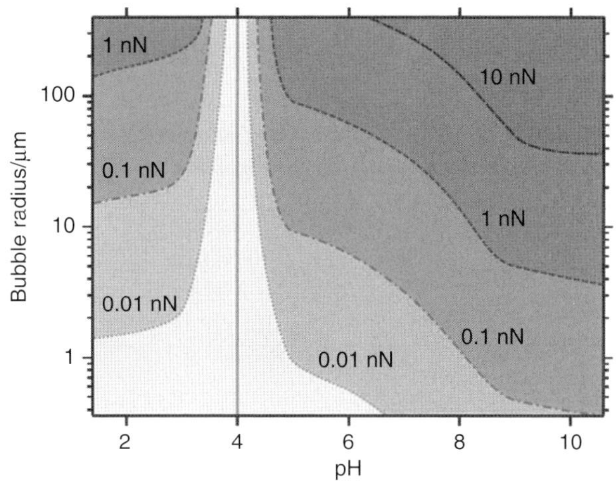

Figure 16.5 Maps of stability for interactions of two air bubbles in aqueous solutions. The plot shows the critical force required for the bubble coalescence during their axisymmetric collision, in the absence of hydrodynamic effects. Reprinted from *Journal of Colloid and Interface Science*, 371, R.F. Tabor, F. Grieser, R.R. Dagastine, D.Y.C. Chan, *Measurement and analysis of forces in bubble and droplet systems using AFM*, 1–14 (2012). (Source: Tabor et al. 2013. Reproduced with permission of Elsevier.)

also explains coalescence in extensional flows in microfluidic channels (Bremond et al., 2008). The analysis was later expanded by Tabor et al. (2012) to predict the stability or coalescence between two droplets or bubbles, for which the disjoining pressure in the thin film separating them can be calculated or predicted. The model can be used to determine behavior of two drops or bubbles over a range of drive conditions. The stability maps for interaction of two air bubbles in aqueous solutions are shown in Figure 16.5.

The theoretical modeling based on the Stokes-Reynolds drainage and the augmented Young-Laplace equation captures very well the qualitative features observed in the bubble-plate, droplet-plate, bubble-bubble and droplet-droplet AFM experiments. However, in all the cases a no-slip (immobile interface) hydrodynamic boundary condition at the air/water and hydrocarbon/water must have been imposed in order to obtain the quantitative agreement between the AFM experiments and the theory. Since a slip (mobile interface) is expected in the cases of air/water and hydrocarbon/water the most frequently used argument for such a no-slip boundary condition is that there must be some surfactant molecules present in the system. Even if present in very small quantities, surfactant molecules will adsorb at the fluid interface resulting in its immobilization. However, for bubbles or droplets in ultra-pure water or salt solutions, this seems to be a conventional explanation.

16.6 OUTLOOK FOR THE APPLICATION OF AFM IN FOOD TEXTURE STUDIES

The abovementioned studies on droplet and bubble interactions, which have used the AFM to great effect, have revealed a number of insights into these systems that would not be apparent to food scientists from traditional studies of food rheology. A number of these will be summarized below, and the implications for controlling/modifying emulsion and foam texture are discussed, along with a description of future studies that need to be undertaken with the AFM to further capitalize on these insights:

16.6.1 Deformation of droplets is dependent on interfacial properties – not bulk rheology

A number of the AFM studies have highlighted this critical aspect of droplet deformation. It is not the properties of the intervening liquid film (and therefore the bulk rheology of the suspension – which is controlled to a great degree by the rheology of the continuous phase) that determine how much a droplet will deform when pressed against another droplet or a solid surface (i.e. oral surfaces). Rather, it is the interfacial properties of the droplet (surface tension, interfacial elasticity) that determine how much a droplet will deform under load.

The reason that this is so critical an insight for food texture is that the oral surfaces experience droplet flow most sensitively as the droplets are squeezed between the tongue and the other oral surfaces. It is this regime of interaction in which the flow and sensory properties of emulsions become tangled with their ability to act as mixed-phase lubricants (as in metal-working fluids) (Dresselhuis et al., 2007). The deformability of emulsion droplets will significantly influence the sliding friction in such systems, and thus textural attributes like creaminess (van Vliet et al., 2009). One cannot alter the deformability (and thus sliding friction) by making alterations to the continuous aqueous phase, or indeed by making alterations to the identity of the oil used as the dispersed phase. The largest influence on the sliding friction will be the stretchability of the oil/water interface.

There are some techniques in physical chemistry that can access this aspect of droplet properties (interfacial rheometry), but none apart from the AFM will allow the researcher to make a direct connection between that property and a measurement of sliding friction. To date there has been only one successful attempt to measure forces during sliding droplet flow with the AFM (see Figure 16.6). Unfortunately, due to an inability to measure the absolute separation or contact area between the droplet and the surface, the measurement could only detect differences in liquid hydrodynamics (Vakarelski et al., 2010). For AFM to make a significant contribution to this problem, and thus make direct connections between droplet deformability and sliding friction (and all the textural attributes that there are encoded within that physical observable), drop profile imaging through optical methods like interferometry are required in addition to AFM force measurement. Such a system has been developed, and future studies of sliding droplets with AFM will hopefully become a useful means to make rational decisions about emulsion formulation to control texture.

16.6.2 Coalescence of bubbles or droplets is driven by van der waals forces

Whereas the forces experienced in sliding droplets rely on surface properties, the actual event of coalescence for two bubbles or two droplets in an aqueous medium (under the usual high salt conditions experienced in food systems) can be controlled by bulk properties (dielectric constants) of the system. In food emulsions and foams, the salt content

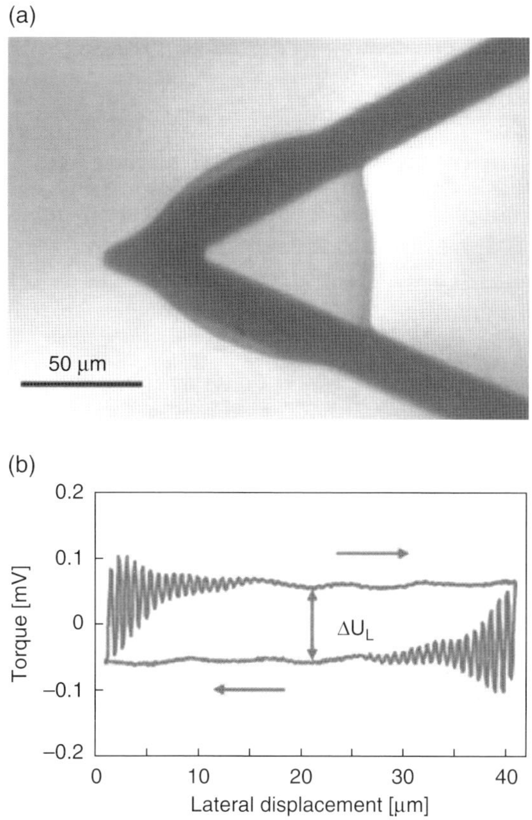

Figure 16.6 (a): top-view microscopic image of a triangular cantilever with an attached oil droplet, and (b): Typical hydrodynamic friction-force loop of lateral piezo displacement vs torque for an oil droplet (diameter = 90 μm) under normal load of 40 nN and a scan velocity of 800 μm/s. (Source: Vakarelski et al. 2010. Reproduced with permission of American Chemical Society.)

of the aqueous phases diminishes the reach of the electrostatic forces: the double layer is compressed and the two bodies do not sense the charge on each other's surfaces until they are very close. This close contact allows the van der Waals interaction to dominate and dictate whether the two droplets/bubbles will attract one another enough to cause rupture of the intervening aqueous thin film.

Van der Waals forces are not surface properties (see above), and rely on the dielectric constant of the oil/gas and water. In binary systems, these forces are always attractive. Although gas bubbles do not have a large variance in dielectric constant for different gases, oil droplets have a much wider range of dielectric constants, and therefore the likelihood is greater that coalescence could be tailored by selecting oils with different bulk properties. The fine tuning of coalescence in this manner could hold the key to controlled emulsion properties that are advantageous for texture and stability: emulsions that are just stable during storage, but that easily coalesce and break in the altered environment of the oral cavity. This could enhance the sensory perceptions associated with mouth coating (silky, velvety) (van Vliet et al., 2009). Detailed measurement of the forces experienced during droplet-droplet coalescence using the AFM would provide a window into this area of coalescence fine-tuning.

16.6.3 Detachment of droplets from surfaces

The previous two examples are based on measurements made between deformable systems with the AFM, and lead to thoughts about studies that can and should be performed in the future to give valuable information about tailored texture, and specifically about the initial droplet-surface and droplet-droplet interactions that occur in the oral cavity. They also depend on measurements that are made on droplet/bubble approach – the first stage of any AFM colloid probe measurement. One aspect of the technique that has yet to be highlighted here is the ability of the technique to measure the detachment forces upon removal of contact between a droplet/bubble and a surface.

Pull-off forces are relevant for studies of food texture due to the sensory properties 'after-feel' texture (van Vliet, 2009). The ease with which oil can be removed from the oral surfaces is critical for determining the residence time of thin oil films within the mouth, and thus the 'after-feel' texture. This droplet removal is a complex phenomenon, involving dilution of the liquid phase in the mouth, rinsing with saliva, and continued rubbing of the tongue against the oral surfaces. Although important, there have been no studies on how much force is required to remove an oil droplet from a solid surface as a function of system variables that would mimic the conditions in the oral cavity. Measurements of droplet retraction from solid surfaces using the AFM would provide information on the adhesion of droplets. Measurements of sliding droplets with increased sliding speeds would give an indication of the forces necessary to encourage the detachment due to altered hydrodynamic conditions (Vakarelski et al., 2010). As with the previous examples cited above, such measurements would provide critical information into textural attributes of food emulsions.

16.6.4 A new stabilisation mechanism for three component systems (bubbles and droplets in water)

Conventional interpretations of DLVO theory prepare people to accept that van der Waals forces are always attractive. However, when one considers the 'fine print', it becomes clear that this only holds exclusively for binary systems. In many systems of relevance for food, such as dairy foams, where oil droplets and air bubbles are both present (in addition to some solid particles), there are examples of heterogeneous droplet/bubble interactions for which the van der Waals interaction becomes repulsive. The importance of this realization only truly becomes relevant in systems for which there is high salt content – when the electrostatic interactions become minor in altering the interaction forces between deformable bodies in the food system.

There have been no studies to date, in the AFM literature or otherwise, that have investigated the influence of repulsive van der Waals forces on heteroaggregation in food systems. Given the importance of partial coalescence and droplet-air bubble interactions in food products, and the subsequent strength of the network structures formed in many food products (which will control textural properties), it is important to look in detail at what might be controllable by understanding and manipulating the contribution of repulsive van der Waals interactions to these systems.

16.7 CONCLUSION AND GENERAL OUTLOOK

The value of colloid probe atomic force microscopy for food science and food texture is clear. The methodology gives access to direct measurement of bubble and droplet interactions that are connected to aspects of sensory perception that are important for all food products consisting of dispersed oil and air within liquid and solid matrices. The technique is demanding experimentally, but the quality of the data, and the ability to make links between the individual droplet/bubble interactions and macroscopic scale measurements of food product rheology and textural properties, makes colloid probe AFM a powerful tool for food product design and optimization.

As with any technique in physical chemistry and food science, AFM needs to be coupled with other methods of analysis to ensure that the information gained is translated from one scale of measurement to another, and to gain additional insights into the nature of interfacial interactions. With AFM, coupling measurements with bulk rheology allows one to translate the information from single objects to multiple objects. In addition to bulk rheology measurements, interfacial rheology and dynamic surface tension measurement (using bubble/droplet tensiometers) can connect bubble and droplet interaction studies with molecular level information on surfactant, polymer, protein, and particle adsorption at liquid interfaces (oil–water, air-water) (Kotsmar et al., 2009). Also, parallel methods for studying single droplet/bubble interactions with interfaces (solid, oil, or air), such as high speed video capture techniques, allow one to explore single droplet/bubble interactions in different force regimes and size ranges (Kuznicki et al., 2013).

Such a multi-technique approach is especially useful in dealing with the complex multi-component systems so prevalent in food science. The advantages of AFM argue strongly for its inclusion in the suite of methods deployed by food scientists to address future challenges in food texture design and optimization. Issues such as influencing oral tribology, controlling both storage AND in-mouth emulsion stability, and directing novel product design and formulation can all be addressed using the information gained from the technique. The methodology is being applied now by a number of groups in physical chemistry, and it is likely that its uptake will increase for food scientists as the power of the colloid probe AFM approach becomes more evident from research published in the scientific literature.

References

Assemi, S., Nguyen, A.V., and Miller, J.D. (2008) Direct measurement of particle–bubble interaction forces using atomic force microscopy. *International Journal of Mineral Processing* 89(1–4), 65–70.

Aston, D.E., and Berg, J.C. (2002) Thin-film hydrodynamics in fluid interface-atomic force microscopy. *Ind. Eng. Chem. Res.* 41, 389–396.

Attard, P., and Miklavcic, S.J. (2001) Effective spring constant of bubbles and droplets. *Langmuir* 17, 8217–8223.

Attard, P., and Miklavcic, S.J. (2002) Effective spring description of a bubble or a droplet interacting with a particle. *J. Colloid Interface Sci.* 247, 255–257.
Attard, P., and Miklavcic, S.J. (2003) Effective spring constant of bubbles and droplets. [Erratum to document cited in CA136:249665]. *Langmuir* 19, 2532.
Beattie, J.K., and Djerdjev, A.M. (2004) The pristine oil/water interface: surfactant-free hydroxide-charged emulsions. *Angew. Chemie Int. Ed.* 43, 3568–3571.
Bremond, N., Thiam, A.R., and Bibette, J. (2008) Decompressing emulsion droplets favors coalescence. *Phys. Rev. Lett.* 100, 024501-1-024501-4.
Butt, H.J. (1991) Measuring electrostatic, van der Waals, and hydration forces in electrolyte solutions with an atomic force microscope. *Biophys. J.* 60, 1438–1444.
Butt, H.-J., (1992) Electrostatic interaction in scanning probe microscopy when imaging in electrolyte solutions. *Nanotechnology* 3, 60–68.
Carambassis, A., Jonker, L.C., Attard, P., and Rutland, M.W. (1998) Forces measured between hydrophobic surfaces due to a submicroscopic bridging bubble. *Phys. Rev. Lett.* 80(24), 5357–5360.
Chapman, D.L. (1913) A contribution to the theory of electrocapillarity. *Phil. Mag.* 25, 475–481.
Debye, P., and Hückel, E. (1923) The theory of electrolytes. I. Lowering of freezing point and related phenomena *Physikalische Zeitschrift* 24, 185–206.
Chan, D.Y.C., Dagastine, R.R., and White, L.R. (2001) Forces between a rigid probe particle and a liquid interface: I. The Repulsive Case *J. Colloid Interface Sci.* 236(1), 141–154.
Clark, S.C., Walz, J.Y., and Ducker, W.A. (2004) Atomic force microscopy colloid-probe measurements with explicit measurement of particle-solid separation. *Langmuir* 20(18), 7616–7622.
Cleveland, J.P., Manne, S., Bocek, D., and Hansma, P.K. (1993) A nondestructive method for determining the spring constant of cantilevers for scanning force microscopy. *Rev. Sci. Instrum.* 64, 403–405.
Connor, J.N., and Horn, R.G. (2001) Measurement of aqueous film thickness between charged mercury and mica surfaces: a direct experimental probe of the Poisson-Boltzmann distribution. *Langmuir* 17, 7194–7197.
Connor, J.N., and Horn, R.G. (2003) Extending the surface force apparatus capabilities by using white light interferometry in reflection. *Review of Scientific Instruments* 74, 4601–4606.
Connor, J.N., and Horn, R.G. (2003) The influence of surface forces on thin film drainage between a fluid drop and a flat solid. *Faraday Discussions* 123, 193–206.
Creux, P., Lachaise, J., Graciaa, A., Beattie, J.K., and Djerdjev, A. (2009) Strong specific hydroxide ion binding at the Pristine Oil/Water and Air/Water Interfaces. *J. Phys. Chem. B* 113, 14146–14150.
Dagastine, R.R., and White, L.R. (2002) Forces between a rigid probe particle and a liquid interface: II. The general case. *J. Colloid Interface Sci.* 247(2), 310–320.
Dagastine, R.R., Chau, T.T., Chan, D.Y.C., Stevens, G.W., and Grieser, F. (2005) Interaction forces between oil–water particle interfaces—Non-DLVO forces. *Faraday Discuss.* 129, 111–124.
Derjaguin, B., Landau, L. (1941) Theory of the stability of strongly charged lyophobic sols and of the adhesion of strongly charged particles in solutions of electrolytes. *Acta Physico Chimica URSS* 14, 633–662.
Dresselhuis, D.M., de Hoog, E.H.A., Cohen Stuart, M.A., and van Aken, G.A. (2007) Tribology as a tool to study emulsion behaviour in the mouth. In *Food Colloids: Self-Assembly and Material Science*, The Royal Society of Chemistry, pp. 451–461.
Ducker, W.A., Senden, T.J., and Pashley, R.M. (1992) Measurement of forces in liquids using a force microscope. *Langmuir* 8(7), 1831–1836; (b) Ducker, W.A., Xu, Z.G., and Israelachvili, J.N. (1994) *Langmuir* 10 (9), 3279–3289.
Dzyaloshinskii, I.E., Lifshitz, E.M., and Pitaerskii, L.P. (1961) The general theory of van der Waals forces. *Adv. Phys.* 10, 165–209.
Englert, A.H., Krasowska, M., Fornasiero, D., Ralston, J., and Rubio, J. (2009) Interaction force between an air bubble and a hydrophilic spherical particle in water, measured by the colloid probe technique. *International Journal of Mineral Processing* 92, 121–127.
Farrokhpay, S., Morris, G.E., Fornasiero, D., and Self, P. (2006) Titania pigment particles dispersion in water-based paint films. *Journal of Coatings Technology, Research* 3(4), 275–283.
Fielden, M.L., Hayes, R.A., and Ralston, J. (1996) Surface and capillary forces affecting air bubble-particle interactions in aqueous electrolyte. *Langmuir* 12(15), 3721–3727.
Gillies, G., Prestidge, C.A., and Attard, P. (2001) Determination of the separation in colloid probe atomic force microscopy of deformablebodies. *Langmuir* 17, 7955–7956.
Gillies, G., Prestidge, C.A., and Attard, P. (2002) An AFM study of the deformation and nanorheology of cross-linked PDMS droplets. *Langmuir* 18, 1674–1679.

Gouy, M. (1910) Sur la constitution de la charge électrique à la surface d'un électrolyte. *J. Phys. (Paris)* 9, 457–468.

Gromer, A. and Gunning, A.P. (2010) *Interactions Between Oil Droplets Probed by Force Spectroscopy with the MFP-3D™ AFM.*

Hartley, P.G., Griesser, F., Mulvaney, P., and Stevens, G.W. (1999) Surface forces and deformation at the oil–water interface probed using AFM force measurement. *Langmuir* 15(21), 7882–7289.

Hogg, R., Healy, T.W., and Fuerstenau, D.W. (1966) Mutual coagulation of colloidal dispersions *Trans. Faraday Soc.* 62, 1638–1651.

Hutter, J.L., and Bechhoefer, J. (1993) Calibration of atomic-force microscope tips. *Rev. Sci. Instrum.* 64(7), 1868–1873.

Ishida, N., and Higashitani, K. (2006) Interaction forces between chemically modified hydrophobic surfaces evaluated by AFM—The role of nanoscopic bubbles in the interactions. *Miner. Eng.* 19(6–8), 719–725.

Israelachvili, J. N. (1992) *Intramolecular and Surface Forces*. Academic Press: London.

Israelachvili, J.N., and Tabor, D. (1972) Measurement of van der Waals dispersion forces in the range 1.5 to 130 nm. *Proc. R. Soc. Lond. A.* 331(1584), 19–38.

Ivanov, I.B. (1980) Effect of surface mobility on the dynamic behavior of thin liquid films. *Pure Appl. Chem.* 52, 1241–1262.

Ivanov, I.B., and Kralchevsky, P.A. (1988) *Thin Liquid Films. Fundamentals and Applications*. Marcel Dekker: New York, Vol. 29.

Johnson, S.B., Drummond, C.J., Scales, P.J., and Nishimura, S. (1995) Comparison of techniques for measuring the electrical double layer properties of surfaces in aqueous solution: Hexadecyltrimethylammonium Bromide self-assembly structures as a model system. *Langmuir* 11, 2367–2375.

Kotsmar, C., Pradines, V., Alahverdjieva, V.S., Aksenenko, E.V., Fainerman, V.B., Kovalchuk, V.I., Krägel, J., Leser, M.E., Noskov, B.A., and Miller, R. (2009) Thermodynamics, adsorption kinetics and rheology of mixed protein-surfactant interfacial layers. *Advances in Colloid and Interface Science* 150(1), 41–54.

Krasowska, M. Carnie, S.L., Fornasiero, D., and Ralston, J. (2011) Ultrathin wetting films on hydrophilic titania surfaces: equilibrium and dynamic behavior. *J. Phys. Chem. C* 115(22), 11065–11076.

Kuznicki, N.P., Krasowska, M., Sellaperumage, P.M.F., Xu, Z., Masliyah, J., Ralston, J., and Popescu, M.N. (2013) Cascade partial coalescence phenomena at electrolyte-oil interfaces and determination of bounds for the surface potential. *Soft Matter*, accepted for publication.

Larson, I., Drummond, C.J., Chan, D.Y.C., and Grieser, F. (1993) Direct force measurements between titanium dioxide surfaces. *J. Am. Chem. Soc.* 115, 11885–11890.

Larson, I., Drummond, C.J., Chan, D.Y.C., and Grieser, F. (1997) Direct force measurements between Silica and Alumina. *Langmuir* 13, 2109–2112.

Levy, R. and Maaloum, M. (2002) Measuring the spring constant of atomic force microscope cantilevers: thermal fluctuations and other methods. *Nanotechnology*, 13, 33–37.

Lockie, H.E., Manica, R., Stevens, G.W., Grieser, F., Chan, D.Y.C., and Dagastine, R.R. (2011) Precision AFM measurements of dynamic interactions between deformable drops in aqueous surfactant and surfactant-free solutions. *Langmuir* 27, 2676–2685.

Manor, O., Vakarelski, I.U., Stevens, G.W., Grieser, F., Dagastine, R.R., and Chan, D.Y.C. (2008) Dynamic forces between bubbles and surfaces and hydrodynamic boundary conditions. *Langmuir* 24(20), 11533–11543.

McClements, D.J. (2004) Colloidal interactions. In: *Food Emulsions: Principles, Practice, and Techniques*, CRC Press.

McKee, C.T., Clark, S.C., Walz, J.Y., and Ducker, W.A. (2005) Relationship between scattered intensity and separation for particles in an evanescent field. *Langmuir* 21(13), 5783–5789.

Milling, A., and Biggs, S. (1995) Direct measurement of the depletion force using an atomic force microscope. *J. Colloid Interface Sci.* 170(2), 604–606.

Neumeister, J.M., and Ducker, W.A. (1994) Lateral, normal, and longitudinal spring constants of atomic force microscopy cantilevers. *Rev. Sci. Instrum.* 65, 2527–2531.

Nguyen, A.V., Evans, G.M., Nalaskowski, J., and Miller, J.D. (2004) Hydrodynamic interaction between an air bubble and a particle: atomic force microscopy measurements. *Exp. Therm. Fluid Sci.* 28, 387–394.

Nguyen, A.V., Nalaskowski, J., Miller, J.D. (2003) A study of the bubble–particle interaction using atomic force microscopy. *Minerals Engineering* 16 (11), 1173–1181.

Rentsch, S., Pericet-Camara, R., Papastavrou, G., and Borkovec, M. (2006) Probing the validity of the Derjaguin approximation for heterogeneous colloidal particles *Phys. Chem. Chem. Phys.* 8, 2531–2538.

Rutland, M.W., and Senden, T.J. (1993) Adsorption of the poly(oxyethylene) nonionic surfactant C12E5 to silica: a study using atomic force microscopy. *Langmuir* 9(2), 412–418.

Sader, J.E., Larson, I., Mulvaney, P., and White, L.R. (1995) Method for the calibration of atomic force microscope cantilevers. *Rev. Sci. Instrum.* 66, 3789–3798.

Tabor, R.F., Grieser, F., Dagastine, R.R., and Chan, D.Y.C. (2012) Measurement and analysis of forces in bubble and droplet systems using AFM. *Journal of Colloid and Interface Science* 371 (1), 1–14.

Tabor, R.F., Lockie, H., Mair, D., Manica, R., Chan, D.Y.C., Grieser, F., and Dagastine, R. R. (2011) Combined AFM–Confocal microscopy of oil droplets: absolute separations and forces in nanofilms. *J. Phys. Chem. Lett.* 2, 961–965.

Tabor, R.F., Wu, C., Lockie, H., Manica, R., Chan, D.Y.C., Grieser, F., and Dagastine, R.R. (2011) Homo- and hetero-interactions between air bubbles and oil droplets measured by atomic force microscopy. *Soft Matter* 7, 8977–8983.

Toikka, G., and Hayes, R.A. (1997) Direct measurement of colloidal forces between mica and silica in aqueous electrolyte. *J. Colloid Interface Sci.* 191(1), 102–109.

Vakarelski, I.U., Dagastine, R.R., Chan, D.Y.C., Stevens, G.W., Higashitani, K., and Grieser, F. (2010) Lateral hydrodynamic interactions between an emulsion droplet and a flat surface evaluated by frictional force microscopy. *Langmuir* 26(11), 8002–8007.

Vakarelski, I., Toritani, A., Nakayama, M., and Higashitani, K. (2003) Effects of particle deformability on interaction between surfaces in solutions. *Langmuir* 19(1), 110–117.

Vakarelski, I.U., Lee, J., Dagastine, R.R., Chan, D.Y.C., Stevens, G.W., Grieser, F. (2008) Bubble colloidal AFM probes formed from ultrasonically generated bubbles. *Langmuir* 24(3), 603–605.

Vakarelski, I.U., Manica, R., Tang, X., O'Shea, S.J., Stevens, G.W., Grieser, F., Dagastine, R.R., and Chan, D.Y.C. (2010) Dynamic interactions between microbubbles in water *Proc. Natl. Acad. Sci. USA* 107, 11177–11182.

van Vliet, T., van Aken, G.A., de Jongh, H.H.J., and Hamer, R.J. (2009) Colloidal aspects of texture perception. *Advances in Colloid and Interface Science* 150(1), 27–40.

Vervey, G., and Overbeek, J.T.G. (1948) *Theory of the Stability of Lyophobic Colloids* Elsevier: Amsterdam.

Vingerhoeds, M.H., Blijdenstein, T.B.J., Zoet, F.D., and Van Aken, G.A. (2005) Emulsion flocculation induced by saliva and mucin. *Food Hydrocolloids* 19 (5), 915–922.

Zauscher, S., and Klingenberg, D.J. (2000) Normal forces between cellulose surfaces measured with colloidal probe microscopy. *J. Colloid Interface Sci.* 229 (2), 497–510.

17 Importance of understanding mouth behavior when optimizing product texture now and in the future

Melissa Jeltema, PhD, Jacqueline H. Beckley and Jennifer Vahalik

The Understanding & Insight Group, LLC, Denville, New Jersey, USA

17.1 INTRODUCTION

17.1.1 Texture fundamental knowledge

It is well known that texture is important to product liking and preferences. As with taste, liking of textural attributes (e.g., thickness, firmness) follows a parabola (inverted U) (Moskowitz, 2007). Optimum liking is at a textural level that matches the top of the parabola and liking decreases as the textural attribute is increased or decreased. The peak of the curve differs within food type and by individual. Additionally, it is well known that texture can be a major reason for food rejection (Drewnowski, 1997). This textural knowledge was slow in coming relative to other attribute discoveries and lagged behind flavor research.

Research on the texture of foods did not seriously begin until the 1950s and 60's when Alina Szczesniak began working on developing a texture lexicon and texture profiling. In 1963, Szczesniak published several foundational articles on the measurement and classification of food texture (Szczesniak, 1963a; Szczesniak, 1963b). In 1969, Szczesniak was a Founding Editor of the *Journal of Texture Studies*. The primary stated goal of the *Journal of Texture Studies* when started was to be a 'primary source on all of the sciences that relate to sensory perception and the physical characterization of texture, and it focuses on how the tactile senses are used to determine the acceptability of food and other consumer products' (Daubert et al., 2013). This work began a focus on texture that was previously lacking.

Through her early work, Szczesniak found that individuals have low texture awareness, seeing texture as an integral part of the nature of food; and therefore individuals have a limited ability to verbalize or distinguish textural characteristics (Szczesniak and Kahn, 1971). When describing food, texture is often not mentioned. However, it is one of the strongest drivers in food aversion (Scott and Downey, 2007).

If texture is such a strong driver of food aversion, why is texture awareness so low? Chen and Engelen (2012) suggest that individuals bring to each food a certain texture expectation. If that expectation is met, then there is no focus on texture. It is when that expectation is not met, that there then is a reason for food rejection.

Food Texture Design and Optimization, First Edition. Edited by Yadunandan Lal Dar and Joseph M. Light.

Possibly due to this perceived inability of individuals to adequately notice or describe their textural experiences, and the complexity of texture perception, a preponderance of research has focused on describing and measuring textural attributes sensorially and then relate the sensory textural attributes to liking. Much of the research to understand the textural drivers of liking has utilized highly trained panels to describe texture and then statistically relating those textural characteristics to consumer liking. In many cases, texture is analyzed in conjunction with other sensory attributes for flavor (taste and aroma). This is because flavor and texture are both drivers of food acceptance and the perception of one can change the perception of the other (Chen and Engelen, 2012).

A variety of statistical techniques are used to determine which sensory variables drive liking of samples. One technique often used is Preference Mapping. This technique creates maps that overlay consumer liking on the sensory and product space.

Since it is understood that not all consumers are similar in their preferences, Cluster Analysis is often first used to form groups of consumers who are similar in their liking of the products. A multidimensional representation of products, based on their sensory properties, is usually obtained by Principle Components Analysis (PCA). The consumer data is then fit on the sensory space. The goal is to determine the optimum sensory region for the consumer segments as well as to indicate which products are closest to the optimum space.

Other statistical techniques commonly employed to determine which sensory variables relate to liking are ANOVA and PLS (Partial Least Squares Regression). A combination of these techniques have been used on a wide variety of food types including (to name a few) candy (Kalviainen et al., 2000), cheese (Murray and Delahunty, 2000), poultry (Sow and Grongnet, 2010), and liquid dairy (Richardson-Harmon et al., 2000).

Currently, some marketing research companies have adapted sensory data in their market segmentation studies. Questions asked can range from degree of liking of individual flavor, texture, and aroma words to degree of liking of specific foods. Consumers are segmented using these attributes and then socio-demographics or psychographics are used to describe the segments.

While a significant amount of research has been done to measure texture, to group consumers based on their texture preferences and to measure which textural attributes may drive liking, none of this research has focused on understanding what drives differences in textural rejection or preference. Without this understanding, product developers rely on mathematical models to drive product formulation, without ever having a true consumer understanding of why products are succeeding or failing.

17.1.2 Oral processing

Separately, there is a growing body of research on oral processes to understand their effects on sensory perception. Differences in food manipulation and mastication have been found to affect sensory sensations. This research has highlighted some important findings, showing that chewing behavior varies by individual, but is consistent within an individual (Laussauzay et al., 2000; Poe et al., 2011).

Brown and Baxton (2000) identified four different groups of people based on their efficiency in reducing the size of almonds and gum. This research also showed that individuals manifested different mechanisms for oral breakdown so that at any point, different groups of individuals would experience the samples differently. In this research, Brown and Baxton suggested that individual differences in the ability to manipulate and manage the product in the mouth may be a key driver in liking and personal preferences. However, the only link found in their research was a correlation between chewing force and preference.

Engelen and van Doorn also found that there were large individual differences in eating styles. In this research on semi-solid foods (custard and mayonnaise), they asked subjects to describe or draw what they did with food in their mouths. Based on this self-description, they identified four styles which they named simple, taster, manipulator, and tonguer. Differences were in how the individual used the tongue, palate, and teeth (Chen and Engelen, 2012).

Supporting the possible relationship between chewing behavior and preference, de Wijk et al., (2003) found that, for a given product, individuals achieved the highest sensory perception when they manipulated the product as they desired. Further research has also shown that perception of attributes changed over time and were associated with specific oral movements, suggesting that individuals use oral movements to enhance specific sensations (de Wijk et al., 2006; de Wijk et al., 2008). It has been suggested that individuals use an oral processing style that is balanced in terms of efficiency and perception and that the processing style is also dependent on the food (depending on the pleasantness of the food, Chen and Engelen, 2012). Therefore, the research suggests that individuals may use the chewing behavior that best provides the flavors and flavor intensity they are seeking (chewing a food differently may help people get the tastes they want).

Additionally it has been shown that chewing behavior is modulated by personality (Rey et al., 2007). This suggests that there may be genetic components to chewing behavior.

17.1.3 A new paradigm for understanding texture

Our research suggests that there is a definite relationship between chewing behavior and food liking and satisfaction. In this chapter we will explain a new way of understanding the drivers of texture preference and how to use that information in optimizing products. Our research suggests that Brown and Braxton were correct in their belief that the ability to manage food in the mouth is a key driver of preference. Our results show that individuals have a preferred way to manipulate food in their mouths (Mouth Behavior) and that this behavior determines the food textures they prefer. *Therefore, it is not texture which is the key driver of liking, but textures that fit with the preferred mouth behavior.* Further, our findings indicate that individuals fall into four mouth behavior groups which we have named Crunchers, Chewers, Suckers, and Smooshers.

17.2 UNDERSTANDING MOUTH BEHAVIOR

In order to use mouth behavior in product optimization research, it is first important to understand how the theory behind mouth behavior was conceived and how the typing tool (to assist mouth behavior classification) was developed. For the purposes of this chapter, we will focus on one (of many) examples to illustrate the importance of mouth behavior on product research and design.

17.2.1 Developing the insight

The existence of different mouth behavior groups was first discovered during reflection on quantitative findings and qualitative product research. These areas of inquiry began in 2001 and continued through the remainder of that decade.

17.2.1.1 An example of the observation that led to the insight

A product that was meant to be held in the mouth for a long time was given to research participants in order to evaluate interest. These individuals were generally not interested in the product because most did not want to suck on it for a long time and were unsure what to do with it in their mouths. This begged the question of what people do want to do with products in their mouths. Further observations showed that individuals used products differently in their mouths, and that the way they were interacting with the products seemed to drive product acceptability.

17.2.1.2 Exploration of the observations

Further exploratory qualitative research was conducted to understand the apparent differences in the ways individuals interacted with food and snacks. Individuals were asked to respond to a variety of statements aimed at understanding how they preferred to manipulate food in their mouths. They were asked to sort the statements into three groups: 'This is exactly like me,' 'This is somewhat like me,' 'This is not like me.' Some of the questions are shown below:

- I like to suck on hard candy until it fully dissolves.
- I usually break up hard candy quickly and swallow it.
- I prefer hard crunchy cookies to soft chewy cookies.
- I prefer soft creamy candies to hard candies.

What was of interest to us during this exploration was the complete lack of awareness (unarticulated need) of these behaviors by the individuals themselves until prompted through the card sort technique. Review of early research into psychology and perception suggests that the fundamental importance of the behavior may have been part of the lack of awareness (Hochberg,1964; Alpern et Al.,1967).

17.2.2 Hypothesizing mouth behavior groups

Based on these qualitative studies conducted over several years (more than 250 hours of observation and listening), it was hypothesized that there were four major mouth behavior groups. These four groups are: 1) Crunchers, 2) Chewers, 3) Suckers, and 4) Smooshers. These groups fell into two major modes of mouth actions. Crunchers and Chewers liked to use their teeth to break down foods. Crunchers were more forceful in

Table 17.1 Examples of products of choice by mouth behavior. (Source: The Understanding & Insight Group LLC.)

Types of Products Chosen	Mouth Behavior Classification Group
Peanut Brittle, Sonic™ ice, whole almonds, highly toasted granola	Crunchers
Gummy bears, caramels, soft Chips Ahoy! ™ cookies, Krackle ™ chocolate bars	Chewers
Caramel crèmes, creamy yogurt, Reeses Cups, Twinkies™	Smooshers
Jolly Ranchers, Hard candies, Werther's Butterscotch pieces	Suckers

their bite and preferred foods that broke up (fractured) on biting. Chewers liked foods that could be chewed longer (length of time varied – there did seem to be short Chewers and long Chewers) and did not fracture on biting. Suckers and Smooshers preferred to manipulate food between the tongue and roof of the mouth. They differed primarily in the hardness of preferred foods. Suckers liked harder foods (like hard candies and items that they could hold in their mouths) that could be sucked on for a long time. Smooshers preferred soft foods, such as creamy candies (like the wrapped candy called 'Cowtails') or puddings that did not require much mouth activity but would spread throughout the mouth and could be held in the mouth for a long time.

It is important to note that just because a person falls into a preferred group, does not mean that he/she rejects other mouth behaviors. While a person may generally be a 'Cruncher,' and often chooses foods that can be crunched, he/she may also have some foods that he/she prefers to 'smoosh.' However, in general, foods that allow the preferred mouth behavior will bring a higher level of satisfaction than foods which do not facilitate that mouth behavior. While subjects are aware of their choices, they are unaware of the reasons that these food choices bring a higher level of pleasure (unexpressed).

17.2.3 Qualitative validation of mouth behavior groups

Our qualitative validation came from many, varied studies, where a wide variety of products were available for people to choose. Across these studies we found that the particular product choice was clearly driven by their primary mouth behavior. While this finding was qualitative, they were numerous. Examples of this are shown in Table 17.1.

17.2.4 Developing the tool for classification

While we found that individuals could be consistently typed during qualitative interviews using the card sort technique and follow-up in-depth discussions on the reasoning for those choices, we needed to develop a typing tool that could reproduce the assessments made during in-depth interviews. This proved to be difficult in the survey format. Reasons for this are many and include:

1. *Individuals are not aware of how they use food in their mouths.* Most people do not consciously think about how they interact with food – and therefore don't often know what they do. 'It would be easier if you gave me a food and then asked me what I was doing.' This unconscious awareness of such a primary need highlights the difficulties inherent in trying to understand behavior, emotion, and perception.

2. *Survey questions are often interpreted differently than intended.* The individual may be thinking about a product grouping that is different than you intended. For example, one smoosher said 'I only smoosh soft foods that can't be chewed' because she was thinking about meat, not something like a French fry.
3. *Individual behavior is not always consistent.* While a question such as 'do you prefer crunchy hard cookies or soft chewy cookies' may generally work, a Cruncher may say to you –'Yes, most foods I like are crunchy, but with cookies I like a soft one that I can chew and smoosh.' Or they may say, 'I choose a soft cookie because a hard one is dry and probably not as fresh.' Since surveys do not allow follow-up questions to resolve inconsistencies, accurately typing an individual through survey questions alone, was difficult.
4. *Individuals modify foods to meet their mouth behavior requirements.* When a food does not have a desirable texture, individuals will automatically modify their use of the product. For example, ice cream is much more likely to be eaten right out of the freezer by a Cruncher, while the other mouth behavior groups (particularly Smooshers) are more likely to let it soften before eating it. The same is found with cereals. While a Smoosher may eat a crunchy cereal, it is usually after it has been softened by the milk. On the other hand, individuals may modify the texture by adding other foods that will modify the texture. For example, Chewers may add chewy fruits like raisins to a cereal or Crunchers, when eating a very soft cereal like oatmeal may add nuts and/or fruit. This modification is particularly evident when an individual chooses to eat something less desirable (for reasons such as perceived health).

The reasons above illustrate how difficult it is to determine a person's mouth behavior from typical survey questions. Individuals have a low awareness of texture to begin with, and they are completely unaware of how they manipulate products in their mouths. Additionally they will eat, for many reasons, products that are not optimal in texture, but will either modify their behavior or the product themselves to make them more acceptable.

This observation was first made in the 1980s when work was being done by a cereal company on extruded cereals (Beckley, unpublished data). The observation was made that research participants would vary widely in their cereal preparation. Some participants used only a very small amount of milk (so as not to have soggy cereal) while others used a lot of milk and let the cereal get soft before eating it. This led to creating methodology for 'bowl life testing'. Bowl life testing was a form of observational and over-time evaluation of texture and flavor properties of the cereal. It was an early attempt to combine the evaluation of texture with other sensory properties to understand why a specific cereal might or might not be rejected by someone with a particular texture preference (crunchy, crispy, or soggy) and to establish metrics for specific cereal texture over time when consumed with specific milk types.

The following illustrates the types of questions used in the last few years to type individuals for mouth behavior. The types of questions used and reasons for inconsistencies are shown in Table 17.2.

The survey tool we first developed was found to clearly identify approximately 75% of those sampled, with the other 25% being somewhat questionable as to which group they belonged.

Table 17.2 Examples of questions in the survey tool and reasons for inconsistencies. (Source: The Understanding & Insight Group LLC.)

Type of question	How well did question work in typing person[a]	Notes on reasons for inconsistencies
Ranking on choice of behavior – suck, smoosh, chew, crunch	Fairly well	Individuals are often not aware of what they do with foods in their mouths – it is subconscious
Chewy vs. Crunchy Cookie	Not well	Crunchers often picked a chewy cookie because it was felt to be fresher and higher quality
Chew cookie completely vs. chew and then smoosh	Fairly well	Some Chewers will smoosh certain foods, and a soft cookie is sometimes one of them
Soft creamy vs hard candy	So so	Those who do not like food stuck in their teeth will not pick a soft creamy candy unless they are thinking of chocolate
Let hard candy dissolve in mouth or break it up as soon as possible	Fairly well	Hard candies are sometimes not broken up by Crunchers because they can stick in the teeth
Chew chocolate candies or hold them in the mouth	Not well	Chocolate is a product that many individuals will use differently than normal – more will hold and smoosh than expected
Crispy crunchy vs. thick and creamy snack	Fairly well	Smooshers may pick crisp because they are thinking of chips, even when they actually smoosh them the mouth

[a] Qualitative scale: Not well, So so, Fairly well, Well

17.2.5 Refining the tool

Based on this, another approach was taken. Instead of basing results on simple differentiating statements, it was felt that an approach that indicated patterns would be less susceptible to error. Therefore, a pictorial (image) solution was generated. Images have been found to be a good way of accessing the unconscious mind. 'Rather than stating its meanings to us in specific logical terms, the unconscious mind portrays its messages to us in images and symbols' and 'Dynamics within the unconscious can be known through imagery that may be visual, verbal, auditory…' (Progoff, 1963).

Subjects were shown four groupings of foods and asked which was 'most like you.' This indicated their 'preferred' mouth behavior because the foods shown were those that most easily allowed that behavior. The use of multiple foods removed the difficulty that arose with a particular food or flavor being disliked or handled differently in the mouth. Subjects were also asked if any of the groups were 'not like me at all' which would indicate a rejection of a given mouth behavior. The JBMB™ (Jeltema/Beckley Mouth Behavior) Classification Tool is shown in Figure 17.1.

To validate this tool, a stepwise confirmatory procedure was put in place. Twenty interviews were first conducted using the graphic tool. This was followed by the traditional survey tool for mouth behavior. Lastly, subjects were interviewed to understand inconsistencies (those whose survey question answers were not consistent with any given mouth behavior or which were unexpected due to the pictorial mouth behavior

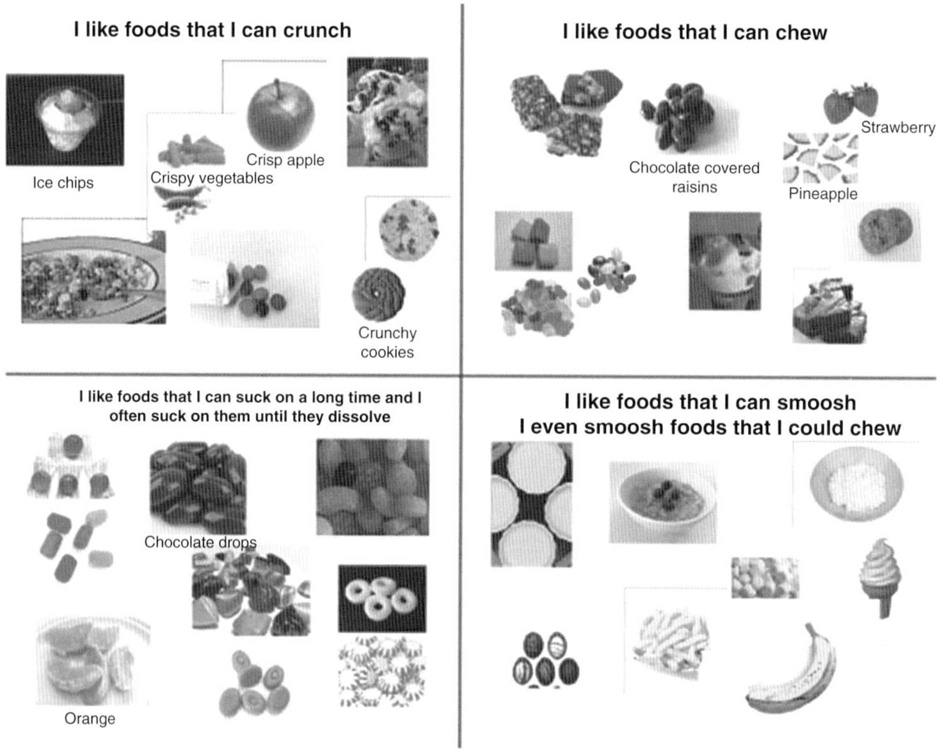

Figure 17.1 Graphic Mouth Behavior Tool (JBMB™)[a]. (Source: The Understanding & Insight Group LLC.)

chosen). This was done to determine their true mouth behavior, following the belief that these in-depth interviews provided the most accurate assessment of mouth behavior. In all 20 cases, the graphical solution was found to match the determination by interview and was deemed to be the correct mouth behavior. This indicated that the pictorial determination of mouth behavior was much less susceptible to error.

17.3 CATEGORIZATION OF MOUTH BEHAVIOR

17.3.1 Prevalence of mouth behavior groups in the U.S.

A survey was conducted among a demographically accurate distribution of 500 individuals across the U.S. The proportion of each mouth behavior group is shown in Table 17.3.

While it is important to know the type of mouth behavior most desirable to individuals, it is equally important to know what mouth behaviors are rejected. Those are shown in Table 17.4.

As indicated earlier, the original example discussed on mouth behavior centered around evaluation of a product that was meant to simply be sucked on. From this data, it is clear that this product would not have a large following. Only 8% of the population prefers products that will be sucked on for a long time, while 45% reject products where the design limits them to this mouth behavior. This of course does not imply that only

Table 17.3 U.S. population mouth behavior categorization. (Source: The Understanding & Insight Group LLC.)

Mouth Behavior	Chewer	Cruncher	Sucker	Smoosher
Percent (%)	43	33	8	16

Table 17.4 U.S. population – mouth behavior rejection. Source: The Understanding & Insight Group LLC.

Mouth Behavior	Chewer	Cruncher	Sucker	Smoosher
Percent %	10	16	45	29

8% of the population will eat a hard candy. Individuals who are not Suckers will simply choose to break it up a lot sooner.

17.3.2 Reported behavior by mouth activity group

As indicated previously, traditional survey questions were often not answered as expected. However, while none of the traditional survey questions were 100% accurate in separating behavioral groups, expected patterns did emerge. Since both Chewers and Crunchers have a similar mouth action (using teeth to break food), it was not surprising that they were more similar to each other in how they answered mouth behavior questions than they were to Suckers or Smooshers. A similar pattern was seen with Suckers and Smooshers (answering more questions similarly).

There were questions where a particular mouth behavior group answered the question significantly differently than other groups. Examples of questions that were different from at least two of the three groups are shown in Table 17.5. This further validates the existence of mouth behavior groups.

While there were fewer questions for Chewers that were significantly different from all other groups, there were many questions that were different from one or two of the groups and many questions that differentiated Crunchers from Chewers. For example, Chewers were more likely than Crunchers to say:

- I like chewy cookies, let cereal soften, let ice cream soften, prefer chewy pieces in ice cream, choose soft chewy snacks, pick soft dried fruit.
- I don't like granola, don't like hard dried fruit, you can't hear me crunch as I eat.

Similarly, there were other questions that differentiated Chewers from Suckers and Smooshers.

While many of these questions were able to show significant differences between groups, there were no questions where 100% of a mouth behavior group agreed. The highest agreement was in the 80s, (for example, 89% of Crunchers indicated that they would rather have a crisp or crunchy snack over a creamy snack). Other questions, while still showing significant differences, might be in the 30s (Crunchers preferring granola cereals over flakes or soft cereals). This again illustrates the difficulty using simple traditional survey research questions to type individuals who use complex behaviors.

Table 17.5 Patterns of behavior shown from survey questions[a,b]. (Source: The Understanding & Insight Group LLC.)

Chewers	Crunchers	Suckers	Smooshers
Prefer products can chew[a]	Hard crunchy cookies over soft chewy[a]	Prefer hard candy over soft[a]	Let cereal get soft or eat soft cereals like oatmeal[a]
Chewy candy over hard candy[b]	Hard granola bars over soft[a]	Like chocolate hard enough to suck on[a]	Prefer soft creamy candies over hard candy[a]
Would choose dried fruit that is chewy[b]	Eat ice cream right out of freezer[a]	Like to suck a long time on candy[a]	Prefer thick creamy snacks over crispy[a]
Like chocolate with good chewing texture[b]	Like apples that are crisp[b]	Always have hard candy around[a]	Flavored ice cream with no pieces[a]
Prefer cereals like Cheerios or flakes[b]	Like raw vegetables[b]	Like mints with some burn[a]	Chewing gum hurts jaw[a]
Don't prefer chocolates hard enough to suck on[b]	Prefer ice cream with crunchy pieces[a]	Like high carbonation in drinks[a]	Like food that is soft and spreads through mouth[a]
	Don't like to play with food in mouth[a]		Smoosh foods that could chew[b]

[a] Significantly different than all other groups;
[b] Significantly different from two of the three groups.

17.3.2.1 Mouth behavior and product choice

CPG (consumer packaged goods) companies know that 'one size does not fit all', and therefore have a wide variety of line extensions for any given product, some based on flavor and others on texture. Confection is one area where there are many texture options. However, the reasons for these segments of consumers have not been investigated. The following illustrate how consumers make choices in a product to optimize the texture that best matches their mouth behavior.

Chocolate is an example of a product that is eaten differently, based on the person eating the chocolate. Some individuals chew chocolate, while others tend to hold it in their mouths and let it melt. As you would expect, Mouth Behavior groups reflect these differences. There were significant differences in how these groups responded to a series of questions on chocolate.

- Suckers – like chocolate that is hard enough to suck on - like to alternate chewing and sucking on chocolates.
- Smooshers – like chocolate that melts fast.
- Crunchers – like chocolate that contains nuts.
- Chewers – like chocolate that has a good chewing texture.

This illustrates how consumers consistently choose food options based on their mouth behavior. Understanding, not only how consumers eat foods differently, but also how they are making these choices is crucial to Product Development.

17.3.2.2 Influence of mouth behavior on desired sensation levels

It has been pointed out that individuals achieve the highest food sensations by utilizing complex mouth actions (tongue, teeth, palate, etc.), and that this is done to maximize pleasantness (de Wijk, 2003). However, it has also been suggested that individuals balance chewing efficiency and perception in their choice of oral processing style (Chen and Engelen, 2012).

Then how do individuals, like Suckers and Smooshers maximize their food sensations? Do different groups seek out products that are different in food sensation levels or do different groups want different levels of sensation?

While we have minimal data on this, our survey data suggests individuals may seek (or avoid) different levels of sensation. Our data suggests that Suckers are more likely to enjoy high levels of carbonation in their drinks and to enjoy mints with 'burn'. This may indicate that these individuals use the increased trigeminal (burn) to help boost sensations. On the other hand, (maybe because they hold food longer in their mouths) both Suckers and smooshers were more likely to indicate a dislike of hot food or cold drinks.

17.4 IMPLICATIONS OF MOUTH BEHAVIOR

This research has shown that people differ in the ways that they like to manipulate food in their mouths and that these preferences drive texture preference and food choice. It should not be interpreted from these results that this implies that people would only choose to eat foods that easily allow a certain mouth behavior. It is more an indication that foods which are more often chosen when there is a choice are more 'delightful' or 'satisfying.'

Individuals are believed to adapt the food they are eating. For example, if a Cruncher is eating a soft food such as chocolate or a soft granola bar, they will more often choose a product with nuts or chips. A Chewer who is eating cereal will often add chewy ingredients like raisins or allow the cereal to soften slightly before eating it.

While there are foods that need to be chewed, such as meats, vegetables, and so on, we hypothesize that a person brings their mouth behavior preference to the food being eaten. For example, a Cruncher will more forcibly chew a meat than will a Chewer. They will probably choose different meats or prefer different meats cooked to different internal temperatures. A Sucker may prefer riper stone fruits that can be sucked on to release the juice before needing to be chewed. Interestingly, Suckers indicated that they were often dissatisfied with foods that they ate. Thirty percent said 'I don't usually buy the foods that I prefer, it's more about what my family will eat and I just deal with it'. This could reflect that there are fewer foods available that truly 'delight' Suckers.

People were also found to enjoy more than one mouth behavior. For example, a Cruncher may also enjoy 'smooshing' and have some foods that they sometimes (or partially) smoosh (like chocolates or ice cream). Only mouth behaviors that are 'rejected' will reflect themselves in foods that are never or rarely chosen (if that food requires the use of that mouth behavior). For example, those individuals who reject 'smooshing' will rarely eat semi-liquid foods that can't be chewed.

Not explored here were two other groups who were noted in the qualitative research. The first is a small group who seemed to want food to be in their mouths only a very short time. Given the option they would rather 'drink' their food or 'take a pill'. Interestingly, Smooshers were more likely to agree with this statement than any other group. Unfortunately, as of this writing, we do not understand why they would provided this response. The second group is a 'fiddling' group. Individuals may be 'fiddlers' in addition to their primary mouth behavior. 'Fiddlers' can be 'body fiddlers' (ex. always tapping feet or tapping or clicking pens) or 'mouth fiddlers' (spend a lot of time playing with food in their mouths, enjoy twirling or chewing on straws). Mouth Fiddlers probably exist in some, but not all mouth behavior groups (unlikely in smooshers). Suckers who spend the most time with food in their mouths were most likely to like to play with food (in the mouth).

The implications of this research on product design and development are profound. Products that allow multiple mouth behaviors will be more universally liked, while those that do not allow multiple behaviors, may only attract a certain type of person. No one food will be optimum for all mouth behaviors. Designers and developers need to understand these groups and use that knowledge to develop optimum products.

Additionally, there are implications for the growing number of older people. Qualitatively, those older individuals who have experienced dental issues indicated that they could no longer enjoy many of the foods that were preferred when they were younger. For example, Crunchers were not able to choose hard crunchy foods and were therefore forced to eat softer foods. While older persons have been identified as showing lower pleasure for food because of reduced ability to taste food, inability to use the preferred mouth behavior probably also drives reduced interest in food. Research conducted on elderly people with dentures showed that palatal coverage interferes with oral perception (Kremer et al., 2007). Those individuals with dentures rated food as less creamy. The reduced perception of texture was thought to be due to impairment of chewing and mouth movements. However, Kremer also found that compensatory changes in flavor or texture did not improve pleasantness. Our research indicates that, in the normal course of eating, individuals do exhibit unconscious behaviors to improve the texture of foods. This area of inquiry needs further study.

A third area of important inquiry is related to current issues in developed countries around obesity. Rolls (Rolls, et al., 2006; Rolls et al., 1998) created a model that has been found to be very good at assisting in weight loss (the volumetric diet) as has the Weight Watchers approach to foods (Comarow, 2012). We believe that once an individual understands their specific mouth behavior requirements, they may be able to enhance their compliance with diet regimes and may actually be able to avoid certain weight gain scenarios (eating foods that are low in calories but not satisfying because they do not allow the right mouth behavior).

While we do not currently have data to support this, we believe that, while mouth behavior may correlate with gender and personality, it is more basic, and is something we are born with (genetic). This hypothesis is supported by evidence that food cravings (Moskowitz et al., 2002; Beckley and Moskowitz, 2010), taste sensitivity and chewing behavior (Michon et al., 2009) differ by gender.

Additionally, we believe that our mouth behaviors will be found to relate to chewing behaviors measured analytically. We also believe that it will be found that an individual's mouth behavior does allow that person to achieve their desired level of sensory stimulation. The creation of the JBMB™ Classification Tool as a reliable device for understanding individual mouth behavior provides a basis for other researchers to explore these hypotheses for the better understanding of human behavior and food interactions.

17.5 USING MOUTH BEHAVIOR IN THE DESIGN OF PRODUCTS

Given how new the use of Mouth Behavior is in the area of product design and development, this section has been created to show proposed models of use for this new discovery. The two examples have been designed to show how incorporating mouth behavior in new product research could be used to improve our understanding of texture and the textural requirements of different groups of individuals.

17.5.1 Example one: potato chips case study example

Recently there has been a lot of attention given to developing products that are 'crispy/crunchy'. These terms are often used synonymously and products are often advertised as both crispy and crunchy. This emphasis seems to be driven by the belief that 'crunchy' products are liked by most consumers. However, developing crispy/crunchy products without understanding how mouth behavior group affects perception and liking (of both crispy and crunchy separately) is likely to lead to a product that is not well designed for anyone.

The following is an example and because of the small number of consumers is not meant to be definitive. More research would be needed before any hard conclusions could be drawn.

Five different chips were given to a small group of consumers who fell into different mouth behavior groups Crunchers(4), Chewers(3), Smoosher(1). The chips were chosen to represent a range of crunchiness. Chips included were: Pringles Original, Lays Original, Ruffles Original, Kettle Original, and Cape Cod Original. Each person was asked to map the chips on degree of crunchiness, crispiness, and degree that the texture was 'delightful'.

The ratings on crunchiness were very consistent across individuals; however ratings of crispiness varied. Some individuals rated crispiness and crunchiness similarly, some did not think the chips varied much on crispiness, and some rated crispiness very differently from crunchiness (thinking that crispy chips fractured easily and were not too soft or too hard).

As expected, texture preference varied and tended to follow mouth behavior group. The most preferred and reasons for preference were as follows:

- Smoosher – preferred the Lays product. Pringles dissolved too soon, causing a lot of salivation forcing you to swallow it. The Lays could be smooshed without difficulty because it didn't break down too soon. The others were too hard to smoosh.
    - *Note – some Smooshers may avoid chips altogether, while others may choose to use a different mouth activity style when they eat chips – depending on which chip it is and whether alternate eating styles are acceptable to them.*
- Crunchers – Of the chips presented, three of the four preferred the Kettle, which was felt to be the most crunchy. The fourth, preferred the Lays, because the Kettle crunch was too hard and it did not fracture easily enough - wasn't crisp). While the Kettle was the best of the chips given, most of the Crunchers would rather have Crunchy Cheetos, because the crunch was believed to be better (fractured correctly).

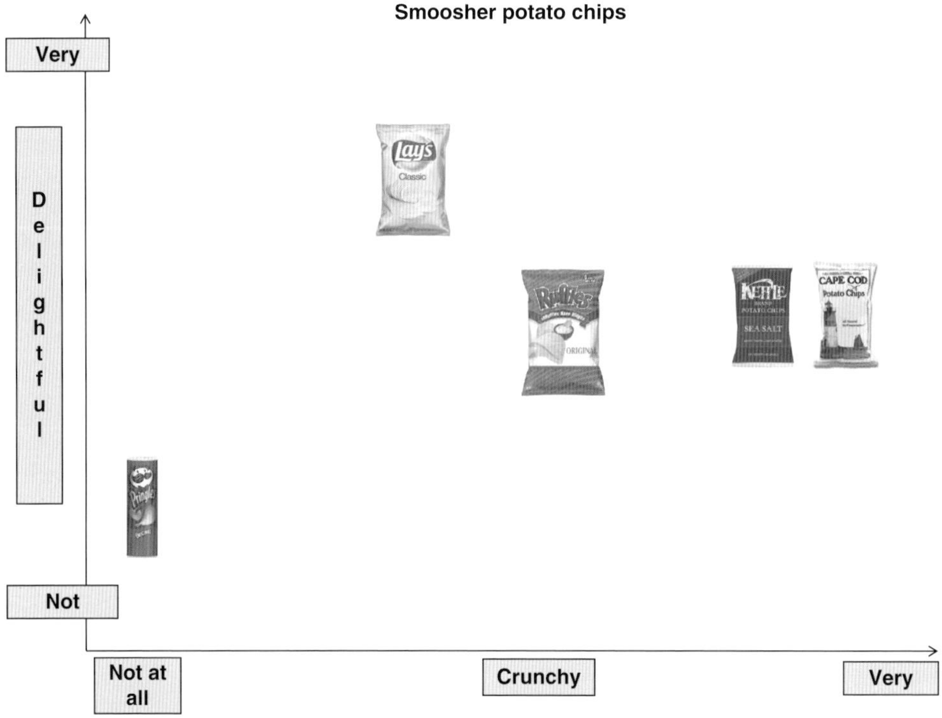

Figure 17.2 Example evaluation of chips by Smoosher. (Source: Pringles™, Lays™, Ruffles™, Cape Cod™, Kettle™.)

- Chewers – All of the Chewers preferred the Ruffles. It chewed well. Two of the three Chewers would not eat a Kettle, the third liked Kettle. Interestingly, this person was a 'long Chewer', while the others were 'short Chewers'
  - *Note – further research would be needed to indicate whether short and long Chewers needed to be considered separately*

Example maps are shown in Figures 17.2 through 17.4. For illustration purposes only the crunchy scale is shown; however, to truly understand what makes a 'good crunch', perceptions of crispiness and how chips 'fracture' is crucial. Interestingly, based on these few individuals, a perfect chip for Crunchers may not currently exist.

How to read the figures: The package graphic is placed in the space on the biplot representing how crunchy and delightful a particular chip was for the group. As a result, in Figure 17.2, for the Smoosher, Pringles was felt to be low in delightful and very low in crunchy, while Cape Cod was very high in crunchy while middle of the range in delightful. Lays was more delightful for the Smoosher because it could be smooshed rather than crunched. In figures where a product is shown more than once – this is indicative of variable responses and shows how the responses differed among individuals. In research, these are often shown as ellipses outlining the areas in which a product fell (an ellipse would circle an area that would include responses for a given product).

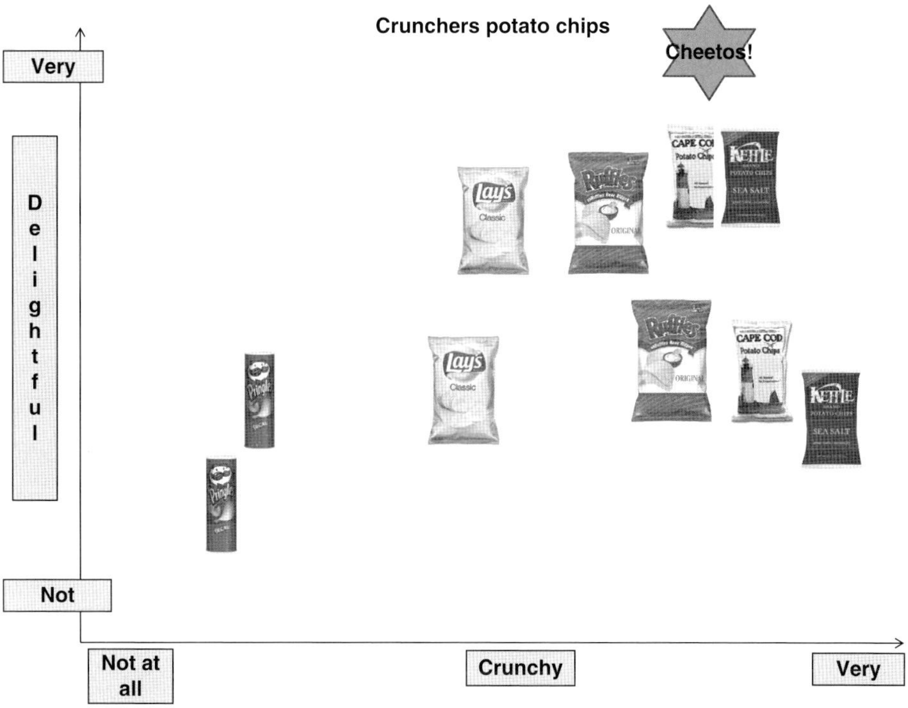

Figure 17.3 Example evaluation of chips by Crunchers. (Source: Pringles™, Lays™, Ruffles™, Cape Cod™, Kettle™.)

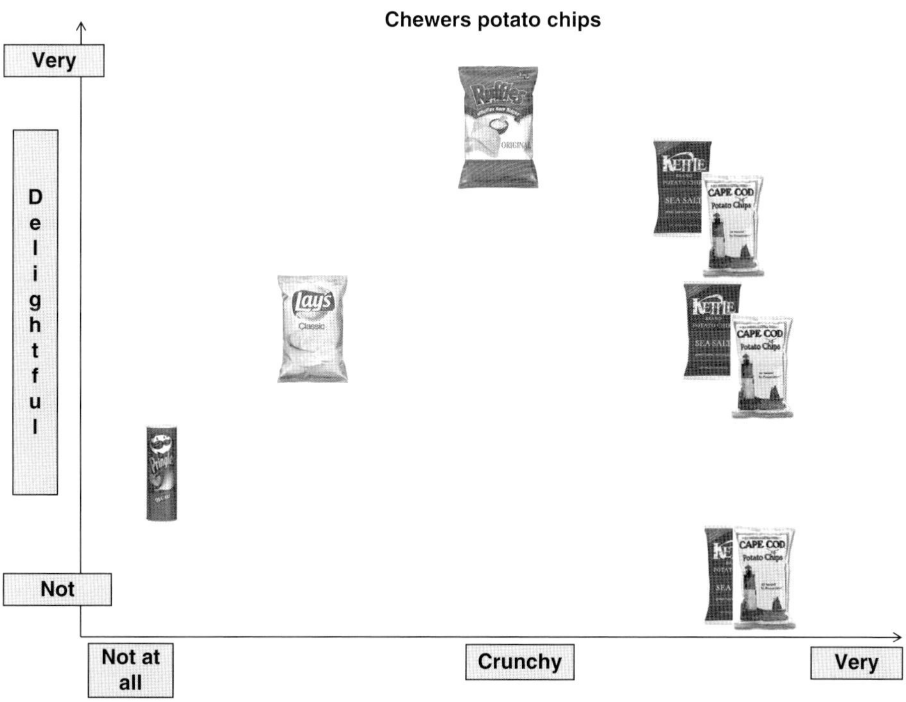

Figure 17.4 Example evaluation of chips by Chewers. (Source: Pringles™, Lays™, Ruffles™, Cape Cod™, Kettle™.)

17.5.2 Example two: yogurt case study

The following case study is about developing a new yogurt. *All data in this case study has been created to illustrate this example and should not be thought of as real values for yogurt.* This case study will highlight the importance of incorporating mouth behavior into your research inquiry, even in a product, like yogurt, where the importance is not obvious (as it is a semi–liquid product).

17.5.2.1 Background objective

A dairy company wanted to successfully compete in the growing yogurt market. They were aware that yogurt was often viewed by people as a healthier option to other high calorie foods and snacks or as a milk replacement, but they were interested in creating a yogurt product that would be viewed as a desirable product in its own right, not just as a replacement. To this end, they wanted to make it not just healthy, but also very 'satisfying' and 'craveable'. They believed that they had several design parameters worth pursuing. They wanted to understand the relationships between thickness and creaminess on satisfaction and craveability. They also wanted to understand how fruit texture and size of fruit pieces affected acceptance.

Knowing that there were probably differences in perception between different mouth behaviors, they had formed some initial hypotheses. Hypotheses:

1. Smooshers would want a very creamy but possibly thinner yogurt than Chewers or Crunchers. They would also want no fruit pieces or very blended fruit.
2. Chewers and Crunchers would want thicker yogurts that allowed chewing, along with larger, firmer fruit pieces.
3. While Chewers might be delighted by larger fruit pieces, there was a question around whether Crunchers would need a nuts or crunchy cereal added to be truly delighted. This would be part of the enquiry during the interviews.
4. They had no clear hypotheses around Suckers.
5. They were also interested in understanding whether the optimal flavor intensity would vary by mouth behavior group.

17.5.2.2 Developing the sensorial design space

In order to assess their hypotheses a design space was created that varied in the dimensions of interest, including thickness, creaminess, flavor intensity, and fruit size and firmness (design spaces are discussed in Lewis et al., 2010). They decided to first understand fruit size, flavor and thickness and then follow up with optimization on creaminess. To that end they created the sensory design space shown in Figure 17.5.

How to read the figures: The figures indicate the products that were made for the research and where each product would be positioned on the attributes of interest. For example, in Figure 17.5, eighteen products were made. Six were low on thickness, six were moderate on thickness, and six were high on thickness. There were also three

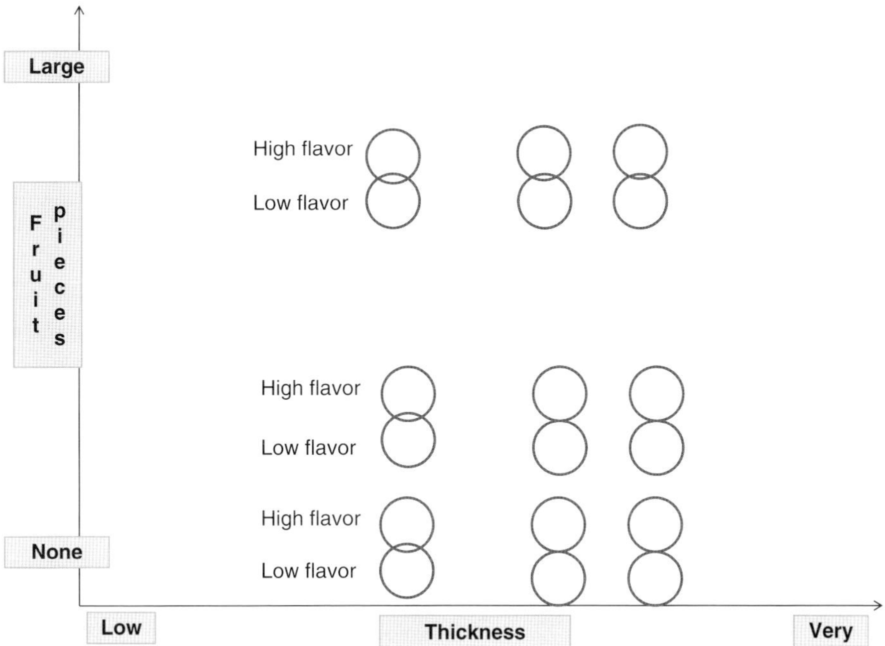

Figure 17.5 Sensory design space to understand thickness, fruit flavor and fruit size. (Source: The Understanding & Insight Group LLC.)

levels of fruit size. At each thickness, six products were made that varied in fruit size. Then at each combination of thickness and fruit size, two flavor levels were made.

17.5.2.3 Conducting the research

Individuals who consume yogurt were recruited based on their mouth behavior type. Separate sessions were held with each mouth behavior group. In this research, while a complete design space is available, not all options are provided to individuals. Instead a methodology called Rapid Product Navigation (Lewis et al., 2010) is utilized. This is a person driven navigation process, where responses from individuals on a product drive the selection of the next product to be evaluated. In our example (Figure 17.6), the team decided to first evaluate thickness, follow that with flavor intensity and then fruit size. In the group of Chewers, the first product presented to participants was the middle level of thickness, high flavor intensity, no fruit pieces (A1). This product was deemed by the group to be not thick enough, so product A2 was presented and was preferred. Based on this, no additional products of lower thickness were presented. The next evaluation (B) was at the higher level of thickness, still at no fruit pieces, but evaluating the lower flavor level. For Chewers, the higher flavor level was preferred. The last evaluation involved fruit size. The Chewers were presented with the products at the high thickness, high flavor level, varying in fruit size (C). This group preferred the larger fruit size. This matched the hypothesis that Chewers would want a product that enabled them to chew – both the thickness and fruit size supported their ability to chew.

Evaluation followed the following scheme: A1,A2,B2,C2,C3

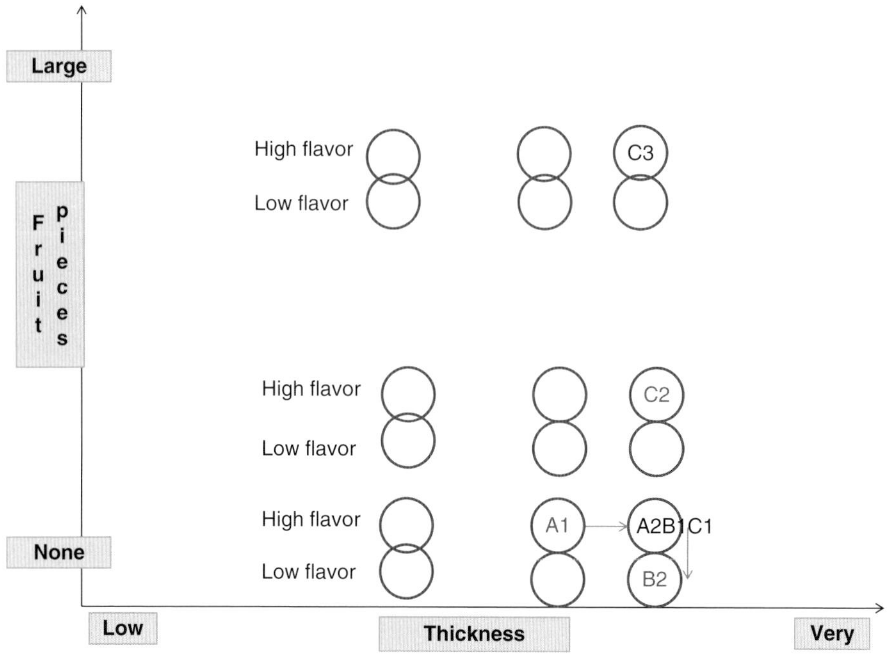

Figure 17.6 Scheme for evaluation of products among Chewers. (Source: The Understanding & Insight Group LLC.)

17.5.2.4 Results and use

By conducting the research among the different mouth activity groups, the company was able to understand whether more than one product was necessary to satisfy the different groups. They could then strategically decide how many and which ones they should commercialize. Were there some options that were generally liked by multiple groups but not a delighter of any? If so, should they go with that option or should they commercialize a few that are highly desirable by some groups but rejected by others?

17.6 CONCLUSION

In order to develop products that will be successful in the marketplace, it is essential that we understand the reasons for product acceptance and rejection. However, this is not enough. Most products are playing in very competitive spaces. To be successful the products must not only be acceptable, but must 'delight' the person. Developing products with the knowledge of how texture acts to aid in this will be a critically important tool for the future.

We would hope that it is clear that this is a very important area of scientific enquiry, and is fundamental in understanding of product acceptance/rejection.

The authors would like to pose several questions. 1) How could something that is this important to food choice be so unrecognized by the individual. 2) How is it possible that the whole area of consumer/sensory research has not to date identified the importance of mouth behavior.

There remains much that needs to be explored. Outlined in the discussions are additional areas of inquiry, including the role of genetics in determining mouth behavior, and the relationship between mouth behavior and flavor perception. Additionally, the authors have hypothesized the use of mouth behavior to aid in key problems such as weight reduction and poor food acceptance by the elderly.

While the typing tool has been found to reliably differentiate individuals and segment them into one of four groups, it has not been expanded to measure the small group who 'would prefer to have nothing in their mouths' or to separately type mouth or body fiddlers. In addition, it has been recognized that Chewers differ in the amount of time they prefer to have food in the mouth. It may be found that 'short' and 'long' Chewers are different enough in texture preferences that they need to be considered separately when developing optimized food products.

The authors look forward to others becoming involved in this research and expanding the knowledge and use of mouth behavior in understanding food processing, textural preferences, impact of genetics, and so on.

References

Alpern, M., Lawrence, M., and Wolsk, D. (1964) *Sensory Process. Basic Concepts in Psychology Series*. Belmont, CA. Wadsworth Publishing Company.

Beckley, J. (2013) Personal communication to lead author regarding the creation of a method for 'bowl life.'

Beckley, J., and Moskowitz, H. (2002) Databasing the person mind: The Crave It! Drink It! Buy It! & Healthy You! Databases. Institute of Food Technologist, Anaheim.

Brown, W. E., and Braxton, D. (2000) Dynamics of food breakdown during eating in relation to perceptions of texture and preference: a study on biscuits. *Food Quality and Preference*, 11, 259–267.

Chen, J. and Engelen, L. (2012) *Food Oral Processing*. Wiley- Blackwell, West Sussex, UK, 159–176.

Comarow, A. (2012) Best Diets – US News and World Reports. US News and World Reports, January 3, 2012.

Daubert, R., Chen, J. and Lundin, L. (2013) Editors Description of Journal Purpose. *Journal of Texture Studies*, Wiley Online Library, ISSN 1745–4603.

de Wijk, R. A., Engelen, L., and Prinz, J. F. (2003) The role of intra-oral manipulation in the perception of sensory attributes. *Appetite*, 40, 1–7.

de Wijk, R. A., Wulfert, F. and Prinz, J.F. (2006) Oral processing assessed by M-mode ultrasound imaging varies with food attribute. *Physiology & Behavior*, 89(1), 15–21.

de Wijk, R. A., Polet, I. A., Bult, J. H. and Prinz, J. F. (2008) Vibromyography of oral processing varies with type of semi-solid food and with sensory judgments. *Physiology & Behavior*, 95(3), 521–526.

Drewnowski, A. (1997) Taste preferences and food intake. Annual Review of Nutrition Human Nutrition Program, School of Public Health, The University of Michigan, Ann Arbor, MI, 17, 237–253.

Hochberg, J. (1967)*Perception. Foundations of Modern Psychology Series*. Englewood Cliffs, N.J. Prentice-Hall. Inc.

Kalviainine, N., Schlich, P., and Tuorila, H. (2000) Consumer Texture Preferences: Effect of age, gender and previous experience. *Journal of Texture Studies* 31, 593–607.

Kremer, S., Bult, J. H. F., Mojet, J., and Kroeze, J. H. A. (2007) Food Perception with age and its relationship to pleasantness. *Chemical Senses*, 32, 591–622.

Lassauzay, C., Peyron, M. A., Albuisson, E., Dransfield, E., and Woda, A.(2000) Variability of the masticatory process during chewing of elastic model foods. *European Journal of Oral Science*, 108, 484–492.

Lewis, J. N., Teratanavat, R., Beckley, J. and Jeltema, M. A. (2010) 'Using a consumer driven Rapid Product Navigation Process to Develop an Optimal Product'. *Food Quality and Preference*, 21, 1052–1058.

Michon, C., O'Sullivan, M. G., Delahunty, C. M., and Kerry, J. P. (2009) The investigation of gender-related sensitivity differences in food perception. *Journal of Sensory Studies*, 24, 922–937.

Moskowitz, H., Beckley, J., and Adams, J.(2002) What makes people crave fast food? *Nutrition Today*, 37, 6, 237–242.

Moskowitz, H. (2007) Sensory intensity versus hedonic functions: Classical Psychophysical approaches. *Journal of Food Quality*, 5(2), 109–137.

Murray, J. M. and Delahunty, C. M. (2000) Mapping consumer preference for the sensory and packaging attributes of Cheddar Cheese. *Food Quality and Preference*, 11, 419–435.

Poe, J. M., Kieser, J. A., Gallo, L. M., Tesenyi, A. J., Herbison, P. and Farella, M. (2011) Time-Frequency analysis of chewing activity in the natural environment. *Journal of Dental Research*. 90, 1206–1210.

Progoff, I. (1963) *The Symbolic and the Real*. McGraw, New York, NY. 91, 208.

Rey, A., Gonzalez, R., Martinez-de-Juan, J. B., and Mulet, A. (2007) EMG assessment of chewing behavior for food evaluation: Influence of personality characteristics. *Food Quality and Preference*, 18, 585–595.

Richardson-Harmon, N. J., Stevens, R., Walker, S., Gamble, J. Miller, M., Wong, M., and McPherson, A. (2000) Mapping consumer perception of creaminess and liking for liquid products. *Food Quality and Preference*, 11, 239–246.

Rolls, B. J., Roe, L. S. and Meengs, J. S. (2006) Reductions in portion size and energy density of foods are additive and lead to sustained decreases in energy intake over two days. *American Journal of Clinical Nutrition*, 83, 11–17.

Rolls, B. J., Castellanos, V. H., Halford, J. C., Kilara, A., Pelkman, C. L., Smith, G. P. M, and Thorwart, M. L.(1998) Volume of food consumed affects satiety in men. *American Journal of Clinical Nutrition*, 67, 412–420.

Sow, T. M. and Grongnet, J. F. (2010) Sensory characteristics and consumer preference for chicken meat in Guinea. *Poultry Science*, 89, 2281–2292.

Scott, C. L. and Downey, R. G. (2007) Types of food aversion: animal, vegetable, and texture. *The Journal of Psychology: Interdisciplinary and Applied*, 141, 127–134.

Szczesniak, A. S. (1963a) Objective Measurement of Food Texture. *Journal of Food Science*, 28, 410–420.

Szczesniak, A. S. (1963b) Classification of textural characteristic. *Journal of Food Science*, 28,385-389.

Szczesniak, A. S. and Kahn, E. L. (1971) Consumer awareness of and attitudes to food texture. *Journal of Texture Studies*, 2,280-295.

Index

Note: Page references in *italics* refer to Figures; those in **bold** refer to Tables

Food Texture Design and Optimization, First Edition. Edited by Yadunandan Lal Dar and Joseph M. Light.